VOLUME ONE HUNDRED AND TWO

ADVANCES IN BOTANICAL RESEARCH

Soybean Physiology and Genetics

ADVANCES IN BOTANICAL RESEARCH

Series Editor

Jean-Pierre Jacquot
Membre de l'Institut Universitaire de France, Unité Mixte de Recherche INRA, UHP 1136 "Interaction Arbres Microorganismes", Université de Lorraine, Faculté des Sciences, Vandoeuvre, France

VOLUME ONE HUNDRED AND TWO

ADVANCES IN BOTANICAL RESEARCH

Soybean Physiology and Genetics

Volume Editors

HON-MING LAM

Centre for Soybean Research of the State Key Laboratory of Agrobiotechnology and School of Life Sciences, The Chinese University of Hong Kong, Shatin, Hong Kong Special Administrative Region, China

MAN-WAH LI

Centre for Soybean Research of the State Key Laboratory of Agrobiotechnology and School of Life Sciences, The Chinese University of Hong Kong, Shatin, Hong Kong Special Administrative Region, China

ELSEVIER

ACADEMIC PRESS

An imprint of Elsevier

Academic Press is an imprint of Elsevier
125 London Wall, London, EC2Y 5AS, United Kingdom
The Boulevard, Langford Lane, Kidlington, Oxford OX5 1GB, United Kingdom
50 Hampshire Street, 5th Floor, Cambridge, MA 02139, United States
525 B Street, Suite 1650, San Diego, CA 92101, United States

First edition 2022

ISBN: 978-0-323-91219-8
ISSN: 0065-2296

For information on all Academic Press publications
visit our website at https://www.elsevier.com/books-and-journals

Publisher: Zoe Kruze
Acquisitions Editor: Sam Mahfoudh
Developmental Editor: Jhon Michael Peñano
Production Project Manager: James Selvam
Cover Designer: Vicky Pearson

Typeset by STRAIVE, India

Contents

Contributors

Shwu-Pyng T. Chen
Centre for Soybean Research of the State Key Laboratory of Agrobiotechnology and School of Life Sciences, The Chinese University of Hong Kong, Shatin, Hong Kong Special Administrative Region, China

Yinglong Chen
The UWA Institute of Agriculture, and UWA School of Agriculture and Environment, The University of Western Australia, Perth, WA, Australia

Sau-Shan Cheng
Centre for Soybean Research of the State Key Laboratory of Agrobiotechnology and School of Life Sciences, The Chinese University of Hong Kong, Shatin, Hong Kong Special Administrative Region, China

Gyuhwa Chung
Department of Biotechnology, Chonnam National University, Yeosu, Korea

Weiguang Du
Soybean Research Institute, Heilongjiang Academy of Agricultural Sciences, Harbin, China

Kejing Fan
Centre for Soybean Research of the State Key Laboratory of Agrobiotechnology and School of Life Sciences, The Chinese University of Hong Kong, Shatin, Hong Kong Special Administrative Region, China

Brett J. Ferguson
Integrative Legume Research Group, School of Agriculture and Food Sciences, The University of Queensland, Brisbane, Australia

Christine H. Foyer
School of Biosciences, College of Life and Environmental Sciences, University of Birmingham, Edgbaston, United Kingdom

Peter M. Gresshoff
Integrative Legume Research Group, School of Agriculture and Food Sciences, The University of Queensland, Brisbane, Australia

Yuefeng Guan
FAFU-UCR Joint Center for Horticultural Biology and Metabolomics, Fujian Agriculture and Forestry University, Fuzhou, China

Tianfu Han
MARA Key Laboratory of Soybean Biology (Beijing), Institute of Crop Sciences, Chinese Academy of Agricultural Sciences, Beijing, China

April H. Hastwell
Integrative Legume Research Group, School of Agriculture and Food Sciences, The University of Queensland, Brisbane, Australia

Zhihong Hou
Innovative Center of Molecular Genetics and Evolution, School of Life Sciences, Guangzhou University, Guangzhou, China

Yi Ying Huang
Centre for Soybean Research of the State Key Laboratory of Agrobiotechnology and School of Life Sciences; Simon F.S. Li Marine Science Laboratory, School of Life Sciences, The Chinese University of Hong Kong, Shatin, Hong Kong Special Administrative Region, China

Jerome H.L. Hui
Centre for Soybean Research of the State Key Laboratory of Agrobiotechnology and School of Life Sciences; Simon F.S. Li Marine Science Laboratory, School of Life Sciences, The Chinese University of Hong Kong, Shatin, Hong Kong Special Administrative Region, China

Bingjun Jiang
MARA Key Laboratory of Soybean Biology (Beijing), Institute of Crop Sciences, Chinese Academy of Agricultural Sciences, Beijing, China

Candice H. Jones
Integrative Legume Research Group, School of Agriculture and Food Sciences, The University of Queensland, Brisbane, Australia

Fanjiang Kong
Innovative Center of Molecular Genetics and Evolution, School of Life Sciences, Guangzhou University, Guangzhou, China

Yee-Shan Ku
Centre for Soybean Research of the State Key Laboratory of Agrobiotechnology and School of Life Sciences, The Chinese University of Hong Kong, Shatin, Hong Kong Special Administrative Region, China

Karl J. Kunert
Department of Plant and Soil Sciences, FABI, University of Pretoria, Pretoria, South Africa

Hon-Ming Lam
Shenzhen Research Institute, The Chinese University of Hong Kong, Shenzhen; Centre for Soybean Research of the State Key Laboratory of Agrobiotechnology and School of Life Sciences, The Chinese University of Hong Kong, Shatin, Hong Kong Special Administrative Region, China

Man-Wah Li
Shenzhen Research Institute, The Chinese University of Hong Kong, Shenzhen; Centre for Soybean Research of the State Key Laboratory of Agrobiotechnology and School of Life Sciences, The Chinese University of Hong Kong, Shatin, Hong Kong Special Administrative Region, China

Ailin Liu
Centre for Soybean Research of the State Key Laboratory of Agrobiotechnology and School of Life Sciences, The Chinese University of Hong Kong, Shatin, Hong Kong Special Administrative Region, China

Baohui Liu
Innovative Center of Molecular Genetics and Evolution, School of Life Sciences, Guangzhou University, Guangzhou, China

Shuo Liu
College of Natural Resources and Environment, and Institute of Soil and Water Conservation, Northwest A&F University, Yangling, Shaanxi, China

Xiao Liu
Photosynthesis Research Center, Key Laboratory of Photobiology, Institute of Botany, Chinese Academy of Sciences, Beijing, China

Ndiko Ludidi
Department of Biotechnology and the DST/NRF Centre of Excellence in Food Security, University of the Western Cape, Bellville, South Africa

Ching-Yee Luk
Centre for Soybean Research of the State Key Laboratory of Agrobiotechnology and School of Life Sciences, The Chinese University of Hong Kong, Shatin, Hong Kong Special Administrative Region, China

Ming-Sin Ng
Centre for Soybean Research of the State Key Laboratory of Agrobiotechnology and School of Life Sciences, The Chinese University of Hong Kong, Shatin, Hong Kong Special Administrative Region, China

Henry T. Nguyen
Division of Plant Sciences, 25 Agriculture Lab, College of Agriculture, Food and Natural Resources, University of Missouri, Columbia, MO, United States

Kadambot H.M. Siddique
The UWA Institute of Agriculture, and UWA School of Agriculture and Environment, The University of Western Australia, Perth, WA, Australia

Lianjun Sun
College of Agronomy and Biotechnology, China Agricultural University, Beijing, China

Satyanarayana Taddi
Centre for Soybean Research of the State Key Laboratory of Agrobiotechnology and School of Life Sciences; Simon F.S. Li Marine Science Laboratory, School of Life Sciences, The Chinese University of Hong Kong, Shatin, Hong Kong Special Administrative Region, China

Dagang Wang
Nanjing Agricultural University, National Center for Soybean Improvement, National Key Laboratory for Crop Genetics and Germplasm Enhancement, Nanjing, Jiangsu, China

Zhili Wang
Centre for Soybean Research of the State Key Laboratory of Agrobiotechnology and School of Life Sciences, The Chinese University of Hong Kong, Shatin, Hong Kong Special Administrative Region, China

Fuk Ling Wong
Centre for Soybean Research of the State Key Laboratory of Agrobiotechnology and School of Life Sciences, The Chinese University of Hong Kong, Shatin, Hong Kong Special Administrative Region, China

Qinzhen Xu
Photosynthesis Research Center, Key Laboratory of Photobiology, Institute of Botany, Chinese Academy of Sciences, Beijing, China

Wenqiang Yang
Photosynthesis Research Center, Key Laboratory of Photobiology, Institute of Botany, Chinese Academy of Sciences; College of Advanced Agricultural Sciences, University of Chinese Academy of Sciences; Innovative Academy of Seed Design, Chinese Academy of Sciences; Sino-Danish College, University of Chinese Academy of Sciences, Beijing, China

Heng Ye
Division of Plant Sciences, 25 Agriculture Lab, College of Agriculture, Food and Natural Resources, University of Missouri, Columbia, MO, United States

Ho Yin Yip
Centre for Soybean Research of the State Key Laboratory of Agrobiotechnology and School of Life Sciences; Simon F.S. Li Marine Science Laboratory, School of Life Sciences, The Chinese University of Hong Kong, Shatin, Hong Kong Special Administrative Region, China

Zhihui Yuan
College of Agronomy and Biotechnology, China Agricultural University, Beijing, China

Wai-Shing Yung
Centre for Soybean Research of the State Key Laboratory of Agrobiotechnology and School of Life Sciences, The Chinese University of Hong Kong, Shatin, Hong Kong Special Administrative Region, China

Chunyang Zhang
Photosynthesis Research Center, Key Laboratory of Photobiology, Institute of Botany, Chinese Academy of Sciences, Beijing, China

Guohong Zhang
Institute of Dryland Agriculture, Gansu Academy of Agricultural Sciences, Key Laboratory of Northwest Drought Crop Cultivation of Chinese Ministry of Agriculture, Lanzhou, China

Haijian Zhi
Nanjing Agricultural University, National Center for Soybean Improvement, National Key Laboratory for Crop Genetics and Germplasm Enhancement, Nanjing, Jiangsu, China

Preface

The soybean plant originated in East Asia. It was estimated that the domestication of soybean took place more than 5000 years ago. Domestication has altered soybean in many ways to suit our demand for food, feed, and industrial uses. Nowadays, instead of just serving as a food crop, soybean has been playing an irreplaceable role in the global trade and economy. With the intensification of climate change, soybean, as a representative grain legume, also serves as one of the many saviors in low-emission agriculture. Given the importance of soybean, many breeders and scientists have dedicated their efforts to advance our knowledge of this crop in different aspects, with the aim of making further improvements to satisfy future needs.

It has been a decade since the release of the first soybean reference genome. Since then, various genomic researches in soybean have blossomed. High-quality reference genomes, pangenomes, and data from genome resequencing have become more and more readily available. Genome-wide association studies, quantitative trait loci mapping, and genomic selection have all flourished on the foundation of the soybean genomes. Taking advantage of advanced genome editing techniques, the mystery of the soybean genome, sharing similarities with other plant genomes and yet unique in its own right, is being unraveled bit by bit.

Soybean is by nature a highly photosensitive plant. Flowering and maturation at the right time are critical to ensure seed production. Thus, the photo-sensitiveness of soybean has limited the expansion of cultivation as soybean is highly sensitive to the gradual change in daylength at varying latitudes. After years of screening and breeding, accessions adapted to different daylengths have been identified. Some of the genes responsible for these adaptations and the regulatory network governing the flowering and maturation times of soybean have been revealed in recent years.

The root is responsible for nutrient acquisition, anchorage to substratum, stress interactions, nodulation, and nitrogen fixation. These root functions contribute greatly to plant growth, health, and final yield. Root performance is largely determined by its architecture. However, while the root is buried in the soil, the root traits were, in general, selected indirectly by its influence on the aboveground attributes of the plant during soybean domestication. Nowadays, with the advancement in phenotype collecting technologies, studies of soybean root variability have become more accessible. Understanding the genetic variability governing root growth, nutrient acquisition, and responses toward stresses will enable further improvement of the crop.

The soybean root is at the frontline of interactions with most stresses such as drought, salt stress, and root-related diseases. Thus, it has played essential roles in safeguarding the plant from these hazards. Phytophthora root rot is a major disease harming soybean roots. Soybean varieties with resistance against different races of Phytophthora have been developed. Understanding the mechanisms and discovering the genetic factors behind the resistance have become the major strategy against this pathogen.

Photosynthesis provides both energy and building blocks of macromolecules in plant. Efficient photosynthesis in soybean not only fills the seed sink, but it also provides both the carbon backbone and an energy source for nitrogen fixation. Thus, understanding the photosynthetic process, sugar transportation and allocation, and carbon-related metabolisms can inform soybean improvement in carbon cycling and nitrogen fixation.

Soybean provides relatively environmentally friendly protein and oil in comparison to animal sources. It is not only because the plant is a primary producer, but it is also because soybean, as a legume, is able to manufacture its own nitrogen fertilizer through its interactions with rhizobium in the root nodule. Molecules such as phytohormones, genes, peptides, secondary metabolites, and reactive oxygen species (ROS) orchestrate the regulatory process for nodulation and nitrogen fixation. Among these signaling molecules, the CLAVATA3-Endosperm Surrounding Region-related (CLE) peptides have recently attracted quite a lot of attention due to its crucial functions in regulating soybean nodulation, symbiosis with mycorrhizae, and other biological functions. Interestingly, these peptides not only control local functions, but they also serve as long-distance signaling molecules. Understanding the functions of these peptides will provide new information for improving soybean production.

The production of cellular ROS is a dilemma. Normally, ROS is considered harmful as its overproduction is usually associated with environmental stress and could damage cellular structures and disrupt cellular processes. However, recent studies have suggested that regulated ROS production plays important roles in signaling and balancing cellular redox status in stress conditions, nodulation, nitrogen fixation, and nodule senescence. However, research into ROS signaling in soybean is limited. Applying the knowledge from other legumes would be a shortcut for understanding ROS signaling in soybean.

Insects feed on the soybean plant, lowering the yield and quality of seeds. Insects sometimes also serve as vectors for pathogenic microbes to invade the plant. Plants have developed different strategies to defend against insect herbivory such as physical barriers and chemical weapons. So far, a number of insects that inhabit and feed on soybean have been identified. Since the intensive use of chemicals to repel/kill insects and pathogens can lead to adverse consequences for human health and the environment, efforts have been made to delineate the interactions between soybean and its insect pests and to develop new agronomic practices and new soybean cultivars with better resistance.

Soybean is a unique crop even among legumes. Unlike the seeds of major cereals, which are mainly made up of carbohydrates, calories in the soybean seed are mainly entrapped in protein and oil. The protein content in soybean seeds is one of the highest among legume seeds. As soybean constitutes 28% of vegetable oil and 70% of protein meal consumption worldwide, a small change in the seed protein and oil contents is sufficient to bring about a big impact. The quality and quantity of soybean seed protein and oil contents are affected by both environmental and genetic factors. The regulatory network governing the protein and oil accumulation has been gradually pieced together in recent years. There are quite a few breakthroughs in altering the oil composition in soybean seed, making soybean a successful model for genome editing.

Secondary metabolism produces compounds that are beneficial for the plant to better adapt to the environment. Sometimes, these secondary metabolites are also of special interest to humans. Unfortunately, some of these metabolites were bred out of the soybean plant during the domestication process, as they contributed to bitter tastes and other undesirable consumer-centric traits at a time when their health benefits were not known. With the advanced technology in detecting these metabolites and a better understanding of their beneficial properties, it is possible to tailor their enrichment in specific cultivars by making use of the rich genetic variability contained in the wild soybean germplasms.

Seed morphology not only contributes to yield components, but it can also affect the taste, texture, and nutritional value of the final food products. Domestication has altered many aspects of the soybean seed appearance. Due to genetic variability, soybean seeds come in different shapes, sizes, colors, and textures. Some of them are directly associated with the nutritional value, agronomic performance, and yield of the seed. To date, we have a better understanding of the genetic controls on seed morphology and its interactions with other seed traits.

This book summarizes the current understanding of the above characteristics of soybean owing to the dedicated efforts of experts in this field. We are grateful for their contributions. Soybean research has taken some big leaps in recent years. Soybean, being a legume in which most of the biological functions are tied to nitrogen fixation, is a unique model for crop research. Since soybean has been gaining importance in different aspects of society, the improvement of soybean becomes a crucial ambition in the food and industrial supply chains. We hope this book can provide the latest information in various important aspects of soybean research, for both current and prospective researchers.

Hon-Ming Lam
Centre for Soybean Research of the State Key Laboratory of Agrobiotechnology and School of Life Sciences, The Chinese University of Hong Kong, Shatin, Hong Kong, China

Man-Wah Li
Centre for Soybean Research of the State Key Laboratory of Agrobiotechnology and School of Life Sciences, The Chinese University of Hong Kong, Shatin, Hong Kong, China

CHAPTER ONE

Genomic research on soybean and its impact on molecular breeding

Man-Wah Li[a,b,*,†], Bingjun Jiang[c,†], Tianfu Han[c], Guohong Zhang[d], and Hon-Ming Lam[a,b,*]

[a]Shenzhen Research Institute, The Chinese University of Hong Kong, Shenzhen, China
[b]Centre for Soybean Research of the State Key Laboratory of Agrobiotechnology and School of Life Sciences, The Chinese University of Hong Kong, Shatin, Hong Kong Special Administrative Region, China
[c]MARA Key Laboratory of Soybean Biology (Beijing), Institute of Crop Sciences, Chinese Academy of Agricultural Sciences, Beijing, China
[d]Institute of Dryland Agriculture, Gansu Academy of Agricultural Sciences, Key Laboratory of Northwest Drought Crop Cultivation of Chinese Ministry of Agriculture, Lanzhou, China
*Corresponding authors: e-mail address: limanwah@cuhk.edu.hk; honming@cuhk.edu.hk

Contents

† Authors contributed equally.

Advances in Botanical Research, Volume 102
ISSN 0065-2296
https://doi.org/10.1016/bs.abr.2022.02.021

Abstract

Soybean, with its high protein and oil contents as well as nitrogen-fixing ability, is an important component in climate-smart agriculture. Yet, to meet the demands on production and consumption, it is necessary to continue to improve its agronomic traits related to yield, adaptability, and nutritional values. Benefiting from advances in genome sequencing technologies, soybean research has blossomed in the past decade since the release of the first soybean reference genome. Improvements in the quality of soybean reference genomes in recent years have facilitated ever higher levels of soybean research. New features such as open chromatin regions and long non-coding RNAs in the soybean genome have also been revealed. To take advantage of the genetic diversity inherent in the soybean germplasm collections, genotyping-by-sequencing, high-resolution mapping, and subsequent gene functional analyses are now routine processes to precisely identify functional markers and genes for molecular breeding. Molecular breeding techniques allow the efficient stacking or modification of desirable genes, and greatly improve the rate and precision of generating new soybean varieties. In this chapter, a brief summary will be given on the current progress in soybean genomic research and its impact on soybean molecular breeding.

1. Introduction

Archaeological evidence suggested that human consumption of soybean can be dated back to >8,000 years before present (BP) in East Asia (Lee, Crawford, Liu, Sasaki, & Chen, 2011). Since it is not known whether the soybean seeds found in archaeological sites were wild-gathered or field-grown, the actual beginning of soybean domestication is still largely debatable. In general, the undomesticated or wild soybean is classified as *Glycine soja* (*G. soja*) while the domesticated or cultivated soybean is classified as *Glycine max* (L.) Merr. (*G. max*). Despite being classified into different species, wild and cultivated soybeans can cross-breed to produce fertile offspring (Kim et al., 2010; Li et al., 2014), and thus wild soybeans have been regarded as reservoirs of genetic materials for the improvement of cultivated soybeans. Some intermediate accessions between wild and cultivated soybeans could sometimes be regarded as semi-wild soybeans and landraces. The definition is fuzzy and is largely dependent on how they were collected, regardless of their true origins. Genomic analyses suggested that cultivated soybean originated from a *G. soja*/*G. max* lineage that had diverged from other *G. soja* lineages 0.27–0.8 million years ago (MYA). Yet these estimations were based on a small number of soybean genomes and several assumptions. More accurate predictions should be available in upcoming large-scale genomic analyses. While *G. max* was derived from a limited genetic background, the substantial genetic resources contained in wild soybeans could potentially be important for soybean improvement.

Over 70,000 cultivated and 20,000 wild soybean accessions are maintained by different seed banks worldwide (Li et al., 2020; Liu et al., 2020; Valliyodan et al., 2021). To make good use of these germplasm resources, sequencing, analyzing and publishing their genomes would be an essential step. Since the official release of the first soybean reference genome in 2010 (Schmutz et al., 2010), genomic research has been expanding, contributing greatly to the molecular breeding of soybean.

2. A survey of published soybean reference genomes

2.1 A brief overview of the soybean genome

The current soybean genome resulted from two rounds of recent whole-genome duplication (WGD). The first round of whole-genome duplication occurred around 58 MYA, common to all members of Papilionoideae (Schmutz et al., 2010)Kim et al., 2010; Schmutz et al., 2021. The second round happened about 13 MYA and was specific to *Glycine* spp. The resultant soybean genome comprises of 20 pairs of chromosomes which are morphologically identical under the microscope during metaphase (Findley et al., 2010). The genome size of soybean was estimated using flow cytometry to be 1.1–1.15 Gb (Arumuganathan & Earle, 1991; Shultz et al., 2006), which, in general, agrees with the 0.88–1.1 Gb estimate by K-mer frequency, based on sequencing reads obtained in recent genomic studies (Li et al., 2014; Qi et al., 2014). It is predicted that there are 45,000–65,000 protein-coding genes in the soybean genome, of which ~50,000–55,000 are of high confidence (Li et al., 2014; Liu et al., 2020; Qi et al., 2014; Schmutz et al., 2010; Xie et al., 2019). Around 60–70% of these genes exist as paralogues as a consequence of WGDs (Schmutz et al., 2010). Although gene redundancy could be an obstacle for genetic studies, it was found that just altering one gene out of a paralogous pair was normally sufficient to create observable phenotypic changes in plant traits during domestication (Lu et al., 2017; Wang et al., 2016, 2020). Similar to other crops, 50%–60% of the soybean genome is made up of transposable elements (TE) dominated by long terminal repeat (LTR) retrotransposons (Schmutz et al., 2010; Xie et al., 2019). Moreover, the soybean genome also contains ~40,000 long non-coding RNA (lncRNA) gene loci which produce transcripts that are not traditionally considered to be functional or protein-encoding (Lin et al., 2020), constituting a relatively unexplored component of the soybean genome. For regulatory elements, >20,000 open chromatin regions potentially containing transcription factor-binding *cis* elements have been identified in a wild soybean genome (Huang et al., 2021), providing an area of focus for future research on soybean transcription regulation.

2.2 Currently available high-quality reference genomes and their characteristics

2.2.1 First generation of soybean reference genomes

The official release of the Williams 82 (Wm82) reference genome in 2010 (Schmutz et al., 2010) marked the beginning of the genomic era of soybean research. A few *de-novo* genome assemblies or reference-guided genome assemblies were subsequently released (Table 1). These first-generation soybean genome assemblies were mostly built on short sequencing reads from next-generation sequencing platforms, except Wm82 which was constructed from large-scale Sanger sequencing. Limited by the short sequencing read length and the complexity of the soybean genome, these genome assemblies were usually fragmented, with short contig N50 ranging from 20 to 190 Kb. Among these soybean genome assemblies, the most commonly used was Wm82, owing to its better contiguity over the others. These published genome assemblies have been widely used as references for resequencing projects, for variation calling and association mapping (see below and other sections), speeding up the progress of soybean research. However, the short reads were not able to resolve the genomic regions with low complexity or duplications, leading to bubbles in the de Brujin graph and resulting in contigs that were usually relatively short and fragmented. The consequence is that it becomes challenging to anchor the short contigs onto scaffolds and chromosomes, leaving large numbers of gaps, which in turn results in ambiguous gene models spanning these gaps. Such gap-filled genomes are not suitable for structural variation comparisons and make a lot of downstream analyses problematic. Fortunately, this issue is now partially resolved by second-generation soybean genome assemblies.

2.2.2 Second generation of soybean reference genomes

The breakthrough in long-read sequencing has brought soybean genome assemblies to another level. Contig assemblies using long sequencing reads and scaffolding using Hi-C sequencing, occasionally supplemented with optical mapping, have now become the standard for the second-generation soybean genome assembly pipeline. The first second-generation soybean

Table 1 First-generation soybean genome assemblies.

Variety	Genome ID and version information	Cultivated/wild	Year of official release	Assembled size[a] (Mb)	Contig N50 (Kb)	Scaffold N50	Reference
Williams 82	Wm82.a1	Cultivated	2010	955	189.4	47.8 Mb	Schmutz et al. (2010)
Menjing1	W05	Wild	2014	808	24.2	401 Kb	Qi et al. (2014)
Lanxi 1	–	Wild	2014	895	21.7	51 Kb	Qiu et al. (2014)
Williams 82	Wm82.a2	Cultivated	2015	955	182.8	48.6 Mb	Schmutz et al. (2010)
Enrei	–	Cultivated	2015	905	NA	NA	Shimomura et al. (2015)
ZYD04569	GsojaA	Wild	2014	813	9	18.3 Kb	Li et al. (2014)
PI507600	GsojaB	Wild	2014	895	22.2	57.2 Kb	Li et al. (2014)
PI407222	GsojaC	Wild	2014	841	8	17 Kb	Li et al. (2014)
ZYD03247	GsojaD	Wild	2014	985	11	48.7 Kb	Li et al. (2014)
ZYD02878	GsojaE	Wild	2014	920	27	65.1 Kb	Li et al. (2014)
ZYD00401	GsojaF	Wild	2014	886	24.3	52.4 Kb	Li et al. (2014)
PI578344B	GsojaG	Wild	2014	878	19.2	44.9 Kb	Li et al. (2014)

[a]Assembled size without gap.

reference genome released was Zhonghuang 13 (Gmax_ZH13), published in 2018 (Shen et al., 2018). It was quickly updated to version 2 in 2019 with a seven-fold improvement in the contig N50 (Shen et al., 2019). Within three years, around 40 more high-quality soybean reference genomes were released (Table 2). In general, most of the second-generation soybean reference genomes have a contig N50 at the megabase level, a hundred times longer than those of the first-generation assemblies. The drastic improvement in the contiguity of the reference genomes allows better gene model predictions, identification and comparison of large structural variations, and association studies with higher precision (Liu et al., 2020; Xie et al., 2019).

2.2.3 Currently available soybean pan-genomes

A single reference genome cannot effectively represent the whole species. Although there are more than 40 high-quality reference genomes of soybean available, using all of them as references in a single study would involve tedious analyses of mostly redundant sequences. Although whole-genome resequencing enables comparisons between large natural populations (see Section 2.2.4), variation calling relies heavily on the information contained within the reference genome. Therefore, features absent in the reference genome are usually overlooked and structural variations are usually missed.

A pan-genome has the potential to solve the above issues. It is built by collapsing the genomic information of multiple genome assemblies into nonredundant sequences or a set of nonredundant genes (Lei et al., 2021). In brief, the set of sequences or genes shared by all genome assemblies are defined as the core genome or core gene set (Lei et al., 2021). Those shared by some genomes are defined as dispensable or variable genome or genes (Lei et al., 2021). Private sequences and genes are those unique to a specific genome assembly. A gene-based pangenome is easy to build as it is only concerned with the mostly conserved genes or protein-coding sequences, and is especially suitable for analyzing genomes with a distant relationship. A sequence-based pangenome, on the other hand, is usually more computationally demanding and is suitable for analyzing closely related genomes with small numbers of sequence variations. It is more information-intensive as it also considers the non-coding elements in the genomes, which can be important in molecular breeding as well.

So far, there are four pangenomic studies on soybean (Table 3). Compared with those built from wild soybeans or a collection of both wild and cultivated soybeans, a pangenome built solely from cultivated soybeans would have a higher proportion of core genes (Table 3). Although more

Table 2 Second-generation soybean genome assemblies.

Variety	Genome ID and version information	Wild/cultivated	Year of official release	Assembled size (Gb)	Contig N50 (Mb)	Scaffold N50 (Mb)	Reference
Zhonghuang 13	Gmax_ZH13	Cultivated	2018	1.025	3.46	51.87	Shen et al. (2018)
Zhonghuang 13	Gmax_ZH13_v2.0	Cultivated	2019	1.011	22.6	NA	Shen et al. (2019)
Mengjin 1	W05	Wild	2019	1.013	3.3	50.7	Xie et al. (2019)
Williams 82	Wm82v4	Cultivated	2019	0.978	0.419	20.4	Valliyodan et al. (2019)
Lee	Lee v1.1	Cultivated	2019	1.016	0.037	15.0	Valliyodan et al. (2019)
PI 483463	Glycine soja v1.1	Wild	2019	0.985	0.011	4.4	Valliyodan et al. (2019)
Wenfeng 7	IGA1001	Cultivated	2021	0.996	1.74	50.7	Chu et al. (2021)
Hefeng 25	IGA1002	Cultivated	2021	0.987	2.92	49.51	Chu et al. (2021)
GsojaF	IGA1003	Wild	2021	0.975	1.55	48.77	Chu et al. (2021)
Zhonghuang 35	IGA1004	Cultivated	2021	1.001	1.41	50.49	Chu et al. (2021)
Zhonghuang 13	IGA1005	Cultivated	2021	0.988	4.65	50.28	Chu et al. (2021)
Jingyuan	IGA1006	Cultivated	2021	0.995	4.27	50.57	Chu et al. (2021)
Huaxia 3	IGA1007	Cultivated	2021	0.986	6.16	50.7	Chu et al. (2021)
Williams 82	IGA1008	Cultivated	2021	0.993	1.88	49.8	Chu et al. (2021)
PI 562565	SoyW01	Wild	2020	1.021	23.9	52.1	Liu et al. (2020)
PI 549046	SoyW02	Wild	2020	1.007	20.5	21.9	Liu et al. (2020)

Continued

Table 2 Second-generation soybean genome assemblies.—cont'd

Variety	Genome ID and version information	Wild/cultivated	Year of official release	Assembled size (Gb)	Contig N50 (Mb)	Scaffold N50 (Mb)	Reference
PI 578357	SoyW03	Wild	2020	1.015	26.8	52.3	Liu et al. (2020)
Zhutwinning2	SoyL01	Landrace	2020	0.999	23.3	50.6	Liu et al. (2020)
Zi Hua No.4	SoyL02	Landrace	2020	1.012	23.1	52.3	Liu et al. (2020)
Tong Shan Tian E Dan	SoyL03	Landrace	2020	1.040	21.3	51.3	Liu et al. (2020)
58-161	SoyL04	Landrace	2020	1.005	23.0	51.4	Liu et al. (2020)
PI 398296	SoyL05	Landrace	2020	1.060	22.1	51.2	Liu et al. (2020)
Zhang Chun Man Cang Jin	SoyL06	Landrace	2020	0.998	21.5	50.7	Liu et al. (2020)
Feng Di Huang	SoyL07	Landrace	2020	1.005	23.7	50.9	Liu et al. (2020)
Tie Jia Si Li Huang	SoyL08	Landrace	2020	1.000	22.6	51.1	Liu et al. (2020)
Shi Sheng Chang Ye	SoyL09	Landrace	2020	1.028	23.1	51.0	Liu et al. (2020)
Xu Dou No.1	SoyC01	Cultivated	2020	1.004	23.5	51.5	Liu et al. (2020)
Tie Feng No.18	SoyC02	Cultivated	2020	1.011	23.8	51.2	Liu et al. (2020)
Ju Xuan No.23	SoyC03	Cultivated	2020	1.008	23.9	51.3	Liu et al. (2020)

Wan Dou No.28	SoyC04	Cultivated	2020	1.002	22.7	50.9	Liu et al. (2020)
Amsoy	SoyC05	Cultivated	2020	0.992	22.7	50.3	Liu et al. (2020)
Yu Dou No.22	SoyC06	Cultivated	2020	1.008	23.0	52.0	Liu et al. (2020)
Jin Dou No.23	SoyC07	Cultivated	2020	1.009	21.8	51.2	Liu et al. (2020)
Qi Huang No. 34	SoyC08	Cultivated	2020	1.002	22.4	50.9	Liu et al. (2020)
Han Dou No.5	SoyC09	Cultivated	2020	1.004	22.6	50.4	Liu et al. (2020)
PI 548362	SoyC10	Cultivated	2020	1.005	19.8	51.0	Liu et al. (2020)
Ji Dou No.17	SoyC11	Cultivated	2020	1.025	18.8	51.4	Liu et al. (2020)
Dong Nong No.50	SoyC12	Cultivated	2020	1.025	20.0	50.8	Liu et al. (2020)
Hei He No.43	SoyC13	Cultivated	2020	1.010	23.8	51.1	Liu et al. (2020)
Ke Shan No.1	SoyC14	Cultivated	2020	1.007	23.0	51.6	Liu et al. (2020)
Jidou 17	JD17	Cultivated	Preprint	0.965	18.0	NA	Yi et al. (2021)
Fiskeby III (PI 438471)	Fiskeby v1.1	Cultivated	Pre-release	0.992	NA	NA	DOE-JGI (2021)

Table 3 Pangenome studies on soybean.

Number of accessions	Assembled pan-genome size (Mb)	Core genome size (Mb)	Total no. of gene families	Core gene families	Reference
7 wild soybeans	986.3	790.1	59,080	28,716	Li et al. (2014)
3 wild soybeans	–	–	57,492	20,623	Liu et al. (2020)
9 landraces					
15 cultivars					
204 cultivated	1,086	–	54,531	49,431	Torkamaneh, Lemay, and Belzile (2021)
157 wild soybeans	1,213	–	51,414	44,654	Bayer et al. (2021)
723 landraces					
228 cultivars					
2 unclassified					

accessions were involved in building PanSoy (Torkamaneh et al., 2021), due to the lower genetic diversity in cultivated soybean, the size of this pangenome is not any bigger and the number of core gene families is higher than that built from just seven wild soybean accessions (Li et al., 2014).

2.2.4 Large-scale soybean resequencing projects

While it is still costly to build high-quality reference genomes for all soybean accessions, whole-genome resequencing is still an affordable way to capture decent genetic variations for genomic analyses. Since the official release of the Wm82 reference genome in 2010, a number of large-scale soybean resequencing projects have been published (Table 4). In the early phase, resequencing studies mainly focused on the analysis of genetic diversity. They all came to the same long-known but previously unproven conclusion that wild soybeans have higher genetic diversity than cultivated soybeans (Chung et al., 2014; Lam et al., 2010; Qiu et al., 2014; Zhou et al., 2015). Furthermore, domestication- and adaptation-related genomic loci were rediscovered in these studies, improving our understanding of the domestication process. With the reduction in sequencing costs, resequencing

Table 4 List of large-scale soybean resequencing studies.

Number of accessions	Average sequencing depth/mapping depth	Reference genome	Total number of SNPs	Genetic diversity	Reference
17 Wild soybeans	–/5×	Williams 82 (Wm82.a1)	6,318,109	Wild: $\theta_\pi = 2.97 \times 10^{-3}$	Lam et al. (2010)
17 Cultivated soybeans				Cultivated: $\theta_\pi = 1.89 \times 10^{-3}$	
1 Wild soybean	52×/43×	Williams 82 (Wm82.a1)	2,504,985		Kim et al. (2010)
9 Semi-wild soybeans	–/3.0×	Williams 82 (Wm82.a1)	7,704,637[a]	Wild: $\pi = 2.173 \times 10^{-3}$	Qiu et al. (2014)
1 Semi-wild soybean	–/41.4×			Semi-wild: $\pi = 1.416 \times 10^{-3}$	
1 Wild soybean	–/55.0×			Cultivated: $\pi = 1.332 \times 10^{-3}$	
6 Wild soybeans	17×/14×	Williams 82 (Wm82.a1)	9,028,250	Wild: $\pi = 1.08 \times 10^{-3}$	Chung et al. (2014)
10 Cultivated soybeans				Cultivated: $\pi = 0.46 \times 10^{-3}$	
62 Wild soybeans	–/11×	Williams 82 (Wm82.a1) + *G. soja* var. IT182932 genome-specific sequences	9,790,744	Wild: $\pi = 2.94 \times 10^{-3}$	Zhou et al. (2015)
130 Landraces				Landraces: $\pi = 1.40 \times 10^{-3}$	
110 Improved cultivars				Improved cultivars: $\pi = 1.05 \times 10^{-3}$	
28 Brazilian cultivars	–/14.8×	Williams 82 (Wm82.a2.v1)	5,835,185	Not available	dos Santos et al. (2016)
102 Canadian short-season soybean	11x[b]/>1x	Williams 82 (Wm82.a2.v1)	4,071,378	Not available	Torkamaneh et al. (2018)

Continued

Table 4 List of large-scale soybean resequencing studies.—cont'd

Number of accessions	Average sequencing depth/mapping depth	Reference genome	Total number of SNPs	Genetic diversity	Reference
195 Cultivars	-/16.3 ×	Williams 82 (Wm82.a2.v1)	10,116,707	Not available	Kajiya-Kanegae et al. (2021)
3 Wild soybeans					
41 Wild soybeans	12.9 × /-	Gmax_ZH13	31,870,983	Not available	Liu et al. (2020)
632 Landraces					
1354 Improved cultivars					
85 wild soybean	-/>10x	Gmax_ZH13	12,506,103	Not available	Lu et al. (2020)
153 Landraces					
186 Cultivars					
279 Landrace	-/6.03x	Williams 82 (Wm82.a2.v1)	6,341,742	Not available	Li, Li, et al. (2020)
134 Chinese cultivars	-/8 ×	Williams 82 (Wm82.a2.v1)	4,163,977	$\pi = 1.28 \times 10^{-3}$	Qi et al. (2021)
429 Cultivars	15 × /-	Williams 82 (Wm82.a2.v1)	7,869,806	Not available	Valliyodan et al. (2021)
52 Wild soybeans					
51 Landraces	11 × /-	Williams 82 (Wm82.a2.v1)	6,333,721	Not available	Yang et al. (2021)
199 Cultivars					

[a]43 soybean accessions, including those from Lam et al. (2010) and Kim et al. (2010), were used for SNP-calling.
[b]Median sequencing depth/minimum mapped depth.

of larger populations becomes readily accessible. As a consequence, these studies have generated a large number of molecular markers for various uses, including mapping, DNA chip construction, marker-assisted breeding and genomic selection. At the same time, the increased size of populations being resequenced has enabled association studies to identify single-nucleotide polymorphisms (SNPs) associated with the causal locus or gene for a specific trait (see Section 3.3).

3. Quantitative trait loci/association mapping using whole-genome sequencing methods

3.1 Basics of genetic mapping for soybean breeding

The availability of functional genes and their associated markers are the foundation of molecular breeding. Unlike many other model plants, there is only a limited collection of publicly available systematic mutants for soybean due to the high maintenance and labor costs (Brown et al., 2021). Thus, the mutant-based forward-genetics approach for the identification of causal genes or associated markers for certain traits is not favored in soybean research. At the same time, the efficiency of soybean transformation is also not comparable with other model crops. As a result, the pace of using a reverse-genetics approach to determine the gene function also lags far behind the other major crops. This means the availability of markers and genes associated with specific traits for soybean molecular breeding is also limited.

Association mapping, either through quantitative trait locus (QTL) mapping or genome-wide association studies (GWAS), is the common way to identify the genomic element associated with a specific phenotype. Before the publishing of the soybean reference genome, there were three major limitations in soybean association mapping. First of all, the number of available molecular markers was limited, and genotyping them was normally laborious, directly affecting the resolution of the map. There are more than 1,000 documented simple-sequence repeat (SSR) markers in the soybean genome (Song et al., 2004). Considering the polymorphism of these SSR markers in the target population and the labor cost of genotyping them, only a fraction of these markers would normally be used. What makes this worse is that these markers are usually not evenly distributed in the genome. Thus, the physical distance between markers can be up to several megabases. In such a case, the resolution of mapping and efficiency of breeding would be low. Secondly, without the whole-genome sequence, the positions of genetic markers were mostly expressed in relative terms to the rate of recombination (i.e. genetic distances expressed in centimorgans). Thus, the location and range of the mapped genomic region could be very fuzzy. On some occasions, the relative positions of markers in different genetic maps vary, which makes comparison between maps problematic. Finally, sequence and gene information within the mapped region could be completely unknown. Hence the identification of the causal element of the phenotype of interest would require tremendous amounts of follow-up work.

Genome sequencing and the construction of reference genomes have revolutionized association mapping in soybean. By mapping the resequencing reads to the reference genome or making comparisons between reference genomes, one can theoretically identify all the variations between genomes that can be used as markers for mapping or breeding.

Since the cost of sequencing may still be unaffordable for some laboratories, instead of carrying out costly whole-genome resequencing and *de novo* genome assembly, genotyping-by-sequencing (GBS)/reduced representation sequencing (RRS) can be an economical option for soybean research. There are quite a number of variants of GBS/RRS, such as restriction site-associated DNA sequencing (RAD-seq), double-digestion RAD sequencing (ddRAD-seq), specific-locus-amplified fragment sequencing (SLAF-seq), Diversity Arrays Technology sequencing (DArTseq), low pass/shallow whole-genome sequencing (LP-WGS) and so on. By sequencing only specific parts of the genome, the sequencing depth of each marker can thus be higher with the same number of total sequencing reads as whole-genome resequencing. These methods may sacrifice the marker density in return for a lower cost and higher precision. Still, the actual marker density is normally at least an order of magnitude higher than the traditional SSR markers, thus the mapping resolution can be greatly improved. This is especially suitable for biparental populations with a limited number of recombination events.

3.2 Important QTLs identified by whole-genome sequencing

Before the release of the soybean reference genome, QTL mapping was mainly done with SSR or restriction fragment length polymorphism (RFLP) markers. After the release of the Wm82 reference genome, reference-based mapping or sequencing-mapping blossomed. Single-nucleotide polymorphism (SNP) and insertion/deletion (INDEL) markers became more readily available. Up to date, there are over 200 published reference-based mapping studies regarding all sort of soybean traits. To keep it simple, we will use those key studies that include gene function validation as examples to illustrate the effectiveness of reference-based mapping.

3.2.1 Growth and development

Low-depth whole-genome resequencing is prone to sequencing error as well as leaving gaps in the coverage at certain genomic regions. To overcome the low coverage of the genome due to low-depth whole-genome resequencing, "bin" markers (Xie et al., 2010) are normally adopted instead of individual SNPs. The genotype of a bin marker is defined by the genotypes of an array of SNPs in a pre-determined window, such that it can tolerate sequencing error, ambiguous base-calling and missing data due to low sequencing coverage. This strategy is better used in a biparental population with limited recombination events. A recombinant inbred (RI) population between a wild and a cultivated parent was genotyped through low-depth whole-genome resequencing at 1X, for the mapping of QTLs governing plant height and growth habit (Wang et al., 2021). SNPs were converted into bin markers before mapping. With more than 6,000 bin markers, this study identified major QTLs related to plant height and growth habit (Wang et al., 2021). Through optical mapping and a comparison between high-quality second-generation soybean reference genomes, a copy number variation (CNV) in *gibberellin 2-oxidase 8* genes was found within these QTLs. Since each unit of the CNV spans ~50 kb, it would not have been able to be resolved using the first-generation soybean reference genomes. The increase in the copy number of *gibberellin 2 oxidase 8* genes is widespread among soybean cultivars, and it reduces internode elongation and suppresses the trailing growth habit (Wang et al., 2021).

The genetic regulation of growth period is a major focus of soybean genomic research. A number of genes and loci regulating growth period have been identified. By genotyping 308 RI lines originating from a very late-flowering and an extremely early-flowering parent using 2b-RAD (Wang et al., 2020), a variant of ddRAD-seq (Wang, Meyer, McKay, & Matz, 2012), a genetic map consisting of 3,454 markers was constructed. In total, 15 QTL regions were identified, with some of them containing reported genes controlling flowering time (Wang et al., 2020). Specifically, in the QTL *qFT12-2*, a circadian clock-related gene, *GmPRR37*, was identified to control the photoperiodic flowering time that helped soybean adapt to the regional conditions during domestication (Wang et al., 2020). The same genes were also discovered in GWAS (Li et al., 2020; Lu et al., 2020).

The cultivation of soybean at low latitudes usually suffers from low yields due to early flowering under inductive short-day conditions. The adoption of soybean varieties with the long-juvenile (LJ) trait, which have a prolonged vegetative phase under short-day conditions, can significantly improve the yield. PI 159925 and BR121 are two soybean varieties showing the LJ trait while Harosoy is a normal variety sensitive to day-length. QTL mapping using two genetic populations originating from PI 159925 x Harosoy and BR121 x Harosoy identified the same QTL region on chromosome 4 governing the LJ phenotype (Lu, Zhao, et al., 2017). An orthologue of the circadian clock gene, *EARLY FLOWERING 3* (*ELF3*), encoding a component of the evening complex, was found to carry the variations that could explain the LJ phenotype (Lu, Zhao, et al., 2017). Complementation of the loss-of-function variants with a functional copy of *ELF3* can significantly reverse the LJ phenotype (Lu, Zhao, et al., 2017). Further analyses also supported that variants of other evening complex genes may also contribute to soybean adaptation to low latitudes (Lu, Zhao, et al., 2017).

3.2.2 Yield potential

A 100-seed weight study was carried out using 198 RI lines generated from a wild (ZYD7) and a cultivated (HN44) soybean (Lu et al., 2017). Interestingly, an elite RI line, R245, was found having seed weights exceeding those of the parental lines, suggesting that the alleles of some loci from the wild parent ZYD7, which has a lower seed weight than the cultivated one, may contribute to the increased seed weight of the RI line. The 198 RI lines were resequenced at 2X depth for SNP-calling and bin-map construction. Fourteen QTLs associated with the 100-seed weight were mapped. It was found that R245 had 13 alleles from the cultivated parent and one from the wild parent in these 14 seed weight QTLs. Of this particular locus from the wild parent that contributed to a large seed size, analyses identified 2 out of 22 genes bearing polymorphisms between the two parental lines. These two genes encode protein phosphatase 2C (PP2C) and EamA-like (EAL), respectively (Lu, Xiong, et al., 2017). Functional analyses in Arabidopsis showed that the ectopic expression of *PP2C* from ZYD7 could increase the plant and seed sizes (Lu, Xiong, et al., 2017). The analysis of 72 wild soybeans and 94 cultivars showed that only 60% of the cultivars carried the ZYD7 *PP2C* allele (Lu, Xiong, et al., 2017), suggesting that the yield of some of the cultivars could be further improved by introducing the ZYD7 *PP2C* alleles.

3.2.3 Abiotic stress tolerance

It has been known that soybean salt tolerance was controlled by a single dominant gene since 1969 (Abel, 1969). The salt tolerance QTL was later mapped to linkage group N (chromosome 3) of the soybean genome using SSR and RFLP markers (Lee et al., 2004). A decade later, the determinant gene was finally uncovered by one of the pioneer sequencing-based QTL mapping studies in soybean (Qi et al., 2014). The salt tolerance QTL was mapped to a 978-kb region in chromosome 3 of the soybean genome using a bin map constructed from 96 RI lines with a 1X sequencing depth (Qi et al., 2014). It was further narrowed down to a 388-kb region using SNP markers developed from the resequencing of 31 soybean germplasms (Lam et al., 2010; Qi et al., 2021). After analyzing a basket of genomic data, it was found that *GmCHX1* was the determinant gene for salt tolerance (Qi et al., 2014). The protein sequence of GmCHX1 is conserved in all salt-tolerant soybean accessions. Any mutation affecting the protein sequence or reducing its expression could be detrimental (Qi et al., 2014). The identity of GmCHX1 (also known as GmSALT3 or Ncl) was later confirmed by another mapping study using 367 F_5 RI lines and 5,769 $F_{5:6}$ individuals using a non-sequencing-based mapping method (Guan et al., 2014). This information has also led to the breeding of a few new multi-stress-tolerant soybean varieties (Li, Wang, et al., 2020).

Apart from *GmCHX1/GmSALT3/Ncl*, another genomic study has also identified a key gene corresponding to soybean salt tolerance during germination (Zhang et al., 2019). By combining the mapping results from an RI population consisting of 184 lines and a GWAS using 211 soybean accessions, a 560-kb region on chromosome 8 was pinpointed to determine the salt tolerance of soybean at germination (Zhang, Liao, et al., 2019). To identify the causal variations within the QTL region, the parental lines were resequenced for variant-calling against the Wm82 reference genome. After identifying 29 genes with variations between the parents, qRT-PCR was used to detect the genes that were distinctively responsive to salt treatment and differentially expressed between parents. A cation diffusion facilitator-encoding gene (*GmCDF1*) was finally isolated as the potential candidate. Functional analyses suggested that GmCDF1 probably mediates Na^+ and K^+ homeostasis by regulating the expression of ion transporter genes such as *GmSOS1* and *GmNHX1* (Zhang, Liao, et al., 2019).

3.2.4 Biotic stress resistance

Sclerotinia stem rot is a disease that can devastate a soybean crop. Using 149 $F_{5:20}$ RI lines for QTL mapping and 261 soybean accessions for GWAS, a region on chromosome 13 was found to correspond to the sclerotinia stem rot resistance in both populations (Zou et al., 2021). A glutathione S-transferase (GST)-encoding gene (*GmGST*) was differentially expressed between susceptible and resistant lines upon *Sclerotinia sclerotiorum* challenge (Zou et al., 2021). The overexpression of *GmGST* from the resistant lines can improve the resistance toward *S. sclerotiorum* in both resistant and susceptible lines (Zou et al., 2021). At the same time, the CRISPR/cas9-mediated mutation of *GmGST* in the resistant line also reduced its resistance toward the pathogen (Zou et al., 2021).

3.2.5 Seed coat color

The transition from a pigmented to a non-pigmented seed coat was an important trait of domestication of soybean. It has long been known that this transition was due to the tissue-specific RNA silencing of *chalcone synthase* genes in the presence of interfering RNAs generated by the *I* locus on chromosome 8 of the soybean genome (Todd & Vodkin, 1996; Tuteja, Zabala, Varala, Hudson, & Vodkin, 2009). Nevertheless, the causal variation was not uncovered till 2019, when the seed coat color variation of a soybean RI population from a wild and a cultivated parent was also mapped to the*I*locus through QTL mapping (Xie et al., 2019). By comparing the wild soybean reference genome with bacterial artificial chromosome clones corresponding to the*I*locus of cultivated soybean, it was found that there was an inversion within the*I*locus that caused a head-to-head fusion between a *subtilisin* gene and a *chalcone synthase* gene (Xie et al., 2019). The fused gene produced chimeric transcripts that led to the RNA silencing of the *chalcone synthase* gene family, thus resulting in an opaque seed coat. The segmental inversion was also confirmed by optical contigs as well as a high-quality reference genome published later (Valliyodan et al., 2019; Xie et al., 2019). Using PacBio long read sequencing, the variation in the pigmentation of the hilum was revealed to be also controlled by another chimeric fusion in the*I*locus (Kajiya-Kanegae et al., 2021).

3.3 Current findings and limitations of genome-wide association-mapping

The first comprehensive sequencing-based GWAS can be dated back to 2015 (Zhou et al., 2015) where 302 soybean accessions consisting of wild soybeans, landraces, and improved cultivars were sequenced at an average depth of 11X (Zhou et al., 2015). Both known and novel genomic regions associated with domestication-related phenotypes were identified using GWAS (Zhou et al., 2015), demonstrating that sequencing-based GWAS on soybean is feasible. Since then, a number of GWAS have been carried out on yield-related traits such as yield stability (Quero et al., 2021), seed shape (Zhao, Li, et al., 2019), and seed weight (Zhang et al., 2021; Zhao, Dong, et al., 2019). Other GWAS were also performed on plant architecture (Seck, Torkamaneh, & Belzile, 2020), stress performance (Jing et al., 2021; Ravelombola et al., 2020; Steketee, Schapaugh, Carter, & Li, 2020), nitrogen fixation-related traits (Ray et al., 2015) and other traits (Dhanapal et al., 2016). Although these studies may not provide sufficient evidence to pinpoint the causal genes, they have provided valuable information for marker-assisted selections and further studies. A few examples are discussed below to illustrate how GWAS were applied to soybean.

Apart from the previously mentioned *J* locus (Lu, Zhao, et al., 2017; Yue et al., 2017), other genes/loci were also found to contribute to the later-flowering trait in soybean cultivars developed in these regions. A GWAS using 329 soybean accessions, including 165 from low latitudes, identified a locus named *Tof16* on chromosome 16 that controls the flowering time under short-day conditions (Dong et al., 2021). After fine mapping and sequence comparisons, *Tof16* was found to encode LHY1a, a circadian clock gene (Dong et al., 2021). A loss-of-function mutation of *LHY1a* induced by CRISPR/cas9 caused delayed flowering in soybean under short-day conditions, through the derepression of *E1* and the repression of *GmFT2a* and *GmFT5a* (Dong et al., 2021). Loss-of-function haplotypes of *LHY1a* are widely distributed in tropical soybean accessions, indicating its importance in soybean adaptation to low-latitude regions (Dong et al., 2021). In the same study, the 4 *LHY* homologues in the soybean genome were demonstrated to function redundantly in regulating flowering time and yield under short-day conditions (Dong et al., 2021).

Using RAD-seq, 182 wild soybean accessions were genotyped to obtain 72,574 SNPs for GWAS on salt tolerance (Jin et al., 2021). This is one of the few GWAS using only wild soybean accessions. Among the 209 genes associated with the 11 SNPs significantly linked to salt tolerance, nine were directly identified as salt tolerance-related, including a gene corresponding to *GmCHX1/GmSALT3/Ncl* (Jin et al., 2021). For the uncharacterized gene, *GsERD15B* (*Early Responsive to Dehydration 15B*), although there was no polymorphism identified in the protein sequence between salt-tolerant and salt-sensitive accessions, a 7-bp insertion was spotted in its promoter in most of the salt-sensitive ones (Jin et al., 2021), leading to reduced expression.

Overexpression of *GsERD15B* was shown to enhance the salt tolerance of soybean, probably through the activation of other stress-responsive genes (Jin et al., 2021).

Interactions between genetic loci could sometimes make these loci hard to identify. A GWAS involving 809 soybean accessions initially identified 150 loci significantly associated with 57 traits (Fang et al., 2017). However, some of the well-characterized loci, such as *dt2* which corresponds to semi-determinate growth, were not detected in this initial study due to the epistatic effect of the loci (Fang et al., 2017). To solve this problem, the 809 accessions were divided into subgroups according to their genotypes at the trait-associated loci identified in the first round, for a second round of GWAS. By employing this strategy, 95 additional secondary loci were identified, including a locus on chromosome 19 that interacts with *Ln* on chromosome 20 to control leaf area (Fang et al., 2017).

Another study genotyped 200 soybean germplasms by SLAF sequencing to obtain 28,926 high-quality SNPs for GWAS on isoflavone contents (Wu et al., 2020). Of the significantly associated SNPs, a peak was consistently identified on chromosome 8, which was associated with total isoflavone content and daidzein content across all the tested environments (Wu, Li, et al., 2020). Together with the results from biparental mapping and fine mapping, a mitogen-activated protein kinase (MPK)-encoding gene, *GmMPK1*, was identified. The overexpression of *GmMPK1* in low-isoflavone soybean accessions, either by transient expression in hairy root or by stable transformation, could increase the isoflavone contents in hairy root and seeds, respectively (Wu, Li, et al., 2020). The direct link between *GmMPK1* and isoflavone accumulation remains unknown. However, it is suggested that GmMPK1 participates in a MAPK-mediated signal transduction pathway to regulate isoflavone production upon pathogen challenge (Wu, Li, et al., 2020).

On some occasions, SNPs alone cannot provide sufficient resolution for association mapping. Other sequence variations such as INDELs could then be used to improve the marker density. While soybean is one of the most important oil crop, efforts have also been spent on discovering oil content related genes or loci. A seed protein and oil QTL was mapped to a 4-Mb region on chromosome 15 using a biparental RI population (Zhang et al., 2020). To further delimit this QTL, an association study was done using 631 soybean accessions. After analyzing the association between the protein/oil content and the 79,725 SNPs and 17,088 INDELs within the 4-Mb region, the QTL was finally narrowed down to a 34-kb region (Zhang et al., 2020).

Within this 34-kb region, a sugar transporter-encoding gene (*GmSWEET39*) was suggested to be the causative gene. A CC deletion in *GmSWEET39* has led to a truncation of the C-terminus of the resulting protein, which was associated with the low protein-high oil trait (Zhang et al., 2020). Apparently, high oil content was a more favorable trait in cultivated soybean, and the allele with the CC deletion was selected during soybean domestication, elevated from <3% prevalence in wild soybeans to over 85% in improved cultivars (Zhang et al., 2020).

In a GWAS using 219 soybean accessions with seed oil content ranging from 9.64% to 27.69% in six environmental conditions, 110 SNPs were identified to associated with the oil content (Zhang et al., 2019). Collapsing the 110 SNPs yield 3 QTL regions. Within the QTL on chromosome 20, a domestication related oleosin-encoding gene, *GmOLE1*, was identified based on the *Fst* value. Variations in the promoter of GmOLE1 was highly associated with the expression and oil content (Zhang, Zhang, et al., 2019). Ectopic overexpression of *GmOLE1* in soybean not only improved the seed oil content, it also increased overall seed yield per plant, and germination rate of the transgenic seeds (Zhang, Zhang, et al., 2019). The tradeoff was the reduction of protein content and 100-seed weight (Zhang, Zhang, et al., 2019).

The polyunsaturated fatty acid content determines the stability of soybean oil and also nutritional consequence. The major fatty acid in soybean oil in linoleic acid, which is a polyunsaturated fatty acid. A GWAS using 510 Chinese soybean accessions, with a linoleic acid range from 36.22% to 72.18%, identified 612 significant SNPs distributed on 9 chromosomes (Di et al., 2022). Among these region, a APETALA2/ethylene responsive element-binding protein (AP2/EREBP) transcription factor encoding gene on chromosome 4 was stably identified in 3 different conditions (Di et al., 2022). This gene was named *GmWRI4* after its homologue *WRINKLED1* (*AtWRI1*) from Arabidopsis. The expression of GmWRI4 was significantly negatively correlated with the linoleic acid (Di et al., 2022). Low linoleic acid was resulted in transgenic soybean seed overexpressing GmWRI4 due to the repression of fatty acid desaturase encoding gene *GmFAD2-1A* and *GmFAD2-1B* (Di et al., 2022).

Symbiotic nitrogen fixation is a unique feature of legumes that is agriculturally and environmentally important. A total of 5,294,054 markers, including both SNPs and INDELs, from 496 soybean accessions were used for GWAS on nodule numbers (Zhang et al., 2021). Only one locus spanning ~105 kb on chromosome 2 was found significantly passing the

Bonferroni threshold. Although there are 20 protein-coding gene models within the region, the only TIR-NBS-LRR-encoding gene, namely *GmNNL1*, was assumed to be the most probable candidate as TIR-NBS-LRR proteins are usually involved in plant-microbe interactions. Sequence analyses categorized the haplotypes of *GmNNL1* into two classes, either with or without a SINE-like transposon insertion. Genetics and functional analyses suggested that *GmNNL1* without the SINE-like transposon is the dominant functional allele that inhibits nodulation (Zhang, Wang, et al., 2021). The wide distribution of *GmNNL1* with a SINE insertion in both wild and cultivated soybeans suggests that the reduced symbiotic specificity was evolutionarily beneficial. It might give soybean a better nitrogen fixation ability and therefore might have improved its fitness in both natural and field environments.

4. Impacts of genomic research on soybean molecular breeding

4.1 Marker-assisted selection

Discovering the function-linked molecular markers and even identifying the underlying gene are imperative for introducing and merging elite alleles into germplasms to enhance soybean cultivar performance. Many types of molecular marker systems have been successfully developed and widely applied in various researches. For example, before high-throughput sequencing technologies were widely available, researchers and breeders made use of low-density restriction fragment length polymorphism (RFLP), amplified fragment length polymorphism (AFLP), random amplified polymorphic DNA (RAPD), sequence characterized amplified region (SCAR), simple sequence repeat (SSR), and cleaved amplified polymorphic sequence (CAPS). However, in the genomic era, with the release of soybean reference genomes and an increasing amount of genome-wide association studies on diverse traits, high-density molecular markers such as SNPs and INDELs are more frequently employed.

With the help of soybean reference genomes and high-throughput sequencing, many molecular markers have been developed with higher-than-ever linkages to the traits of interest, among which some are the functional markers located within the causal genes. This progress has significantly promote the efficiency of marker-assisted selection. Furthermore, marker-assisted selection together with recurrent backcrossing is an excellent strategy to introduce elite alleles into elite germplasms. A few examples of such successes are highlighted here.

Resistance to pod shattering is a domesticated trait that retains the seeds on the plant until harvest. A major QTL, *qPdh1*, corresponding to pod shattering resistance, were first located in the region between two SSR markers, Sat_366 and Sat_093, and then finally identified as a gene encoding a Dirigent-like protein (Funatsuki et al., 2006, 2014). These linked markers have been used to successfully assist the breeding of several pod-shattering-resistant cultivars, such as "Sachiyutaka A1 gou" (Hajika et al., 2016), "Fukuyutaka A1 gou" (Yamada et al., 2017), "Enreinosora" (Yamada et al., 2017), and "Kotoyutaka A1 gou" (MAFF, Japan).

Similarly, several QTLs associated with soybean cyst nematode (SCN) resistance, such as *Rhg4*, *rhg1s*, *rhg1g*, *rhg2g* and *rhg2s*, were discovered and their mapping information has been used in different breeding projects. For example, "Yukihomare R" and "Suzumaru R" are respectively derived from "Yukihomare" and "Suzumaru" through recurrent back-crossing. Apart from acquiring *rhg2g* from "Gendenshirazu," "Suzumaru R" also acquired *rhg1s* and *Rhg4* from PI 84751, resulting in a cultivar that is resistant to SCN races 1 and 3 (Kurosaki et al., 2017; Suzuki et al., 2017). As for soybean mosaic virus (SMV) resistance, Kato et al. (2016) successfully conferred the resistance against SMV strains C and D on a leading variety, Ohsuzu, by transferred the *Rsv3* marker from a U.S. variety, Harosoy (Kato et al., 2016). Oki et al. (2011) introduced two QTLs, CCW-1 and CCW-2, into "Fukuyutaka" in developing the variety "Fukuminori" with antibiosis resistance (Oki et al., 2011). A null allele of kunitz trypsin inhibitor (KTI) from PI542044 was introgressed into JS97-52, DS9712 and DS9814 to get KTI-free lines (Kumar, Rani, Rawal, & Mourya, 2015; Maranna et al., 2016). To speed up the soybean improvement process, a fast-breeding strategy, integrating off-site generation advancement and marker-assisted selection to shorten the generation time, was established to allow at least four generations per year under natural growing conditions (Fang et al., 2021). In these successful cases, marker-assisted selection significantly promoted the efficiency of breeding. However, it appears that this method works better with the qualitative traits with high heritability, but the benefits to quantitative traits such as yield are still limited.

4.2 Genomic selection

Instead of directly selecting the markers associated with major-effect QTLs or functional genes using marker-assisted selection (MAS), genomic selection (GS) estimates the genomic-estimated-breeding-values (GEBVs) by considering the effects of all markers as a whole. It has been demonstrated that GS using 500 SNP markers could outperform MAS using 22 loci identified via GWAS for seed weight (Zhang, Song, Cregan, & Jiang, 2016). GS requires a training population to supply the genotype and phenotype information to establish a statistical model for the estimation of the marker effects. In turn, the GEBVs of the individuals in the test population can be calculated and cross-validated with the actual phenotype. To obtain more reliable GEBVs, the training cycle can be repeated until satisfactory.

With the emerging SNP and INDEL markers linked to all kinds of traits, GS can be used for both background and foreground selections. GS in soybean is in the early stage and apparently still requires a lot of optimization. For example, for soybean maturity and yield components, the training population size apparently was the most important factor in improving the prediction accuracy, more than the choice of statistical models and marker density. On average, ~9.1% improvement in accuracy can be observed by doubling the population size. However, once the population size exceeded 2,000, the effect on prediction accuracy improvement plateaued (Xavier, Muir, & Rainey, 2016). Similarly, to carry out GS for quantitative resistance toward *Phytophthora sojae*, when the population size was limited, the predictive power was improved with an increase in marker density of up to 1,000 SNPs, and the improvement in prediction accuracy diminished with any further increase in marker number (Rolling, Dorrance, & McHale, 2020). In this case, a simplified BARCSoySNP6K developed from the SoySNP50K assay would be sufficient for GS (Song et al., 2020).

One obstacle for GS is the missing genotypes, which were normally imputed, but they may not reflect the genotype effects. With the emergence of deep-learning technology, a framework based on convolutional neural networks (CNNs) was developed to predict the GEBV of soybean without the imputation of missing genotypes (Liu et al., 2019). The deep-learning model showed an overall improvement over conventional statistic models in predicting the yield, seed moisture and oil contents (Liu et al., 2019). Instead of using the entire set of SNPs, a GS study on SCN tolerance used only the SNPs from GWAS with logarithm of the odds (LOD) scores higher than 2 (Ravelombola et al., 2020). This strategy improved the GS accuracy by two times when compared to another strategy that employed the whole set of SNPs (Ravelombola et al., 2020). Furthermore, the same study also demonstrated that GS accuracy was dependent on the SNP set, training model, and training population size (Ravelombola et al., 2020). Similarly, using just 231 SNPs associated with amino acid contents for GS gave better GEBVs than those generated using 23,279 random SNPs (Qin et al., 2019).

A QTL-allele matrix of seed oil, oleic acid and linolenic acid contents was constructed for the Chinese soybean landrace population (CSLRP) and was used for the GS of optimal crosses with high transgressive potentials (Zhang et al., 2018). Since phenotype is normally determined by the haplotype instead of individual SNPs alone, haplotype-based analyses are expected to be able to better infer the phenotype. This has been confirmed by a GS using preselected markers based on haplotype blocks, which effectively improved the prediction accuracy for grain yield (Ma et al., 2016).

4.3 Random mutagenesis

Random mutagenesis is a traditional way of introducing variations into the soybean genome, usually achieved through physical methods such as irradiating soybean seeds with high-energy radiations or chemical methods such as treating soybean seeds with chemical mutagens to induce changes in their DNA. Other methods could involve the use of random T-DNA or transposon insertions into the genome to alter endogenous gene functions. These treated seeds would then be germinated and screened for the desired traits. These random processes may occasionally create beneficial mutations that improve certain traits of soybean. However, the outcomes of these processes are highly unpredictable, and the causal mutations were largely unknown before the genomic era. Multiple mutations could occur in the same treated seed and result in simultaneous changes in other traits besides the targeted one. Nevertheless, randomly induced mutagenesis could create new alleles that do not exist naturally, and thus, in its own right, is still an important tool in genetic studies and crop improvement especially when the causal gene of a certain trait is still unknown.

A soybean mutant collection was created by fast neutron bombardment of the seeds of soybean cultivar M92-220 (Bolon et al., 2011). Characterization of the genomes of these mutants, by either comparative genomic hybridization or resequencing, identified all sorts of mutations including deletion, tandem duplication, and translocation, etc. (Bolon et al., 2011, 2014). This mutant collection showed a wide range of variations in seed composition such as changes in protein and oil contents, and visible phenotype such as alterations in plant architecture (Bolon et al., 2011). Another soybean mutant collection was produced by ethyl methane sulfonate (EMS)-induced mutagenesis (Espina et al., 2018). The M2 plants also displayed a wide range of phenotypic variations. Such mutants are valuable for genetic studies and crop improvement.

It is relatively straightforward to identify the causal mutation using forward genetic approaches. In a nutshell, a mutant of interest can be crossed with an unrelated soybean variety. The causal mutation can then be identified through a bulked segregant analysis (BSA) of their offsprings. For example, the *chlorophyll-deficient gold yellow trifoliolate leaves* (*gyl*) mutation was mapped by BSA to a region on chromosome 13 containing a chlorophyll synthesis-associated gene (Li et al., 2017). The causal mutation resulting in brown-seededness in a fast-neutron mutant in the yellow-seeded Williams 82 background was identified using comparative genome hybridization (CGH) (Stacey et al., 2016). CGH analyses using five brown-seeded BC_1F_2 plants identified three deletions in common (Stacey et al., 2016). In another study, a fast neutron-induced chromosomal translocation between chromosomes 8 and 13 corresponding to the high-sucrose and low-oil phenotype of soybean seed was identified by BSA, CGH and genome resequencing. Using a genomic approach, this mutation may also be mapped by QTL mapping (Dobbels et al., 2017). This inter-chromosomal translocation may have disrupted a β-ketoacyl-[acyl carrier protein] synthase 1 (KAS1), which is essential for fatty acid biosynthesis, and led to the inefficient synthesis of oil from sucrose in seeds (Dobbels et al., 2017). PE2166 was a mutant of Pungsannamul with elevated α-linolenic acid (Johnson et al., 2021). An RI population was produced by crossing PE2166 with Daepung. A QTL corresponding to elevated α-linolenic acid was mapped to chromosome 5 containing a homeodomain-like transcriptional regulator-encoding gene that bears the causal mutation (Johnson et al., 2021). In cases where the major causal gene of the unmutated phenotype is well characterized, checking the sequence of the causal gene directly may provide the answer to the type and location of the mutation. For example, six mutants were identified to have high seed stearic acid contents compared to the wild type (Lakhssassi et al., 2017). As it is known that the metabolic enzyme stearoyl-acyl carrier protein desaturase (SACPD-C) is responsible for the conversion of stearic acid to oleic acids, all six mutants were verified to carry mutations in the *SACPD-C* gene (Lakhssassi et al., 2017).

One obstacle with generating randomly-induced mutants is that the locations of the mutations are unknown and could occur at multiple locations on the genome. Thus, using a reverse genetic approach to study the function of a given gene using randomly-induced mutants could be confusing. However, with the availability of soybean reference genomes and next-generation sequencing platforms, the high-throughput detection of mutations in targeted genes becomes more feasible (Rigola et al., 2009). In brief, DNAs of the mutants are mixed by multidimensional pooling. Targeted regions are amplified and indexed from different pools. The amplicons are then sequenced on next-generation sequencing platforms to identify the desired mutations. The loss-of-function mutant of *GmDNJ1* was isolated by such a strategy. A study on the *dnj1* mutant demonstrated that *GmDNJ1* played crucial roles in plant growth and heat stress tolerance (Li et al., 2021). A similar technology called TILLING-by-Sequencing (+) (TbyS[+]), which uses magnetic breads to enrich the target amplicon before sequencing, has successfully identified mutants from three desaturase gene families involved in the soybean fatty acid biosynthesis pathway (Lakhssassi et al., 2021). Furthermore, next-generation sequencing can also detect copy number variations in the mutant lines. Using genotyping-by-sequencing technology, copy number variations of up to 50 Mb were detected in a fast-neutron mutant population (Lemay et al., 2019).

The application of the activation tagging technique could induce the expressions of tagged genes while at the same time providing an anchor with a known sequence for later identification of the tagged genes. A variant of the rice *mPing*-based activation tag called *mmPing20* was adopted in soybean research (Johnson et al., 2021). Expression analyses showed that *mmPing20* could actually activate the expressions of nearby genes in the soybean genome (Johnson et al., 2021). The drawback of this method is that the activation of genes requires exogenous DNA. As a result, the resulting mutant is regarded as a genetically-modified organism, which is in general not suitable for commercialization as a food crop.

4.4 Genome editing in soybean

Compared to traditional transgenesis, new crop varieties generated by genome editing are in general more acceptable to the public. The main reason is that although the process still requires the introduction of exogenous DNA into the plant, the foreign DNA can be completely removed afterward, leaving only the edited site in the genome. Genomic research has played a fundamental role in genome editing, especially for those highly duplicated genomes such as that of soybean. Targets for editing are usually determined by homology search, genetic studies, mapping, or association studies. Building good reference genomes with precise annotations followed by in-depth studies is the key for identifying the right targets for editing.

4.4.1 TALEN

Transcription activator-like effector nucleases (TALENs) are sequence-specific DNA endonucleases engineered from TAL effectors, where the transcription activation domain is replaced by an endonuclease domain, specifically the *Fok*I domain, and the DNA-binding domain is a customizable stretch of 34-amino-acid repeats. These repeats vary only at amino acid positions 12 and 13, so that each repeat binds one specific nucleotide on the DNA (Sprink, Metje, & Hartung, 2015). Thus, by designing a unique combination of repeats, TALEN can act like a pair of scissors to make a double-stranded break (DSB) in a specific location on the DNA determined by the corresponding combination of repeats. The DSB is usually repaired via two main pathways including the error-prone nonhomologous end-joining (NHEJ) and the homology-directed repair (HDR) to introduce insertions, deletions, substitutions, or even larger chromosomal rearrangements. So far, TALEN-based genome editing has successfully been applied in diverse plants and crops, such as Arabidopsis, tomato, rice, barley and wheat (Sprink et al., 2015). As for soybean, there were also several successful cases. It has been demonstrated that engineered TALENs can disrupt both copies of *fatty acid desaturase 2* genes (*FAD2-1A* and *FAD2-1B*) to decrease the levels of polyunsaturated fats to increase the stability of soybean oil against oxidation and make it healthier for human consumption (Haun et al., 2014). Another mutation of *FAD3A* was stacked into the *fad2-1a fad2-1b* mutant to further decrease the linoleic acid level from 5.1% in the double mutant to 2.7% in the triple mutant and increase the oleic acid from 77.5% to 82.2% (Demorest et al., 2016). The first gene-edited crop commercialized in the U.S. was also a high-oleic low-linolenic acid soybean engineered using TALEN (Li, Wang, et al., 2020).

4.4.2 ZFN

Zinc-finger nuclease (ZFN) is another synthetic engineered nuclease with a *Fok*I nonspecific endonuclease domain fused to a customized array of engineered zinc fingers, where each finger typically recognizes three nucleotides (Urnov, Rebar, Holmes, Zhang, & Gregory, 2010). Thus, with a well-designed array of zinc fingers, a ZFN can also help edit the genome at a specific location. In soybean, Curtin et al. (2011) first demonstrated that custom-designed ZFNs could carry out transmissible mutations of soybean genes, specifically *DCL4a* and *DCL4b* in their work (Curtin et al., 2011). Not only can they create INDELs, ZFNs have also been demonstrated to integrate a large DNA fragment containing four genes into the soybean genome through NHEJ. However, the rate of successful incorporation in the regenerated plant was only 0.03%. Therefore further improvement in this technique is required (Bonawitz et al., 2019).

4.4.3 CRISPR/Cas9

The CRISPR (clustered regularly interspaced short palindromic repeat)/Cas9 (CRISPR-associated endonuclease 9) system is an emerging DSB technology with great promise. Different from the previous TALEN and ZFN platforms which rely on protein-nucleotide interactions to recognize a specific DNA sequence, the CRISPR/Cas9 system generally uses a simple short guide RNA (gRNA) to recognize a specific DNA sequence followed by a protospacer adjacent motif (PAM) to generate a DSB in the targeted sequence. More specifically, for the *Streptococcus pyogenes* CRISPR/Cas9 system, the PAM consensus sequence is 5'-NGG-3', and the cleavage normally occurs between the third and fourth nucleotides upstream of the PAM. The cleaved DNA will then be repaired by NHEJ or HDR in the presence of a sequence with homology from either an exogenous or an endogenous source.

The history of CRISPR/cas9 genome editing in soybean is short. Due to the low efficiency of stable transformations in whole soybean plants, transient expression methods, such as soybean hairy root transformation, were first utilized to demonstrate that CRISPR/Cas9 system can effectively edit both endogenous and exogenous genes, including the *bar* gene, *GFP*, *GmFEI2*, *GmSHR*, *Glyma06g14180* (a homolog of *TTG1*), *Glyma08g02290* (a homolog of *KUP4*), and *Glyma12g37050* (a homolog of *ETR1*) (Cai et al., 2015; Jacobs, LaFayette, Schmitz, & Parrott, 2015; Sun et al., 2015). Furthermore, HDR gene integration and replacement were also demonstrated to be feasible through cotransforming the target site-specific Cas9-sgRNA and donor DNA constructs by particle bombardment (Li et al., 2015). Since then, CRISPR/Cas9 has been increasingly utilized in various aspects of soybean research and breeding.

The CRISPR/Cas9 system has significantly facilitated the precise identification of the causal genes for important traits, especially when combining its use in the construction of a knock-out mutant by targeted mutagenesis with a conventional overexpression-complementation experiment. For example, *Rj4*, a gene that restricts certain *Bradyrhizobium* strains, was originally reported to be encoded by *Glyma.01g165800*. However, it was later found to be encoded by *Glyma.0165800-D*, the neighboring gene duplicate of *Glyma.01g165800*. Knocking out *Glyma.01G165800-D* by CRISPR/cas9 abolished the function of *Rj4*, i.e. allowing the nodulation by *B. elkanii*, but knocking out *Glyma.01G165800* did not (Tang, Yang, Liu, & Zhu, 2015). This provided concrete evidence to support that *Glyma.01G165800-D* is in fact *Rj4*. A fast neutron-induced six-gene deletion was found to be associated with the short-trichome phenotype (Campbell et al., 2019). Among the six genes, *Glyma.06g145800* encoding a homologue of *Arabidopsis* Constitutive Expression of PR Genes 5 (CPR5) was the strongest candidate (Campbell et al., 2019). A panel of mutants with different numbers of base deletions in *CPR5* were generated by CRISPR/cas9 genome editing. All these mutants showed a significant reduction in the trichome length, with was highly consistent with the fast-neutron mutant phenotype (Campbell et al., 2019). Similarly, the causal gene *GmMS1* (*male-sterility 1*), located in a 39-kb region where five genes were deleted, was demonstrated by CRISPR/cas9 to be *Glyma13G114200* (Jiang et al., 2021). *GmKIX8-1* encodes a KIX domain-containing protein. A CRISPR/cas9 mutant of *GmKIX8-1* showed increased cell proliferation and enlarged organ size. It was shown to be the causative gene for the major seed weight QTL qSw17-1 (Nguyen, Paddock, Zhang, & Stacey, 2021). As mentioned previously, although there are limited mutant collections for soybean, the CRISPR/Cas9 system has greatly facilitated the forward genetics studies on soybean.

In addition to basic research, CRISPR/Cas9 also enables the direct change in elite traits to promote soybean breeding and immediate applications. It is generally agreed that flowering time and maturity are the most important traits for soybean breeding. Research is focused on producing soybeans with times to maturity that are suited to different latitudes. For example, the flowering integrator *GmFT2a* and *GmFT5a* were knocked out to delay flowering time (Cai et al., 2018, 2020). The resulting double mutant *ft2aft5a*, which displays a late-flowering phenotype, is suitable for low-latitude high-temperature environments. On the other hand, to promote the adaptation to high-latitude environments where early flowering is required, CRISPR/Cas9-edited *E1* shortened the time to flowering as desired (Han et al., 2019). While there is a basket of maturity-related genes found in the soybean genome, strategic editing of them could provide more refined controls on soybean ecological adaptations. Furthermore, seven cryptochrome genes (*GmCRYs*) in soybean were edited. Among them, GmCRY1s was found to regulate soybean shade avoidance in response to reduced blue light, through modulating gibberellin metabolism (Lyu et al., 2021), thus providing a new target for improving soybean performance under high-density cultivation in the field.

CRISPR/cas9 has also been applied to improve qualitative traits. The Δ12-fatty acid desaturase II (FAD2) family of genes were targeted by CRISPR/Cas9 to improve the oil profile, similar to the process using TALEN technology (Do et al., 2019; Wu et al., 2020). Gly m Bd 28 K and Gly m Bd 30 K are two major allergens present in soybean. Their nullification by CRISPR/Cas9 has been demonstrated to be a feasible way of producing hypoallergenic soybeans (Sugano et al., 2020).

To potentially improve disease resistance in soybean by introducing variations to existing genes, two regions of tandem-duplicated arrays of NBS-LRR (nucleotide-binding-site leucine-rich-repeat) gene families, Rpp1 and Rps1, were targeted by CRISPR/Cas9 to induce rearrangements and produce novel chimeric paralogues that could lead to new disease resistance specificities (Nagy et al., 2021). CRISPR/cas9 can not only target protein-coding genes in the soybean genome, but it can also be used to edit other regulatory elements. For example, it has been demonstrated that the knockout of microRNA miR169c by CRISPR/cas9 could promote nodulation through the repression of soybean *Nuclear Factor-Y Subunit A* (*GmNFYA*) (Xu et al., 2021).

4.4.4 Potential applications of base-editing using CRISPR/Cas9

The commonly used CRISPR/Cas9 system usually disrupts the gene function by introducing a frame-shift variation. However, the complete knockout of a gene may lead to undesirable pleiotropic effects. In fact, variations at the base level could lead to enhanced or diminished protein functions through the alteration of three-dimensional structures, enzymatic activities or interactions with other molecules. Recent genomic researches have identified rich sequence variations in the soybean genome (Kajiya-Kanegae et al., 2021; Lam et al., 2010; Valliyodan et al., 2021; Yang et al., 2021; Zhou et al., 2015). Instead of incorporating a random combination of these variations through traditional breeding, the precise introduction of a desired sequence variation by base editing would be a much better option. In soybean, a CRISPR/Cas9-mediated base-editing tool was developed through combining the Cas9^{D10A} nickase, the rat cytosine deaminase (APOBEC1) or the uracil glycosylase inhibitor (UGI). This system successfully introduced an edited base at the target position in *GmFT2a* and *GmFT4a* (Cai et al., 2020). The mutation from C to G in *GmFT2a* has led to a significant delay in flowering time when compared to the wild type, but a slightly earlier flowering time when compared to the *ft2a* frameshift-knockout line (Cai, Chen, et al., 2020). This example has sufficiently demonstrated the potential of base editing in soybean. However, the current base-editing tool is still limited and need to be further improved.

5. Perspectives

Although soybean is the major dietary source of plant-based protein, its yield is still much lower than those of other major crops including rice, wheat, and maize (FAOstat [retrieved on Oct 5, 2021, from http://www.fao.org/faostat/en/#data/QCL]). Furthermore, due to population explosion and climate change, food security has become the major challenge in the coming century. Thus, increasing yield, improving yield stability, and enhancing added values are the major goals for soybean breeding. To achieve these goals, genome sequencing has fueled soybean research in the past decade, accumulating vast amounts of information to serve as the basis for soybean breeding.

So far, thousands of soybean accessions have been sequenced in various projects. In most cases, sequencing data are required by publishers of scientific journals to be deposited in public domains, but genetic resources are not. Thus, even though genomic information is available for sophisticated bioinformatic analyses, one cannot test a hypothesis or collect new phenotypic data for analyses without access to the seeds. To get the most out of the published data in order to advance soybean research and breeding, the sharing of soybean germplasm resources between research groups and across national boarders would benefit the whole soybean research community. Taking it a step further, raw phenotypic data should also be systematically shared. In contrast to the volume of available genomic data, phenotypic data are the biggest limiting factor for association studies. The phenotyping of hundreds of soybean accessions is already tedious enough, let alone phenotyping thousands of accessions at once. At the same time, datasets collected across different studies are normally considered not directly comparable among one another. To solve this problem, high-throughput phenotyping systems, either ground-based or unmanned aerial system (UAS)-based, have drawn the attention of researchers (Baek et al., 2020;

Bai, Ge, Hussain, Baenziger, & Graef, 2016; Kaler et al., 2020; Trevisan, Perez, Schmitz, Diers, & Martin, 2020). On the other hand, although directly combining the data from multiple studies for genome-wide association studies (GWAS) is improper, meta-analyses of the results of multiple GWAS have been demonstrated to increase the statistical power and lead to the discovery of new genetic loci (Evangelou & Ioannidis, 2013; Shook et al., 2021).

Simply breeding high-yielding soybean cultivars is apparently not sufficient to fulfill the demand for food in the second half of the 21st century. Serving as a carbon sink and a nitrogen source in a climate-smart agricultural system, new soybean varieties would also need to be highly adaptive to environmental changes and require low agricultural inputs. Although these goals are difficult to attain, new technologies such as machine learning techniques/artificial intelligence have emerged to contribute to both soybean genomic analyses and breeding (Herrero-Huerta, Rodriguez-Gonzalvez, & Rainey, 2020; Yoosefzadeh-Najafabadi, Earl, Tulpan, Sulik, & Eskandari, 2021), giving rise to new possibilities.

Acknowledgments

This work was supported by Hong Kong Research Grants Council Area of Excellence Scheme (AoE/M-403/16), Lo Kwee-Seong Biomedical Research Fund and Guangdong Provincial Department of Science and Technology 2020 Key Areas Research and Development Programs: Breeding for High Yield and High Quality New Soybean Cultivars for Tropics and Subtropics (2020B020220008), the China Agriculture Research System (CARS-04) and the CAAS Agricultural Science and Technology Innovation Project. JY Chu copy-edited the text. Any opinions, findings, conclusions or recommendations expressed in this publication do not reflect the views of the Government of Hong Kong Special Administrative Region or the Innovation and Technology Commission.

References

Abel, G. H. (1969). Inheritance of the capacity for chloride inclusion and chloride exclusion by soybeans. *Crop Science*, *9*, 697–698. https://doi.org/10.2135/cropsci1969.0011183X000900060006x.

Arumuganathan, K., & Earle, E. D. (1991). Nuclear DNA content of some important plant species. *Plant Molecular Biology Reporter*, *9*, 208–218. https://doi.org/10.1007/BF02672069.

Baek, J., Lee, E., Kim, N., Kim, S. L., Choi, I., Ji, H., et al. (2020). High throughput phenotyping for various traits on soybean seeds using image analysis. *Sensors*, *20*, 248. https://doi.org/10.3390/s20010248.

Bai, G., Ge, Y. F., Hussain, W., Baenziger, P. S., & Graef, G. (2016). A multi-sensor system for high throughput field phenotyping in soybean and wheat breeding. *Computers and Electronics in Agriculture*, *128*, 181–192. https://doi.org/10.1016/j.compag.2016.08.021.

Bayer, P. E., Valliyodan, B., Hu, H. F., Marsh, J. I., Yuan, Y. X., Vuong, T. D., et al. (2021). Sequencing the USDA core soybean collection reveals gene loss during domestication and breeding. *Plant Genome*, e20109. https://doi.org/10.1002/tpg2.20109.

Bolon, Y. T., Haun, W. J., Xu, W. W., Grant, D., Stacey, M. G., Nelson, R. T., et al. (2011). Phenotypic and genomic analyses of a fast neutron mutant population resource in soybean. *Plant Physiology*, *156*, 240–253. https://doi.org/10.1104/pp.110.170811.

Bolon, Y. T., Stec, A. O., Michno, J. M., Roessler, J., Bhaskar, P. B., Ries, L., et al. (2014). Genome resilience and prevalence of segmental duplications following fast neutron irradiation of soybean. *Genetics*, *198*, 967–981. https://doi.org/10.1534/genetics.114.170340.

Bonawitz, N. D., Ainley, W. M., Itaya, A., Chennareddy, S. R., Cicak, T., Effinger, K., et al. (2019). Zinc finger nuclease-mediated targeting of multiple transgenes to an endogenous soybean genomic locus via non-homologous end joining. *Plant Biotechnology Journal*, *17*, 750–761. https://doi.org/10.1111/pbi.13012.

Brown, A. V., Conners, S. I., Huang, W., Wilkey, A. P., Grant, D., Weeks, N. T., et al. (2021). A new decade and new data at SoyBase, the USDA-ARS soybean genetics and genomics database. *Nucleic Acids Research*, *49*, D1496–D1501. https://doi.org/10.1093/nar/gkaa1107.

Cai, Y., Chen, L., Liu, X., Guo, C., Sun, S., Wu, C., et al. (2018). CRISPR/Cas9-mediated targeted mutagenesis of *GmFT2a* delays flowering time in soya bean. *Plant Biotechnology Journal*, *16*, 176–185. https://doi.org/10.1111/pbi.12758.

Cai, Y. P., Chen, L., Liu, X. J., Sun, S., Wu, C. X., Jiang, B. J., et al. (2015). CRISPR/Cas9-mediated genome editing in soybean hairy roots. *PLoS One*, *10*, e0136064. https://doi.org/10.1371/journal.pone.0136064.

Cai, Y. P., Chen, L., Zhang, Y., Yuan, S., Su, Q., Sun, S., et al. (2020). Target base editing in soybean using a modified CRISPR/Cas9 system. *Plant Biotechnology Journal*, *18*, 1996–1998. https://doi.org/10.1111/pbi.13386.

Cai, Y. P., Wang, L. W., Chen, L., Wu, T. T., Liu, L. P., Sun, S., et al. (2020). Mutagenesis of *GmFT2a* and *GmFT5a* mediated by CRISPR/Cas9 contributes for expanding the regional adaptability of soybean. *Plant Biotechnology Journal*, *18*, 298–309. https://doi.org/10.1111/pbi.13199.

Campbell, B. W., Hoyle, J. W., Bucciarelli, B., Stec, A. O., Samac, D. A., Parrott, W. A., et al. (2019). Functional analysis and development of a CRISPR/Cas9 allelic series for a *CPR5* ortholog necessary for proper growth of soybean trichomes. *Scientific Reports*, *9*, 14757. https://doi.org/10.1038/s41598-019-51240-7.

Chu, J. S., Peng, B., Tang, K., Yi, X., Zhou, H., Wang, H., et al. (2021). Eight soybean reference genome resources from varying latitudes and agronomic traits. *Scientific Data*, *8*, 164. https://doi.org/10.1038/s41597-021-00947-2.

Chung, W. H., Jeong, N., Kim, J., Lee, W. K., Lee, Y. G., Lee, S. H., et al. (2014). Population structure and domestication revealed by high-depth resequencing of Korean cultivated and wild soybean genomes. *DNA Research*, *21*, 153–167. https://doi.org/10.1093/dnares/dst047.

Curtin, S. J., Zhang, F., Sander, J. D., Haun, W. J., Starker, C., Baltes, N. J., et al. (2011). Targeted mutagenesis of duplicated genes in soybean with zinc-finger nucleases. *Plant Physiology*, *156*, 466–473. https://doi.org/10.1104/pp.111.172981.

Demorest, Z. L., Coffman, A., Baltes, N. J., Stoddard, T. J., Clasen, B. M., Luo, S., et al. (2016). Direct stacking of sequence-specific nuclease-induced mutations to produce high oleic and low linolenic soybean oil. *BMC Plant Biology*, *16*, 225. https://doi.org/10.1186/s12870-016-0906-1.

Dhanapal, A. P., Ray, J. D., Singh, S. K., Hoyos-Villegas, V., Smith, J. R., Purcell, L. C., et al. (2016). Genome-wide association mapping of soybean chlorophyll traits based on canopy spectral reflectance and leaf extracts. *Bmc Plant Biology*, *16*, 174. https://doi.org/10.1186/s12870-016-0861-x.

Di, Q., Piersanti, A., Zhang, Q., Miceli, C., Li, H., & Liu, X. Y. (2022). Genome-wide association study identifies candidate genes related to the linoleic acid content in soybean seeds. *International Journal of Molecular Sciences, 23*, 454. https://doi.org/10.3390/ijms23010454.

Do, P. T., Nguyen, C. X., Bui, H. T., Tran, L. T. N., Stacey, G., Gillman, J. D., et al. (2019). Demonstration of highly efficient dual gRNA CRISPR/Cas9 editing of the homeologous *GmFAD2–1A* and *GmFAD2–1B* genes to yield a high oleic, low linoleic and α-linolenic acid phenotype in soybean. *BMC Plant Biology, 19*, 311. https://doi.org/10.1186/s12870-019-1906-8.

Dobbels, A. A., Michno, J. M., Campbell, B. W., Virdi, K. S., Stec, A. O., Muehlbauer, G. J., et al. (2017). An induced chromosomal translocation in soybean disrupts a KASI ortholog and is associated with a high-sucrose and low-oil seed phenotype. *G3-Genes Genomes Genetics, 7*, 1215–1223. https://doi.org/10.1534/g3.116.038596.

DOE-JGI. (2021). Glycine max Fiskeby v1.1. Retrieved from https://phytozome-next.jgi.doe.gov/info/GmaxFiskeby_v1_1.

Dong, L., Fang, C., Cheng, Q., Su, T., Kou, K., Kong, L., et al. (2021). Genetic basis and adaptation trajectory of soybean from its temperate origin to tropics. *Nature Communications, 12*, 5445. https://doi.org/10.1038/s41467-021-25800-3.

dos Santos, J. V. M., Valliyodan, B., Joshi, T., Khan, S. M., Liu, Y., Wang, J. X., et al. (2016). Evaluation of genetic variation among Brazilian soybean cultivars through genome resequencing. *BMC Genomics, 17*, 110. https://doi.org/10.1186/s12864-016-2431-x.

Espina, M. J., Ahmed, C. M. S., Bernardini, A., Adeleke, E., Yadegari, Z., Arelli, P., et al. (2018). Development and phenotypic screening of an ethyl methane sulfonate mutant population in soybean. *Frontiers in Plant Science, 9*, 394. https://doi.org/10.3389/fpls.2018.00394.

Evangelou, E., & Ioannidis, J. P. A. (2013). Meta-analysis methods for genome-wide association studies and beyond. *Nature Reviews Genetics, 14*, 379–389. https://doi.org/10.1038/nrg3472.

Fang, C., Ma, Y. M., Wu, S. W., Liu, Z., Wang, Z., Yang, R., et al. (2017). Genome-wide association studies dissect the genetic networks underlying agronomical traits in soybean. *Genome Biology, 18*, 161. https://doi.org/10.1186/s13059-017-1289-9.

Fang, Y. D., Wang, L. W., Sapey, E., Fu, S., Wu, T. T., Zeng, H. Y., et al. (2021). Speed-breeding system in soybean: Integrating off-site generation advancement, fresh seeding, and marker-assisted selection. *Frontiers in Plant Science, 12*, 717077. https://doi.org/10.3389/fpls.2021.717077.

Findley, S. D., Cannon, S., Varala, K., Du, J. C., Ma, J. X., Hudson, M. E., et al. (2010). A fluorescence *in situ* hybridization system for karyotyping soybean. *Genetics, 185*, 727–744. https://doi.org/10.1534/genetics.109.113753.

Funatsuki, H., Ishimoto, M., Tsuji, H., Kawaguchi, K., Hajika, M., & Fujino, K. (2006). Simple sequence repeat markers linked to a major QTL controlling pod shattering in soybean. *Plant Breeding, 125*, 195–197. https://doi.org/10.1111/j.1439-0523.2006.01199.x.

Funatsuki, H., Suzuki, M., Hirose, A., Inaba, H., Yamada, T., Hajika, M., et al. (2014). Molecular basis of a shattering resistance boosting global dissemination of soybean. *Proceedings of the National Academy of Sciences of the United States of America, 111*, 17797–17802. https://doi.org/10.1073/pnas.1417282111.

Guan, R. X., Qu, Y., Guo, Y., Yu, L. L., Liu, Y., Jiang, J. H., et al. (2014). Salinity tolerance in soybean is modulated by natural variation in *GmSALT3*. *The Plant Journal, 80*, 937–950. https://doi.org/10.1111/tpj.12695.

Hajika, M., Funatsuki, H., Yamada, T., Takahashi, K., Hishinuma, A., Hirata, K., et al. (2016). Development of a new pod dehiscence-resistant soybean cultivar 'Sachiyutaka A1 gou'. *Bulletin of NARO Institute of Crop Science, 16*, 1–34.

Han, J. N., Guo, B. F., Guo, Y., Zhang, B., Wang, X. B., & Qiu, L. J. (2019). Creation of early flowering germplasm of soybean by CRISPR/Cas9 technology. *Frontiers in Plant Science, 10*, 1446. https://doi.org/10.3389/fpls.2019.01446.

Haun, W., Coffman, A., Clasen, B. M., Demorest, Z. L., Lowy, A., Ray, E., et al. (2014). Improved soybean oil quality by targeted mutagenesis of the *fatty acid desaturase 2* gene family. *Plant Biotechnology Journal, 12*, 934–940. https://doi.org/10.1111/pbi.12201.

Herrero-Huerta, M., Rodriguez-Gonzalvez, P., & Rainey, K. M. (2020). Yield prediction by machine learning from UAS-based multi-sensor data fusion in soybean. *Plant Methods, 16*, 78. https://doi.org/10.1186/s13007-020-00620-6.

Huang, M. K., Zhang, L., Zhou, L. M., Yung, W. S., Li, M. W., & Lam, H. M. (2021). Genomic features of open chromatin regions (OCRs) in wild soybean and their effects on gene expressions. *Genes, 12*, 640. https://doi.org/10.3390/genes12050640.

Jacobs, T. B., LaFayette, P. R., Schmitz, R. J., & Parrott, W. A. (2015). Targeted genome modifications in soybean with CRISPR/Cas9. *BMC Biotechnology, 15*, 16. https://doi.org/10.1186/s12896-015-0131-2.

Jiang, B. J., Chen, L., Yang, C. Y., Wu, T. T., Yuan, S., Wu, C. X., et al. (2021). The cloning and CRISPR/Cas9-mediated mutagenesis of a male sterility gene *MS1* of soybean. *Plant Biotechnology Journal, 19*, 1098–1100. https://doi.org/10.1111/pbi.13601.

Jin, T., Sun, Y. Y., Shan, Z., He, J. B., Wang, N., Gai, J. Y., et al. (2021). Natural variation in the promoter of *GsERD15B* affects salt tolerance in soybean. *Plant Biotechnology Journal, 19*, 1155–1169. https://doi.org/10.1111/pbi.13536.

Jing, Y., Teng, W. L., Qiu, L. J., Zheng, H. K., Li, W. B., Han, Y. P., et al. (2021). Genetic dissection of soybean partial resistance to sclerotinia stem rot through genome wide association study and high throughout single nucleotide polymorphisms. *Genomics, 113*, 1262–1271. https://doi.org/10.1016/j.ygeno.2020.10.042.

Johnson, A., Mcassey, E., Diaz, S., Reagin, J., Redd, P. S., Parrilla, D. R., et al. (2021). Development of mPing-based activation tags for crop insertional mutagenesis. *Plant Direct, 5*, e00300. https://doi.org/10.1002/pld3.300.

Kajiya-Kanegae, H., Nagasaki, H., Kaga, A., Hirano, K., Ogiso-Tanaka, E., Matsuoka, M., et al. (2021). Whole-genome sequence diversity and association analysis of 198 soybean accessions in mini-core collections. *DNA Research, 28*, dsaa032. https://doi.org/10.1093/dnares/dsaa032.

Kaler, A. S., Abdel-Haleem, H., Fritschi, F. B., Gillman, J. D., Ray, J. D., Smith, J. R., et al. (2020). Genome-wide association mapping of dark green color index using a diverse panel of soybean accessions. *Scientific Reports, 10*, 5166. https://doi.org/10.1038/s41598-020-62034-7.

Kato, S., Takada, Y., Shimamura, S., Hirata, K., Sayama, T., Taguchi-Shiobara, F., et al. (2016). Transfer of the *Rsv3* locus from 'Harosoy' for resistance to soybean mosaic virus strains C and D in Japan. *Breeding Science, 66*, 319–327. https://doi.org/10.1270/jsbbs.66.319.

Kim, M. Y., Lee, S., Van, K., Kim, T. H., Jeong, S. C., Choi, I. Y., et al. (2010). Whole-genome sequencing and intensive analysis of the undomesticated soybean (*Glycine soja* Sieb. and Zucc.) genome. *Proceedings of the National Academy of Sciences of the United States of America, 107*, 22032–22037. https://doi.org/10.1073/pnas.1009526107.

Kumar, V., Rani, A., Rawal, R., & Mourya, V. (2015). Marker assisted accelerated introgression of null allele of kunitz trypsin inhibitor in soybean. *Breeding Science, 65*, 447–452. https://doi.org/10.1270/jsbbs.65.447.

Kurosaki, H., Fujita, S., Ohnishi, S., Kosaka, F., Tanaka, Y., Takeuchi, T., et al. (2017). A new soybean variety 'Suzumaru R'. *Bulletin of Hokkaido Research Organization Agricultural Experiment Stations, 101*, 1–13.

Lakhssassi, N., Colantonio, V., Flowers, N. D., Zhou, Z., Henry, J., Liu, S. M., et al. (2017). Stearoyl-acyl carrier protein desaturase mutations uncover an impact of stearic acid in leaf and nodule structure. *Plant Physiology*, *174*, 1531–1543. https://doi.org/10.1104/pp.16.01929.

Lakhssassi, N., Zhou, Z., Cullen, M. A., Badad, O., El Baze, A., Chetto, O., et al. (2021). TILLING-by-sequencing+ to decipher oil biosynthesis pathway in soybeans: A new and effective platform for high-throughput gene functional analysis. *International Journal of Molecular Sciences*, *22*, 4219.

Lam, H. M., Xu, X., Liu, X., Chen, W. B., Yang, G. H., Wong, F. L., et al. (2010). Resequencing of 31 wild and cultivated soybean genomes identifies patterns of genetic diversity and selection. *Nature Genetics*, *42*, 1053–U1041. https://doi.org/10.1038/ng.715.

Lee, G. J., Boerma, H. R., Villagarcia, M. R., Zhou, X., Carter, T. E., Li, Z., et al. (2004). A major QTL conditioning salt tolerance in S-100 soybean and descendent cultivars. *Theoretical and Applied Genetics*, *109*, 1610–1619. https://doi.org/10.1007/s00122-004-1783-9.

Lee, G. A., Crawford, G. W., Liu, L., Sasaki, Y., & Chen, X. X. (2011). Archaeological doybean (*Glycine max*) in East Asia: Does size matter? *PLoS One*, *6*, e26720. https://doi.org/10.1371/journal.pone.0026720.

Lei, L., Goltsman, E., Goodstein, D., AlbertWu, G., Rokhsar, D. S., & Vogel, J. P. (2021). Plant pan-genomics comes of age. *Annual Review of Plant Biology*, *72*, 411–435. https://doi.org/10.1146/annurev-arplant-080720-105454.

Lemay, M. A., Torkamaneh, D., Rigaill, G., Boyle, B., Stec, A. O., Stupar, R. M., et al. (2019). Screening populations for copy number variation using genotyping-by-sequencing: A proof of concept using soybean fast neutron mutants. *BMC Genomics*, *20*, 634. https://doi.org/10.1186/s12864-019-5998-1.

Li, Z. F., Jiang, L. X., Ma, Y. S., Wei, Z. Y., Hong, H. L., Liu, Z. X., et al. (2017). Development and utilization of a new chemically-induced soybean library with a high mutation density. *Journal of Integrative Plant Biology*, *59*, 60–74. https://doi.org/10.1111/jipb.12505.

Li, C., Li, Y. H., Li, Y., Lu, H., Hong, H., Tian, Y., et al. (2020). A domestication-associated gene *GmPRR3b* regulates the circadian clock and flowering time in soybean. *Molecular Plant*, *13*, 745–759. https://doi.org/10.1016/j.molp.2020.01.014.

Li, Z. S., Liu, Z. B., Xing, A. Q., Moon, B. P., Koellhoffer, J. P., Huang, L. X., et al. (2015). Cas9-Guide RNA directed genome editing in soybean. *Plant Physiology*, *169*, 960–970. https://doi.org/10.1104/pp.15.00783.

Li, M. W., Wang, Z. L., Jiang, B. J., Kaga, A., Wong, F. L., Zhang, G. H., et al. (2020). Impacts of genomic research on soybean improvement in East Asia. *Theoretical and Applied Genetics*, *133*, 1655–1678. https://doi.org/10.1007/s00122-019-03462-6.

Li, K. P., Wong, C. H., Cheng, C. C., Cheng, S. S., Li, M. W., Mansveld, S., et al. (2021). GmDNJ1, a type-I heat shock protein 40 (HSP40), is responsible for both Growth and heat tolerance in soybean. *Plant Direct*, *5*, e00298. https://doi.org/10.1002/pld3.298.

Li, Y. H., Zhou, G. Y., Ma, J. X., Jiang, W. K., Jin, L. G., Zhang, Z. H., et al. (2014). De novo assembly of soybean wild relatives for pan-genome analysis of diversity and agronomic traits. *Nature Biotechnology*, *32*, 1045–1052. https://doi.org/10.1038/nbt.2979.

Lin, X., Lin, W. G., Ku, Y. S., Wong, F. L., Li, M. W., Lam, H. M., et al. (2020). Analysis of soybean long non-coding RNAs reveals a subset of small peptide-coding transcripts. *Plant Physiology*, *182*, 1359–1374. https://doi.org/10.1104/pp.19.01324.

Liu, Y. C., Du, H. L., Li, P. C., Shen, Y. T., Peng, H., Liu, S. L., et al. (2020). Pan-genome of wild and cultivated soybeans. *Cell*, *182*, 162–176.e113. https://doi.org/10.1016/j.cell.2020.05.023.

Liu, Y., Wang, D. L., He, F., Wang, J. X., Joshi, T., & Xu, D. (2019). Phenotype prediction and genome-wide association study using deep convolutional neural network of soybean. *Frontiers in Genetics*, *10*, 1091. https://doi.org/10.3389/fgene.2019.01091.

Lu, S., Dong, L., Fang, C., Liu, S., Kong, L., Cheng, Q., et al. (2020). Stepwise selection on homeologous *PRR* genes controlling flowering and maturity during soybean domestication. *Nature Genetics*, *52*, 428–436. https://doi.org/10.1038/s41588-020-0604-7.

Lu, X., Xiong, Q., Cheng, T., Li, Q. T., Liu, X. L., Bi, Y. D., et al. (2017). A *PP2C-1* allele underlying a quantitative trait locus enhances soybean 100-seed weight. *Molecular Plant*, *10*, 670–684. https://doi.org/10.1016/j.molp.2017.03.006.

Lu, S. J., Zhao, X. H., Hu, Y. L., Liu, S. L., Nan, H. Y., Li, X. M., et al. (2017). Natural variation at the soybean *J* locus improves adaptation to the tropics and enhances yield. *Nature Genetics*, *49*, 773–779. https://doi.org/10.1038/ng.3819.

Lyu, X. G., Cheng, Q. C., Qin, C., Li, Y. H., Xu, X. Y., Ji, R. H., et al. (2021). GmCRY1s modulate gibberellin metabolism to regulate soybean shade avoidance in response to reduced blue light. *Molecular Plant*, *14*, 298–314. https://doi.org/10.1016/j.molp.2020.11.016.

Ma, Y. S., Reif, J. C., Jiang, Y., Wen, Z. X., Wang, D. C., Liu, Z. X., et al. (2016). Potential of marker selection to increase prediction accuracy of genomic selection in soybean (*Glycine max* L.). *Molecular Breeding*, *36*, 113. https://doi.org/10.1007/s11032-016-0504-9.

Maranna, S., Verma, K., Talukdar, A., Lal, S. K., Kumar, A., & Mukherjee, K. (2016). Introgression of null allele of Kunitz trypsin inhibitor through marker-assisted backcross breeding in soybean (*Glycine max* L. Merr.). *BMC Genetics*, *17*, 106. https://doi.org/10.1186/s12863-016-0413-2.

Nagy, E. D., Stevens, J. L., Yu, N., Hubmeier, C. S., LaFaver, N., Gillespie, M., et al. (2021). Novel disease resistance gene paralogs created by CRISPR/Cas9 in soy. *Plant Cell Reports*, *40*, 1047–1058. https://doi.org/10.1007/s00299-021-02678-5.

Nguyen, C. X., Paddock, K. J., Zhang, Z., & Stacey, M. G. (2021). *GmKIX8-1* regulates organ size in soybean and is the causative gene for the major seed weight QTL *qSw17-1*. *New Phytologist*, *229*, 920–934. https://doi.org/10.1111/nph.16928.

Oki, N., Komatsu, K., Sayama, T., Ishimoto, M., Takahashi, M., & Takahashi, M. (2011). Genetic analysis of antixenosis resistance to the common cutworm (*Spodoptera litura* Fabricius) and its relationship with pubescence characteristics in soybean (*Glycine max* (L.) Merr.). *Breeding Science*, *61*, 608–617. https://doi.org/10.1270/jsbbs.61.608.

Qi, X., Jiang, B., Wu, T., Sun, S., Wang, C., Song, W., et al. (2021). Genomic dissection of widely planted soybean cultivars leads to a new breeding strategy of crops in the post-genomic era. *The Crop Journal*, *9*, 1079–1087. https://doi.org/10.1016/j.cj.2021.01.001.

Qi, X. P., Li, M. W., Xie, M., Liu, X., Ni, M., Shao, G. H., et al. (2014). Identification of a novel salt tolerance gene in wild soybean by whole-genome sequencing. *Nature Communications*, *5*, 4340. https://doi.org/10.1038/ncomms5340.

Qin, J., Shi, A. N., Song, Q. J., Li, S., Wang, F. M., Cao, Y. H., et al. (2019). Genome wide association study and genomic selection of amino acid concentrations in soybean seeds. *Frontiers in Plant Science*, *10*, 1445. https://doi.org/10.3389/fpls.2019.01445.

Qiu, J., Wang, Y., Wu, S. L., Wang, Y. Y., Ye, C. Y., Bai, X. F., et al. (2014). Genome re-sequencing of semi-wild soybean reveals a complex *soja* population structure and deep introgression. *PLoS One*, *9*, e108479. https://doi.org/10.1371/journal.pone.0108479.

Quero, G., Simondi, S., Ceretta, S., Otero, A., Garaycochea, S., Fernandez, S., et al. (2021). An integrative analysis of yield stability for a GWAS in a small soybean breeding population. *Crop Science*, *61*, 1903–1914. https://doi.org/10.1002/csc2.20490.

Ravelombola, W. S., Qin, J., Shi, A. N., Nice, L., Bao, Y., Lorenz, A., et al. (2020). Genome-wide association study and genomic selection for tolerance of soybean biomass to soybean cyst nematode infestation. *PLoS One*, *15*, e0235089. https://doi.org/10.1371/journal.pone.0235089.

Ray, J. D., Dhanapal, A. P., Singh, S. K., Hoyos-Villegas, V., Smith, J. R., Purcell, L. C., et al. (2015). Genome-wide association study of ureide concentration in diverse maturity group IV soybean [*Glycine max*(L.) Merr.] accessions. *G3-Genes Genomes Genetics*, *5*, 2391–2403. https://doi.org/10.1534/g3.115.021774.

Rigola, D., van Oeveren, J., Janssen, A., Bonne, A., Schneiders, H., van der Poel, H. J. A., et al. (2009). High-throughput detection of induced mutations and natural variation using KeyPoint(TM) technology. *PLoS One*, *4*, e4761. https://doi.org/10.1371/journal.pone.0004761.

Rolling, W. R., Dorrance, A. E., & McHale, L. K. (2020). Testing methods and statistical models of genomic prediction for quantitative disease resistance to *Phytophthora sojae* in soybean [*Glycine max* (L.) Merr] germplasm collections. *Theoretical and Applied Genetics*, *133*, 3441–3454. https://doi.org/10.1007/s00122-020-03679-w.

Schmutz, J., Cannon, S. B., Schlueter, J., Ma, J. X., Mitros, T., Nelson, W., et al. (2010). Genome sequence of the palaeopolyploid soybean. *Nature*, *463*, 178–183. https://doi.org/10.1038/nature08670.

Seck, W., Torkamaneh, D., & Belzile, F. (2020). Comprehensive genome-wide association analysis reveals the genetic basis of root system architecture in soybean. *Frontiers in Plant Science*, *11*, 590740. https://doi.org/10.3389/fpls.2020.590740.

Shen, Y. T., Du, H. L., Liu, Y. C., Ni, L. B., Wang, Z., Liang, C. Z., et al. (2019). Update soybean Zhonghuang 13 genome to a golden reference. *Science China-Life Sciences*, *62*, 1257–1260. https://doi.org/10.1007/s11427-019-9822-2.

Shen, Y. T., Liu, J., Geng, H. Y., Zhang, J. X., Liu, Y. C., Zhang, H. K., et al. (2018). *De novo* assembly of a Chinese soybean genome. *Science China-Life Sciences*, *61*, 871–884. https://doi.org/10.1007/s11427-018-9360-0.

Shimomura, M., Kanamori, H., Komatsu, S., Namiki, N., Mukai, Y., Kurita, K., et al. (2015). The *Glycine max* cv. Enrei genome for improvement of Japanese soybean cultivars. *International Journal of Genomics*, *2015*, 358127. https://doi.org/10.1155/2015/358127.

Shook, J. M., Zhang, J., Jones, S. E., Singh, A., Diers, B. W., & Singh, A. K. (2021). Meta-GWAS for quantitative trait loci identification in soybean. *G3-Genes Genomes Genetics*, *11*, jkab117. https://doi.org/10.1093/g3journal/jkab117.

Shultz, J. L., Kurunam, D., Shopinski, K., Iqbal, M. J., Kazi, S., Zobrist, K., et al. (2006). The soybean genome database (SoyGD): A browser for display of duplicated, polyploid, regions and sequence tagged sites on the integrated physical and genetic maps of *Glycine max*. *Nucleic Acids Research*, *34*, D758–D765. https://doi.org/10.1093/nar/gkj050.

Song, Q. J., Marek, L. F., Shoemaker, R. C., Lark, K. G., Concibido, V. C., Delannay, X., et al. (2004). A new integrated genetic linkage map of the soybean. *Theoretical and Applied Genetics*, *109*, 122–128. https://doi.org/10.1007/s00122-004-1602-3.

Song, Q. J., Yan, L., Quigley, C., Fickus, E., Wei, H., Chen, L. F., et al. (2020). Soybean BARCSoySNP6K: An assay for soybean genetics and breeding research. *Plant Journal*, *104*, 800–811. https://doi.org/10.1111/tpj.14960.

Sprink, T., Metje, J., & Hartung, F. (2015). Plant genome editing by novel tools: TALEN and other sequence specific nucleases. *Current Opinion in Biotechnology*, *32*, 47–53. https://doi.org/10.1016/j.copbio.2014.11.010.

Stacey, M. G., Cahoon, R. E., Nguyen, H. T., Cui, Y. Y., Sato, S., Nguyen, C. T., et al. (2016). Identification of homogentisate dioxygenase as a target for vitamin E biofortification in oilseeds. *Plant Physiology*, *172*, 1506–1518. https://doi.org/10.1104/pp.16.00941.

Steketee, C. J., Schapaugh, W. T., Carter, T. E., & Li, Z. L. (2020). Genome-wide association analyses reveal genomic regions controlling canopy wilting in soybean. *G3-Genes Genomes Genetics*, *10*, 1413–1425. https://doi.org/10.1534/g3.119.401016.

Sugano, S., Hirose, A., Kanazashi, Y., Adachi, K., Hibara, M., Itoh, T., et al. (2020). Simultaneous induction of mutant alleles of two allergenic genes in soybean by using site-directed mutagenesis. *BMC Plant Biology*, *20*, 513. https://doi.org/10.1186/s12870-020-02708-6.

Sun, X. J., Hu, Z., Chen, R., Jiang, Q. Y., Song, G. H., Zhang, H., et al. (2015). Targeted mutagenesis in soybean using the CRISPR-Cas9 system. *Scientific Reports*, *5*, 10342. https://doi.org/10.1038/srep10342.

Suzuki, C., Miyoshi, T., Shirai, S., Yumoto, S., Tanaka, Y., Hagihara, S., et al. (2017). A new soybean variety 'Yukihomare R' introduced resistance for soybean cyst nematode race1 into 'Yukihomare' by marker assisted selection. *Bulletin of Hokkaido Research Organization Agricultural Experiment Stations*, *101*, 33–47.

Tang, F., Yang, S., Liu, J., & Zhu, H. (2015). *Rj4*, a gene vontrolling nodulation specificity in soybeans, encodes a thaumatin-like protein but not the one previously reported. *Plant Physiology*, *170*, 26–32. https://doi.org/10.1104/pp.15.01661.

Todd, J. J., & Vodkin, L. O. (1996). Duplications that suppress and deletions that restore expression from a chalcone synthase multigene family. *Plant Cell*, *8*, 687–699. https://doi.org/10.1105/tpc.8.4.687.

Torkamaneh, D., Laroche, J., Tardivel, A., O'Donoughue, L., Cober, E., Rajcan, I., et al. (2018). Comprehensive description of genomewide nucleotide and structural variation in short-season soya bean. *Plant Biotechnology Journal*, *16*, 749–759. https://doi.org/10.1111/pbi.12825.

Torkamaneh, D., Lemay, M. A., & Belzile, F. (2021). The pan-genome of the cultivated soybean (PanSoy) reveals an extraordinarily conserved gene content. *Plant Biotechnology Journal*, *19*, 1852–1862. https://doi.org/10.1111/pbi.13600.

Trevisan, R., Perez, O., Schmitz, N., Diers, B., & Martin, N. (2020). High-throughput phenotyping of soybean maturity using time series UAV imagery and convolutional neural networks. *Remote Sensing*, *12*, 3617. https://doi.org/10.3390/rs12213617.

Tuteja, J. H., Zabala, G., Varala, K., Hudson, M., & Vodkin, L. O. (2009). Endogenous, tissue-specific short interfering RNAs silence the *chalcone synthase* gene family in *Glycine max* seed coats. *Plant Cell*, *21*, 3063–3077. https://doi.org/10.1105/tpc.109.069856.

Urnov, F. D., Rebar, E. J., Holmes, M. C., Zhang, H. S., & Gregory, P. D. (2010). Genome editing with engineered zinc finger nucleases. *Nature Reviews Genetics*, *11*, 636–646. https://doi.org/10.1038/nrg2842.

Valliyodan, B., Brown, A. V., Wang, J. X., Patil, G., Liu, Y., Otyama, P. I., et al. (2021). Genetic variation among 481 diverse soybean accessions, inferred from genomic re-sequencing. *Scientific Data*, *8*, 50. https://doi.org/10.1038/s41597-021-00834-w.

Valliyodan, B., Cannon, S. B., Bayer, P. E., Shu, S. Q., Brown, A. V., Ren, L. H., et al. (2019). Construction and comparison of three reference-quality genome assemblies for soybean. *Plant Journal*, *100*, 1066–1082. https://doi.org/10.1111/tpj.14500.

Wang, Y., Gu, Y. Z., Gao, H. H., Qiu, L. J., Chang, R. Z., Chen, S. Y., et al. (2016). Molecular and geographic evolutionary support for the essential role of GIGANTEAa in soybean domestication of flowering time. *BMC Evolutionary Biology*, *16*, 79. https://doi.org/10.1186/s12862-016-0653-9.

Wang, X., Li, M. W., Wong, F. L., Luk, C. Y., Chung, C. Y. L., Yung, W. S., et al. (2021). Increased copy number of *gibberellin 2-oxidase 8* genes reduced trailing growth and shoot length during soybean domestication. *Plant Journal, 107*, 1739–1755. https://doi.org/10.1111/tpj.15414.

Wang, S., Meyer, E., McKay, J. K., & Matz, M. V. (2012). 2b-RAD: A simple and flexible method for genome-wide genotyping. *Nature Methods, 9*, 808–810. https://doi.org/10.1038/Nmeth.2023.

Wang, L. W., Sun, S., Wu, T. T., Liu, L. P., Sun, X. G., Cai, Y. P., et al. (2020). Natural variation and CRISPR/Cas9-mediated mutation in *GmPRR37* affect photoperiodic flowering and contribute to regional adaptation of soybean. *Plant Biotechnology Journal, 18*, 1869–1881. https://doi.org/10.1111/pbi.13346.

Wu, D. P., Li, D. M., Zhao, X., Zhan, Y. H., Teng, W. L., Qiu, L. J., et al. (2020). Identification of a candidate gene associated with isoflavone content in soybean seeds using genome-wide association and linkage mapping. *Plant Journal, 104*, 950–963. https://doi.org/10.1111/tpj.14972.

Wu, N., Lu, Q., Wang, P. W., Zhang, Q., Zhang, J., Qu, J., et al. (2020). Construction and analysis of *GmFAD2-1A* and *GmFAD2-2A* soybean fatty acid desaturase mutants based on CRISPR/Cas9 technology. *International Journal of Molecular Sciences, 21*, 1104. https://doi.org/10.3390/ijms21031104.

Xavier, A., Muir, W. M., & Rainey, K. M. (2016). Assessing predictive properties of genome-wide selection in soybeans. *G3-Genes Genomes Genetics, 6*, 2611–2616. https://doi.org/10.1534/g3.116.032268.

Xie, M., Chung, C. Y. L., Li, M. W., Wong, F. L., Wang, X., Liu, A. L., et al. (2019). A reference-grade wild soybean genome. *Nature Communications, 10*, 1216. https://doi.org/10.1038/s41467-019-09142-9.

Xie, W. B., Feng, Q., Yu, H. H., Huang, X. H., Zhao, Q. A., Xing, Y. Z., et al. (2010). Parent-independent genotyping for constructing an ultrahigh-density linkage map based on population sequencing. *Proceedings of the National Academy of Sciences of the United States of America, 107*, 10578–10583. https://doi.org/10.1073/pnas.1005931107.

Xu, H. Y., Li, Y. J., Zhang, K. F., Li, M. J., Fu, S. Y., Tian, Y. Z., et al. (2021). miR169c-NFYA-C-ENOD40 modulates nitrogen inhibitory effects in soybean nodulation. *New Phytologist, 229*, 3377–3392. https://doi.org/10.1111/nph.17115.

Yamada, T., Makita, H., Funatsuki, H., Takahashi, K., Hirata, K., Hishinuma, A., et al. (2017). Causal analysis of yield-increase by introgression of shattering resistance gene *pdh1* in soybean. *Japanese Journal of Crop Science, 86*, 251–257. https://doi.org/10.1626/jcs.86.251.

Yang, C., Yan, J., Jiang, S., Li, X., Min, H., Wang, X., et al. (2021). Resequencing 250 soybean accessions: New insights into genes associated with agronomic traits and genetic networks. *Genomics, Proteomics & Bioinformatics*. https://doi.org/10.1016/j.gpb.2021.02.009.

Yi, X., Liu, J., Chen, S., Wu, H., Liu, M., Xu, Q., et al. (2021). Platinum-grade soybean reference genome facilitates characterization of genetic underpinnings of traits. *Research Square*. https://doi.org/10.21203/rs.3.rs-57915/v1.

Yoosefzadeh-Najafabadi, M., Earl, H. J., Tulpan, D., Sulik, J., & Eskandari, M. (2021). Application of machine learning algorithms in plant breeding: Predicting yield from hyperspectral reflectance in soybean. *Frontiers in Plant Science, 11*, 624273. https://doi.org/10.3389/fpls.2020.624273.

Yue, Y. L., Liu, N. X., Jiang, B. J., Li, M., Wang, H. J., Jiang, Z., et al. (2017). A single nucleotide deletion in *J* encoding GmELF3 confers long juvenility and is associated with adaption of tropic soybean. *Molecular Plant, 10*, 656–658. https://doi.org/10.1016/j.molp.2016.12.004.

Zhang, H. Y., Goettel, W., Song, Q. J., Jiang, H., Hu, Z. B., Wang, M. L., et al. (2020). Dual use and selection of GmSWEET39 for oil and protein improvement in soybean. *PLoS Genetics*, *16*, e1009114. https://doi.org/10.1371/journal.pgen.1009114.

Zhang, Y. H., He, J. B., Wang, H. W., Meng, S., Xing, G. N., Li, Y., et al. (2018). Detecting the QTL-allele system of seed oil traits using multi-locus genome-wide association analysis for population characterization and optimal cross prediction in soybean. *Frontiers in Plant Science*, *9*, 1793. https://doi.org/10.3389/fpls.2018.01793.

Zhang, W., Liao, X. L., Cui, Y. M., Ma, W. Y., Zhang, X. N., Du, H. Y., et al. (2019). A cation diffusion facilitator, GmCDF1, negatively regulates salt tolerance in soybean. *PLoS Genetics*, *15*, e1007798. https://doi.org/10.1371/journal.pgen.1007798.

Zhang, J. P., Song, Q. J., Cregan, P. B., & Jiang, G. L. (2016). Genome-wide association study, genomic prediction and marker-assisted selection for seed weight in soybean (*Glycine max*). *Theoretical and Applied Genetics*, *129*, 117–130. https://doi.org/10.1007/s00122-015-2614-x.

Zhang, B., Wang, M. D., Sun, Y. F., Zhao, P., Liu, C., Qing, K., et al. (2021). Glycine max NNL1 restricts symbiotic compatibility with widely distributed bradyrhizobia via root hair infection. *Nature Plants*, 7, 239. https://doi.org/10.1038/s41477-021-00872-7.

Zhang, W., Xu, W. J., Zhang, H. M., Liu, X. Q., Cui, X. Y., Li, S. S., et al. (2021). Comparative selective signature analysis and high-resolution GWAS reveal a new candidate gene controlling seed weight in soybean. *Theoretical and Applied Genetics*, *134*, 1329–1341. https://doi.org/10.1007/s00122-021-03774-6.

Zhang, D., Zhang, H. Y., Hu, Z. B., Chu, S. S., Yu, K. Y., Lv, L. L., et al. (2019). Artificial selection on GmOLEO1 contributes to the increase in seed oil during soybean domestication. *PLoS Genetics*, *15*, e1008267. https://doi.org/10.1371/journal.pgen.1008267.

Zhao, X., Dong, H. R., Chang, H., Zhao, J. Y., Teng, W. L., Qiu, L. J., et al. (2019). Genome wide association mapping and candidate gene analysis for hundred seed weight in soybean [*Glycine max* (L.) Merrill]. *BMC Genomics*, *20*, 648. https://doi.org/10.1186/s12864-019-6009-2.

Zhao, X., Li, W. J., Zhao, X. Y., Wang, J. Y., Liu, Z. Y., Han, Y. P., et al. (2019). Genome-wide association mapping and candidate gene analysis for seed shape in soybean (*Glycine max*). *Crop & Pasture Science*, *70*, 684–693. https://doi.org/10.1071/Cp19028.

Zhou, Z. K., Jiang, Y., Wang, Z., Gou, Z. H., Lyu, J., Li, W. Y., et al. (2015). Resequencing 302 wild and cultivated accessions identifies genes related to domestication and improvement in soybean. *Nature Biotechnology*, *33*, 408–U125. https://doi.org/10.1038/nbt.3096.

Zou, J. A., Li, W. J., Zhang, Y. T., Song, W., Jiang, H. P., Zhao, J. Y., et al. (2021). Identification of *glutathione transferase* gene associated with partial resistance to *Sclerotinia* stem rot of soybean using genome-wide association and linkage mapping. *Theoretical and Applied Genetics*, *134*, 2699–2709. https://doi.org/10.1007/s00122-021-03855-6.

CHAPTER TWO

Regulation of flowering and maturation in soybean

Zhihong Hou[†], Baohui Liu[†], and Fanjiang Kong*[,†]
Innovative Center of Molecular Genetics and Evolution, School of Life Sciences, Guangzhou University, Guangzhou, China
*Corresponding author: e-mail address: kongfj@gzhu.edu.cn

Contents

Abstract

Precise timing of flowering and maturation is critical for crop adaptation and productivity in specific environments. Soybean (*Glycine max*) is a facultative short-day plant and is highly sensitive to photoperiod, which limits the geographical range for cultivating specific soybean varieties. However, soybean cultivars overall have a wide range of geographical adaptation. Adaptation to different latitudes in soybean is controlled by maturity genes *E1* to *E11* and *J*, and quantitative trait loci (QTLs) including *Time of Flowering 5* (*Tof5*), *Tof11*, *Tof12*, and *Tof16*. The interactions of these genes and QTLs allow soybean to adapt to different daylengths. Photoperiod sensitivity must be reduced to adapt to

† These authors contributed equally to this work.

Advances in Botanical Research, Volume 102
ISSN 0065-2296
https://doi.org/10.1016/bs.abr.2022.02.007

long-day conditions, while later flowering allows for higher yields under short-day conditions. This chapter provides an overview of recent research in photoperiodic flowering of soybean and discusses current challenges and future directions.

1. Introduction

Time to flowering and maturation influences the adaptability, productivity, and quality of seed plants. Plants integrate flowering signals from a range of different endogenous and exogenous cues to transit from vegetative growth to flowering to seed set, and finally plant maturation. In soybean, the vegetative stages (V) in soybean can be divided into emergence (VE), cotyledon (VC), first-node (V1), second-node (V2), third-node (V3), and … nth-node (Vn), and the reproductive stages (R) can be divided into R1 to R8 stages (Table 1 and Fig. 1) (Fehr et al., 1971). In general, days to flowering are recorded at the R1 stage (first open flower appeared), and maturation is recorded at the R8 stage (days from emergence to when 95% of pods had a mature color) in soybean (Fehr et al., 1971).

Studies of the model plant *Arabidopsis thaliana* and other taxa have identified six major pathways that regulate flowering time, including the autonomous pathway, the aging pathway, the gibberellin (GA) pathway, the photoperiod pathway, the vernalization pathway, and the ambient temperature pathway (Fornara, de Montaigu, & Coupland, 2010; Roux, Touzet, Cuguen, & Le Corre, 2006). These pathways, alone or in combination, form a complex network that regulates flowering at the appropriate time based on changes in ambient conditions.

Photoperiodic regulation is one of the major flowering-time mechanisms (Song, Shim, Kinmonth-Schultz, & Imaizumi, 2015). Garner and Allard discovered "photoperiodism" in 1920, describing how daylength influences flowering time in many plant species (Garner & Allard, 1920). In the photoperiod pathway, photoreceptors perceive light signals and control the expression of circadian rhythm genes; these genes regulate flowering time by inducing downstream genes related to floral organ development and flowering (Guo, Yang, Mockler, & Lin, 1998). Flowering plants are classified as long-day (LD), short-day (SD), or neutral-day plants according to their photoperiodic flowering behavior. LD plants such as Arabidopsis exhibit precocious flowering under LD conditions but delayed flowering or no flowering under SD conditions (Guo et al., 1998). In contrast, SD plants, such as soybean and rice (*Oryza sativa*), display precocious flowering

Table 1 Description of vegetative stages and reproductive stages.

Stage no.	Abbreviated stage title	Description
VE	Emergence	Cotyledons above the soil surface.
VC	Cotyledon	Unifoliolate leaves expand sufficiently so the leaf edges are not touching.
V1	First-node	Fully developed leaves at unifoliolate nodes.
V2	Second-node	Fully developed trifoliolate leaf at node above the unifoliolate nodes.
V3	Third-node	Three nodes on the main steam with fully developed leaves beginning with the unifoliolate node.
Vn	nth-node	n number of nodes on the main stem with fully developed leaves beginning with the unifoliolate nodes.
R1	Beginning flowering	One open flower at any node on the main stem.
R2	Full bloom	Open flower at one of the two uppermost nodes on the main stem.
R3	Beginning pod	Pods are 5 mm long on one of the four uppermost nodes on the main stem.
R4	Full pod	Pods are 2 cm long on one of the four uppermost nodes on the main stem.
R5	Beginning seed	Seed is 3 mm long in a pod at one of the four uppermost nodes on the main stem.
R6	Full seed	Pods contain a green seed that fills the pod cavity in one of the four uppermost nodes on the main stem.
R7	Beginning maturity	One normal pod on the main stem has reached mature pod color.
R8	Full maturity	95% of the pods that have reached their mature pod color.

Information about developmental stages is adapted from Fehr, W. R., Caviness, C. E., Burmood, D. T., & Pennington, J. S. (1971). Stage of development descriptions for soybeans. *Crop Science*, 11, 929–931.

under SD conditions but delayed flowering or no flowering under LD conditions (Huber, Rufty, & Kerr, 1984).

Soybean is a facultative SD plant and is highly sensitive to photoperiod, which limits the geographical range for cultivation of specific cultivars (Garner & Allard, 1920; Gong, 2020; Miranda, Scaboo, Cober, Denwar, & Bilyeu, 2020; Watanabe, Harada, & Abe, 2012). However, the diversity in

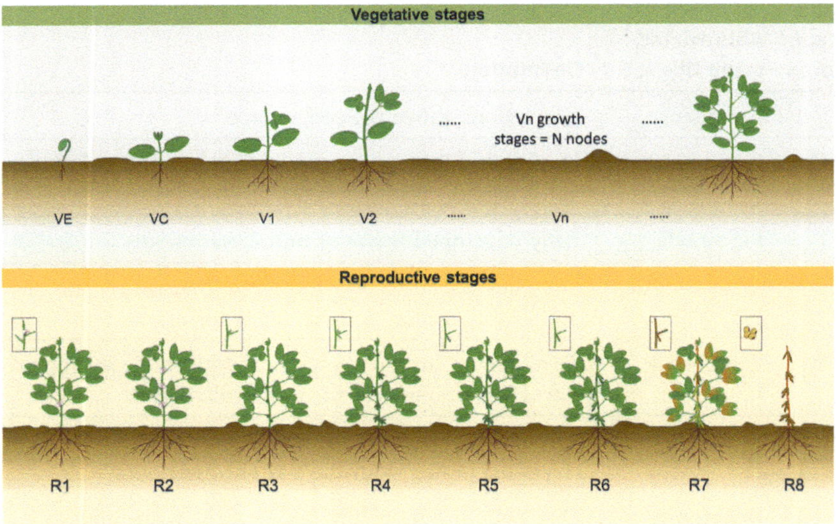

Fig. 1 Stages of soybean development. Soybean undergoes vegetative and reproductive growth stages from seedling emergence to maturity. The vegetative stages (V) in soybean can be divided into emergence (VE), cotyledon (VC), first-node (V1), second-node (V2), third-node (V3), and ...nth-node (Vn). The reproductive stages (R) can be divided into R1 to R8 stages. *Drawings about developmental stages are adapted from Fehr, W. R., Caviness, C. E., Burmood, D. T., & Pennington, J. S. (1971). Stage of development descriptions for soybeans.* Crop Science, *11, 929–931.*

geographical adaptation among different cultivars allows soybean to be grown over a wide range of latitudes, from equatorial regions to at least 50°N and 35° S, owing to variation in the major loci and quantitative trait loci (QTLs) controlling flowering time and maturity (Watanabe et al., 2012; Wong, Singh, & Bhalla, 2013; Zhang et al., 2020a).

The timing of flowering and maturity are important considerations for maximizing soybean yield, and photoperiod sensitivity is a key factor for local adaptation. For example, at high latitudes, the long photoperiod delays flowering and maturation of photoperiod-sensitive varieties. Soybean varieties bred for high latitudes adapted to LD conditions by reducing their photoperiod sensitivity. In contrast, at low latitudes, photoperiod-sensitive varieties flower and mature earlier because of the shorter photoperiod. Thus, in SD conditions, delayed flowering and extended reproductive periods are required to produce and maintain sufficient vegetative growth to support higher yields. The photoperiod adaptation of soybean is controlled by maturity genes and several QTLs (Table 2).

Table 2 Genes involved in photoperiodic flowering in soybean.

Locus	Gene name	Gene number	Gene function	References
E1	*E1*	*Glyma.06G207800*	Inhibits flowering	Bernard (1971); Xia et al. (2012a)
	E1la	*Glyma.04G156400*	Inhibits flowering	Xu et al. (2015)
	E1lb	*Glyma.04G143300*	Inhibits flowering	Watanabe et al. (2011); Xu et al. (2015)
E2	*GI*	*Glyma.10G221500*	Inhibits flowering	Bernard (1971); Watanabe et al. (2011)
E3	*phyA3*	*Glyma.19G224200*	Inhibits flowering	Buzzell (1971); Watanabe et al. (2009)
E4	*phyA2*	*Glyma.20G090000*	Inhibits flowering	Buzzell & Voldeng (1980); Liu et al. (2008)
E5	*Inexistence*			McBlain & Bernard (1987); Dissanayaka et al. (2016)
E6	*ELF3*	*Glyma.04G050200*	Promotes flowering	Bonato & Vello, (1999); Fang et al. (2021b)
E7	*Unknown*		Inhibits flowering	Cober & Voldeng (2001); Molnar, Rai, Charette, & Cober (2003)
E8	*Unknown*		Inhibits flowering	Cober, Molnar, Charette, & Voldeng (2010)
E9/ LJ16.1	*FT2a*	*Glyma.16G150700*	Promotes flowering	Kong et al. (2010); Zhao et al. (2016); Cai et al. (2018, 2020); Li et al. (2021)
E10	*FT4*	*Glyma.08G363100*	Inhibits flowering	Zhai et al. (2014); Samanfar et al. (2017)
LJ16.2	*FT5a*	*Gyma.16G044100*	Promotes flowering	Kong et al. (2010); Cai et al. (2020); Li et al. (2021)
	FT1a	*Glyma.18G298900*	Inhibits flowering	Guo et al. (2015); Liu et al. (2018)
	FT1b	*Glyma.18G299000*	Inhibits flowering	Guo et al. (2015)

Continued

Table 2 Genes involved in photoperiodic flowering in soybean.—cont'd

Locus	Gene name	Gene number	Gene function	References
	FT2b	*Glyma.16G151000*	Promotes flowering	Fan et al. (2014); Chen, Cai, et al. (2020)
	FT3a	*Glyma.16G044200*	Promotes flowering	Fan et al. (2014)
	FT3b	*Glyma.19G108100*	Promotes flowering	Fan et al. (2014)
	FT5b	*Glyma.19G108200*	Promotes flowering	Fan et al. (2014)
	FT6	*Glyma.08G363200*	Inhibits flowering	Fan et al. (2014); Wang et al. (2015)
E11	Unknown		Promotes flowering	Wang et al. (2019)
J	*ELF3*	*Glyma.04G050200*	Promotes flowering	Ray, Hinson, Mankono, & Malo (1995); Lu et al. (2017); Yue et al. (2017)
Tof5	*FUL2a*	*Glyma.05G018800*	Promotes flowering	Dong et al. (2021a)
Tof11	*PRR3a*	*Glyma.U034500*	Inhibits flowering	Li, Liu, Lam, & Gendron (2019b); Lu et al. (2020)
Tof12	*PRR3b*	*Glyma.12G073900*	Inhibits flowering	Li, Liu, et al. (2019b); Li et al. (2019a), Li et al. (2020b); Wang et al. (2020); Lu et al. (2020)
Tof16	*LHY1a*	*Glyma.16G017400*	Promotes flowering	Lu et al. (2020); Dong et al. (2021b)
	LHY1b	*Glyma.07G048500*	Promotes flowering	Dong, Fang, et al. (2021b)
Dt1	*TFL1b*	*Glyma.19G194300*	Inhibits flowering	Yue et al. (2021a)

In this chapter, we summarize recent research on photoperiodic flowering of soybean, highlighting the progress in the identification of flowering- and maturity-related genes, flowering gene networks, and determinants of latitudinal adaptation in soybean.

2. Flowering and maturity genes in soybean

In soybean, forward genetic approaches have identified the major genes *E1* to *E11* and *J*, and several QTLs, such as *Time of Flowering 5* (*Tof5*), *Tof11*, *Tof12*, and *Tof16*, which control time to flowering and maturity (Bernard, 1971; Bonato & Vello, 1999; Buzzell, 1971; Buzzell & Voldeng, 1980; Cober et al., 2010; Cober & Voldeng, 2001; Dong, Cheng, et al., 2021a; Dong, Fang, et al., 2021b; Kong et al., 2014; Lu et al., 2020; McBlain & Bernard, 1987; Ray et al., 1995; Wang et al., 2019; Zhai et al., 2014). Dominant alleles at *E1*, *E2*, *E3*, *E4*, *E7*, *E8*, *E10*, *Tof11*, and *Tof12* confer late flowering, whereas dominant alleles at *E6*, *E9*, *E11*, *J*, *Tof5*, and *Tof16* confer early flowering (Table 2). In addition to the above 16 genes/loci, several soybean orthologs of Arabidopsis flowering genes have been characterized via reverse-genetic approaches.

2.1 *E*-series genes

Since Owen (1927) identified a series of genes closely related to photoperiod and maturation in 1927 and defined them as *E* and *e* series genes, 11 *E*-series genes have been revealed, *E1* to *E11*.

Among these *E*-series genes, *E1* was the first locus shown to affect soybean flowering and maturation and has the strongest effect on flowering time (Tsubokura et al., 2014; Upadhyay, Ellis, Summerfield, Roberts, & Qi, 1994). *E1* is a legume-specific gene and a major regulator of the photoperiod response, functioning as a critical point of integration within the soybean flowering pathway, where it represses flowering (Upadhyay et al., 1994; Xia et al., 2012a). *E1* encodes a putative transcription factor with a bipartite nuclear localization signal and a 20.3-kDa B3 distantly related domain (Xia, Watanabe, et al., 2012a). *E1* is highly expressed in leaves but is expressed at low levels in other tissues (Xia, Watanabe, et al., 2012a). The expression pattern of *E1* differs under LD and SD conditions. *E1* is significantly inhibited under SD conditions but shows a bimodal pattern under LD conditions (Xia, Watanabe, et al., 2012a). Therefore, *E1* expression represses flowering under LD conditions, while its limited expression in SD conditions promotes flowering (Xia, Watanabe, et al., 2012a; Zhai et al., 2015). There are three major natural loss-of-function variants of *E1* in soybean: the leaky allele ($e1^{as}$) and two null alleles ($e1^{fs}$ and $e1^{nl}$), which cause earlier flowering and maturity under LD conditions (Tsubokura et al., 2014; Xia, Watanabe, et al., 2012a; Xu et al., 2013). In addition, *E1* has two homologs in soybean, *E1-like-a* (*E1La*) and *E1Lb* (Xia, Watanabe, et al., 2012a; Xu et al., 2015). Their expression patterns are similar to that of *E1* under LD and SD conditions, and both genes function as flowering inhibitors (Xia, Watanabe, et al., 2012a; Xu et al., 2015). Under LD conditions, *E1* and *E1L* inhibit flowering by down-regulating *FT2a* and *FT5a*, two soybean orthologues of Arabidopsis *FLOWERING LOCUS T* (*AtFT*) genes that lead to early flowering (Xia, Watanabe, et al., 2012a; Xu et al., 2015).

E2 is a soybean ortholog (*GIa*) of Arabidopsis *GIGANTEA* (*AtGI*), which plays multiple roles in the circadian clock and flowering (Watanabe et al., 2011). *E2* has two known alleles, the functional dominant *E2* and the recessive *e2* (Watanabe et al., 2011). The *E2* allele encodes a full-length protein of 1170 amino acids, and the *e2* allele is a nonsense allele encoding a truncated protein of 521 amino acids (Watanabe et al., 2011). The *e2* allele promotes early flowering through up-regulation of *FT2a* (Watanabe et al., 2011). Insertion and deletion of 36 bases in the eighth intron (*E2-in* and *E2-dl*) were observed at the *E2* locus (Tsubokura et al., 2014). The soybean genome has two additional *GI* homologs, *GIb* and *GIc*, which reside on chromosomes 20 and 16, respectively (Wang et al., 2016); however, their effects remain unknown.

E3 and *E4* affect flowering time, maturity, and photoperiod sensitivity in soybean. *E3* and *E4* encode homologs of the Arabidopsis photoreceptor phytochrome A (AtphyA), phyA3 and phyA2, respectively (Liu & Abe, 2010; Liu et al., 2008; Watanabe et al., 2009). There are four homologs of *phyA* in soybean: *phyA1*, *phyA2*, *phyA3*, and *phyA4* (Li et al., 2014). Functional *E3* and *E4* delay flowering and maturity. There are six types of *E3* alleles: *E3-Mi*, *E3-Ha*, *e3-tr*, *e3-Mo*, *e3-ns*, and *e3-fs*. E3-Mi and E3-Ha are normal phyAs. *e3-Mo* encodes phyA with reduced protein function. The protein encoded by *e3-tr* does not function as a phyA (Watanabe et al., 2009). Among the *E*-series genes, *E4* was the first to be cloned (Liu et al., 2008). There are five types of *E4* mutants: *e4-SORE-1*, *e4-kam*, *e4-kes*, e*4-oto*, and *e4-tsu*. A *Ty1/copia*-like retrotransposon inserted into exon 1 of the recessive *E4* allele *e4-SORE-1* weakens the function of *E4* in flowering repression (Liu et al., 2008). Different red: far-red (R:FR) light ratios cause different responses to soybean flowering (Cober, Tanner, & Voldeng, 1996). *E3* regulates flowering under LD conditions with high R:FR quantum ratios, and *E4* regulates flowering under LD conditions with low R:FR ratios (Cober et al., 1996; Cober & Voldeng, 2001). The *e3e3* recessive homozygote showed early flowering at high R:FR ratios in LD conditions (Buzzell, 1971), while the *e4e4* recessive homozygote was less sensitive to photoperiod at low R:FR ratios in LD conditions (Cober et al., 1996). Therefore, LD soybean cultivars must contain the *e3e3* recessive combination for early flowering. In addition, *E3* and *E4* have redundant functions in photomorphogenesis and flowering at low R:FR ratios (Liu et al., 2008). *E3* has a larger effect than *E4*, and *E3* has an epistatic effect on *E4* (Saindon, Voldeng, Beversdorf, & Buzzell, 1989).

The effect of *E5* on time to flowering and maturity is similar to that of *E2* (McBlain & Bernard, 1987), and the *e5* allele was observed in the late-flowering cultivar Harosoy (McBlain & Bernard, 1987). Furthermore, none of the QTLs found in three test crosses (Harosoy-*E5* × Clark-*e2*, Harosoy × Clark-*e2E5*, and Harosoy-*E5Dt2* × Clark-*e2*) corresponded to *E5* (Dissanayaka et al., 2016). Therefore, *E5* may not exist, and the effect of the *E5* locus on flowering and maturity of soybean may be attributable to the functional difference of the *E2* allele (Dissanayaka et al., 2016).

In contrast to the *E1–E4* loci, *E6* is primarily involved in promoting flowering and maturity under SD conditions. Recessive *e6* allele inhibits flowering (Bonato & Vello, 1999; Ray et al., 1995). This late-flowering trait has commonly been referred to as "long juvenile" (LJ) (Cober, 2011; Sinclair & Hinson, 1992). Genetic analysis indicated that *E6* is tightly linked to *J*, the ortholog of Arabidopsis *EARLY FLOWERING 3* (*AtELF3*) (Lu et al., 2017). Positional cloning, sequence analysis, and transgenic complementation confirmed that *E6* is a novel allele of *J*, and the mutant allele was named $e6^{PG}$ (Fang et al., 2021b). The recessive $e6^{PG}$ allele contains a *Ty1/copia*-like retrotransposon in the fourth exon of *J* and is a rare allele (Fang et al., 2021b).

E7 and *E8* were identified in genetic analyses of early-maturing cultivars with the genotype $e1^{as}/e2/e3/e4$ (Cober et al., 2010; Cober & Voldeng, 2001). Recessive alleles at both loci confer earlier flowering and maturity under LD conditions (Cober et al., 2010; Cober & Voldeng, 2001). *E7* is located in linkage group (LG) C2 (chromosome 6) and is linked to *E1* (Cober & Voldeng, 2001; Liu et al., 2010; Molnar et al., 2003). *E8* is located in LG C1 (chromosome 4), where *E1La* and *E1Lb* are located (Cober et al., 2010). However, the molecular mechanisms of the maturity loci *E7* and *E8* are still unknown.

E9 and *E10* correspond to *FT2a* and *FT4*, respectively, which are orthologs of *Arabidopsis FT* gene (Kong et al., 2010; Liu et al., 2007; Samanfar et al., 2017; Zhao et al., 2016). *E9* was fine-mapped to a 245-kb interval between markers M5 and M7 on chromosome 16 (Kong et al., 2014). The functional *E9* allele results in early flowering and maturity, and the nonfunctional *e9* allele contains a *Ty1/copia*-like retrotransposon,

SORE-1, inserted in the first intron, which confers late flowering and maturity (Kong et al., 2014; Zhao et al., 2016). *E10* functions as a flowering repressor (Samanfar et al., 2017).

E11 is located on chromosome 7 and confers early flowering and maturity in soybean (Wang et al., 2019). The gene was fine-mapped to an ~138-kb interval. The three most likely candidate genes are *Glyma.07G048500*, *Glyma.07G049000*, and *Glyma.07G049200* (Wang et al., 2019). *Glyma.07G048500* is one of the four *LATE ELONGATED HYPOCOTYL* (*LHY*) genes in soybean (Wang et al., 2019). Clustered regularly interspaced short palindromic repeats (CRISPR)/CRISPR-associated protein 9 (Cas9)-mediated targeted mutagenesis of all four *LHY* genes delays flowering and reduces soybean plant height (Cheng et al., 2019; Lu et al., 2020). Moreover, a single mutant of *Glyma.07G048500* showed significantly delayed flowering time and maturity (Dong, Fang, et al., 2021b).

2.2 Circadian clock genes

The circadian clock is an endogenous mechanism for maintaining a 24-h rhythm via central oscillators that synchronize with diurnal changes in the environment (Bendix, Marshall, & Harmon, 2015; Greenham & McClung, 2015). In Arabidopsis, the circadian clock consists of at least three interlocking loops (Locke et al., 2006; Zeilinger, Farré, Taylor, Kay, & Doyle, 2006). The morning-expressed genes include *CIRCADIAN CLOCK-ASSOCIATED 1* (*CCA1*)/*LHY* and *PSEUDORESPONSE REGULATOR 7* (*PRR7*)/*PRR9*, and evening-phased genes include *TIMING OF CAB EXPRESSION 1* (*TOC1*), *GI*, and the evening complex (EC) genes *EARLY FLOWERING 4/3* and *LUX ARRHYTHMO* (*ELF4/ELF3/LUX*) (Huang & Nusinow, 2016; Oakenfull & Davis, 2017). In soybean, numerous circadian clock homologs have been linked to flowering and maturity, including *E2* (*GIa*), *J* (*ELF3*), *LUX1*, *LUX2*, *Tof11* (*PRR3a*), *Tof12* (*PRR3b*), and *Tof16* (*LHY1a*) (Bu et al., 2021; Dong, Fang, et al., 2021b; Li et al., 2020b; Li, Dong, et al., 2019a; Li, Liu, et al., 2019b; Lu et al., 2017, 2020; Watanabe et al., 2011).

J gene controlling the LJ trait is a homolog of the circadian clock gene *ELF3* (Lu et al., 2017; Ray et al., 1995; Yue et al., 2017), which is a component of the EC (Nusinow et al., 2011). *J* promotes flowering time and maturity under SD conditions (Lu et al., 2017). J physically interacts with the *E1* promoter and suppresses its transcription, thereby allowing *FT2a* and *FT5a* to be highly expressed to accelerate flowering under SD conditions (Lu et al., 2017). Many natural variants of *J* have been identified and loss-of-function *j* alleles are restricted to the low-latitude accessions; moreover, some of these variants have been utilized to develop elite soybean cultivars with adaption to tropical regions (Dong, Fang, et al., 2021b; Lu et al., 2017; Yue et al., 2017).

Two soybean *LUX* homologs, *LUX1* and *LUX2*, like *J*, are components of the EC (Bu et al., 2021; Fang et al., 2021a). LUX1 and LUX2 interact with J to form the EC in soybean. *LUX1* and *LUX2* are functionally redundant, but together play critical roles in the regulation of soybean flowering (Bu et al., 2021). The single *lux1* and *lux2* mutations generated by CRISPR-Cas9 confer no significant effect on flowering time under SD and LD conditions (Bu et al., 2021). However, the *lux1 lux2* double mutant shows extremely late flowering, and this phenotype is highly similar to the famous photoperiod-sensitive tobacco (*Nicotiana tabacum*) mutant "Maryland Mammoth" that was discovered by Garner and Allard (1920) (Bu et al., 2021). Therefore, the author named the *lux1 lux2* double mutant "Guangzhou Mammoth." In addition, there was no significant difference in flowering time between LD and SD conditions in the *lux1 lux2* double mutant (Bu et al., 2021). Complete impairment of the EC abolishes soybean photoperiod sensitivity, indicating that the EC is critical for soybean photoperiodism. LUX1 and LUX2 physically interact with J, and the J–LUX complex may directly bind to the *E1* and *E1L* promoters and repress their activity (Bu et al., 2021). The *EC* and *E1* family genes may therefore constitute key targets for customized breeding of soybean varieties with precise flowering time (Bu et al., 2021). ELF3-LUX, the soybean EC component, controls flowering time through co-regulation of gene expression, which contributes to the development of soybean varieties suitable for growing at low latitudes.

Two other circadian clock genes, *Tof11* and *Tof12*, named *Gp11* and *qFT12-1/Gp12*, respectively, were identified by whole-genome resequencing and genome-wide association studies (GWAS) and by positional cloning of QTLs (Li et al., 2020b; Li, Dong, et al., 2019a; Li, Liu, et al., 2019b; Lu et al., 2020; Wang et al., 2020). They are homologs of *Arabidopsis AtPRR3*, *PRR3a*, and *PRR3b* (Lu et al., 2020). *Tof12* and *Tof11* function independently to regulate flowering time and maturity, but with some degree of redundancy (Lu et al., 2020). In LD conditions, Tof12 and Tof11 promote *E1* expression through LHY homologs and delay flowering and maturity (Lu et al., 2020).

Tof16 is a homolog of *LHY* (Dong, Fang, et al., 2021b). Soybean has four *LHY/CCA1* homologs (*LHY1a*, *LHY1b*, *LHY2a*, and *LHY2b*) and their amino acid sequences are highly similar (Cheng et al., 2019; Dong, Fang, et al., 2021b; Lu et al., 2020). *Tof16* encodes LHY1a. Analysis of two F_7 near-isogenic lines (NILs) carrying either the functional *Tof16* allele (NIL-*Tof16*) or the non-functional *tof16-1* alleles (NIL-*tof16-1* or NIL-*tof16-2*) in different genetic backgrounds showed that non-functional *tof16* significantly delays flowering time and maturity, and improves yield under LD and SD conditions (Dong, Fang, et al., 2021b). *LHY* homologs have both redundant and divergent functions in controlling flowering time, maturity, and grain yield in soybean. Researchers investigated the phenotype of CRISPR/Cas9-generated 15 homozygous mutational combinations of the four *LHY* genes and found that all mutants except *lhy2a*, *lhy2b*, and *lhy2a/2b* exhibited delayed flowering time and maturity under natural SD and LD conditions. All mutants containing the *lhy1a* mutation had higher yields than those without the mutation, suggesting that *lhy1a* plays a key role in controlling flowering time and maturity under SD conditions (Dong, Fang, et al., 2021b).

2.3 The soybean phosphatidylethanolamine-binding protein (PEBP) family

Two phosphatidylethanolamine-binding protein (PEBP) family members, FT and TERMINAL FLOWER 1 (TFL1), control flowering. TFL1 and FT have a high degree of sequence similarity, indicating that they likely share common biochemical characteristics (Ahn et al., 2006). *FT* encodes a putative florigen protein and functions as an integrator to regulate flowering (Corbesier et al., 2007; Turck, Fornara, & Coupland, 2008); its function is highly conserved among different species (Amasino, 2010; Hayama & Coupland, 2004). FT is transported from leaves to shoot apical meristems (SAMs) through the phloem and functions as a long-distance signal to induce floral initiation (Corbesier et al., 2007; Jaeger & Wigge, 2007; Mathieu, Warthmann, Küttner, & Schmid, 2007; Notaguchi et al., 2008).

Ten *FT* homologs have been characterized in soybean, which include five homoeologous pairs: *FT1a/b*, *FT2a/b*, *FT3a/b*, *FT4*, *FT5a/b*, and *FT6* (Kong et al., 2010; Li et al., 2014; Wu, Sedivy, Price, Haider, & Hanzawa, 2017). Among these *FT* genes, *FT1a* and *FT4* repress flowering and are expressed at higher levels under LD compared to SD conditions (Liu et al., 2018; Zhai et al., 2014). *FT4* was reported to be the most likely candidate gene at a newly identified maturity locus, *E10* (Samanfar et al., 2017).

When ectopically expressed under the control of the Cauliflower mosaic virus (CaMV) 35S promoter, *FT2a/b*, *FT3a/b*, and *FT5a/b* promoted flowering of the Arabidopsis ecotype Columbia (Col-0), suggesting that these genes are floral inducers (Fan et al., 2014; Guo et al., 2015; Kong et al., 2010; Thakare, Kumudini, & Dinkins, 2011). *FT2b*, which has high sequence similarity to *FT2a*, promotes soybean flowering under LD conditions, but not under SD conditions (Chen, Cai, et al., 2020).

FT2a and *FT5a* mainly promote flowering time in soybean (Kong et al., 2010). Analysis of loss-of-function mutants produced by genome editing further indicated that *FT2a* and *FT5a* play different roles in floral induction (Cai et al., 2018, 2020; Li et al., 2021). The single *ft2a* and *ft5a* mutants show a slight impact on flowering under LD and SD conditions, but *ft2a ft5a* double mutants showed significantly late flowering under LD and SD conditions (Li et al., 2021). The *ft2a* mutants mainly have delayed flowering in SD conditions, while *ft5a* mutants primarily have delayed flowering in LD conditions (Cai et al., 2020; Li et al., 2021). The *ft2a* and *ft5a* mutations delay flowering by inhibiting the floral meristem identity gene *APETALA1* (*AP1*) under SD conditions (Li et al., 2021).

Determinate stem 1 (*Dt1*), a shoot meristem identity gene, is an ortholog of *TFL1* (Liu et al., 2010, 2016; Tian et al., 2010). *Dt1* is a major controller of stem growth habit. It regulates the stem growth habit in soybean, controlling the number of nodes and pods per plant and thus ultimately determining soybean yield (Liu et al., 2010; Tian et al., 2010). Control of the flowering transition determines adaptability and yield of soybean, and *Dt1* also controls soybean flowering (Yue et al., 2021a). *Dt1* is not expressed in SD conditions. However, under LD conditions, *Dt1* is induced by E3 and E4 (Xu et al., 2013). Dt1 interacts with the bZIP transcription factor FDc1 to directly bind to the *AP1* promoter and inhibit its expression, thus inhibiting flowering and stem apical meristem growth (Yue et al., 2021a). FT5a competes with Dt1 for binding to FDc1 to promote *AP1* expression. AP1 in turn directly binds to the *Dt1* promoter and inhibits *Dt1* expression, thus promoting flowering, thereby forming a negative feedback regulatory pathway in which the DT1–AP1 interaction controls flowering and apical growth of soybean (Yue et al., 2021a).

2.4 Other genes in the soybean flowering pathway

Soybean contains many other homologs of Arabidopsis flowering genes, such as *CONSTANS*-like (*COL*), *AP1*, *AGAMOUS-like1* (*AGL1*),

FRUITFULL (*FUL*), *NIGHT LIGHT-INDUCIBLE AND CLOCK-REGULATED 2* (*LNK2*), *RELATED TO ABI3/VP1* (*RAV*), and *GUANYLATE-BINDING PROTEIN 1* (*GBP1*).

In LD-flowering Arabidopsis, the nuclear protein CONSTANS (AtCO) plays a central role in photoperiodic flowering (Turck et al., 2008). There are 26 *CO*-like (*COL*) genes in soybean. Among them, two pairs of homoeologous genes, *COL1a/1b* and *COL2a/2b*, encode proteins with the highest amino acid similarity to Arabidopsis CO (Wu et al., 2014). Analyses of *COL1a* overexpression and artificially induced missense *col1b* mutants showed that *COL1a* and *COL1b* delay flowering under LD conditions, but not in SD conditions (Cao, Li, Lu, et al., 2015). Moreover, *COL1* RNA interference (RNAi) transgenic soybean plants also flowered significantly earlier than wild-type plants (Wu et al., 2019). The lack of *COL2b* may moderate the repression of flowering by cool temperatures (Zhang et al., 2020b).

AP1 is a MADS-box transcription factor that functions as a Class A gene in the ABCE model of flower organ development, mediating the formation of the carpels, stamens, petals, and sepals of Arabidopsis flowers (Mandel, Gustafson-Brown, Savidge, & Yanofsky, 1992). There are four *AP1* homologs in soybean: *AP1a*, *AP1b*, *AP1c*, and *AP1d* (Chen, Nan, et al., 2020). Under SD conditions, the *ap1* quadruple mutant generated by CRISPR-Cas9 gene editing exhibited delayed flowering, and overexpression of *AP1a* resulted in early flowering (Chen, Nan, et al., 2020); the same phenotype was observed in LD conditions (Yue et al., 2021a).

Soybean *AGAMOUS-like1* (*AGL1*), another MADS-box gene, promotes flowering through the photoperiod pathway. *AGL1* overexpression not only promoted flowering, but also led to early maturation and affected petal development under SD and LD conditions. Furthermore, *AGL1* promotes flowering more under LD than SD conditions (Zeng et al., 2018).

Tof5, which controls flowering time and maturity, was recently identified via GWAS (Dong, Cheng, et al., 2021a). *Tof5*, also called *FULc* and *FUL2a*, is a homolog of *Arabidopsis FUL* (Cao, Li, Wang, et al., 2015; Dong, Cheng, et al., 2021a; Sun et al., 2021). A null $tof5^{CR}$ mutant produced by CRISPR/Cas9 in soybean showed significantly delayed flowering and maturity under natural LD conditions, and *Tof5* overexpression resulted in early flowering and maturity under natural LD conditions (Dong, Cheng, et al., 2021a). Molecular analysis showed that *E1* directly regulates the expression of *Tof5*, and *Tof5* directly activates the expression of *FT2a* and *FT5a* to promote soybean flowering (Dong, Cheng, et al., 2021a).

In addition, overexpression of *FULc* in transgenic Arabidopsis caused significantly earlier flowering time, and *FULc* overexpression rescued the late-flowering phenotype of Arabidopsis *Atful* mutants under SD conditions (Sun et al., 2021). *FULa*, another *FUL* homolog in soybean, functions in flowering and maturation (Jia et al., 2015), and *FULa* overexpression enhanced vegetative growth with more biomass accumulation, and ultimately increased soybean yield (Yue et al., 2021b).

Analysis of variations in coding and non-coding regions of *GBP1* genes in 278 soybean accessions showed that shorter growth periods might be induced by higher *GBP1* expression. In addition, RNAi of *GBP1* resulted in a longer growth period under different daylengths and reduced photoperiod sensitivity. Under SD conditions, *GBP1* leads to early flowering by regulating *FT2a* and *FT5a* to activate the expression of *FULc* (Zhao et al., 2018).

Four homologs of *A. thaliana LNK2* were identified in soybean and named *LNK2* (Li et al., 2020a). The quadruple mutant of *LNK2* obtained by CRISPR/Cas9 flowered significantly earlier than the wild type under LD conditions. Further analysis showed that *E1* expression was significantly decreased in the quadruple mutant, while *FT2a* expression was increased, suggesting that *LNK2* may regulate soybean flowering by activating *E1* and inhibiting *FT* (Li et al., 2020a).

Overexpressing *RAV* inhibited flowering time and maturity, whereas its inhibition resulted in earlier flowering time and maturity under SD and LD conditions. *RAV* may act downstream of *E3* and *E4*. *RAV* downregulated *FT5a* transcription, thereby delaying soybean growth and extending the vegetative and reproductive phases (Wang et al., 2021b).

3. Molecular mechanisms for the photoperiodic regulation of flowering in soybean

The basic framework and mechanisms of photoperiodic flowering regulation in the model plants Arabidopsis and rice are well understood. Arabidopsis is a LD plant, and its photoperiod regulation is dominated by the CO–FT module. Rice, a SD plant, has a Heading date 1 (Hd1) (CO)–Hd3a (FT) pathway similar to that of Arabidopsis and a rice-specific Ghd7–Ehd1–RFT1 (FT) regulatory pathway (Doi et al., 2004; Matsubara, Hori, Ogiso-Tanaka, & Yano, 2014; Song et al., 2015; Xue et al., 2008).

Soybean shares many homologous genes with Arabidopsis and rice. The key flowering gene network in soybean is conserved (Cao et al., 2017; Lin,

Liu, Weller, Abe, & Kong, 2021; Zhang et al., 2021). The major maturity gene *E2* is homologous to *GI* (*GIa*), and *E3* and *E4* are homologs of the photoreceptor *phyA*. In addition, *FT2a* and *FT5a* have conserved roles of the soybean *FT* homologs in flowering (Wu et al., 2017). The roles and regulation of microRNAs (miRNAs) are also generally conserved in soybean (Cao et al., 2017). Although photoperiodic flowering in soybean is regulated by many signaling pathways (Cao et al., 2017; Lin et al., 2021; Xia et al., 2012b; Zhang et al., 2021), phyA–E1 is the main regulatory pathway. This pathway begins with phyA and eventually manifests in changes in *FT* expression, and E1 is an important core transcription factor in this pathway. The phyA proteins E3 and E4 are the major photoreceptors of soybean and play crucial roles in regulating photoperiodic flowering (Cober et al., 1996; Xu et al., 2015). They are activated by light signals under LD conditions and up-regulate *E1* transcription. In turn, *E1* regulates *FT2a* and *FT5a* expression (Xia, Watanabe, et al., 2012a). *FT2a* and *FT5a* are key flowering integrators in soybean that synergistically regulate flowering in the photoperiod pathway (Kong et al., 2010). FT2a and FT5a migrate from leaves to the apical meristem and then interact with the bZIP transcription factor FD-like 19 (FDL19) to up-regulate several downstream flowering-related genes (such as *AP1*, *SUPPRESSOR OF OVEREXPRESSION OF CONSTANS 1* [*SOC1*], and *LEAFY* [*LFY*]) in both a redundant and differential manner (Cao, Li, Lu, et al., 2015; Chen, Nan, et al., 2020; Nan et al., 2014). Soybean, as a photoperiod-sensitive SD plant, has formed different regulatory networks to adapt to different LD and SD conditions.

3.1 Molecular mechanisms for adaptation to long days

Flowering in soybean is controlled by three modules under LD conditions: GI–CO, phyA–E1, and miRNA-dependent modules (Cao et al., 2017; Lin et al., 2021; Zhang et al., 2021).

In Arabidopsis, *AtCO* is a key integrator in the photoperiod pathway that induces *AtFT* expression (Song, 2016; Suárez-López et al., 2001). The two pairs of soybean *COL* genes, *COL1a/1b* and *COL2a/2b*, have different expression patterns (Cao, Li, Lu, et al., 2015; Fan et al., 2014) and belong to a branch linked to *CO*, *COL1*, and *COL2* in Arabidopsis. The expression patterns of *COL1a/1b* are more similar to that of *AtCOL2* than to *AtCO*, whereas *COL2a/2b* expression patterns are closer to that of *AtCO* (Fan et al., 2014; Wu et al., 2014). *COL1a* and *COL1b* suppress flowering in soybean under LD conditions. *COL1a*-overexpressing transgenic lines showed delayed flowering, and *col1b* mutants and *COL1* RNAi transgenic plants showed early flowering in soybean under LD conditions (Cao, Li, Lu, et al., 2015; Wu et al., 2019). *COL1a/b* expression upregulates *E1* (Wu et al., 2019). However, *COL1a/1b* had no effect on flowering under SD conditions (Cao et al., 2015a). Overexpression of *COL1a* down-regulates *FT2a, FT5a,* and *E2* under LD conditions (Cao, Li, Lu, et al., 2015). *COL1a* expression was upregulated in *E1* and *E2* NILs, suggesting that *COL1a, E1,* and *E2* may function in a negative feedback loop (Cao, Li, Lu, et al., 2015). The expression of *E2* is up-regulated by *E3* and *E4,* but down-regulated by *E1* (Cao, Li, Lu, et al., 2015).

AtmiR156 regulates the SQUAMOSA PROMOTER-BINDING PROTEIN-LIKE (AtSPL) transcription factor family to control flowering time in Arabidopsis (Zheng, Ye, Sang, & Wu, 2019). Flowering time is delayed by miR156 overexpression in soybean under LD conditions, but not under SD conditions (Cao, Li, Wang, et al., 2015). miR156b overexpression inhibits *SPL3* and *SPL9* expression in leaves, thus regulating the expression of flowering-related genes such as the MADS-box genes *FUL* and *SOC1* in shoot apices (Cao, Li, Wang, et al., 2015). In addition, miR156b is positively regulated by *E1, E2, E3,* and *E4* in soybean (Cao, Li, Wang, et al., 2015). miR156b overexpression down-regulates miR172 and up-regulates *TOE4a,* a soybean ortholog of *TARGET OF EAT 1* (*TOE1*) (Cao, Li, Wang, et al., 2015). Flowering time is delayed by *TOE4a* overexpression under LD and SD conditions (Zhao et al., 2015). The expression of *FT2a* and *FT5a* in the leaves of wild-type and *TOE4a*-overexpressing plants was low and similar to each other (Zhao et al., 2015). However, *FT2a* and *FT5a* expressions in shoot apices are up-regulated in the wild type but strongly down-regulated in miR156b-overexpressing plants, suggesting that the delayed flowering is likely caused by the inhibition of *FT* genes in shoot apices (Cao, Li, Wang, et al., 2015; Zhao et al., 2015).

The E3/E4–E1–FT module forms a flowering regulation pathway in soybean. Under LD conditions, E3 and E4 cooperate to delay flowering, and *FT5a* and *FT2a* are controlled by *E1. E3* and *E4* are induced under LD conditions. Furthermore, E3 and E4 induce *E1* expression. E3 and E4 show functional differentiation in *E1* regulation: E3 strongly induces *E1* expression, while E4 weakly induces *E1* (Cao, Li, Lu, et al., 2015). *E1* inhibits *FT2a* and *FT5a* expression to delay flowering (Kong et al., 2010; Thakare et al., 2011; Xia, Watanabe, et al., 2012a; Xu et al., 2015). In addition, E1 induces two flowering inhibitors, *FT1a* and *FT4* (Liu et al., 2018; Wang et al., 2015). *E1* and its two homologs, *E1La* and

E1Lb, function similarly in flowering (Xu et al., 2015). *E1la* and *E1lb* transcriptions are also controlled by E3 and E4 (Xu et al., 2015). E1La and E1Lb up-regulate the expression of *FT5a* and *FT2a* and lead to early flowering (Xu et al., 2015). In addition, circadian clock genes, such as *Tof11* (*PRR3a*), *Tof12* (*PRR3b*), and *LHY* coordinately control *E1* expression under LD conditions (Lu et al., 2020). E3 and E4 promote *Tof11* and *Tof12* expression, and Tof11 and Tof12 then directly bind to the promoters of *LHY* genes to suppress their transcription. LHY proteins bind to the *E1* promoter to suppress its transcription and in turn release its transcriptional suppression of *FT2a* and *FT5a*, ultimately promoting flowering and earlier maturity (Dong, Fang, et al., 2021b; Lu et al., 2020). The molecular regulatory network of the soybean photoperiod module E3E4–Tof11/Tof12–LHY–E1–FT was revealed under LD conditions. In addition, *Tof5* was shown to be involved in the E3E4–E1–FT module, further clarifying this pathway. Under LD conditions, *Tof5* expression is downregulated by E3 and E4. E1 also directly represses the transcription of *Tof5*. Tof5 physically associates with the *FT2a* and *FT5a* promoters to promote their expression, ultimately resulting in earlier flowering and earlier maturity (Dong, Cheng, et al., 2021a).

3.2 Molecular mechanisms for adaptation to short days

Soybean flowering is regulated by the E3/E4–E1–FT module under SD conditions. Under SD conditions, E3 and E4 have significantly reduced functionality, meanwhile *E3* and *E4* mutations only slightly affect flowering time (Xia, Watanabe, et al., 2012a; Lu et al., 2017). *E1* expression is inhibited under SD conditions, but *FT2a* and *FT5a* are up-regulated (Kong et al., 2010; Xu et al., 2013). This results in early flowering and early maturation in soybean. The *J* gene acts as a repressor to inhibit *E1* expression. Mutation of *J* leads to up-regulation of *E1* under SD conditions, resulting in delayed flowering (Lu et al., 2017). J protein directly binds to the *E1* promoter near the LUX-binding motif to suppress *E1* transcription. E1 represses *FT2a* and *FT5a* transcription to delay flowering (Lu et al., 2017). Therefore, *E1* suppression by J releases its inhibitory effect on *FT2a* and *FT5a* to promote flowering. When *J* function is lost, *E1* is not suppressed. J protein associates with LUX1 and LUX2 to form a soybean EC, which directly binds to the promoters of *E1* and its two homologous genes *E1La* and *E1Lb*, thereby releasing transcriptional inhibition of *FT2a* and *FT5a* by *E1* and promoting flowering (Bu et al., 2021).

In addition, *Tof16* expression is downregulated by E3 and E4. Tof16 directly binds to the *E1* promoter to suppress *E1* expression and release *FT2a* and *FT5a* transcriptional inhibition (Dong, Fang, et al., 2021b).

4. Soybean adaptation to different latitudes

Soybean flowering time and maturity are the key factors determining soybean adaptation. Cultivated soybean was domesticated from its wild progenitor (*Glycine soja* Sieb. & Zucc.) more than 5000 years ago in temperate regions of China between 32° and 40°N (Carter, Nelson, Sneller, & Cui, 2004; Hymowitz, 1970; Li et al., 2008). The cultivated soybean originated from the Huang-Huai-Hai region of China, but when it was introduced into higher latitudes, flowering was delayed due to the longer daylengths, which resulted in the growth period exceeding the growth season and soybean plants not fully maturing before the winter. When introduced into lower latitudes with shorter daylengths, soybean yield decreased significantly due to reduced vegetative growth and earlier flowering. Understanding the distribution and the degree of fixation of flowering time and maturity alleles is important for improving soybean adaptation.

4.1 Flowering genes involved in soybean adaptation to high latitudes

Photoperiod insensitivity is an important feature for adaptation of soybean cultivars to high latitudes, which is mainly achieved through accumulation of alleles that reduce photoperiod sensitivity.

The maturity loci *E1*, *E3*, and *E4* play major roles in regulating photoperiod sensitivity and adaptation to high latitudes (Abe et al., 2003; Cober et al., 1996; Liu & Abe, 2010; Xu et al., 2013). Therefore, different allelic combinations of *E1*, *E3*, and *E4* are the basis of soybean adaptation in high latitude areas (Jiang et al., 2014) and have been widely used in breeding. The first type of LD insensitivity was achieved by deletion of both *E3* and *E4* genotypes. The second is the dysfunctional *e1e3e4* genotype. The third is the *e1e3E4* genotype. These three genotypes reduce soybean sensitivity to long days to varying degrees (Cober et al., 1996; Xu et al., 2013; Xu et al., 2015).

Tof11 and *Tof12* play important roles in soybean domestication and adaptation to high latitudes. Almost all improved soybean cultivars carry *tof12*, regardless of their origin (Lu et al., 2017). Analysis of *Tof12* coding sequence variations in 1295 soybean materials (including wild soybeans, landraces, and improved cultivars) revealed 25 *Tof12* haplotypes, including four distinct loss-of-function alleles (*tof12-1* to *tof12-4*). Among the four *tof12* loss-of-function alleles, *tof12-1* was strongly selected and rapidly fixed in cultivated soybean, leading to earlier flowering and maturity. The other three *tof12* loss-of-function alleles appeared at low frequency in wild lineages and landraces. The *tof12-1* mutation has been central to the breeding of modern soybean varieties. For *Tof11*, 11 putative loss-of-function mutations were identified. The loss-of-function *tof11-1* mutation occurred after *tof12-1* and was selected in the genetic background of *tof12-1*, thus further shortening the flowering and growth period of cultivated soybeans and improving adaptation. This was the first study to systematically report the molecular mechanisms of anthesis gene evolution and selection during crop domestication (Lu et al., 2020).

The genes mentioned above are important for high-latitude adaptation in cultivated soybeans. $Tof5^{H2}$ and loss-of-function alleles of *E3* played an important role in adaptation of wild soybean to high-latitude regions (Dong, Cheng, et al., 2021a). Different alleles of *Tof5* improved the adaptability of wild soybean and cultivated soybean (Dong, Cheng, et al., 2021a). The genetic basis of high-latitude adaptation was revealed for the first time in wild soybean. Three *Tof5* haplotypes were found in 372 soybean accessions (including 199 domesticated and 173 wild soybean accessions), named $Tof5^{H1}$, $Tof5^{H2}$, and $Tof5^{H3}$. These *Tof5* alleles underwent parallel selection during the adaptation of cultivated and wild soybean to high latitudes. In cultivated soybean, the $Tof5^{H1}$ allele underwent strong artificial selection, while in wild soybean, the $Tof5^{H2}$ allele underwent natural selection, promoting the high-latitude adaptation of cultivated and wild soybean, respectively (Dong, Cheng, et al., 2021a). Analysis of natural variants of *E1*, *E3*, and *E4*, the key genes determining photoperiod sensitivity of cultivated soybean, in 257 wild accessions showed that only loss-of-function alleles of *E3* contributed to high-latitude adaptation of wild soybean. This suggests that the genetic basis for adaptation of wild and cultivated soybean to high latitudes differs. The loss-of-function alleles of *E3* and $Tof5^{H2}$ alleles occurred independently in the adaptation of wild soybean to high latitudes (Dong, Cheng, et al., 2021a). These studies provide a theoretical basis for the adaptation of wild soybean to high latitudes during natural selection.

4.2 Flowering genes involved in soybean adaptation to low latitudes

In contrast to the lack of or reduced photoperiod sensitivity of soybean varieties to LD conditions, suppression of flowering under SD conditions is indispensable for growth at low latitudes. The adaptation of soybean to low latitudes is achieved by the LJ trait, which delays flowering and maturity under SD conditions. In low latitudes such as Brazil, the introduction of LJ delayed flowering and maturation and significantly increased grain yield. The LJ trait was first introduced into soybean varieties in central and western Brazil in the 1970s, which allowed subsequent extension of soybean cultivation to tropical regions (Hartwig & Kiihl, 1979; Spehar, 1995).

The adaptation of soybean to low latitudes or tropical regions largely depends on the natural loss of function of the flowering enhancer *J* (Lu et al., 2017). Analysis of 302 resequenced accessions (including 62 wild soybeans, 130 landraces, and 110 improved cultivars) (Zhou et al., 2015) and 125 non-resequenced accessions collected from low-latitude regions in Brazil, South and Southeast Asia, and southern China showed that at least eight mutational alleles of *J* existed in soybean cultivars adapted to low latitudes (Lu et al., 2017). These mutational alleles were named *j-1* to *j-8*. The geographical origin of the eight mutant alleles was limited to low-latitude regions, but they were not found in landraces originally domesticated from China (Lu et al., 2017). At low latitudes, the presence of mutant *j* alleles increased soybean yield by 30–50% compared with wild-type *J* (Lu et al., 2017). The mutant allele $e6^{PG}$, which carries a *Ty1/copia*-like retrotransposon insertion, was identified to be an allele of *J* and was named *j-9* (Fang et al., 2021b), extending the number of *j* alleles to nine. Five *j* alleles were selected during modern soybean breeding in North and South America (*j-1*, *j-2*, *j-4*, *j-5*, and *j-9*), and four occurred naturally in soybean landraces and were selected in Japan, South Asia, and Southeast Asia (*j-3*, *j-6*, *j-7*, and *j-8*) (Fang et al., 2021b; Lu et al., 2017). Two new *j* alleles were recently discovered (Dong, Fang, et al., 2021b). *j-11* is a weak loss-of-function allele identified by a transient transfection assay in Arabidopsis protoplasts (Dong, Fang, et al., 2021b). *j-11* first appeared in Huanghuai wild soybean and later in domesticated landraces, but the allele was widely selected in soybean varieties at low latitudes. However, *j-11* was not sufficient for soybean adaptation to the tropics. Therefore, *j-1*, *j-3*, *j-6*, and *j-10* variants continue to occur in the genetic background of *j-11*, which facilitate adaptation to tropical areas and increase soybean yield. This shows that there is a phenomenon of gradual selection between *j-11* and the other alleles including *j-1*, *j-3*, *j-6*, and *j-10* (Dong, Fang, et al., 2021b). In summary, the soybean *J* gene is a major source of adaptation to low-latitude regions (Dong, Fang, et al., 2021b; Lu et al., 2017).

Tof16 also plays an important role in extending the LJ trait and adaptation to low latitudes (Dong, Fang, et al., 2021b). Four loss-of-function alleles of

Tof16 (named *tof16-1* to *tof16-4*) were found in 1624 resequenced soybean accessions. *tof16-2* is a weak loss-of-function allele, and the other three are null alleles. Analysis of haplotype origin indicated that the *tof16-2* allele first appeared in wild soybean in the Huanghuai region and was then introduced into local varieties. *tof16-1* and *tof16-4* commonly occur in the genetic background of *tof16-2. tof16-1* and *tof16-4* mainly occur in Brazil. *tof16-2, tof16-1*, and *tof16-4* underwent gradual selection during adaptation to low-latitude environments. Therefore, these loss-of-function *tof16* alleles were selected in different regions to enhance soybean adaptation to low-latitude environments. Analysis of the contribution of natural variants of *Tof16* and *J* to soybean adaptation at low latitudes found that loss-of-function alleles of *Tof16* were only concentrated in soybean varieties in Brazil, while the loss-of-function alleles of *J* were mainly concentrated in soybean varieties in southern China, Southeast Asia, and Brazil. These results suggest that selection of *Tof16* and *J* mutations may have occurred independently. In summary, the natural variations of *Tof16* and *J* promoted soybean adaptation from its temperate origins to tropical regions (Dong, Fang, et al., 2021b).

FT2a and *FT5a*, which contribute to the LJ trait, also play important roles in adaptation to tropical regions (Cai et al., 2020; Li et al., 2021). Resequencing of soybean populations adapted to low latitudes showed that *FT2a* and *FT5a* variants had different geographical origins and played different roles in the spread of soybean to tropical regions (Cai et al., 2020; Li et al., 2021), and this role represented a new regulatory pathway independent of the classical LJ locus *J*. These studies provide a new strategy for improving the adaptation and yield of soybean in tropical environments and provide an important strategy for breeding other crops by integrating natural variations of flowering hormone–related genes to improve environmental adaptation and yield.

5. Concluding remarks

Photoperiod controls the flowering transition and maturity in soybean. Soybean is a photoperiod-sensitive SD plant; SD conditions induce early flowering, while LD conditions delay flowering. The photoperiod adaptation of soybean is controlled by a series of maturity genes. The interaction between these maturity genes and the circadian clock–related flowering time forms the regulatory network of soybean photoperiodic flowering (Fig. 2). Soybean possesses an E3/E4-E1-FT photoperiod regulation pathway with legume-specific *E1* (there are no *E1* homologs in

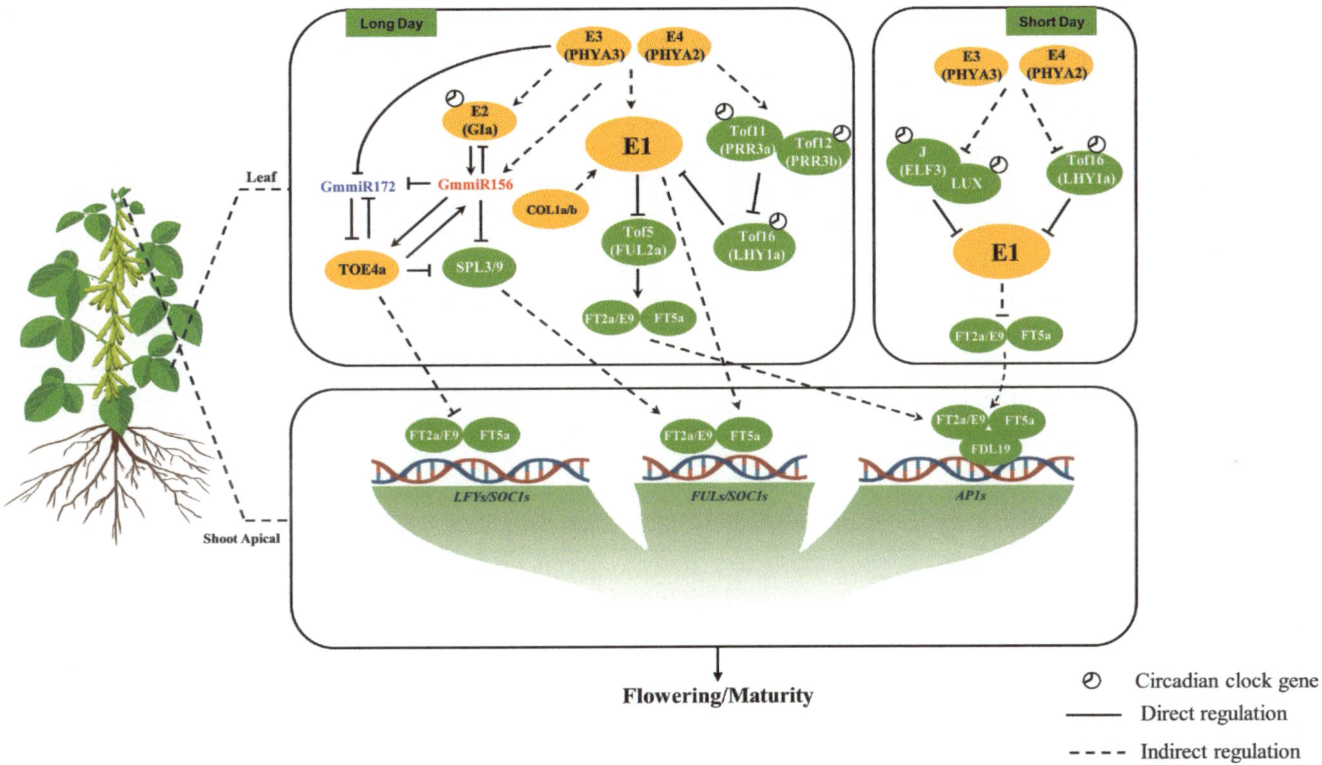

Fig. 2 See figure legend on next page.

Arabidopsis or rice) as the core. *E3* and *E4* sense light signals to mediate the expression of *E1*, which mediates soybean flowering by regulating downstream florigen genes *FT2a* and *FT5a* (Nan et al., 2014; Xia, Watanabe, et al., 2012a; Xu et al., 2015; Zhai et al., 2014). However, the photoperiod response of soybean is very complex and requires coordination of multiple genes to achieve flowering. The detailed regulatory mechanisms are still poorly understood. Many questions need to be resolved to clearly understand how photoperiod regulates soybean flowering. For example, how do E3 and E4 regulate *E1* expression in plants grown under LD conditions? Why are E3 and E4 functions greatly inhibited under SD conditions? Is J protein active under LD conditions? The regulation of the flowering core transcription factor AtCO in Arabidopsis is very precise. In addition to regulating *AtCO* transcription level, the regulation of protein level is also important (Jang et al., 2008; Valverde et al., 2004). As a core transcription factor controlling flowering in soybeans, the regulation of E1 protein level is also expected to be important, but the function of E1 protein level remains a mystery. Further studies are needed to elucidate the molecular mechanisms underlying photoperiodic flowering in soybean. The functions of clock-related genes in soybean have yet to be fully identified. Although the functions of J, LHY and LUX have been preliminarily elucidated, the regulatory

Fig. 2 Photoperiodic flowering regulatory mechanisms in soybean. On the left panel, under LD conditions, *FT2a* and *FT5a* expression is regulated by the PHYA–E1, GI–CO, and miRNA-dependent modules. E3 and E4 induce the expression of *E1*. Tof11 and Tof12 directly bind to the promoters of *LHY* genes to suppress their transcription. LHYs bind to the *E1* promoter to suppress its expression. E1 can directly represses the transcription of *Tof5*. Tof5 physically associates with the *FT2a* and *FT5a* promoters to induce their expression to delay flowering time and maturity. The GI–CO and miRNA-dependent modules are also influenced by photoperiod mediated by E3 and E4. *TOE4a* inhibits flowering by down-regulating *FT2a* and *FT5a* expression. *TOE4a* expression is possibly controlled by *miR172*, which is under the control of *miR156b*. miR156b may delay flowering by negatively regulating *SPL3* and *SPL9*. This suppresses the expression of floral identity genes (*SOC1s*, *LFYs*, *FULs*, *AP1s*). As a result, flowering is delayed under LD conditions. Under SD conditions, the functions of E3 and E4 are greatly weakened, and the induction of *E1* also decreases, thereby weakening the inhibition of *FT2a* and *FT5a*. *J* is suppressed by E3 and E4, and J protein physically binds to the *E1* promoter to suppress *E1* transcription. In addition, *Tof16* expression is down-regulated by E3 and E4. Tof16 directly binds to the *E1* promoter to suppress *E1* expression, thus releasing the E1-mediated inhibition of *FT2a* and *FT5a* and promoting flowering and earlier maturity. Yellow, repressors of flowering; green, promoters. The solid and dotted lines represent direct and indirect regulation, respectively. ↓ Activation and ⊥ suppression of gene expression.

relationship of E2 remains unclear. The interactions between the loci that have been reported so far in relation to fertility need to be further identified. The functions of more soybean homologs of Arabidopsis flowering-related genes are also need to be studied to improve our understanding of photoperiod flowering in soybeans. Molecular characterization of homologs of flowering genes in Arabidopsis has enriched our understanding of the photoperiodic flowering pathway in soybean. Therefore, the model of photoperiodic flowering in Arabidopsis can be used as a guide to study soybean photoperiodic flowering. Similarly, insight gleaned from soybean studies is expected to accelerate the understanding of flowering pathways in other legumes. Studies of the unique photoperiodic flowering mechanism of soybean are helpful for analyzing the adaptive mechanism of soybean at the molecular level, thus providing a theoretical basis for the promotion of high-quality and high-yield soybean.

Soybean is an important source of protein and oil for human diets. Demand for soybean is increasing rapidly. Therefore, it is urgent to expand the global soybean planting area and improve soybean yield and quality. To further understand the adaptability of soybean to different latitudes, it is necessary to clarify not only the molecular mechanism of the interrelationship between genes, but also the origin, distribution, and molecular evolution of alleles at specific flowering time genes of soybean. Soybean originated in the Huang-Huai-hai region of China. Natural variations in *E1*, *E3*, *E4*, *Tof11*, *Tof12*, and *Tof5* have contributed to soybean adaptation to high-latitude regions. The natural variants of *J*, *Tof16*, *FT2a*, and *FT5a* with LJ traits helped soybean adapt to low-latitude areas. Variants of *Tof5* improved the adaptation of wild and cultivated soybeans, while the other genes were primarily involved in the adaptation of cultivated soybeans. Wild soybean was gradually domesticated by artificial selection. Due to long-term selection during domestication, some key genes of important agronomic traits were fixed in cultivated soybean, resulting in a genetic bottleneck. Thus, genetic diversity in cultivated soybeans is much lower than in wild soybeans (Hyten et al., 2006; Lam et al., 2010). This reduced variation may have resulted in the loss of some genes, or alleles, that are important for adaptation to different environments. Thus, wild soybeans may contain useful genetic resources that can be reintroduced into cultivated soybeans through breeding. In addition, our understanding of the origin, distribution, and molecular evolution of these specific flowering time alleles in soybean has improved in recent years, due to the emergence of large-scale genome-wide diversity analyses. Combinations of these approaches will help us better understand the adaptability of soybean, providing the tools to further optimize flowering time adaptability and yield.

Photoperiod not only affects flowering and maturity of soybean, but also node number, seed size, root nodule number, plant type development, disease resistance, and stress tolerance. Therefore, genes controlling photoperiodic flowering of soybean may also affect other phenotypes. For example, *J* positively regulates the expression of downstream salt-stress genes *WRKY12*, *WRKY27*, *WRKY54*, *NAC*, and *SIN1* to improve salt tolerance in soybean (Cheng et al., 2020). A pair of homologous genes, *LHY1a* and *LHY1b*, control the drought response (Wang et al., 2021a). However, the pleiotropic effects of these flowering genes on other agronomic or phenotypic traits have not been well explored. Therefore, we need to continue to elucidate other functions of these genes, as well as their functional diversification and evolution in domestication and modern breeding to further improve adaptability and yield of soybeans.

Acknowledgments

This work was supported by the National Natural Science Foundation of China (Grant Nos. 32090064 and 31725021 to F.K., 31930083 to B.L), and also supported by the Major Program of Guangdong Basic and Applied Research (Grant no. 2019B030302006 to F.K. and B.L.).

Author contributions

Zhihong Hou wrote the first draft of this manuscript. Baohui Liu and Fanjiang Kong provided scientific inputs and contributed to the writing. Fanjiang Kong coordinated the writing effort and prepared the final manuscript.

Conflicts of interest

The authors declare no conflict of interest.

References

Abe, J., Xu, D. H., Miyano, A., Komatsu, K., Kanazawa, A., & Shimamoto, Y. (2003). Photoperiod-insensitive Japanese soybean landraces differ at two maturity loci. *Crop Science*, *43*, 1300–1304.

Ahn, J. H., Miller, D., Winter, V. J., Banfield, M. J., Lee, J. H., Yoo, S. Y., et al. (2006). A divergent external loop confers antagonistic activity on floral regulators FT and TFL1. *The EMBO Journal*, *25*, 605–614.

Amasino, R. (2010). Seasonal and developmental timing of flowering. *The Plant Journal: for cell and molecular biology*, *61*, 1001–1013.

Bendix, C., Marshall, C. M., & Harmon, F. G. (2015). Circadian clock genes universally control key agricultural traits. *Molecular Plant*, *8*, 1135–1152.

Bernard, R. L. (1971). Two major genes for time of flowering and maturity in soybeans. *Crop Science, 11*, 242–244.

Bonato, E. R., & Vello, N. A. (1999). *E6*, a dominant gene conditioning early flowering and maturity in soybeans. *Genetics and Molecular Biology Science, 22*, 229–232.

Bu, T., Lu, S., Wang, K., Dong, L., Li, S., Xie, Q., et al. (2021). A critical role of the soybean evening complex in the control of photoperiod sensitivity and adaptation. *Proceedings of the National Academy of Sciences of the United States of America, 118*, e2010241118.

Buzzell, R. I. (1971). Inheritance of a soybean flowering response to fluorescent-daylength conditions. *Canadian Journal of Genetics and Cytology, 13*, 703–707.

Buzzell, R. I., & Voldeng, H. D. (1980). Inheritance of insensitivity to long daylength. *Soybean Genetics Newsletter, 7*, 26–29.

Cai, Y., Chen, L., Liu, X., Guo, C., Sun, S., Wu, C., et al. (2018). CRISPR/Cas9-mediated targeted mutagenesis of *GmFT2a* delays flowering time in soya bean. *Plant Biotechnology Journal, 16*, 176–185.

Cai, Y., Wang, L., Chen, L., Wu, T., Liu, L., Sun, S., et al. (2020). Mutagenesis of *GmFT2a* and *GmFT5a* mediated by CRISPR/Cas9 contributes for expanding the regional adaptability of soybean. *Plant Biotechnology Journal, 18*, 298–309.

Cao, D., Li, Y., Lu, S., Wang, J., Nan, H., Li, X., et al. (2015a). *GmCOL1a* and *GmCOL1b* function as flowering repressors in soybean under long-day conditions. *Plant & Cell Physiology, 56*, 2409–2422.

Cao, D., Li, Y., Wang, J., Nan, H., Wang, Y., Lu, S., et al. (2015b). GmmiR156b overexpression delays flowering time in soybean. *Plant Molecular Biology, 89*, 353–363.

Cao, D., Takeshima, R., Zhao, C., Liu, B., Abe, J., & Kong, F. (2017). Molecular mechanisms of flowering under long days and stem growth habit in soybean. *Journal of Experimental Botany, 68*, 1873–1884.

Carter, T. E., Nelson, R., Sneller, C. H., & Cui, Z. (2004). *Soybeans: Improvement, production and uses* (3rd ed.). Madison: American Society of Agronomy (Chapter 8).

Chen, L., Cai, Y., Qu, M., Wang, L., Sun, H., Jiang, B., et al. (2020a). Soybean adaption to high-latitude regions is associated with natural variations of *GmFT2b*, an ortholog of *FLOWERING LOCUS T*. *Plant, Cell & Environment, 43*, 934–944.

Chen, L., Nan, H., Kong, L., Yue, L., Yang, H., Zhao, Q., et al. (2020b). Soybean *AP1* homologs control flowering time and plant height. *Journal of Integrative Plant Biology, 62*, 1868–1879.

Cheng, Q., Dong, L., Su, T., Li, T., Gan, Z., Nan, H., et al. (2019). CRISPR/Cas9-mediated targeted mutagenesis of *GmLHY* genes alters plant height and internode length in soybean. *BMC Plant Biology, 19*, 562.

Cheng, Q., Gan, Z., Wang, Y., Lu, S., Hou, Z., Li, H., et al. (2020). The soybean gene *J* contributes to salt stress tolerance by up-regulating salt-responsive genes. *Frontiers in Plant Science, 11*, 272.

Cober, E. R. (2011). Long juvenile soybean flowering responses under very short photoperiods. *Crop Science, 51*, 140–145.

Cober, E. R., Molnar, S. J., Charette, M., & Voldeng, H. D. (2010). A new locus for early maturity in soybean. *Crop Science, 50*, 524–527.

Cober, E. R., Tanner, J. W., & Voldeng, H. D. (1996). Soybean photoperiod-sensitivity loci respond differentially to light quality. *Crop Science, 36*, 606–610.

Cober, E. R., & Voldeng, H. D. (2001). Low R: FR light quality delays flowering of *E7E7* soybean lines. *Crop Science, 41*, 1823–1826.

Corbesier, L., Vincent, C., Jang, S., Fornara, F., Fan, Q. Z., Searle, I., et al. (2007). FT protein movement contributes to long-distance signaling in floral induction of *Arabidopsis*. *Science, 316*, 1030–1033.

Dissanayaka, A., Rodriguez, T. O., Di, S., Yan, F., Githri, S. M., Rodas, F. R., et al. (2016). Quantitative trait locus mapping of soybean maturity gene *E5*. *Breeding Science*, *66*, 407–415.

Doi, K., Izawa, T., Fuse, T., Yamanouchi, U., Kubo, T., Shimatani, Z., et al. (2004). *Ehd1*, a B-type response regulator in rice, confers short-day promotion of flowering and controls *FT-like* gene expression independently of *Hd1*. *Genes & Development*, *18*, 926–936.

Dong, L., Cheng, Q., Fang, C., Kong, L., Yang, H., Hou, Z., et al. (2021a). Parallel selection of distinct *Tof5* alleles drove the adaptation of cultivated and wild soybean to high latitudes. *Molecular Plant*, *14*, 1–14.

Dong, L., Fang, C., Cheng, Q., Su, T., Kou, K., Kong, L., et al. (2021b). Genetic basis and adaptation trajectory of soybean from its temperate origin to tropics. *Nature Communications*, *12*, 5445.

Fan, C., Hu, R., Zhang, X., Wang, X., Zhang, W., Zhang, Q., et al. (2014). Conserved CO-FT regulons contribute to the photoperiod flowering control in soybean. *BMC Plant Biology*, *14*, 9.

Fang, X., Han, Y., Liu, M., Jiang, J., Li, X., Lian, Q., et al. (2021a). Modulation of evening complex activity enables north-to-south adaptation of soybean. *Science China Life Sciences*, *64*, 179–195.

Fang, C., Liu, J., Zhang, T., Su, T., Li, S., Cheng, Q., et al. (2021b). A recent retrotransposon insertion of J caused *E6* locus facilitating soybean adaptation into low latitude. *Journal of Integrative Plant Biology*, *63*, 995–1003.

Fehr, W. R., Caviness, C. E., Burmood, D. T., & Pennington, J. S. (1971). Stage of development descriptions for soybeans. *Crop Science*, *11*, 929–931.

Fornara, F., de Montaigu, A., & Coupland, G. (2010). SnapShot: Control of flowering in *Arabidopsis*. *Cell*, *141*, 550.

Garner, W. W., & Allard, H. A. (1920). Effect of the relative length of day and night and other factors of the environment on growth and reproduction in plants. *Journal of Agricultural Research*, *18*, 553–606.

Gong, Z. (2020). Flowering phenology as a core domestication trait in soybean. *Journal of Integrative Plant Biology.*, *62*, 546–549.

Greenham, K., & McClung, C. R. (2015). Integrating circadian dynamics with physiological processes in plants. *Genetics*, *16*, 598–610.

Guo, G., Xu, K., Zhang, X., Zhu, J., Lu, M., Chen, F., et al. (2015). Extensive analysis of *GmFTL* and *GmCOL* expression in northern soybean cultivars in field conditions. *PLoS One*, *10*, e0136601.

Guo, H., Yang, H., Mockler, T. C., & Lin, C. (1998). Regulation of flowering time by *Arabidopsis* photoreceptors. *Science*, *279*, 1360–1363.

Hartwig, E. E., & Kiihl, R. A. S. (1979). Identification and utilization of a delayed flowering character in soybeans for short-day condtions. *Field Crops Research*, *2*, 145–151.

Hayama, R., & Coupland, G. (2004). The molecular basis of diversity in the photoperiodic flowering responses of Arabidopsis and rice. *Plant Physiology*, *135*, 677–684.

Huang, H., & Nusinow, D. A. (2016). Into the evening: Complex interactions in the *Arabidopsis* circadian clock. *Trends in Genetics*, *32*, 674–686.

Huber, S. C., Rufty, T. W., & Kerr, P. S. (1984). Effect of photoperiod on photosynthate partitioning and diurnal rhythms in sucrose phosphate synthase activity in leaves of soybean (*Glycine max* L. [Merr.]) and tobacco (*Nicotiana Tabacum* L.). *Plant Physiology*, *75*, 1080–1084.

Hymowitz, T. (1970). On the domestication of the soybean. *Economic Botany*, *24*, 408–421.

Hyten, D. L., Song, Q., Zhu, Y., Choi, I. Y., Nelson, R. L., Costa, J. M., et al. (2006). Impacts of genetic bottlenecks on soybean genome diversity. *Proceedings of the National Academy of Sciences of the United States of America*, *103*, 16666–16671.

Jaeger, K. E., & Wigge, P. A. (2007). FT protein acts as a long-range signal in *Arabidopsis*. *Current Biology*, *17*, 1050–1054.

Jang, S., Marchal, V., Panigrahi, K. C., Wenkel, S., Soppe, W., Deng, X. W., et al. (2008). *Arabidopsis* COP1 shapes the temporal pattern of CO accumulation conferring a photoperiodic flowering response. *The EMBO Journal*, *27*, 1277–1288.

Jia, Z., Jiang, B., Gao, X., Yue, Y., Fei, Z., Sun, H., et al. (2015). *GmFULa*, a *FRUITFULL* homolog, functions in the flowering and maturation of soybean. *Plant Cell Reports*, *34*, 121–132.

Jiang, B., Nan, H., Gao, Y., Tang, L., Yue, Y., Lu, S., et al. (2014). Allelic combinations of soybean maturity Loci *E1*, *E2*, *E3* and *E4* result in diversity of maturity and adaptation to different latitudes. *PLoS One*, *9*, e106042.

Kong, F., Liu, B., Xia, Z., Sato, S., Kim, B. M., Watanabe, S., et al. (2010). Two coordinately regulated homologs of *FLOWERING LOCUS T* are involved in the control of photoperiodic flowering in soybean. *Plant Physiology*, *154*, 1220–1231.

Kong, F., Nan, H., Cao, D., Li, Y., Wu, F., Wang, J., et al. (2014). A new dominant gene *E9* conditions early flowering and maturity in soybean. *Crop Science*, *54*, 2529–2535.

Lam, H. M., Xu, X., Liu, X., Chen, W., Yang, G., Wong, F. L., et al. (2010). Resequencing of 31 wild and cultivated soybean genomes identifies patterns of genetic diversity and selection. *Nature Genetics*, *42*, 1053–1059.

Li, Z., Cheng, Q., Gan, Z., Hou, Z., Zhang, Y., Li, Y., et al. (2020a). Multiplex CRISPR/Cas9-mediated knockout of soybean *LNK2* advances flowering time. *The Crop Journal*, *9*, 767–776.

Li, Y., Dong, Y., Wu, H., Hu, B., Zhai, H., Yang, J., et al. (2019a). Positional cloning of the flowering time QTL *qFT12-1* reveals the link between the clock related *PRR* homolog with photoperiodic response in soybeans. *Frontiers in Plant Science*, *10*, 1303.

Li, X., Fang, C., Xu, M., Zhang, F., Lu, S., Nan, H., et al. (2017). Quantitative trait locus mapping of soybean maturity gene *E6*. *Crop Science*, *57*, 1–8.

Li, X., Fang, C., Yang, Y., Lv, T., Su, T., Chen, L., et al. (2021). Overcoming the genetic compensation response of soybean florigens to improve adaptation and yield at low latitudes. *Current Biology*, *31*, 3755–3767.

Li, Y., Guan, R., Liu, Z., Ma, Y., Wang, L., Li, L., et al. (2008). Genetic structure and diversity of cultivated soybean (*Glycine max* (L.) Merr.) landraces in China. *Theoretical and Applied Genetics*, *117*, 857–871.

Li, C., Li, Y. H., Li, Y., Lu, H., Hong, H., Tian, Y., et al. (2020b). A domestication-associated gene *GmPRR3b* regulates the circadian clock and flowering time in soybean. *Molecular Plant*, *13*, 745–759.

Li, M. W., Liu, W., Lam, H. M., & Gendron, J. M. (2019b). Characterization of two growth period QTLs reveals modification of *PRR3* genes during soybean domestication. *Plant & Cell Physiology*, *60*, 407–420.

Li, Y. H., Zhou, G., Ma, J., Jiang, W., Jin, L. G., Zhang, Z., et al. (2014). De novo assembly of soybean wild relatives for pan-genome analysis of diversity and agronomic traits. *Nature Biotechnology*, *32*, 1045–1052.

Lin, X., Liu, B., Weller, J. L., Abe, J., & Kong, F. (2021). Molecular mechanisms for the photoperiodic regulation of flowering in soybean. *Journal of Integrative Plant Biology*, *63*, 981–994.

Liu, B., & Abe, J. (2010). QTL mapping for photoperiod insensitivity of a Japanese soybean landrace Sakamotowase. *The Journal of Heredity*, *101*, 251–256.

Liu, B., Fujita, T., Yan, Z., Sakamoto, S., Xu, D., & Abe, J. (2007). QTL mapping of domestication-related traits in soybean (*Glycine max*). *Annals of Botany*, *100*, 1027–1038.

Liu, W., Jiang, B., Ma, L., Zhang, S., Zhai, H., Xu, X., et al. (2018). Functional diversification of *flowering locus T* homologs in soybean: *GmFT1a* and *GmFT2a/5a* have opposite roles in controlling flowering and maturation. *The New Phytologist*, *217*, 1335–1345.

Liu, B., Kanazawa, A., Matsumura, H., Takahashi, R., Harada, K., & Abe, J. (2008). Genetic redundancy in soybean photoresponses associated with duplication of the phytochrome A gene. *Genetics, 180*, 995–1007.

Liu, B., Watanabe, S., Uchiyama, T., Kong, F., Kanazawa, A., Xia, Z., et al. (2010). The soybean stem growth habit gene *Dt1* is an ortholog of Arabidopsis *TERMINAL FLOWER1*. *Plant Physiology, 153*, 198–210.

Liu, Y., Zhang, D., Ping, J., Li, S., Chen, Z., & Ma, J. (2016). Innovation of a regulatory mechanism modulating semi-determinate stem growth through artificial selection in soybean. *PLoS Genetics, 12*, e1005818.

Locke, J. C., Kozma-Bognár, L., Gould, P. D., Fehér, B., Kevei, E., Nagy, F., et al. (2006). Experimental validation of a predicted feedback loop in the multi-oscillator clock of *Arabidopsis thaliana*. *Molecular Systems Biology, 2*, 59.

Lu, S., Dong, L., Fang, C., Liu, S., Kong, L., Cheng, Q., et al. (2020). Stepwise selection on homeologous *PRR* genes controlling flowering and maturity during soybean domestication. *Nature Genetics, 52*(4), 428–436.

Lu, S., Zhao, X., Hu, Y., Liu, S., Nan, H., Li, X., et al. (2017). Natural variation at the soybean *J* locus improves adaptation to the tropics and enhances yield. *Nature Genetics, 49*(5), 773–779. https://doi.org/10.1038/ng.3819.

Mandel, M. A., Gustafson-Brown, C., Savidge, B., & Yanofsky, M. F. (1992). Molecular characterization of the *Arabidopsis* floral homeotic gene *APETALA1*. *Nature, 360*, 273–277.

Mathieu, J., Warthmann, N., Küttner, F., & Schmid, M. (2007). Export of FT protein from phloem companion cells is sufficient for floral induction in *Arabidopsis*. *Current Biology, 17*, 1055–1060.

Matsubara, K., Hori, K., Ogiso-Tanaka, E., & Yano, M. (2014). Cloning of quantitative trait genes from rice reveals conservation and divergence of photoperiod flowering pathways in Arabidopsis and rice. *Frontiers in Plant Science, 5*, 193.

McBlain, B. A., & Bernard, R. L. (1987). A new gene affecting the time of flowering and maturity in soybeans. *Journal of Heredity, 78*, 160–162.

Miranda, C., Scaboo, A., Cober, E., Denwar, N., & Bilyeu, K. (2020). The effects and interaction of soybean maturity gene alleles controlling flowering time, maturity, and adaptation in tropical environments. *BMC Plant Biology, 20*, 65.

Molnar, S. J., Rai, S., Charette, M., & Cober, E. R. (2003). Simple sequence repeat (SSR) markers linked to *E1*, *E3*, *E4*, and *E7* maturity genes in soybean. *Genome, 46*, 1024–1036.

Nan, H., Cao, D., Zhang, D., Li, Y., Lu, S., Tang, L., et al. (2014). GmFT2a and GmFT5a redundantly and differentially regulate flowering through interaction with and upregulation of the bZIP transcription factor GmFDL19 in soybean. *PLoS One, 9*, e97669.

Notaguchi, M., Abe, M., Kimura, T., Daimon, Y., Kobayashi, T., Yamaguchi, A., et al. (2008). Long-distance, graft-transmissible action of Arabidopsis *FLOWERING LOCUS T* protein to promote flowering. *Plant & Cell Physiology, 49*, 1645–1658.

Nusinow, D. A., Helfer, A., Hamilton, E. E., King, J. J., Imaizumi, T., Schultz, T. F., et al. (2011). The ELF4-ELF3-LUX complex links the circadian clock to diurnal control of hypocotyl growth. *Nature, 475*, 398–402.

Oakenfull, R. J., & Davis, S. J. (2017). Shining a light on the Arabidopsis circadian clock. *Plant, Cell & Environment, 40*, 2571–2585.

Owen, F. V. (1927). Inheritance studies in soybeans. II. Glabrousness, color of pubescence, time of maturity, and linkage relations. *Genetics, 12*, 519–529.

Ray, J. D., Hinson, K., Mankono, J. E. B., & Malo, M. F. (1995). Genetic control of a long-juvenile trait in soybean. *Crop Science, 35*, 1001–1006.

Roux, F., Touzet, P., Cuguen, J., & Le Corre, V. (2006). How to be early flowering: An evolutionary perspective. *Trends in Plant Science, 11*, 375–381.

Saindon, G., Voldeng, H. D., Beversdorf, W. D., & Buzzell, R. I. (1989). Genetic control of long daylength response in soybean. *Crop Science, 29*, 1436–1439.
Samanfar, B., Molnar, S. J., Charette, M., Schoenrock, A., Dehne, F., Golshani, A., et al. (2017). Mapping and identification of a potential candidate gene for a novel maturity locus, *E10*, in soybean. *Theoretical and Applied Genetics, 130*, 377–390.
Sinclair, T. R., & Hinson, K. (1992). Soybean flowering in response to the long-juvenile trait. *Crop Science, 32*, 1242–1248.
Song, Y. H. (2016). The effect of fluctuations in photoperiod and ambient temperature on the timing of flowering: Time to move on natural environmental conditions: Time to move on natural environmental conditions. *Molecules and Cells, 39*, 715–721.
Song, Y. H., Shim, J. S., Kinmonth-Schultz, H. A., & Imaizumi, T. (2015). Photoperiodic flowering: Time measurement mechanisms in leaves. *Annual Review of Plant Biology, 66*, 441–464.
Spehar, C. R. (1995). Impact of strategic genes in soybean on agricultural development in the Brazilian tropical savannah. *Field Crops Research, 41*, 141–146.
Suárez-López, P., Wheatley, K., Robson, F., Onouchi, H., Valverde, F., & Coupland, G. (2001). *CONSTANS* mediates between the circadian clock and the control of flowering in *Arabidopsis*. *Nature, 410*, 1116–1120.
Sun, J., Wang, M., Zhao, C., Liu, T., Liu, Z., Fan, Y., et al. (2021). *GmFULc* is induced by short days in soybean and may accelerate flowering in transgenic *Arabidopsis thaliana*. *International Journal of Molecular Sciences, 22*, 10333.
Thakare, D., Kumudini, S., & Dinkins, R. D. (2011). The alleles at the *E1* locus impact the expression pattern of two soybean *FT-like* genes shown to induce flowering in Arabidopsis. *Planta, 234*, 933–943.
Tian, Z., Wang, X., Lee, R., Li, Y., Specht, J. E., Nelson, R. L., et al. (2010). Artificial selection for determinate growth habit in soybean. *Proceedings of the National Academy of Sciences of the United States of America, 107*, 8563–8568.
Tsubokura, Y., Watanabe, S., Xia, Z., Kanamori, H., Yamagata, H., Kaga, A., et al. (2014). Natural variation in the genes responsible for maturity loci *E1*, *E2*, *E3* and *E4* in soybean. *Annals of Botany, 113*, 429–441.
Turck, F., Fornara, F., & Coupland, G. (2008). Regulation and identity of florigen: FLOWERING LOCUS T moves center stage. *Annual Review of Plant Biology, 59*, 573–594.
Upadhyay, A. P., Ellis, R. H., Summerfield, R. J., Roberts, E. H., & Qi, A. (1994). Characterization of photothermal flowering responses in maturity isolines of soyabean [*Glycine max* (L.) Merrill] cv. Clark. *Annals of Botany, 74*, 87–96.
Valverde, F., Mouradov, A., Soppe, W., Ravenscroft, D., Samach, A., & Coupland, G. (2004). Photoreceptor regulation of CONSTANS protein in photoperiodic flowering. *Science, 303*, 1003–1006.
Wang, K., Bu, T., Cheng, Q., Dong, L., Su, T., Chen, Z., et al. (2021a). Two homologous *LHY* pairs negatively control soybean drought tolerance by repressing the abscisic acid responses. *The New Phytologist, 229*, 2660–2675.
Wang, Y., Gu, Y., Gao, H., Qiu, L., Chang, R., Chen, S., et al. (2016). Molecular and geographic evolutionary support for the essential role of *GIGANTEAa* in soybean domestication of flowering time. *BMC Evolutionary Biology, 16*, 79.
Wang, F., Nan, H., Chen, L., Fang, C., Zhang, H., et al. (2019). A new dominant locus, *E11*, controls early flowering time and maturity in soybean. *Molecular Breeding, 39*, 70.
Wang, L., Sun, S., Wu, T., Liu, L., Sun, X., Cai, Y., et al. (2020). Natural variation and CRISPR/Cas9-mediated mutation in *GmPRR37* affect photoperiodic flowering and contribute to regional adaptation of soybean. *Plant Biotechnology Journal, 18*, 1869–1881.
Wang, Y., Xu, C., Sun, J., Dong, L., Li, M., Liu, Y., et al. (2021b). *GmRAV* confers ecological adaptation through photoperiod control of flowering time and maturity in soybean. *Plant Physiology, 187*, 361–377.

Wang, Z., Zhou, Z., Liu, Y., Liu, T., Li, Q., Ji, Y., et al. (2015). Functional evolution of phosphatidylethanolamine binding proteins in soybean and Arabidopsis. *The Plant Cell, 27*, 323–336.

Watanabe, S., Harada, K., & Abe, J. (2012). Genetic and molecular bases of photoperiod responses of flowering in soybean. *Breeding Science, 61*, 531–543.

Watanabe, S., Hideshima, R., Xia, Z., Tsubokura, Y., Sato, S., Nakamoto, Y., et al. (2009). Map-based cloning of the gene associated with the soybean maturity locus *E3*. *Genetics, 182*, 1251–1262.

Watanabe, S., Xia, Z., Hideshima, R., Tsubokura, Y., Sato, S., Yamanaka, N., et al. (2011). A map-based cloning strategy employing a residual heterozygous line reveals that the *GIGANTEA* gene is involved in soybean maturity and flowering. *Genetics, 188*, 395–407.

Wong, C. E., Singh, M. B., & Bhalla, P. L. (2013). The dynamics of soybean leaf and shoot apical meristem transcriptome undergoing floral initiation process. *PLoS One, 8*, e65319.

Wu, F., Kang, X., Wang, M., Haider, W., Price, W. B., Hajek, B., et al. (2019). Transcriptome-enabled network inference revealed the *GmCOL1* feed-forward loop and its roles in photoperiodic flowering of soybean. *Frontiers in Plant Science, 10*, 1221.

Wu, F., Price, B. W., Haider, W., Seufferheld, G., Nelson, R., & Hanzawa, Y. (2014). Functional and evolutionary characterization of the *CONSTANS* gene family in short-day photoperiodic flowering in soybean. *PLoS One, 9*, e85754.

Wu, F., Sedivy, E. J., Price, W. B., Haider, W., & Hanzawa, Y. (2017). Evolutionary trajectories of duplicated *FT* homologues and their roles in soybean domestication. *The Plant Journal: Cell and Molecular Biology, 90*, 941–953.

Xia, Z., Watanabe, S., Yamada, T., Tsubokura, Y., Nakashima, H., Zhai, H., et al. (2012a). Positional cloning and characterization reveal the molecular basis for soybean maturity locus *E1* that regulates photoperiodic flowering. *Proceedings of the National Academy of Sciences of the United States of America, 109*, E2155–E2164.

Xia, Z., Zhai, H., Liu, B., Kong, F., Yuan, X., Wu, H., et al. (2012b). Molecular identification of genes controlling flowering time, maturity, and photoperiod response in soybean. *Plant Systematics and Evolution, 298*, 1217–1227.

Xu, M., Xu, Z., Liu, B., Kong, F., Tsubokura, Y., Watanabe, S., et al. (2013). Genetic variation in four maturity genes affects photoperiod insensitivity and PHYA-regulated post-flowering responses of soybean. *BMC Plant Biology, 13*, 91.

Xu, M., Yamagishi, N., Zhao, C., Takeshima, R., Kasai, M., Watanabe, S., et al. (2015). The soybean-specific maturity gene E1 family of floral repressors controls night-break responses through down-regulation of *FLOWERING LOCUS T* orthologs. *Plant Physiology, 168*, 1735–1746.

Xue, W., Xing, Y., Weng, X., Zhao, Y., Tang, W., Wang, L., et al. (2008). Natural variation in *Ghd7* is an important regulator of heading date and yield potential in rice. *Nature Genetics, 40*, 761–767.

Yue, L., Li, X., Fang, C., Chen, L., Yang, H., Yang, J., et al. (2021a). FT5a interferes with the Dt1-AP1 feedback loop to control flowering time and shoot determinacy in soybean. *Journal of Integrative Plant Biology, 63*, 1004–1020.

Yue, Y., Liu, N., Jiang, B., Li, M., Wang, H., Jiang, Z., et al. (2017). A single nucleotide deletion in *J* encoding GmELF3 confers long juvenility and is associated with adaption of tropic soybean. *Molecular Plant, 10*, 656–658.

Yue, Y., Sun, S., Li, J., Yu, H., Wu, H., Sun, B., et al. (2021b). *GmFULa* improves soybean yield by enhancing carbon assimilation without altering flowering time or maturity. *Plant Cell Reports, 40*(10), 1875–1888.

Zeilinger, M. N., Farré, E. M., Taylor, S. R., Kay, S. A., & Doyle, F. J., 3rd. (2006). A novel computational model of the circadian clock in *Arabidopsis* that incorporates PRR7 and PRR9. *Molecular Systems Biology, 2*, 58.

Zeng, X., Liu, H., Du, H., Wang, S., Yang, W., Chi, Y., et al. (2018). Soybean MADS-box gene *GmAGL1* promotes flowering via the photoperiod pathway. *BMC Genomics*, *19*, 51.

Zhai, H., Lü, S., Liang, S., Wu, H., Zhang, X., Liu, B., et al. (2014). *GmFT4*, a homolog of *FLOWERING LOCUS T*, is positively regulated by *E1* and functions as a flowering repressor in soybean. *PLoS One*, *9*, e89030.

Zhai, H., Lü, S., Wu, H., Zhang, Y., Zhang, X., Yang, J., et al. (2015). Diurnal expression pattern, allelic variation, and association analysis reveal functional features of the *E1* gene in control of photoperiodic flowering in soybean. *PLoS One*, *10*, e0135909.

Zhang, L., Liu, W., Tsegaw, M., Xu, X., Qi, Y., Sapey, E., et al. (2020a). Principles and practices of the photo-thermal adaptability improvement in soybean. *Journal of Integrative Agriculture*, *19*, 295–310.

Zhang, M., Liu, S., Wang, Z., Yuan, Y., Zhang, Z., Liang, Q., et al. (2021). Progress in soybean functional genomics over the past decade. *Plant Biotechnology Journal*, 1–27.

Zhang, J., Xu, M., Dwiyanti, M. S., Watanabe, S., Yamada, T., Hase, Y., et al. (2020b). A soybean deletion mutant that moderates the repression of flowering by cool temperatures. *Frontiers in Plant Science*, *11*, 429.

Zhao, X., Cao, D., Huang, Z., Wang, J., Lu, S., Xu, Y., et al. (2015). Dual functions of GmTOE4a in the regulation of photoperiod-mediated flowering and plant morphology in soybean. *Plant Molecular Biology*, *88*, 343–355.

Zhao, L., Li, M., Xu, C., Yang, X., Li, D., Zhao, X., et al. (2018). Natural variation in GmGBP1 promoter affects photoperiod control of flowering time and maturity in soybean. *The Plant Journal: Cell and Molecular Biology*, *96*, 147–162.

Zhao, C., Takeshima, R., Zhu, J., Xu, M., Sato, M., Watanabe, S., et al. (2016). A recessive allele for delayed FLOWERING at the soybean maturity LOCUS *E9* is a leaky allele of *FT2a*, a *FLOWERING LOCUS T* ortholog. *BMC Plant Biology*, *16*, 20.

Zheng, C., Ye, M., Sang, M., & Wu, R. (2019). A regulatory network for miR156-SPL module in *Arabidopsis thaliana*. *International Journal of Molecular Sciences*, *20*, 6166.

Zhou, Z., Jiang, Y., Wang, Z., Gou, Z., Lyu, J., Li, W., et al. (2015). Resequencing 302 wild and cultivated accessions identifies genes related to domestication and improvement in soybean. *Nature Biotechnology*, *33*, 408–414.

CHAPTER THREE

Root physiology and morphology of soybean in relation to stress tolerance

Yinglong Chen[a,*], Zhili Wang[b], Heng Ye[c], Shuo Liu[d], Henry T. Nguyen[c], Hon-Ming Lam[b], and Kadambot H.M. Siddique[a]

[a]The UWA Institute of Agriculture, and UWA School of Agriculture and Environment, The University of Western Australia, Perth, WA, Australia
[b]Centre for Soybean Research of the State Key Laboratory of Agrobiotechnology and School of Life Sciences, The Chinese University of Hong Kong, Shatin, Hong Kong Special Administrative Region, China
[c]Division of Plant Sciences, 25 Agriculture Lab, College of Agriculture, Food and Natural Resources, University of Missouri, Columbia, MO, United States
[d]College of Natural Resources and Environment, and Institute of Soil and Water Conservation, Northwest A&F University, Yangling, Shaanxi, China
*Corresponding author: e-mail address: yinglong.chen@uwa.edu.au

Contents

Abstract

The significance of root system architecture (RSA) to the productivity and adaptation of crops such as soybean (*Glycine* max (L.) Merr.) is well recognized. RSA is an essential trait in the selection of genotypes with improved resource acquisition and adaptation to environmental stresses. High-throughput phenotyping, imaging of root growth and structure, genomic sequencing and association analysis techniques are useful tools

Advances in Botanical Research, Volume 102
ISSN 0065-2296
https://doi.org/10.1016/bs.abr.2022.02.005

in soybean breeding programs that aim to improve plant performance and productivity. This chapter discusses genetic variability and plasticity in soybean root system morphology, root architecture and root anatomic traits that are relevant to increased soil resource use efficiency and better adaptation to specific soil environments. We consider the importance of soybean/rhizobia and rhizosphere interactions and highlight current challenges and future directions in improving soybean root physiology and morphology for better adaptation to abiotic stresses.

1. Introduction

Soybean is one of the five most important crops worldwide with global production of over 340 million metric tons in 2016 (FAOSTAT, 2017). The global import and export values of soybean have outstripped all other major crops such as wheat, rice, and maize. However, the potential for global production is currently restricted by climatic factors, particularly drought, because of the increasing variability in air temperatures and rainfall and limited agricultural water resources (Kunert et al., 2016). Drought stress is predicted to worsen as a result of climate change, thereby reducing soybean yields (Li et al., 2015; Trenberth et al., 2013). The ability of soybean roots to extract water from larger soil volumes is critical for maintaining yield stability as soil water becomes depleted. Deep root systems potentially help to sustain yield under drought conditions (Costa et al., 2000; Xiong et al., 2021). Moreover, root morphology, including root biomass, root length, root surface area, average root diameter and root volume (Vamerali et al., 2003), determine the ability of crops to obtain nutrients and water.

Roots are the major organs responsible for foraging and acquiring nutrients and water from the soil. The root system is a complex three-dimensional (3D) structure, exhibiting specific spatial and temporal configurations. Root system architecture (RSA) refers to the in situ spatial distribution of the root system within the soil (Hinsinger et al., 2011; Lynch, 1995, 2007; Manschadi, Manske, & Vlek, 2013). RSA plays a vital role in the exploration of soil zones and the acquisition of water and nutrients (Chen, Djalovic, & Rengel, 2015; Gregory et al., 2009; Lynch & Brown, 2012).

Root phenotyping of soybean germplasm is hindered because of the absence of appropriate technologies that allow the accurate observation, quantification and collection of important root trait data. The technology gap is exacerbated by the genetic complexity of the traits, as well as the phenotypic expression and morphometric nature of trait expression, in addition to environmental interactions including the availability of water and nutrients and interactions with other plants and microbes (Falk et al., 2020). Genetic and phenotypic diversity must be explored and characterized by soybean breeders prior to selection for RSA traits, Therefore, accurate and efficient quantification of root architecture traits and diversity as well as associated physiological processes is a perquisite for the fulfillment of breeding objectives.

RSA is plastic and dynamic, allowing plants to optimize acquisition of important soil resources in response to environmental cues (Zhu, Ingram, Benfey, & Elich, 2011). Roots foster a wide range of plant–microbe interactions, such as nodulation in soybean and other legumes, which can be indirectly selected traits in breeding programs, particularly for non-tuber or root crops (York, Galindo-Castañeda, Schussler, & Lynch, 2015). The following discussion provides an overview of current knowledge concerning root morphology, architecture and anatomic traits in soybean. We discuss current developments in root phenotyping and summarize the genetic control of root-related traits, highlighting the importance of root plasticity in plant responses to abiotic stresses, such as drought, flooding, and salinity.

2. Root morphology and architecture

2.1 Soybean root system

Soybean is a dicotyledonous species. As such, the root system consists of primary or tap root and several orders of lateral roots. Root development is defined as primary root growth and the proliferation of secondary roots. The elaboration of the root system defines RSA, which is the spatial distribution of the roots in the soil. The primary or tap root is the first organ produced by germinating seeds by hypophysis (Smet, Lau, Mayer, & Jürgens, 2010). The thick soybean primary root develops lateral roots from primordia that are produced from the pericycle cells (Lucas et al., 2013). Root nodules are formed by symbiotic interactions between the roots and soil rhizobia. Soybean has determinate nodules with a spherical shape because of the absence of a persistent meristem.

Soybean root development has been explored under controlled environment, greenhouse and field environments. Soybean roots develop in three stages: (i) rapid root growth occurs during vegetative stage. At this stage, primary root tip extension is gradual, lateral root primordia are formed but their growth depends on environmental cues; (ii) more lateral roots and root branches are formed during reproductive growth. At this stage, primary root tip extension is rapid and root length is substantially increased. The depth of the roots is linked to soil water availability; (iii) at maturity, root tip extension slows and senescence begins. However, some lateral root penetration occurs to ensure seed maturity and to guarantee high yields (Kaspar, Stanley, & Taylor, 1978; Torrion et al., 2012).

RSA is defined by several parameters, including the depth of the primary roots, root surface area, root length, number of lateral roots, and root volume. RSA is partly genetically determined and partly determined by the soil composition. Hence, RSA is defined by the interaction between genetic controls and environmental factors. Soybean RSA can be classified into three phenotypic classes: (i) a shallow root phenotype with more lateral roots in the topsoil and lateral roots horizontally displayed; (ii) a deep root phenotype with a longer primary root and a greater angle to the lateral roots; and (iii) an intermediate root phenotype, which is between the shallow and the deep root phenotypes. The physiological properties of the plant are determined to a large extent by RSA. For example, field experiments have shown that soybean genotypes with the shallow root phenotype are more drought-sensitive, while the deep-rooted genotypes are more drought-escaping and the intermediate root type is more drought-tolerant with higher yields under limiting water conditions (Fenta et al., 2014). The growth rate of the tap root and lateral root density together determine the ability of the root system to absorb water, particularly under drought conditions (Hoogenboom, Huck, & Peterson, 1987; Kaspar, Taylor, & Shibles, 1984). Hence, RSA and its dynamic responses to environmental conditions directly influences shoot growth and ultimately crop yields.

2.2 Genotypic variation in root architecture traits

Soybean breeders have consistently selected for improved shoot and yield traits because shoot parameters are easier to determine than root traits. Root system traits have not been investigated until recently because of limitations in accurate root trait measurements and an inability to correctly assess root architecture, morphology, topology, and distribution within the soil. Root responses to environmental stimuli and root growth over time have largely been measured under controlled environment conditions. Researchers have generally applied one of the following three strategies for root phenotyping: (i) controlled environment methods, (ii) greenhouse-based methods, and (iii) minimally controlled field testing (Chen, Djalovic, &

Siddique, 2018; Falk, Jubery, Mirnezami, et al., 2020). While controlled laboratory studies are amenable to large scale phenotyping and throughput, field-based methods are essential to understand root/environment interactions. Field studies are often labor-intensive and costly. Improved technologies are therefore required to bridge the gap between lab and field studies.

Traditional methods for measuring RSA in field-grown plants are difficult and often involve the destruction of the soil habitat (Chen et al., 2014) through excavation of the root system (Jackson, 1995; Trachsel, Kaeppler, Brown, & Lynch, 2011). Other methods that involve excavating soil core or trench profiles (Kücke, Schmid, & Spiess, 1995) are also destructive, tedious, and also time-consuming. Crucially, the accuracy of results obtained by such methods is questionable (Smit et al., 2000). However, root excavation and washing in order to measure root length and architecture remains the most common and simplest method of studying roots in the field. Trench profiles can also provide quantitative information on RSA (Chen et al., 2018) but they provide measurements for only a small fraction of the root system (Pierret, Kirby, & Moran, 2003). Moreover, such techniques are often unsuitable for investigations of root development (Liedgens, Richner, Stamp, & Soldati, 2000) and information on fine roots and root hairs is often lost (Smit et al., 2000). Rhizotrons, rhizoboxes or minirhizotrons are often used for *in situ* root measurements. Such techniques allow extrapolation of demographic methods to field environments, facilitating measurements of developmental stages from root initiation to senescence. However, such results cannot easily be translated into more useful data, such as root length densities or biomass (Chen et al., 2018). The ability to study larger sample sizes in such systems may provide better opportunities to study RSA (Chen et al., 2015; Falk et al., 2020).

A novel semi-hydroponic phenotyping platform developed by Chen, Dunbabin, Diggle, Siddique, and Rengel (2011) has been used to study root plasticity and responses to changes in water and nutrient supply. This high-throughput phenotyping system was first used to characterize the variability in root architectural traits in the narrow-leafed lupin (Chen, Dunbabin, Diggle, Siddique, & Rengel, 2012). Subsequently, it has been used to phenotype root traits in chickpea (*Cicer arietinum* L.) (Chen, Ghanem, & Siddique, 2017; Chen, Zhou, & Siddique, 2020), maize (*Zea mays* L.) (Qiao et al., 2019), wheat (*Triticum aestivum* L.) (Chen, Palta, Prasad, & Siddique, 2020; Halder, Liu, Chen, Yan, & Siddique, 2021) and barley (*Hordeum vulgare* L.) (Wang et al., 2021). Recently, thus system was used to detail root system characteristics and trait-trait correlations in

Fig. 1 Layout of soybean plants grown in a semi-hydroponic phenotyping platform at 19 days after sowing (DAS) (A), and example images of root systems of eight selected genotypes at 39 DAS (B). Root angle (RA), root width (RW) and maximal root depth (MRD) of the soybean root system grown in a semi-hydroponic phenotyping platform (C), were measured at 39 DAS. White bar = 10 cm (Liu et al., 2021).

171 soybean genotypes developed for use in the Yangtze and Huaihe River districts of eastern China (Fig. 1; Liu et al., 2021). Large variations in root morphological traits were reported between the soybean genotypes. For example, total root length ranged from 208 to 1663 cm per plant, and root dry weight ranged from 19.4 to 251 mg per plant. Three root phenotypes of three genotypes (#054, #055, and #071) identified in the initial phenotyping study were confirmed by later studies in deep rhizoboxes (Fig. 2). A consistent ranking of genotypes with regard to root traits for early growth stages (39 days after sowing), and between different growth stages (38, 81 and 160 days after sowing) was obtained in the two growth systems.

A low-cost and high-resolution root phenotyping system with an imaging platform was recently established to study soybean root morphology at the seedling stage (Falk, Jubery, Mirnezami, et al., 2020). A root category

Fig. 2 Soil-filled rhizoboxes (25 cm long × 5 cm wide × 150 cm deep) used to validate root traits of three selected soybean plants following the phenotyping study in the semi-hydroponic phenotyping platform. The rhizoboxes were placed on steel stands a 30° angle in a temperature-controlled glasshouse (A), and visible roots inside the glass wall after removing the black cover sheets were mapped on the glass walls and transparent plastic film weekly till the maturity (B). *Photos provided by Y. Chen.*

(iRoot) system was produced that linked RSA to genotype in soybean, allowing the characterization of different soybean accessions into different clusters or groups (Falk, Jubery, O'Rourke, et al., 2020).

Comparing to model plant *Arabidopsis*, soybean root system is more complicate. Soybean genotypes with contrasting RSA and anatomic features may differ in response to grow environments. However, limited by technique and facilities, root phenotyping in large number of soybean germplasms remains a challenge. For further perspective, researchers should explore high-throughput phenotyping techniques to identity root trait variability and adaptation to varying environments.

2.3 Root anatomical traits

Soybean root anatomy is related to stress tolerance. The anatomical features of the mature region of the root might be related to stress tolerance. Soybean roots are more complex than Arabidopsis roots and have more cell layers (Fig. 3). The outmost epidermal layer consists of root hair cells and cells without root hairs. The thick cortex region consists of multilayers of cells, some of which can form aerenchyma that improves tolerance to hypoxia and other stresses (Silva, Batista, & Lobato, 2021). For example, the formation of aerenchyma in the root cortical region increases in roots exposed to salinity (Silva et al., 2021).

The endodermis of the root steel surrounds the pericycle and the vasculature, which is made up of xylem and phloem cells. Cell division in the pericycle is essential for the formation of the lateral root primordia (Lucas et al., 2013). Soybean root anatomy changes in response to stress (Makbul, Güler, Durmus, & Güven, 2011). For example, the number and size of the metaxylem determines water uptake capacity under water-limited conditions (Prince et al., 2017). Moreover, the numbers new metaxylem cells are decreased under drought and are replaced by pith cells (Mangena, 2018).

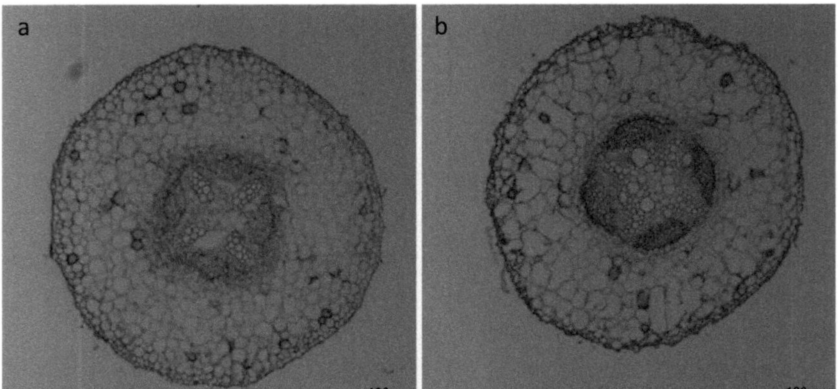

Fig. 3 Cross sections of soybean root sampled 4 (A) and 7 (B) days after germination. *Photos provided by H.M. Lam.*

3. Regulation of soybean root development and architecture

3.1 Signaling regulation of soybean root development

The genes and molecular mechanisms that govern soybean RSA in natural environments are largely unknown. In contrast, many studies have focused on soybean root and nodule formation, which is controlled by a complex interplay of hormone, peptide, and sugar signaling systems. Auxin fulfills many roles in plant development (Michniewicz et al., 2019; Ogura et al., 2019; Overvoorde, Fukaki, & Beeckman, 2010). It plays a fundamental role in soybean root and nodule development. Soybean root and nodule development requires auxin signaling. For example, soybean nodulation and root development are regulated by *GmYUC2a*, which is involved in auxin biosynthesis (Wang et al., 2019). Moreover, knock-down of auxin receptor *GmTIR1/AFB3* in soybean significantly reduces the number of curly root hairs and represses nodulation (Cai et al., 2017). Similarly, the GmARF proteins negatively regulate soybean nodulation (Wang et al., 2015).

Cytokinins play a negative role in lateral root development antagonizing auxin signaling. They are synthetized in the root apical meristem and transported to the shoots (Werner, Motyka, Strnad, & Schmülling, 2001). Cytokinins suppress the gravity responses of lateral roots and allow radical expansion (Waidmann et al., 2019). The application of low concentrations of cytokinin to soybean roots promotes nodulation. The overexpression of *GmRAV1*, which is a positive regulator of cytokinin signaling, promotes soybean root regeneration (Zhang et al., 2019).

Other hormones, such as ethylene, abscisic acid, gibberellin and jasmonate, also affect root growth (Gou et al., 2010; Huang, Liu, Liu, & Song, 2017; Qin & Huang, 2018; Rowe, Topping, Liu, & Lindsey, 2016). For example, the application of GA_3 to soybean results in shorter, thicker roots, which might be related to increases in ethylene (Leite, Rosolem, & Rodrigues, 2003).

MicroRNAs, which are small and noncoding RNA that negatively regulate the expression of target genes are involved in the regulation of root development and nodulation. Overexpression of miR483, miR1512 and miR1515 in soybean hairy roots significantly increases nodule numbers (Li, Deng, Wu, Subramanian, & Yu, 2010). Some microRNAs are auxin responsive. For example, miR167 negatively regulates *GmARF8*, which plays a functional role in nodulation via auxin signaling

(Wang et al., 2015). Overexpression of miR393 in soybean roots decreases lateral root density and increases auxin-hyposensitivity but has no significant effect on nodulation. In contrast, miR160 overexpression increases lateral root density and increases auxin-hypersensitivity but inhibits nodulation (Turner et al., 2013).

Small peptides are also important in the regulation of root development. For example, the ROOT MERISTEM GROWTH FACTOR 1 (RGF1) peptide is required for root stem cell maintenance (Matsuzaki, Ogawa-Ohnishi, Mori, & Matsubayashi, 2010); Similarly, phytosulfokine (PSK) and PLANT PEPTIDE CONTAINING SULFATED TYROSINE 1 (PSY1) promote primary root growth through modifying root cell expansion in the elongation zone (Amano, Tsubouchi, Shinohara, Ogawa, & Matsubayashi, 2007). The RAPID ALKLINIZATION FACTOR (RALF) peptide inhibits root hair growth in Arabidopsis by suppressing cell elongation (Haruta, Sabat, Stecker, Minkoff, & Sussman, 2014; Pearce, Moura, Stratmann, & Ryan, 2001). Small peptides also play a role in controlling nodule numbers. The expression of an Arabidopsis CLAVATA peptide like receptor kinase GmNARK in soybean promoted lateral root formation and regulated nodulation through modulation of shoot to root communication (Searle et al., 2003). Overexpression of *GmNIC1*, which encodes a nitrate-induced CLE peptide led to decreased nodulation in a GmNARK-dependent manner (Lim, Lee, & Hwang, 2011; Reid, Ferguson, & Gresshoff, 2011).

Sucrose plays an important role in root formation (Ogawa, Ando, Toyofuku, & Kawashima, 2009), while high concentrations of glucose inhibit root meristem expansion by reducing root auxin levels (Yuan, Xu, Zhang, Guo, & Lu, 2014). High nitrite concentration inhibit nodulation, but nitrate-induced inhibition can be alleviated by the addition of sucrose (Fujikake et al., 2003).

3.2 Genetic regulation of soybean root system architecture

Quantitative trait loci (QTLs) for RSA have been reported for many crops, including maize, rice and wheat (Ju et al., 2018; Phung et al., 2016). Several QTLs associated with fibrous rooting, root length and later root number have been identified in bi-parental soybean populations, with donor parents derived from an "old" cultivar or landrace (Abdel-Haleem, Lee, & Boerma, 2011; Manavalan et al., 2015). Five QTLs associated with fibrous rooting and root surface area were identified on chromosomes Gm01 (Satt383),

Gm03 (Satt339), Gm04 (Sct_191), Gm08 (Satt429), and Gm20 (Sat_299) by Abdel-Haleem et al. (2011). Other QTLs, (SA_Gm06, TRL_Gm06, RDL3_Gm07 and TRL_Gm08) were associated with soybean root length (Manavalan et al., 2015). QTL LRN_Gm08 was also associated with lateral root number. A major QTL on chromosome16 associated with soybean primary root length was recently identified (Chen, Kumawat, Yan, Fan, & Xu, 2021). Overall, these findings indicate the potential of using "older" soybean varieties and landraces for the improvement of RSA. Wild soybeans are also an important source of valuable alleles, which can be applied in marker-assisted selection to improve RSA in cultivated soybeans.

Many populations of wild and cultivated soybeans have been used to identity novel genetic components regulating RSA. Four QTLs associated with surface area and root volume were identified in a population of *Glycine max* × *Glycine soja*, with favorable alleles identified in the wild parent (PI 407162) (Prince et al., 2015). Four QTLs related to root surface area, root length and root thickness were also derived from the wild parents (Prince et al., 2015). A QTL on Chromosome 7, which is a positive allele from a wild parent, was recently associated with soybean root length, surface area and volume (Prince et al., 2020).

Recently, 252 phenotypically diverse soybean germplasm sources were screened in a GWAS approach. Glycine Lateral Root 1 (*GLR1*) was associated with increased lateral root numbers (Prince et al., 2019). In a further study using 289 different soybean germplasms, 67 loci were identified, including more than 200 SNPs, for different soybean RSA traits under field-grown conditions (Dhanapal, York, Hames, & Fritschi, 2021). Due to the complexity of root related traits, the influence of a single gene is highly dependent on the QTL effects on this trait. Despite these advances, few causal genes which function in soybean root development have been identified.

Soybean root related phenotypes are complexly controlled by multiple factors. Molecular signals such as phytohormone and sugar are involved, and some of these signals are synthetic in shoot (or root) and transported to root (or shoot) via vascular system. However, how genetic controlling on these long-distance signals from distant shoot and working on root development is still unclear. In future, forward genetic approaches using high quality genetic population with larger population size and higher genetic marker density should be widely applied to explore genetic controlling on soybean root development.

4. Root plasticity and adaptation to edaphic stress

RSA is important for nutrient acquisition, drought and flood tolerance, and also lodging resistance (Falk, Jubery, Mirnezami, et al., 2020). Phenotypic diversity, which has been acquired during plant evolution (Mickelbart, Hasegawa, & Bailey-Serres, 2015), is important in stress resilience (Fig. 4).

4.1 Drought stress

More than 50 years ago, Kramer (1969) stated that one essential characteristic of drought resistance is a "deep, wide-spreading, much-branched root system." The benefits of deep branched root system for drought tolerance have been reported for various crops, including rice (Babu et al., 2003; Bernier et al., 2009; Kato, Abe, Kamoshita, & Yamagishi, 2006; MacMillan, Emrich, Piepho, Mullins, & Price, 2006; Nguyen, Babu, & Blum, 1997; Steele, Price, Shashidhar, & Witcombe, 2006; Uga et al., 2013), maize (Giuliani et al., 2005; Hammer et al., 2009; Hund, Reimer, & Messmer, 2011; Landi et al., 2010; Tuberosa et al., 2002), barley (Forster, Thomas, & Chloupek, 2005), wheat (Manschadi, Christopher, Devoil, & Hammer, 2006; Wasson et al., 2012), chickpea (Chen et al., 2012; Varshney et al., 2011), and also soybean (Hudak & Patterson, 1995; Pantalone, Rebetzke, Burton, & Carter, 1996; Sadok & Sinclair, 2011). In addition, the diameter and distribution of

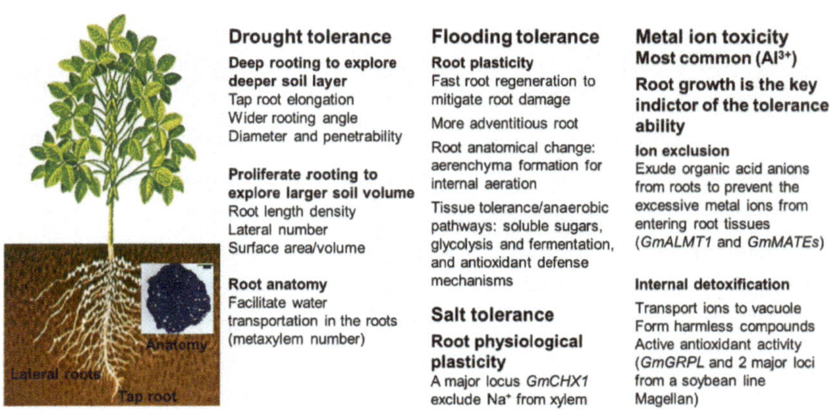

Fig. 4 Root plasticity in root system architecture and anatomy for adaptation to major abiotic stresses in soybean.

the metaxylem vessels, which govern root conductivity, are also beneficial for drought tolerance (Purushothaman et al., 2013).

DEEPER ROOTING 1 (*DRO1*) doubles rice yields under drought stress due to the ability of developing a deep root system (Uga et al., 2013). Thicker roots also tend to penetrate more deeply through soil-layers allowing deep rooting (Yu, Ray, O'Toole, & Nguyen, 1995; Zhao et al., 2020). Proliferate rooting allows higher water uptake efficiencies under water deficits. Chickpea and soybean lines with higher root densities have higher yields and better drought tolerance in water limited environments (Jaganathan et al., 2015).

Drought stress adversely affects root and shoot growth (Fathi & Tari, 2016). Root traits, such as root length, root biomass, rooting depth, root density, fibrous root systems, prolific root systems, rooting depth and osmotic adjustments, are considered to be important traits in drought-resistant soybean genotypes (Khatun et al., 2021; Meister, Rajani, Ruzicka, & Schachtman, 2014; Nielsen, Lynch, & Weiss, 1997; Osmont, Sibout, & Hardtke, 2007; Salazar-Henao, Vélez-Bermúdez, & Schmidt, 2016). Improved root systems allow improved soil penetration and better water acquisition (Hoogenboom et al., 1987). Drought-tolerant grain legume lines have greater rooting depth than sensitive lines (Ye et al., 2018). A drought-tolerant soybean landrace (PI 416937) had more fibrous roots (Hudak & Patterson, 1996). Drought-tolerant soybeans penetrate deeper into the soil than drought-sensitive varieties (Garay & Wilhelm, 1983). Genetic diversity in metaxylem number correlates with drought tolerance in soybean (Prince et al., 2018).

Natural variation in RSA has been used to improve drought tolerance in soybean (Carter, 1989). A drought-tolerant soybean landrace (PI 416937) developed more fibrous roots allowing greater exploration of soil water (Hudak & Patterson, 1996).

Several soybean accessions with promising RSA for extensive fibrous rooting, root length or narrow root angle have been identified in the USDA Germplasm Collection (Falk, Jubery, Mirnezami, et al., 2020; Prince et al., 2018). A genome-wide association study (GWAS) provided evidence that a rare allele that promotes lateral root growth has the potential to improve drought tolerance (Prince et al., 2018). One single nucleotide polymorphism (SNP) in the identified allele caused increased lateral rooting. Further screening of the USDA Germplasm Collection revealed that the frequency of favorable alleles at this locus is about 0.5% (Prince et al., 2018). An examination of RSA traits by combining 35,448 SNPs with an imaging phenotyping platform was conducted in 292 diverse soybean accessions to decipher genetic diversity in RSA (Falk, Jubery, Mirnezami, et al., 2020). An ideotype-based approach was applied to develop informative root categories using these 292 soybean accessions. Eight genotype- and phenotype-based clusters were identified. Such studies demonstrate the efficacy of combinations of genetic and phenotypic analyses and machine learning in identifying key RSA traits for plant breeding (Carter, 1989).

4.2 Flooding

Flooding is the second largest cause of economic loss in soybean (Valliyodan et al., 2014). Yield losses due to flooding are caused by multiple factors including reduced root growth, shoot growth, nodulation, nitrogen fixation, photosynthesis, biomass accumulation, stomatal conductance and nutrient uptake (Oosterhuis, Scott, Hampton, & Wullschleter, 1990; VanToai, Beuerlein, Schmitthenner, & St. Martin, 1994). Flooding stress involves submergence or water-logging. Submergence often occurs in wetland crops (e.g., rice), while water-logging is common in dryland crops, including soybean. Waterlogging during the reproductive growth phase can reduce soybean yields by much as 27% (Linkemer, Board, & Musgrave, 1998; Tamang, Magliozzi, Maroof, & Fukao, 2014). Plants have developed tolerance mechanisms to the hypoxia that results from flooding. Metabolism shifts from aerobic to anaerobic pathways, that can increase the availability of soluble sugars. Aerenchyma formation, greater glycolytic activity and fermentation as well as increased antioxidant defenses are common responses flooding (Ahmed et al., 2013). The formation of adventitious roots has also been observed in many flooded plant species, including soybean (Kim et al., 2015; Valliyodan et al., 2014). Plant hormones also play key roles in root development and growth under flooding. Ethylene, which promotes root hair formation (Zhang, Lynch, & Brown, 2003), accumulates under flooding conditions and activates flooding-adapted growth responses in deep-water rice (Visser & Voesenek, 2005). Ethylene production increases in water-logged soybean roots, with a significantly higher increases in flooding-tolerant lines than sensitive lines (Kim et al., 2015).

Genetic variations in flooding tolerance have also been observed in soybean. The introduction of flooding tolerance associated genomic regions into soybean cultivars prevented substantial yield losses caused by flooding (Rhine, Steven, Shannon, Wrather, & Sleper, 2010; Scott, DeAngulo, Daniels, & Wood, 1989; Shannon et al., 2005; VanToai et al., 2010). Genetic mapping in soybean populations has also identified genomic regions associated with flooding tolerance (Cornelious et al., 2005; Cornelious, Chen, Hou, Shi, & Shannon, 2006; Githiri, Watanabe, Harada, & Takahashi, 2006; Nguyen et al., 2012; VanToai et al., 2001; Ye et al., 2018). Root plasticity was found to play a crucial role in waterlogging tolerance in soybean. The major waterlogging tolerance QTL *qWT_Gm03* regulated total root length, the number of root tips and the number of adventitious roots under waterlogging stress (Ye et al., 2018).

4.3 Salinity

Almost 20% of cultivated area in the world is affected by salinity. The area of cultivated land affected by salt is predicted to double by 2050 (Sairam & Tyagi, 2004). As a medium salt-sensitive crop, soybean can be seriously affected by salinity stress with a 20–50% reduction in yield (Patil et al., 2016). Thus, it is important to study the salt-tolerant mechanism and identify salt-tolerant soybean germplasm. Under salinity stress, soybean performed a series of phenotype changes, such as leaf drooping, increased leaf temperature and decreased photosynthetic rate (Liu et al., 2019). Roots as the first organ to encounter the salt signal are also negatively affected by the salinity stress. It is showed that soybean root growth could be inhibited by reduction the activity of root apical meristem under salinity stress and miRNA-mediated regulatory networks in root apexes play a vital role in this process (Sun et al., 2016).

Importantly, it is an effective way by identifying QTL/gene(s) to breed elite cultivars with greater salt tolerance capability. A major salt tolerance gene, *GmCHX1,* which regulates root plasticity in soybean improves salt tolerance (Qi et al., 2014). The expression of *GmCHX1* is induced in soybean roots by salt stress. The gene encodes a sodium ion transporter protein, which assists Na^+ exclusion from the above-ground tissues (Guan et al., 2014; Qi et al., 2014).

4.4 Metal ion toxicity

Metal ion toxicity stress, particularly caused by excessive aluminum (Al) concentrations in acid soils, is also a significant growth-limiting factor threatening crop production. Unfortunately, about 38% of the farmland in Southeast Asia, 31% of Latin America, and 20% of the arable lands in East Asia, sub-Saharan Africa, and North America suffers from Al toxicity (Magalhaes et al., 2007; Wood, Sebastian, & Scherr, 2000). Al toxicity in acidic soils significantly inhibits root elongation and lateral root initiation

(Haling, Richardson, Culvenor, Lambers, & Simpson, 2010; Sun, Tian, Chen, & Zhang, 2010) and consequently reduces crop yields, or even results in death (Sasaki, Yamamoto, & Matsumoto, 1996; Ma & Furukawa, 2003). Root plasticity to Al toxicity is generally used as an indicator of Al tolerance in plants. Root plasticity indicators include relative primary root length, relative primary root elongation rate, and relative root dry weight (Khabaz-Saberi & Rengel, 2010; Korir, Zhang, Wu, Zhao, & Gai, 2013; Nguyen et al., 2003; Sharma, Sharma, & Lightfoot, 2011). Mechanisms of how plants alleviate Al toxicity include Al^{3+} exclusion and internal detoxification. Al-tolerant *Arabidopsis* and rice plants can exude organic acid anions (malate and citrate) from roots to prevent the excessive Al^{3+} from entering root tissues (Delhaize, Gruber, & Ryan, 2007; Delhaize & Ryan, 1995; Hoekenga et al., 2006; Ligaba, Maron, Shaff, Kochian, & Piñeros, 2012; Liu, Magalhaes, Shaff, & Kochian, 2009; Yokosho, Yamaji, & Ma, 2011). Detoxification mechanisms can alleviate Al toxicity in rice and soybean, for example, by sequestrating Al^{3+} in vacuoles or by forming harmless complexes with organic ligands (Famoso et al., 2011; Huang, Yamaji, Chen, & Ma, 2012; Larsen, Cancel, Rounds, & Ochoa, 2007; Lei, Yokosho, Yamaji, Fujii-Kashino, & Ma, 2017; Li et al., 2014, 2021; Xia, Yamaji, Kasai, & Ma, 2010). Two soybean genes, *GmALMT1* and *GmMATEs,* confer Al tolerance by regulating malate and citrate exudation from roots (Liang et al., 2013; Ma et al., 2018; Wang et al., 2019; Zhou et al., 2019). Moreover, *GmGRPL,* enhances antioxidant activity and auxin and ethylene pathways in soybean roots to alleviate Al toxicity and maintaining growth (Chen et al., 2018). The US soybean cultivar (Magellan) is Al tolerant. Three QTLs associated with Al tolerance have been mapped from a recombinant inbred line (RIL) population derived from Magellan (Li et al., 2021). Two of these natural variations on soybean Chrs. 13 and 20 were confirmed in a near-isogenic background and were found to epistatically regulate Al tolerance (Li et al., 2021).

5. Current challenges and future directions

The significance of soybean RSA to productivity is recognized by breeders, agronomists, and researchers. Root traits conferring efficiency in resource acquisition and adaptation to edaphic stresses, particularly in drying soil environments, are essential to the success of soybean breeding programs. As global climate change, the production of soybean is struggled with multiabiotic stresses. As the first organ to sense different stress in the soil, roots play a vital role in modulating plant growth and mediating plants' response when plants encounter external stimuli. However, there are some challenges in the studying of soybean root physiology and root system architecture and attentions will be paid to the following areas in the future research:

(i) It is essential to explore high-throughput methods together with phenomics, data analytics, and modeling simulations to decipher the genetics of RSA trait expression in order to realize the potential of root-driven breeding. Such approaches will enhance our understanding of the relationships between root structure and function, the complexity of root–soil interactions, as well as increasing the success of targeting for improved soybean performance and productivity in breeding programs.

(ii) The adaptation of soybean to hostile soil and other environmental stress can be enhanced by breeding genotypes with suitable root traits. Soybean genotypes with deeper rooting can capture water from deep soil profile under limited rainfall and thus perform better. Although some studies have pointed and identified some important genes and pathways which are involved and regulate soybean to respond to different stress, the regulation network still need to complete to widen our understanding and knowledge, especially for soybean improvement. In future, we should focus on the deeper mechanism exploration and combine forward genetic approaches with molecular biotechnology and multiomics to promote soybean molecular breeding and improve soybean production under extreme environments.

(iii) Root study in soybean needs to integrate root interactions with rhizosphere microorganisms, in particular nitrogen-fixing bacteria (rhizobia) and arbuscular mycorrhizal fungi. An improved understanding of the consequences of the interplay between soybean root systems, rhizobia, and other root-associated microbiomes for overall productivity is essential in order to ensure the success of future plant improvement strategies.

(iv) Many experimental research were conducted under controlled environments due to unavailability of research methods in examining root growth and performance and infeasibility in carrying out large scale experiments under field conditions. Therefore, there is a need to develop suitable tools for non-constructive observation and sampling methods of root systems of the field-grown soybean plants. On the other hand, development of simulation modeling capable of reconstruction of 3D root structure to link studies conducted under controlled environments and those under field conditions.

Acknowledgments

This study was partially supported by Australian Research Council (FT210100902), National Natural Science Foundation of China (42077055), and Hong Kong Research Grants Council Area of Excellence Scheme (AoE/M-403/16).

References

Abdel-Haleem, H., Lee, G. J., & Boerma, R. H. (2011). Identification of QTL for increased fibrous roots in soybean. *Theoretical and Applied Genetics*, *122*, 935–946.

Ahmed, F., Rafii, M. Y., Ismail, M. R., Juraimi, A. S., Rahim, H. A., Asfaliza, R., et al. (2013). Waterlogging tolerance of crops: Breeding, mechanism of tolerance, molecular approaches, and future prospects. *BioMed Research International*, *2013*, 963525.

Amano, Y., Tsubouchi, H., Shinohara, H., Ogawa, M., & Matsubayashi, Y. (2007). Tyrosine-sulfated glycopeptide involved in cellular proliferation and expansion in *Arabidopsis*. *Proceedings of the National Academy of Sciences of the United States of America*, *104*, 18333–18338.

Babu, R. C., Nguyen, B. D., Chamarerk, V., Shanmugasundaram, P., Chezhian, P., Jeyaprakash, P., et al. (2003). Genetic analysis of drought resistance in rice by molecular markers: Association between secondary traits and field performance. *Crop Science*, *43*, 1457–1469.

Bernier, J., Kumar, A., Venuprasad, R., Spaner, D., Verulkar, S., Mandal, N. P., et al. (2009). Characterization of the effect of a QTL for drought resistance in rice, *qtl12.1*, over a range of environments in the Philippines and eastern India. *Euphytica*, *166*, 207–217.

Cai, Z., Wang, Y., Zhu, L., Tian, Y., Chen, L., Sun, Z., et al. (2017). GmTIR1/GmAFB3-based auxin perception regulated by miR393 modulates soybean nodulation. *New Phytologist*, *215*, 672–686.

Carter, T. E., Jr. (1989). Breeding for drought tolerance in soybean: Where do we stand? In A. J. Pascale (Ed.), *Proceeding world soybean conference IV* (pp. 1001–1008). Buenos Aires: Argentina.

Chen, Y., Djalovic, I., & Rengel, Z. (2015). Phenotyping for root traits. In J. Kumar, A. Pratap, & S. Kumar (Eds.), *Phenomics in crop plants: Trends, options and limitations* (pp. 102–128). New Delhi: Springer.

Chen, Y., Djalovic, I., & Siddique, K. H. M. (2018). Advances in understanding grain legume physiology: Understanding root architecture, nutrient uptake and response to abiotic stress. In S. Sivasankar, D. Bergvinson, P. Gaur, S. Kumar, S. Beebe, & M. Tamò (Eds.), *Advances in breeding and cultivation techniques: Vol. 1. Achieving sustainable cultivation of grain legumes* (pp. 11–27). Cambridge: Burleigh Dodds Science Publishing.

Chen, Y., Dunbabin, V. M., Diggle, A. J., Siddique, K. H. M., & Rengel, Z. (2011). Development of a novel semi-hydroponic phenotyping system for studying root architecture. *Functional Plant Biology*, *38*, 355–363.

Chen, Y., Dunbabin, V. M., Diggle, A. J., Siddique, K. H. M., & Rengel, Z. (2012). Assessing variability in root traits of wild *Lupinus angustifolius* germplasm: Basis for modelling root system structure. *Plant and Soil*, *354*, 141–155.

Chen, Y., Ghanem, M. E., & Siddique, K. H. M. (2017). Characterising root trait variability in chickpea (*Cicer arietinum* L.) germplasm. *Journal of Experimental Botany*, *68*, 1987–1999.

Chen, H., Kumawat, G., Yan, Y., Fan, B., & Xu, D. (2021). Mapping and validation of a major QTL for primary root length of soybean seedlings grown in hydroponic conditions. *BMC Genomics*, *22*, 132.

Chen, Y., Palta, J., Clements, J., Buirchell, B., Siddique, K. H. M., & Rengel, Z. (2014). Root architecture alteration of narrow-leafed lupin and wheat in response to soil compaction. *Field Crops Research*, *165*, 61–70.

Chen, Y., Palta, J., Prasad, P. V. V., & Siddique, K. H. M. (2020). Phenotypic variability in bread wheat root systems at the early vegetative stage. *BMC Plant Biology*, *20*, 185.

Chen, L., Ren, F., Zhou, L., Wang, Q. Q., Zhong, H., & Li, X. B. (2012). The *Brassica napus* calcineurin B-like 1/CBL-interacting protein kinase 6 (CBL1/CIPK6) component is involved in the plant response to abiotic stress and ABA signaling. *Journal of Experimental Botany*, *63*, 6211–6222.

Chen, Y., Zhou, T., & Siddique, K. H. M. (2020). Method for characterization of root traits in chickpea germplasm for legume genomics and breeding. In M. Jain, & R. Garg (Eds.), *Methods in molecular biology: 2107. Legume Genomics* (pp. 269–275). New York: Humana.

Cornelious, B., Chen, P., Chen, Y., de Leon, N., Shannon, J. G., & Wang, D. (2005). Identification of QTLs underlying water-logging tolerance in soybean. *Molecular Breeding*, *16*, 103–112.

Cornelious, B., Chen, P., Hou, A., Shi, A., & Shannon, J. G. (2006). Yield potential and water-logging tolerance of selected near-isogenic lines and recombinant inbred lines from two southern soybean populations. *Journal of Crop Improvement*, *16*, 97–111.

Costa, C., Dwyer, L. M., Hamilton, R. I., Hamel, C., Nantais, L., & Smith, D. L. (2000). A sampling method for measurement of large root systems with scanner-based image analysis. *Agronomy Journal*, *92*, 621–627.

Delhaize, E., Gruber, B. D., & Ryan, P. R. (2007). The roles of organic anion permeases in aluminium resistance and mineral nutrition. *FEBS Letters*, *581*, 2255–2262.

Delhaize, E., & Ryan, P. R. (1995). Aluminum toxicity and tolerance in plants. *Plant Physiology*, *107*, 315–321.

Dhanapal, A. P., York, L. M., Hames, K. A., & Fritschi, F. B. (2021). Genome-wide association study of topsoil root system architecture in field-grown soybean [*Glycine max* (L.) Merr.]. *Frontiers in Plant Science*, *11*, 590179.

Falk, K. G., Jubery, T. Z., Mirnezami, S. V., Parmley, K. A., Sarkar, S., Singh, A., et al. (2020). Computer vision and machine learning enabled soybean root phenotyping pipeline. *Plant Methods*, *16*, 5.

Falk, K. G., Jubery, T. Z., O'Rourke, J. A., Singh, A., Sarkar, S., Ganapathysubramanian, B., et al. (2020). Soybean root system architecture trait study through genotypic, phenotypic, and shape-based clusters. *Plant Phenomics*, *2020*. 1925495.

Famoso, A. N., Zhao, K., Clark, R. T., Tung, C. W., Wright, M. H., Bustamante, C., et al. (2011). Genetic architecture of aluminum tolerance in rice (*Oryza sativa*) determined through genome-wide association analysis and QTL mapping. *PLoS Genetics*, *7*, e1002221.

FAOSTAT. (2017). *FAO stat*. Rome: Italy. (http://www.fao.org/faostat).

Fathi, A., & Tari, D. B. (2016). Effect of drought stress and its mechanism in plants. *International Journal of Life Sciences*, *10*, 1–6.

Fenta, B. A., Beebe, S. E., Kunert, K. J., Burridge, J. D., Barlow, K. M., Lynch, J. P., et al. (2014). Field phenotyping of soybean roots for drought stress tolerance. *Agronomy*, *4*, 418–435.

Forster, B. P., Thomas, W. T. B., & Chloupek, O. (2005). Genetic controls of barley root systems and their associations with plant performance. *Aspects of Applied Biology*, *73*, 199–204.

Fujikake, H., Yamazaki, A., Ohtake, N., Sueyoshi, K., Matsuhashi, S., Ito, T., et al. (2003). Quick and reversible inhibition of soybean root nodule growth by nitrate involves a decrease in sucrose supply to nodules. *Journal of Experimental Botany*, *54*, 1379–1388.

Garay, A. F., & Wilhelm, W. W. (1983). Root system characteristics of two soybean isolines undergoing water stress conditions. *Agronomy Journal*, *75*, 973–977.

Githiri, S. M., Watanabe, S., Harada, K., & Takahashi, R. (2006). QTL analysis of flooding tolerance in soybean at an early vegetative growth stage. *Plant Breeding*, *125*, 613–618.

Giuliani, S., Sanguineti, M. C., Tuberosa, R., Bellotti, M., Salvi, S., & Landi, P. (2005). *Root-ABA1*, a major constitutive QTL, affects maize root architecture and leaf ABA concentration at different water regimes. *Journal of Experimental Botany*, *56*, 3061–3070.

Gou, J., Strauss, S., Tsai, C. J., Fang, K., Chen, Y., Jiang, X., et al. (2010). Gibberellins regulate lateral root formation in *Populus* through interactions with auxin and other hormones. *The Plant Cell*, *22*, 623–639.

Gregory, P. J., Bengough, A. G., Grinev, D., Schmidt, S., Thomas, W. T. B., Wojciechowski, T., et al. (2009). Root phenomics of crops: Opportunities and challenges. *Functional Plant Biology*, *36*, 922–929.

Guan, R., Qu, Y., Guo, Y., Yu, L., Liu, Y., Jiang, J., et al. (2014). Salinity tolerance in soybean is modulated by natural variation in *GmSALT3*. *The Plant Journal*, *80*, 937–950.

Halder, T., Liu, H., Chen, Y., Yan, G. J., & Siddique, K. H. M. (2021). Identification of candidate genes for root traits using genotype-phenotype association analysis of near-isogenic lines in hexaploid wheat (*Triticum aestivum* L.). *International Journal of Molecular Sciences*, *22*, 3579.

Haling, R. E., Richardson, A. E., Culvenor, R. A., Lambers, H., & Simpson, R. J. (2010). Root morphology, root-hair development and rhizosheath formation on perennial grass seedlings is influenced by soil acidity. *Plant and Soil*, *335*, 457–468.

Hammer, G. L., Dong, Z., McLean, G., Doherty, A., Messina, C., Schussler, J., et al. (2009). Can changes in canopy and/or root system architecture explain historical maize yield trends in the U.S. Corn Belt? *Crop Science*, *49*, 299–312.

Haruta, M., Sabat, G., Stecker, K., Minkoff, B. B., & Sussman, M. R. (2014). A peptide hormone and its receptor protein kinase regulate plant cell expansion. *Science*, *343*, 408–411.

Hinsinger, P., Brauman, A., Devau, N., Gérard, F., Jourdan, C., Laclau, J. P., et al. (2011). Acquisition of phosphorus and other poorly mobile nutrients by roots: Where do plant nutrition models fail? *Plant and Soil*, *348*, 29–61.

Hoekenga, O. A., Maron, L. G., Piñeros, M. A., Cançado, G. M. A., Shaff, J., Kobayashi, Y., et al. (2006). *AtALMT1*, which encodes a malate transporter, is identified as one of several genes critical for aluminum tolerance in *Arabidopsis*. *Proceedings of the National Academy of Sciences of the United States of America*, *103*, 9738–9743.

Hoogenboom, G., Huck, M. G., & Peterson, C. M. (1987). Root growth rate of soybean as affected by drought stress. *Agronomy Journal*, *79*, 607–614.

Huang, H., Liu, B., Liu, L., & Song, S. (2017). Jasmonate action in plant growth and development. *Journal of Experimental Botany*, *68*, 1349–1359.

Huang, C. F., Yamaji, N., Chen, Z., & Ma, J. F. (2012). A tonoplast-localized half-size ABC transporter is required for internal detoxification of aluminum in rice. *Plant Journal*, *69*, 857–867.

Hudak, C. M., & Patterson, R. P. (1995). Vegetative growth analysis of a drought resistant soybean plant introduction. *Crop Science*, *35*, 464–471.

Hudak, C. M., & Patterson, R. P. (1996). Root distribution and soil moisture depletion pattern of a drought-resistant soybean plant introduction. *Agronomy Journal*, *88*, 478–485.

Hund, A., Reimer, R., & Messmer, R. (2011). A consensus map of QTLs controlling the root length of maize. *Plant and Soil*, *344*, 143–158.

Jackson, L. E. (1995). Root architecture in cultivated and wild lettuce (*Lactuca* spp.). *Plant, Cell & Environment*, *18*, 885–894.

Jaganathan, D., Thudi, M., Kale, S., Azam, S., Roorkiwal, M., Gaur, P. M., et al. (2015). Genotyping-by-sequencing based intra-specific genetic map refines a "QTL-hotspot" region for drought tolerance in chickpea. *Molecular Genetics and Genomics*, *290*, 559–571.

Ju, C., Zhang, W., Liu, Y., Gao, Y., Wang, X., Yan, J., et al. (2018). Genetic analysis of seedling root traits reveals the association of root trait with other agronomic traits in maize. *BMC Plant Biology*, *8*, 171.

Kaspar, T. C., Stanley, C. D., & Taylor, H. M. (1978). Soybean root growth during the reproductive stages of development. *Agronomy Journal, 70*, 1105–1107.
Kaspar, T. C., Taylor, H. M., & Shibles, R. M. (1984). Taproot-elongation rates of soybean cultivars in the glasshouse and their relation to field rooting depth. *Crop Science, 24*, 916–920.
Kato, Y., Abe, J., Kamoshita, A., & Yamagishi, J. (2006). Genotypic variation in root growth angle in rice (*Oryza sativa* L.) and its association with deep root development in upland fields with different water regimes. *Plant and Soil, 287*, 117–129.
Khabaz-Saberi, H., & Rengel, Z. (2010). Aluminum, manganese, and iron tolerance improves performance of wheat genotypes in waterlogged acidic soils. *Journal of Plant Nutrition and Soil Science, 173*, 461–468.
Khatun, M., Sarkar, S., Era, F. M., Islam, A. K. M. M., Anwar, M. P., Fahad, S., et al. (2021). Drought stress in grain legumes: Effects, tolerance mechanisms and management. *Agronomy, 11*, 2374.
Kim, Y. H., Hwang, S. J., Waqas, M., Khan, A. L., Lee, J. H., Lee, J. D., et al. (2015). Comparative analysis of endogenous hormones level in two soybean (*Glycine max* L.) lines differing in waterlogging tolerance. *Frontiers. Plant Science, 6*, 714.
Korir, P. C., Zhang, J., Wu, K., Zhao, T., & Gai, J. (2013). Association mapping combined with linkage analysis for aluminum tolerance among soybean cultivars released in yellow and Changjiang River valleys in China. *Theoretical and Applied Genetics, 126*, 1659–1675.
Kramer, P. J. (1969). *Plant and soil water relationships: A modern synthesis* (p. 482). New York: McGrawHillpp.
Kücke, M., Schmid, H., & Spiess, A. (1995). A comparison of four methods for measuring roots of field crops in three contrasting soils. *Plant and Soil, 172*, 63–71.
Kunert, K. J., Vorster, B. J., Fenta, B. A., Kibido, T., Dionisio, G., & Foyer, C. H. (2016). Drought stress responses in soybean roots and nodules. *Frontiers in Plant Science, 7*, 1015.
Landi, P., Giuliani, S., Salvi, S., Ferri, M., Tuberosa, R., & Sanguineti, M. C. (2010). Characterization of *root-yield-1.06*, a major constitutive QTL for root and agronomic traits in maize across water regimes. *Journal of Experimental Botany, 61*, 3553–3562.
Larsen, P. B., Cancel, J., Rounds, M., & Ochoa, V. (2007). *Arabidopsis* ALS_1 encodes a root tip and stele localized half type ABC transporter required for root growth in an aluminum toxic environment. *Planta, 225*, 1447–1458.
Lei, G. J., Yokosho, K., Yamaji, N., Fujii-Kashino, M., & Ma, J. F. (2017). Functional characterization of two half-size ABC transporter genes in aluminium-accumulating buckwheat. *New Phytologist, 215*, 1080–1089.
Leite, V. M., Rosolem, C. A., & Rodrigues, J. D. (2003). Gibberellin and cytokinin effects on soybean growth. *Scientia Agricola, 60*, 537–541.
Li, H., Deng, Y., Wu, T., Subramanian, S., & Yu, O. (2010). Misexpression of miR482, miR1512, and miR1515 increases soybean nodulation. *Plant Physiology, 153*, 1759–1770.
Li, J. Y., Liu, J., Dong, D., Jia, X., McCouch, S. R., & Kochian, L. V. (2014). Natural variation underlies alterations in Nramp aluminum transporter (*NRAT1*) expression and function that play a key role in rice aluminum tolerance. *Proceedings of the National Academy of Sciences of the United States of America, 111*, 6503–6508.
Li, H., Wang, X., Liang, Q., Lyu, X., Li, S., Gong, Z., et al. (2021). Regulation of phosphorus supply on nodulation and nitrogen fixation in soybean plants with dual-root systems. *Agronomy-Basel, 11*, 2354.
Li, R., Zeng, Y., Xu, J., Wang, Q., Wu, F., Cao, M., et al. (2015). Genetic variation for maize root architecture in response to drought stress at the seedling stage. *Breeding Science, 65*, 298–307.
Liang, C., Piñeros, M. A., Tian, J., Yao, Z., Sun, L., Liu, J., et al. (2013). Low pH, aluminum, and phosphorus coordinately regulate malate exudation through *GmALMT1* to improve soybean adaptation to acid soils. *Plant Physiology, 161*, 1347–1361.

Liedgens, M., Richner, W., Stamp, P., & Soldati, A. (2000). A rhizolysimeter facility for studying the dynamics of crop and soil processes: Description and evaluation. *Plant and Soil, 223*, 89–99.

Ligaba, A., Maron, L., Shaff, J., Kochian, L., & Piñeros, M. (2012). Maize ZmALMT2 is a root anion transporter that mediates constitutive root malate efflux. *Plant, Cell & Environment, 35*, 1185–1200.

Lim, C. W., Lee, Y. W., & Hwang, C. H. (2011). Soybean nodule-enhanced CLE peptides in roots act as signals in GmNARK-mediated nodulation suppression. *Plant and Cell Physiology, 52*, 1613–1627.

Linkemer, G., Board, J. E., & Musgrave, M. E. (1998). Waterlogging effects on growth and yield components in late-planted soybean. *Crop Science, 38*, 1576–1584.

Liu, S., Begum, N., An, T., Zhao, T., Xu, B., Zhang, S., et al. (2021). Characterization of root system architecture traits in diverse soybean genotypes using a semi-hydroponic system. *Plants, 10*, 2781.

Liu, J., Magalhaes, J. V., Shaff, J., & Kochian, L. V. (2009). Aluminum-activated citrate and malate transporters from the MATE and ALMT families function independently to confer Arabidopsis aluminum tolerance. *The Plant Journal, 57*, 389–399.

Liu, A., Xiao, Z., Li, M. W., Wong, F. L., Yung, W. S., Ku, Y. S., et al. (2019). Transcriptomic reprogramming in soybean seedlings under salt stress. *Plant, Cell & Environment, 42*, 98–114.

Lucas, M., Kenobi, K., von Wangenheim, D., Voss, U., Swarup, K., De Smet, I., et al. (2013). Lateral root morphogenesis is dependent on the mechanical properties of the overlaying tissues. *Proceedings of the National Academy of Sciences of the United States of America, 110*, 5229–5234.

Lynch, J. P. (1995). Root architecture and plant productivity. *Plant Physiology, 109*, 7–13.

Lynch, J. P. (2007). Roots of the second green revolution. *Australian Journal of Botany, 55*, 493–512.

Lynch, J. P., & Brown, K. M. (2012). New roots for agriculture: Exploiting the root phenome. *Philosophical Transactions of the Royal Society, B: Biological Sciences, 367*, 1598–1604.

Ma, J. F., & Furukawa, J. (2003). Recent progress in the research of external Al detoxification in higher plants: A minireview. *Journal of Inorganic Biochemistry, 97*, 46–51.

Ma, Q., Yi, R., Li, L., Liang, Z., Zeng, T., Zhang, Y., et al. (2018). *GsMATE* encoding a multidrug and toxic compound extrusion transporter enhances aluminum tolerance in *Arabidopsis thaliana*. *BMC Plant Biology, 18*, 212.

MacMillan, K., Emrich, K., Piepho, H. P., Mullins, C. E., & Price, A. H. (2006). Assessing the importance of genotype x environmental interaction for root traits in rice using a mapping population II: Conventional QTL analysis. *Theoretical and Applied Genetics, 113*, 953–964.

Magalhaes, J. V., Liu, J., Guimaraes, C. T., Lana, U. G. P., Alves, V. M. C., Wang, Y. H., et al. (2007). A gene in the multidrug and toxic compound extrusion (MATE) family confers aluminum tolerance in sorghum. *Nature Genetics, 39*, 1156–1161.

Makbul, S., Güler, N. S., Durmus, N., & Güven, S. (2011). Changes in anatomical and physiological parameters of soybean under drought stress. *Turkish Journal of Botany, 35*, 369–377.

Manavalan, L. P., Prince, S. J., Musket, T. A., Chaky, J., Deshmukh, R., Vuong, T. D., et al. (2015). Identification of novel QTL governing root architectural traits in an interspecific soybean population. *PLoS One, 10*, e0120490.

Mangena, P. (2018). Water stress: Morphological and anatomical changes in soybean (*Glycine max* L.) plants. In V. Andjelkovic (Ed.), *Plant, abiotic stress and responses to climate change* (pp. 9–31). London: IntechOpen.

Manschadi, A. M., Christopher, J., Devoil, P., & Hammer, G. L. (2006). The role of root architectural traits in adaptation of wheat to water-limited environments. *Functional Plant Biology*, *33*, 823–837.

Manschadi, A. M., Manske, G. G. B., & Vlek, P. L. G. (2013). Root architecture and resource acquisition: Wheat as a model plant. In A. Eshel, & T. Beeckman (Eds.), *Plant roots: The hidden half* (pp. 249–259). New York: Marcel Dekker.

Matsuzaki, Y., Ogawa-Ohnishi, M., Mori, A., & Matsubayashi, Y. (2010). Secreted peptide signals required for maintenance of root stem cell niche in *Arabidopsis*. *Science*, *329*, 1065–1067.

Meister, R., Rajani, M. S., Ruzicka, D., & Schachtman, D. P. (2014). Challenges of modifying root traits in crops for agriculture. *Trends in Plant Science*, *19*, 779–788.

Michniewicz, M., Ho, C. H., Enders, T. A., Floro, E., Damodoran, S., Gunther, L. K., et al. (2019). TRANSPORTER OF IBA1 links auxin and cytokinin to influence root architecture. *Developmental Cell*, *50*, 599–609.

Mickelbart, M. V., Hasegawa, P. M., & Bailey-Serres, J. (2015). Genetic mechanisms of abiotic stress tolerance that translate to crop yield stability. *Nature Reviews Genetics*, *16*, 237–251.

Nguyen, H. T., Babu, R. C., & Blum, A. (1997). Breeding for drought resistance in rice: Physiology and molecular genetics considerations. *Crop Science*, *37*, 1426–1434.

Nguyen, B. D., Brar, D. S., Bui, B. C., Nguyen, T. V., Pham, L. N., & Nguyen, H. T. (2003). Identification and mapping of the QTL for aluminum tolerance introgressed from the new source, Oryza Rufipogon Griff., into indica rice (*Oryza sativa* L.). *Theoretical and Applied Genetics*, *106*, 583–593.

Nguyen, V. T., Vuong, T. D., VanToai, T., Lee, J. D., Wu, X., Mian, M. A. R., et al. (2012). Mapping of quantitative trait loci associated with resistance to *phytophthora sojae* and flooding tolerance in soybean. *Crop Science*, *52*, 2481–2493.

Nielsen, K. L., Lynch, J. P., & Weiss, H. N. (1997). Fractal geometry of bean root systems: Correlations between spatial and fractal dimension. *American Journal of Botany*, *84*, 26–33.

Ogawa, A., Ando, F., Toyofuku, K., & Kawashima, C. (2009). Sucrose metabolism for the development of seminal root in maize seedlings. *Plant Production Science*, *12*, 9–16.

Ogura, T., Goeschl, C., Filiault, D., Mirea, M., Slovak, R., Wolhrab, B., et al. (2019). Root system depth in *Arabidopsis* is shaped by *EXOCYST70A3* via the dynamic modulation of auxin transport. *Cell*, *178*, 400–412.

Oosterhuis, D. M., Scott, H. D., Hampton, R. E., & Wullschleter, S. D. (1990). Physiological response of two soybean [*Glycine max* (L.) Merr] cultivars to short-term flooding. *Environmental and Experimental Botany*, *30*, 85–92.

Osmont, K. S., Sibout, R., & Hardtke, C. S. (2007). Hidden branches: Developments in root system architecture. *Annual Review of Plant Biology*, *58*, 93–113.

Overvoorde, P., Fukaki, H., & Beeckman, T. (2010). Auxin control of root development. *Cold Spring Harbor Perspectives in Biology*, *2*, a001537.

Pantalone, V. R., Rebetzke, G. J., Burton, J. W., & Carter, T. E. (1996). Phenotypic evaluation of root traits in soybean and applicability to plant breeding. *Crop Science*, *36*, 456–459.

Patil, G., Do, T., Vuong, T. D., Valliyodan, B., Lee, J. D., Chaudhary, J., et al. (2016). Genomic-assisted haplotype analysis and the development of high-throughput SNP markers for salinity tolerance in soybean. *Scientific Reports*, *6*, 19199.

Pearce, G., Moura, D. S., Stratmann, J., & Ryan, C. A. (2001). RALF, a 5-kDa ubiquitous polypeptide in plants, arrests root growth and development. *Proceedings of the National Academy of Sciences of the United States of America*, *98*, 12843–12847.

Phung, N. T. P., Mai, C. D., Hoang, G. T., Truong, H. T. M., Lavarenne, J., Gonin, M., et al. (2016). Genome-wide association mapping for root traits in a panel of rice accessions from Vietnam. *BMC Plant Biology*, *16*, 64.

Pierret, A., Kirby, M., & Moran, C. (2003). Simultaneous X-ray imaging of plant root growth and water uptake in thin-slab systems. *Plant and Soil, 255*, 361–373.

Prince, S. J., Murphy, M., Mutava, R. N., Durnell, L. A., Valliyodan, B., Shannon, J. G., et al. (2017). Root xylem plasticity to improve water use and yield in water-stressed soybean. *Journal of Experimental Botany, 68*, 2027–2036.

Prince, S. J., Song, L., Qiu, D., dos Santos, J. V. M., Chai, C., Joshi, T., et al. (2015). Genetic variants in root architecture-related genes in a *Glycine soja* accession, a potential resource to improve cultivated soybean. *BMC Genomics, 16*, 132.

Prince, S. J., Valliyodan, B., Ye, H., Yang, M., Tai, S., Hu, W., et al. (2018). Understanding genetic control of root system architecture in soybean: Insights into the genetic basis of lateral root number. *Plant, Cell & Environment, 42*, 212–229.

Prince, S. J., Vuong, T. D., Wu, X., Bai, Y., Lu, F., Kumpatla, S. P., et al. (2020). Mapping quantitative trait loci for soybean seedling shoot and root architecture traits in an inter-specific genetic population. *Frontiers in Plant Science, 11*, 1284.

Purushothaman, R., Zaman-Allah, M., Mallikarjuna, N., Pannirselvam, R., Krishnamurthy, L., & Gowda, C. L. L. (2013). Root anatomical traits and their possible contribution to drought tolerance in grain legumes. *Plant Production Science, 16*, 1–8.

Qi, X., Li, M. W., Xie, M., Liu, X., Ni, M., Shao, G., et al. (2014). Identification of a novel salt tolerance gene in wild soybean by whole-genome sequencing. *Nature Communications, 5*, 4340.

Qiao, S., Fang, Y., Wu, A., Xu, B., Zhang, S., Deng, X., et al. (2019). Dissecting root trait variability in maize genotypes using the semi-hydroponic phenotyping platform. *Plant and Soil, 439*, 75–90.

Qin, H., & Huang, R. (2018). Auxin controlled by ethylene steers root development. *International Journal of Molecular Sciences, 19*, 3656.

Reid, D. E., Ferguson, B. J., & Gresshoff, P. M. (2011). Inoculation- and nitrate-induced CLE peptides of soybean control NARK-dependent nodule formation. *Molecular Plant-Microbe Interactions, 24*, 606–618.

Rhine, M. D., Steven, G., Shannon, G., Wrather, A., & Sleper, D. (2010). Yield and nutritional responses to waterlogging to soybean cultivars. *Irrigation Science, 28*, 135–142.

Rowe, J. H., Topping, J. F., Liu, J., & Lindsey, K. (2016). Abscisic acid regulates root growth under osmotic stress conditions via an interacting hormonal network with cytokinin, ethylene and auxin. *New Phytologist, 211*, 225–239.

Sadok, W., & Sinclair, T. R. (2011). Crops yield increase under water-limited conditions: Review of recent physiological advances for soybean genetic improvement. *Advances in Agronomy, 113*, 313–337.

Sairam, R. K., & Tyagi, A. (2004). Physiology and molecular biology of salinity stress tolerance in plants. *Current Science, 86*, 407–421.

Salazar-Henao, J. E., Vélez-Bermúdez, I. C., & Schmidt, W. (2016). The regulation and plasticity of root hair patterning and morphogenesis. *Development, 143*, 1848–1858.

Sasaki, M., Yamamoto, Y., & Matsumoto, H. (1996). Lignin deposition induced by aluminium in wheat (*Triticum aestivum*) roots. *Physiologia Plantarum, 96*, 193–198.

Scott, H. D., DeAngulo, J., Daniels, M. B., & Wood, L. S. (1989). Flood duration effects on soybean growth and yield. *Agronomy Journal, 81*, 631–636.

Searle, I. R., Men, A. E., Laniya, T. S., Buzas, D. M., Iturbe-Ormaetxe, I., Carroll, B. J., et al. (2003). Long-distance signaling in nodulation directed by a CLAVATA1-like receptor kinase. *Science, 299*, 109–112.

Shannon, J. G., Stevens, W. E., Wiebold, W. J., McGraw, R. L., Sleper, D. A., & Nguyen, H. T. (2005). Breeding soybeans for improved tolerance to flooding. In *Procedure of 30th Soybean Research Conference, American Seed Trade Association, Chicago, IL, December 7th, 2005*.

Sharma, A. D., Sharma, H., & Lightfoot, D. A. (2011). The genetic control of tolerance to aluminium toxicity in the 'Essex' by 'Forrest' recombinant inbred line population. *Theoretical and Applied Genetics*, *122*, 687–694.

Silva, B. R. S., Batista, B. L., & Lobato, A. K. S. (2021). Anatomical changes in stem and root of soybean plants submitted to salt stress. *Plant Biology*, *23*, 57–65.

Smet, I. D., Lau, S., Mayer, U., & Jürgens, G. (2010). Embryogenesis – The humble beginnings of plant life. *Plant Journal*, *61*, 959–970.

Smit, A. L., Bengough, A. G., Engels, C., van Noordwijk, M., Pellerin, S., & van de Geijn, S. C. (2000). *Root methods: A handbook*. Heidelberg: Springer.

Steele, K. A., Price, A. H., Shashidhar, H. E., & Witcombe, J. R. (2006). Marker-assisted selection to introgress rice QTLs controlling root traits into an Indian upland rice variety. *Theoretical and Applied Genetics*, *112*, 208–221.

Sun, P., Tian, Q. Y., Chen, J., & Zhang, W. H. (2010). Aluminium-induced inhibition of root elongation in *Arabidopsis* is mediated by ethylene and auxin. *Journal of Experimental Botany*, *61*, 347–356.

Sun, Z., Wang, Y., Mou, F., Tian, Y., Chen, L., Zhang, S., et al. (2016). Genome-wide small RNA analysis of soybean reveals auxin-responsive microRNAs that are differentially expressed in response to salt stress in root apex. *Frontiers in Plant Science*, *6*, 1273.

Tamang, B. G., Magliozzi, J. O., Maroof, M. A. S., & Fukao, T. (2014). Physiological and transcriptomic characterization of submergence and reoxygenation responses in soybean seedlings. *Plant, Cell and Environment*, *37*, 2350–2365.

Torrion, J. A., Setiyono, T. D., Cassman, K. G., Ferguson, R. B., Irmak, S., & Specht, J. E. (2012). Soybean root development relative to vegetative and reproductive phenology. *Agronomy Journal*, *104*, 1702–1709.

Trachsel, S., Kaeppler, S. M., Brown, K. M., & Lynch, J. P. (2011). Shovelomics: High throughput phenotyping of maize (*Zea mays* L.) root architecture in the field. *Plant and Soil*, *341*, 75–87.

Trenberth, K. E., Dai, A., van der Schrier, G., Jones, P. D., Barichivich, J., Briffa, K. R., et al. (2013). Global warming and changes in drought. *Nature Climate Change*, *4*, 17–22.

Tuberosa, R., Salvi, S., Sanguineti, M. C., Landi, P., MacCaferri, M., & Conti, S. (2002). Mapping QTLs regulating morpho-physiological traits and yield: Case studies, shortcomings and perspectives in drought-stressed maize. *Annals of Botany*, *89*, 941–963.

Turner, M., Nizampatnam, N. R., Baron, M., Coppin, S., Damodaran, S., Adhikari, S., et al. (2013). Ectopic expression of miR160 results in auxin hypersensitivity, cytokinin hyposensitivity, and inhibition of symbiotic nodule development in soybean. *Plant Physiology*, *162*, 2042–2055.

Uga, Y., Sugimoto, K., Ogawa, S., Rane, J., Ishitani, M., Hara, N., et al. (2013). Control of root system architecture by *DEEPER ROOTING1* increases rice yield under drought conditions. *Nature Genetics*, *45*, 1097–1102.

Valliyodan, B., Van Toai, T. T., Alves, J. D., Goulart, P. D. P., Lee, J. D., Fritschi, F. B., et al. (2014). Expression of root-related transcription factors associated with flooding tolerance of soybean (*Glycine max*). *International Journal of Molecular Sciences*, *15*, 17622–17643.

Vamerali, T., Saccomani, M., Bona, S., Mosca, G., Guarise, M., & Ganis, A. (2003). A comparison of root characteristics in relation to nutrient and water stress in two maize hybrids. *Plant and Soil*, *255*, 157–167.

VanToai, T. T., Beuerlein, A. F., Schmitthenner, S. K., & St. Martin, S. K. (1994). Genetic variability for flooding tolerance in soybeans. *Crop Science*, *34*, 1112–1115.

VanToai, T. T., Hao, T. T. C., Hue, N. T. N., Nguyen, H. T., Shannon, J. G., & Rahman, M. A. (2010). Flooding tolerance of soybean [*Glycine max* (L.) Merr.] germplasm from Southeast Asia under field and screen-house environments. *The Open Agriculture Journal*, *4*, 38–46.

VanToai, T. T., Martin, S. K., St. Chase, K., Boru, G., Schnipke, V., Schmitthenner, A. F., et al. (2001). Identification of a QTL associated with tolerance of soybean tosoil waterlogging. *Crop Science*, *41*, 1247–1252.

Varshney, R. K., Pazhamala, L., Kashiwagi, J., Gaur, P. M., Krishnamurthy, L., & Hoisington, D. (2011). Genomics and physiological approaches for root trait breeding to improve drought tolerance in chickpea (*Cicer arietinum* L.). In A. C. DeOliveira, & P. K. Varshney (Eds.), *Root genomics* (pp. 233–250). Dordrecht: Springer.

Visser, E. J. W., & Voesenek, L. A. C. J. (2005). Acclimation to soil flooding-sensing and signal-transduction. *Plant and Soil*, *274*, 197–214.

Waidmann, S., Rosquete, M. R., Schöller, M., Sarkel, E., Lindner, H., LaRue, T., et al. (2019). Cytokinin functions as an asymmetric and anti-gravitropic signal in lateral roots. *Nature Communications*, *10*, 3540.

Wang, J., Chen, Y., Zhang, Y., Zhang, Y., Ai, Y., Feng, Y., et al. (2021). Phenotyping and validation of root morphological traits in barley (*Hordeum vulgare* L.). *Agronomy*, *11*, 1583.

Wang, Y., Li, K., Chen, L., Zou, Y., Liu, H., Tian, Y., et al. (2015). MicroRNA167-directed regulation of the auxin response factors *GmARF8a* and *GmARF8b* is required for soybean nodulation and lateral root development. *Plant Physiology*, *168*, 984–999.

Wang, Y., Yang, W., Zuo, Y., Zhu, L., Hastwell, A. H., Chen, L., et al. (2019). *GmYUC2a* mediates auxin biosynthesis during root development and nodulation in soybean. *Journal of Experimental Botany*, *70*, 3165–3176.

Wasson, A. P., Richards, R. A., Chatrath, R., Misra, S. C., Prasad, S. V. S., Rebetzke, G. J., et al. (2012). Traits and selection strategies to improve root systems and water uptake in water-limited wheat crops. *Journal of Experimental Botany*, *63*, 3485–3498.

Werner, T., Motyka, V., Strnad, M., & Schmülling, T. (2001). Regulation of plant growth by cytokinin. *Proceedings of the National Academy of Sciences of the United States of America*, *98*, 10487–10492.

Wood, S., Sebastian, K., & Scherr, S. J. (2000). *Pilot analysis of global ecosystems: Agroecosystems*. Washington, DC: World Resources Institute.

Xia, J., Yamaji, N., Kasai, T., & Ma, J. F. (2010). Plasma membrane-localized transporter for aluminum in rice. *Proceedings of the National Academy of Sciences of the United States of America*, *107*, 18381–18385.

Xiong, R., Liu, S., Considine, M. J., Siddique, K. H. M., Lam, H. M., & Chen, Y. (2021). Root system architecture, physiological and transcriptional traits of soybean (*Glycine max* L.) in response to water deficit: A review. *Physiologia Plantarum*, *172*, 405–418.

Ye, H., Roorkiwal, M., Valliyodan, B., Zhou, L., Chen, P., Varshney, R. K., et al. (2018). Genetic diversity of root system architecture in response to drought stress in grain legumes. *Journal of Experimental Botany*, *69*, 3267–3277.

Yokosho, K., Yamaji, N., & Ma, J. F. (2011). An Al-inducible MATE gene is involved in external detoxification of Al in rice. *The Plant Journal*, *68*, 1061–1069.

York, L. M., Galindo-Castañeda, T., Schussler, J. R., & Lynch, J. P. (2015). Evolution of US maize (*Zea mays* L.) root architectural and anatomical phenes over the past 100 years corresponds to increased tolerance of nitrogen stress. *Journal of Experimental Botany*, *66*, 2347–2358.

Yu, L. X., Ray, J. D., O'Toole, J. C., & Nguyen, H. T. (1995). Use of wax–petrolatum layers for screening rice root penetration. *Crop Science*, *35*, 684–687.

Yuan, T., Xu, H., Zhang, K., Guo, T., & Lu, Y. (2014). Glucose inhibits root meristem growth via *ABA INSENSITIVE 5*, which represses PIN1 accumulation and auxin activity in *Arabidopsis*. *Plant, Cell & Environment*, *37*, 1338–1350.

Zhang, Y. J., Lynch, J. P., & Brown, K. M. (2003). Ethylene and phosphorous availability have interacting yet distinct effects on root hair development. *Journal of Experimental Botany*, *54*, 2351–2361.

Zhang, K., Zhao, L., Yang, X., Li, M., Sun, J., Wang, K., et al. (2019). *GmRAV1* regulates regeneration of roots and adventitious buds by the cytokinin signaling pathway in *Arabidopsis* and soybean. *Physiologia Plantarum, 165*, 814–829.

Zhao, L. J., Cui, J. J., Cai, Y. Y., Yang, S. N., Liu, J. G., Wang, W., et al. (2020). Comparative transcriptome analysis of two contrasting soybean varieties in response to aluminum toxicity. *International Journal of Molecular Sciences, 21*, 4316.

Zhou, Y., Wang, Z., Gong, L., Chen, A., Liu, N., Li, S., et al. (2019). Functional characterization of three MATE genes in relation to aluminum-induced citrate efflux from soybean root. *Plant and Soil, 443*, 121–138.

Zhu, J., Ingram, A. P., Benfey, N. P., & Elich, T. (2011). From lab to field, new approaches to phenotyping root system architecture. *Current Opinion in Plant Biology, 14*, 310–317.

CHAPTER FOUR

Inheritance and prevention of soybean root rot

Haijian Zhi* **and Dagang Wang**
Nanjing Agricultural University, National Center for Soybean Improvement, National Key Laboratory for Crop Genetics and Germplasm Enhancement, Nanjing, Jiangsu, China
*Corresponding author: e-mail address: zhj@njau.edu.cn

Contents

Abstract

Soybean root rot is one of the most important diseases in soybean, and it is caused by a variety of pathogens. In this chapter, we introduce the pathogenic species, harmful symptoms, and biological characteristics of soybean root rot, especially the Phytophthora root rot. In addition, the inheritance patterns, the markers linked to resistance genes, and the molecular mechanisms of Phytophthora root rot were discussed. Moreover, the effective and reasonable control techniques were also proposed.

1. Biological properties of soybean root rot

Soybean is an important economic crop in the world, the seeds are rich of high-quality protein and editable vegetable oil that can meet a variety of food and industrial needs. Soybean root rot is a devastating soybean disease occurred worldwide, which have been identified in almost all soybean cultivation areas including the major soybean producing countries such as Brazil, America, Japan, Argentina, and China (Schmitthenner, 1985).

Advances in Botanical Research, Volume 102
ISSN 0065-2296
https://doi.org/10.1016/bs.abr.2022.03.003

Soybean root rot can occur at any developmental stage of soybean plants. The disease is mainly soil-borne and seed-borne. It causes the under-development of root system, retardation of shoot growth, and reduction of pod setting.

In water-saturated soil, swarm spore can infect soybean plants throughout the growing season, causing decay of soybean roots and stems, and eventually death of the whole plants. Root rot can lead to an annual average of 10–30% loss of soybean yield, with 60% loss in some cases, or even failure to produce seeds in the whole infected areas (Jiang et al., 2017).

Soybean root rot is mainly caused by *Phytophthora sojae*, Fusarium spp., Pythium spp. and *Rhizoctonia solani* (Rizvi & Yang, 1996). And *Phytophthora sojae* (*P. sojae*) is the most widely reported and studied pathogens of soybean root rot in the world. The occurrence of *P. sojae* was first reported in Indiana, America and it caused severe reduction in soybean production and the yield loss was estimated to be $1–2 billion per year (Kamoun, Furzer, Jones, et al., 2015; Kaufmann & Gerdemann, 1958; Qiu, Li, Ye, et al., 2021).

P. sojae is a soil-borne pathogen that spreads primarily through soil and water flows. The oospores can survive in the soil for many years and the oospores will germinate into sporangia until meeting favorable conditions. When exposed to water, zoospores are released to actively infect plant roots. *P. sojae* can infect soybean at all growth stages, from seed to adult plants (Tyler, 2007). The symptoms include seed rot, damping-off, browning, and sometimes death of adult soybean plants.

Phytophthora root rot was first reported in 1955, since then, at least 55 races have been reported (Jackson, Kirkpatrick, & Rupe, 2004). Tyler (2007) reported that the disease virulence pathotypes are very complex, and many virulence combinations occur because of mutations and sexual outcrossing. Yang, Wang, Guo, et al. (2019) has reported that pathotypes in *P. sojae* field population in China were complicated too.

There were 213 unique virulence pathotypes identified among the 873 isolates collected in northern regions of the United States (Dorrance, Kurle, Robertson, et al., 2016). Based on the analysis, Dorrance et al. (2016) reported that the complexity of pathological types of *P. sojae* population in the United States has increased.

In South Korea, Phytophthora root rot was first discovered in soybean fields in Hongseong, Cheongyang, and Chungnam in 1996 (Jee, Kim, & Cho, 1998). Recently, Kang, Kang, Jang, et al. (2019) reported two new *P. sojae* isolates identified in two geographical regions in South Korea and confirmed their identification based on both morphological features and molecular evidence by PCR.

2. Resistance identification in soybean to Phytophthora root rot

The use of resistant soybean cultivars is the most economical, effective, and environmentally friendly method to control the pathogen (Sahoo, Abeysekara, Cianzio, et al., 2017). There is a large amount of soybean germplasm resources with a vast genetic diversity in China, and those resistant to *P. sojae* have been identified. Zhu, Huo, Wang, et al. (2006) evaluated 120 soybean cultivars from 11 provinces in China by challenging them with 10 different isolates of *P. sojae*. They identified 110 resistant cultivars or lines of which showed varying levels of resistance to 1–10 strains. The 120 cultivars displayed 57 different reaction types when tested on those 10 strains (Zhu et al., 2006).

Zhang, Sun, Wang, et al. (2014) identified 34 *P. sojae* resistant cultivars from Henan province by inoculating them with 26 *P. sojae* pathotypes. In addition, Zaoshu18 is an elite soybean cultivar, which was grown in the Huanghuaihai region of China during the 1980s. Many studies have shown that Zaoshu18 exhibited excellent resistance to Phytophthora root rot (Zhang et al., 2014; Zhu et al., 2006). A systematic screening by Zhong, Sun, Yao, et al. (2018b) further revealed that Zaoshu18 was resistant to 10 of the 14 *P. sojae* isolates. In a further study, Zhong, Yao, Sun, et al. (2019) found that Xiu94-11 was resistant to all 14 *P. sojae* isolates and exhibited the broadest resistance spectrum among the Phytophthora root rot resistant cultivars. These studies indicate that Zaoshu18 harbors a mechanism that show a broad-spectrum resistance to multiple isolates, which merits further attention.

Yang, Ye, Wang, et al. (2019) reported that soybean cultivars Wandou21020, Xu9302-A, Kedou10, and Lidi055 showed resistance to eight *P. sojae* isolates and 53 cultivars were resistant to seven isolates. Jiang, Cheng, Cai, et al. (2020) found that the Guizao1 cultivar was resistant to *P. sojae* PNJ4 and PNJ1. As such, Kang et al. (2019) identified that six cultivars, including Daepung2, Cheongja, and Pungsannamul, exhibited resistant reaction to one or two isolates in South Korea. Therefore, these cultivars could be deployed in these areas to control locally prevalent *P. sojae*.

3. Inheritance and molecular markers in soybean to Phytophythora root rot

There are two types of resistance mechanism to Phytophythora root rot in soybean: One is controlled by single gene (locus) with complete resistance to specific races, and the other one is controlled by multi-genes, which are quantitative trait loci, with broad but partial resistance to all races of *P. sojae* (Sugimoto, Kato, Yoshida, et al., 2012).

The first *Rps* (*resistance to P. sojae*) gene (*Rps1a*) was identified in soybean in the 1950s and widely deployed in the USA (Bernard, Smith, Kaufmann, et al., 1957). Thus far, more than 33 *Rps* genes have been identified and mapped on nine chromosomes of the soybean genome, namely chromosomes 2, 3, 7, 10, 13, 14, and 16–19 (Li, Lin, Wang, et al., 2016; Li, Sun, Zhong, et al., 2017; Niu, Guo, Sun, et al., 2017; Qin, Song, Shi, et al., 2017; Sahoo et al., 2017; Sugimoto et al., 2012; Yang, Zheng, Wang, et al., 2020; Zhong, Sun, Yao, et al., 2018a).

A further study reveals the resistant loci in the genotypes. For example, Yang et al. (2020) showed that cultivars Dongnong 60 and Henong 72 contained *Rps1a*, while Hedou 19, Henong 76, 75-3, Wandou 21,020, Zheng 196, Wandou 28, Heinong 71 and Wandou 29 all contained *Rps1c*. The cultivars Jidou 12, Henong 72, Heinong 71 and Wandou 29 contained *Rps1k*. The cultivar Henong 72 contained both *Rps1a* and *Rps1k*, while Wandou 29 and Heinong 71 contained both *Rps1c* and *Rps1k*.

According to the investigation, most of the tested cultivars from Henan province were postulated to contain *Rps1b*. Some of those from Heilongjiang, Shandong, and Henan provinces contained *Rps3a* (Yang, Ye, et al., 2019). Some soybean cultivars or lines carrying multiple *Rps* genes, including *Rps1a*, *Rps1c*, and *Rps1k*, have been used widely in commercial breeding programs to reduce yield losses caused by *P. sojae* (Sugimoto et al., 2012).

Genetic studies have been dedicated attempting to uncover the linked marker or genes for marker-assisted breeding program. The resistant loci on chromosome 3 gain more attention than others because of the larger resistance effects over other loci. Among the identified *Rps* genes on chromosome 3, *Rps1k* was mapped to a 125-kb region and further cloned. *Rps1k* encodes a nucleotide binding site 1eucine rich repeat (NBS-LRR) protein that is typical resistance protein (Gao & Bhattacharyya, 2008). Recently, a novel *Rps* gene, *RpsHC18*, was also mapped on chromosome 3 through a combination of QTL-seq and traditional genetic mapping

(Zhong et al., 2018a). Genetic analyses indicated that the resistance to *P. sojae* in Qichadou 1 is controlled by a single dominant gene, *RpsQ*. It was mapped to a 118-kb region between BARCSOYSSR_03_0165 and InDel281 on soybean chromosome 3, and co-segregated with Insert11, Insert144 and SNP276 (Li et al., 2017).

The resistance locus *RpsHN* from Meng8206 was mapped to a 278.7 kb genomic region flanked by the SSR markers SSRSOYN-25 and SSRSOYN-44 and may reside at nucleotide position 4,227,863 and 4,506,526 bp (Niu et al., 2017). *RpsUN1* was localized to the region between 4,020,587 and 4,171,402 bp, flanked by two SSR markers, BARCSOYSSR_03_0233 and BARCSOYSSR_03_0246, based on the Glyma2.0 soybean gene annotation database of the Williams 82 genome sequence (Li et al., 2016). *RpsWY* gene was located in bin 401 between 4,466,230 and 4,502,773 bp on chromosome 3 using RILs-based linkage mapping (Cheng, Ma, Ren, et al., 2017). *RpsYD29* was mapped to a 204.8-kb region, and two NBS-LRR-type genes, *Glyma03g04030.1* and *Glyma03g04080.1*, were proposed as candidate genes (Zhang, Xia, Wang, et al., 2013). The closely-linked markers may be directly used to facilitate the transfer of such important resistance to elite cultivars and it is expected that the underlying genes would be uncovered soon.

Zhong et al. (2019) suggested that the resistance of Xiu94-11 is controlled by a single dominant gene, which was named *RpsX*. *RpsX* was mapped between the InDel marker InDelxz6 and the SSR marker BARCSOYSSR_03_0175, with genetic distances of 0.4 and 0.7 cM, respectively (Zhong et al., 2019). Genetic analyses indicated that resistance to *P. sojae* PNJ4 in Guizao 1 was controlled by a single locus (Jiang et al., 2020). Jiang et al. (2020) identified and finely mapped a novel *Rps* locus (*RpsGZ*) that can confer resistance to *P. sojae* PNJ4 and PNJ1 on chromosome 3 with a 367,371 bp interval between 4,003,401 and 4,370,772 bp. Further investigations are needed to uncover the closely linked markers or genes that can facilitate the breeding.

Genetics for the loci on other chromosomes were also investigated. Yao, Wang, Wu, et al. (2010) revealed that Zaoshu18 resistance to *P. sojae* was controlled by a single dominant gene, *RpsZS18*, which mapped on chromosome 2 in a preliminary study. Further, *RpsZS18* was mapped between SSR markers ZCSSR33 and CSSR46 with genetic distances of 0.9 and 0.5 cM, respectively (Zhong et al., 2018b).

These resistance genes act by recognizing the products of avirulence (*Avr*) genes of the pathogen in a traditional gene-for-gene relationship. Arsenault-Labrecque, Sonah, Lebreton, et al. (2018) sequenced 31 *P. sojae* isolates collected from Canadian fields and analyzed the sequences and variants of the seven most common *Avr* genes: 1a, 1b, 1c, 1d, 1k, 3a, and 6. In addition, Yang, Wang, et al. (2019) recently reported new haplotypes of the *Avr1c* gene from isolates recovered in Chinese fields.

4. The function and mechanism of soybean resistance gene to Phytophythora root rot

To date, attempts have been made to increase the resistance of soybean to *P. sojae* infection by expressing foreign genes conferring resistance to fungal pathogens. The *Rps1k* gene, which is associated with broad-spectrum and durable resistance, has been cloned, and the tandemly arranged NBS-LRR genes *Rps1k-1* and *Rps1k-2* were functionally validated conferring complete resistance to *P. sojae* (Gao & Bhattacharyya, 2008; Gao, Narayanan, Ellison, et al., 2005).

Based on the mapped results, a best gene encoding serine/threonine protein kinase 1eucine rich repeat (STK-LRR) was predicted as the candidate gene for *RpsQ* (Li et al., 2017). Gene *Glyma.03g27200* encoded a protein with a typical STK-LRR structure, this gene could be induced by *P. sojae* infection and was regarded as a plausible candidate gene for *RpsQ* (Li et al., 2017). The STK-LRR resistance gene encodes an extracellular receptor-like protein kinase with an extracellular LRR and an intracellular STK.

Real-time RT-PCR analysis showed that three genes (*Glyma.03g04260*, *Glyma.03g04300*, and *Glyma.03g04340*) were induced by *P. sojae* isolate HeN08-35 (Niu et al., 2017). Genes *Glyma.03g04260* and *Glyma.03g04300* encoded nucleotide-binding adaptor shared by APAF-1, R proteins and CED-4 (NB-ARC) domain-containing disease resistance protein. The third gene encodes STK, which has been identified to be involved in plant disease resistance. These genes were predicted as likely involved candidate genes for *RpsHN* (Niu et al., 2017).

According to fine mapping, whole-genome resequencing and quantitative real-time PCR assay, Zhong et al. (2018b) identified that an EF-hand calcium-binding domain containing protein encoding gene (*Glyma.02g245700*), a phosphofructokinase (pfk) B encoding gene (*Glyma.02g245800*), and a gene with no functional annotation (*Glyma.02g246300*), are putative candidate genes showing Phytophthora root rot resistance. The expression level of *Glyma.02g245800* was significantly higher in resistant Zaoshu18 than it in the susceptible cultivar Williams at all six tested time points post infection (Zhong et al., 2018b). In another study, using genotyping-by-sequencing and real-time RT-PCR assays of possible candidate genes revealed that *Glyma.03g05300* was likely involved in Phytophthora root rot resistance (Jiang et al., 2020).

Overexpression of Gly m 4L in susceptible "Dongnong 50" soybean showed enhanced resistance to *P. sojae* (Fan, Jiang, Wu, et al., 2015). Under *P. sojae* stress, Fan et al. (2015) speculated two potential modes of action by of Gly m 41. In case Gly m 4L protein in soybean could be targeted to non-classical secretory pathway and secreted into the intercellular space, then it performed antimicrobial function by directly degrading pathogenic RNA to inhibit hyphal growth or zoospore release. In the other case Gly m 4L protein could not be targeted to non-classical secretory pathway, then it performed antimicrobial functional role through degrading RNA of haustorium, which is formed and stick into cytoplasm of susceptible soybean with *P. sojae* infection, to limit pathogen invasion (Fan et al., 2015).

A novel ethylene response factor (ERF) in soybean, namely GmERF5, was recently identified, and overexpressing it in soybean upregulated the pathogenesis-related *PR10*, *PR1-1*, and *PR10-1* genes, leading to enhanced *P. sojae* resistance compared with the wild type (Dong, Cheng, Wu, et al., 2015). GmERF5 could interact with soybean basic helix–loop–helix transcription factor (GmbHLH) and eukaryotic translation initiation factor (GmEIF) both in yeast cells and *in planta*. Dong et al. (2015) speculated that GmERF5 and GmbHLH cooperatively regulate the response to *P. sojae* resistance. GmEIF might be involved in the regulation of GmERF5 transcription.

A novel isoflavone reductase (IFR) in soybean, namely GmIFR, encodes a NAD(*P*)H-dependent oxidoreductase and it can enhance resistance to *P. sojae* when overexpressed it in soybean (Cheng, Li, Dong, et al., 2015). Authors speculated that overexpression of GmIFR in transgenic soybean plants can increase the content of glyceollins or GmIFR itself behave as an antioxidant to scavenge reactive oxygen species, which might improve resistance to *P. sojae* in soybean.

Harpin proteins in many plant pathogenic bacteria were identified and shown to enhance disease and insect resistance in crop plants. Overexpression of the harpin-encoding *hrpZm* gene significantly enhances Phytophthora root rot tolerance in soybean by eliciting resistance responses mediated by multiple defense signaling pathways (Du, Yang, Zhang, et al., 2018). Consistent with the expression of the defense genes, the activities of several defense-related enzymes, such as phenylalanine ammonia lyase (PAL), polyphenoloxidase (PPO), peroxidase and superoxide dismutase also increase significantly in transgenic lines compared to the negative control plants after *P. sojae* inoculation (Du et al., 2018).

Cheng, Dong, Gao, et al. (2018) uncovered that *GmPIB1* (*P. sojae*-inducible bHLH transcription factor; *Glyma.01g129700*) transcript levels were much higher in *P. sojae*-resistant soybean cultivar "L77–1863" than in the susceptible cultivar "Williams." Overexpressing *GmPIB1* in soybean hairy roots increased resistance to *P. sojae*, whereas RNA interference (RNAi) of this gene in soybean hairy roots increased susceptibility to this pathogen (Cheng et al., 2018). GmPIB1 bound directly to the promoter of peroxidase gene *GmSPOD1* and inhibited its expression, leading to improved resistance to *P. sojae* (Cheng et al., 2018). GmSPOD1, encodes a key enzyme for reactive oxygen species production. As GmPIB1-modulated genes, the expression of *GmSPOD1* was down-regulated in hairy roots overexpressing *GmPIB1* and up-regulated in *GmPIB1* RNA interference lines (Fig. 1).

The enzyme 4-coumarate: CoA ligase (4CL) has been extensively studied with regard to plant responses to pathogens. Overexpression of *GmPI4L* (*P. sojae*-inducible CoA ligase gene) enhanced resistance to *P. sojae* in transgenic soybean plants possibly by enhancing the content of daidzein, genistein, glycitein and glyceollins in soybean (Chen, Fang, Zhang, et al., 2019). Meanwhile, the expression of *GmPI4L* was also induced by treatments with salicylic acid, gibberellic, and methyl jasmonate (Chen et al., 2019).

Compared with non-transformed plants, the transgenic plants showed milder *P. sojae* symptoms and had higher survival rates. Harpin proteins encoded by the hypersensitive response and pathogenicity (hrp) genes are produced by several gram-negative plant pathogenic bacteria (Tampakaki et al., 2010) The harpinXooc-encoding *hrf2* gene expression in transgenic soybean significantly enhanced the resistance to *P. sojae* by eliciting multiple defense responses mediated by different signaling pathways (Niu, Yang, Zhang, et al., 2019).

The broad complex, Tramtrack, Bric-à-brac/poxvirus and zinc finger (BTB/POZ) domain is a widely distributed structural motif found in an array of proteins involved in various biological processes. Soybean BTB/POZ protein interacts with soybean like heterochromatin protein1 (GmLHP1), which it might play crucial roles in the response to *P. sojae* (Zhang, Gao, Li, et al., 2019). GmLHP1 negatively regulates the response of soybean to *P. sojae* by reducing salicylic acid levels and repressing salicylic acid-marker gene *GmPR1* expression (Zhang, Gao, Sun, et al., 2021). Together, the nuclear localization of GmLHP1 is required for the

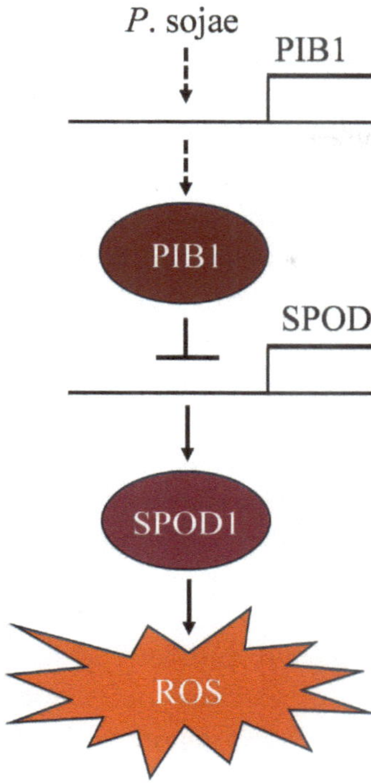

Fig. 1 Model of the GmPIB1-mediated response to P. sojae. GmPIB1 expression is induced by P. sojae. GmPIB1 inhibits GmSPOD1 transcription by binding to the E-box element in its promoter. The suppression of GmSPOD1 expression leads to decreased intracellular ROS levels. *This figure cited to Cheng, Q., Dong, L., & Gao, T., et al. (2018). The bHLH transcription factor GmPIB1 facilitates resistance to Phytophthora sojae in Glycine max. Journal of Experimental Botany, 69 (10), 2527–2541.*

GmLHP1-mediated negative regulation of immunity, salicylic acid levels and the suppression of WRKY transcription factor gene *GmWRKY40* expression. Finally, GmBTB/POZ releases GmLHP1-regulated *GmWRKY40* suppression and increases resistance to *P. sojae* in GmLHP1-OE hairy roots (Zhang et al., 2019, 2021).

5. The function of *P. sojae* effector

Research revealed that the transcript levels of the RxLR (Arg-x-Leu-Arg) effector gene *PsAvh52* were strongly upregulated during the early stages (0.5–6 hpi) of *P. sojae* infection and *PsAvh52* could suppress cell death induced by effectors and pathogen-associated molecular patterns (PAMPs) in soybean (Li, Wang, Jing, et al., 2018; Ma, Song, Zhu, et al., 2015).

Boevink, Wang, McLellan, et al. (2016) identified an RxLR effector from *P. infestans*, PexRD24, that enhances plant susceptibility by attenuating jasmonic acid and salicylic acid-mediated transcriptional responses of the host plant. PexRD24 interacts with three host protein phosphatase 1 catalytic (PP1c) isoforms, causing their re-localization from the nucleolus to the nucleoplasm (Boevink et al., 2016). Moreover, soybean binding immunoglobulin proteins (GmBiPs), which are targets of the *P. sojae* RxLR effector, negatively regulate plant defense responses against *P. sojae* infection (Jing, Guo, Li, et al., 2016).

PsAvh262 interacts with soybean proteins that control endoplasmic reticulum stress in order to modulate plant resistance (Jing et al., 2016). *P. sojae* effector *PsAvh23* enhances plant susceptibility through reducing the levels of General Control Non-depressive 5 (GCN5)-mediated lysine 9 of histone H3 (H3K9) acetylation to suppress plant defense gene expression (Kong, Qiu, Kang, et al., 2017). Li et al. (2018) demonstrated that soybean transacetylase protein (GmTAP1) is a susceptibility factor that is hijacked by *PsAvh52* in order to promote epigenetic modifications that enhance the susceptibility of soybean to *P. sojae* infection. Therefore, *PsAvh52* acts early in infection to promote H3K9 acetylation, stimulating the expression of susceptibility genes, and then *PsAvh23* acts to reverse the action of *PsAvh52*, and also to block any host-mediated increases in H3K9 acetylation that may support the expression of plant defense genes (Kong et al., 2017; Li et al., 2018).

PsAvr3c is an effector identified from oomycete plant pathogen *P. sojae* that causes soybean root and stem rot disease. Research found that *PsAvr3c* reprograms host pre-mRNA splicing to subvert plant immunity, to promote disease (Huang, Gu, Zhang, et al., 2017). *P. sojae* effector *PsAvr3c* could be recognized by soybean cultivars that carry the cognate resistant gene *Rps3c*. Huang, Chen, Lu, et al. (2019) found that substitution of serine for glycine at position 174 in PsAvr3c resulted in evasion of *Rps3c*-mediated immunity.

Zhang, Huang, Ochola, et al. (2018) revealed that both *PsAvh27b* (*PsAvr3c* paralog from *P. sojae*) and *ProbiAvh89* (*PsAvr3c* ortholog from *P. cinnamomi* var. robiniae) interact with soybean spliceosomal complex proteins (GmSKRPs) in vitro, and stabilize GmSKRP1 in vivo. However, *PparvAvh214* (orthologs from *Phytophthora parvispora*) cannot interact with GmSKRPs proteins. As an effector of the *PsAvr3c* family, *PparvAvh214* may have diverged during *Phytophthora* evolution (Zhang et al., 2018). *PsAvr3c* family members display differences in promoting *Phytophthora* infection in a SKRP-dependent manner. It is reported that the conserved motifs at the C-terminus of Avr are critical for their interactions with *Rps* genes. Yang, Wang, et al. (2019), Yang, Ye, et al. (2019) identified that C-terminal of Avr1c is critical for its recognition by *Rps1c*.

Members of the carbohydrate esterase (CE) family 10 (carboxylesterases, arylesterases, sterol esterases and acetylcholine esterases, collectively known as CE10), are thought to be important for penetration process of *P. sojae*. Research revealed that CE10-coding genes are important for both the biotrophic and necrotrophic stages of the *P. sojae* infection process (Grams & Ospina-Giraldo, 2019). Grams and Ospina-Giraldo (2019) found that all CE10-coding genes experienced an increase in transcription during infection when compared to non-infecting *P. sojae* mycelium further supports the suggested role of CE10-coding genes in *P. sojae* pathogenesis.

Protein phosphatase 2A (PP2A) is a subgroup of widely conserved serine/threonine phosphatases and plays diverse roles in transcription, translation, differentiation, cell cycle, and signal transduction (Mumby & Walter, 1993). Qiu et al. (2021) found that the deletion of a regulatory B-subunit of protein phosphatase 2A (*PsPP2Ab1*) in *P. sojae* leads to severe defects in vegetative hyphal growth, soybean infection, and loss of the ability to produce sporangia.

In response to *P. sojae* infection in different lines and at different time points, 90 differentially accumulated metabolites (DAMs) were identified after *P. sojae* infection. Levels of 50 metabolites differed between the resistant line and the susceptible line. 28 DAMs did not only differentially accumulate between the resistant line and the susceptible line but also differentially accumulated after *P. sojae* infection (Zhu, Zhou, Li, et al., 2018).

6. Prevention and future perspectives

The most economical and effective way to protect soybeans against *P. sojae* infection is by breeding for dominant resistance to *P. sojae* (*Rps*) genes (Sugimoto et al., 2012). There are abundant soybean germplasm resources in China for breeding programs by using a variety of techniques. The hypocotyl inoculation method was commonly used in resistance evaluation assays. Jiang et al. (2017) developed a method, in which, *P. sojae* inoculum was mixed into soil and the resistance was evaluated by the survival rate (%) of soybean seedlings (Mixing agar homogenate containing *P. sojae* into the soil, MAPS). Arsenault-Labrecque et al. (2018) developed a simple hydroponic assay to study resistance in soybean.

Genetic engineering utilizing alien genes that confer disease resistance offers an alternative to conventional breeding methods for improving resistance against plant pathogens.

Efficient crop rotation has a significant effect on the prevention and control of soybean root rot (Ye, Zheng, & Wang, 2020). In additional, continuous cropping should be avoided. High-quality and disease-resistant soybean varieties should be chosen and appropriate late sowing, especially in cold areas should be implemented. At the same time, applying more phosphate fertilizer and organic fertilizer, strengthening field management, rational cultivation, implementing deep loosening, improving soil aeration conditions, improving soil preparation quality, and timely drainage to avoid stagnant water, preventing underground pests can be helpful in preventing the disease (Ye et al., 2020).

In addition, treatment of seeds with seed coating agent can delay the occurrence of disease, especially in areas where rain or irrigation occurs shortly after planting, the use of metalaxyl-M and Multi-fungi can effectively reduce the incidence of disease and increase yield. Dressing seeds with a seed coating agent containing thiram fungicides with the dosage of 1:50–1:70 can reduce the incidence of disease by more than 20% and have a certain control effect on the occurrence of soybean root rot. Soybean seed may also be coated with 6.25% metalaxyl and fludioxonil, 25% thiamethoxam, metalaxyl and fludioxonil, 27% thiamethoxam, fludioxonil and difenoconazole to effectively control soybean root rot (Ye et al., 2020).

Wagner, Norris, Chatterjee, et al. (2018) found that antagonistic activity toward oomycetes was significantly more common for aquatic isolates than for soil isolates. According to the collection of 330 *Pseudomonas* strains isolated from soil and freshwater habitats, 118 exhibited antagonistic activity against at least one oomycete species, and 16 strains were inhibitory to all pathogens. One water-derived strain, 06C 126, was predicted to express a siderophore and exhibited diverse antagonistic profiles when tested on nutrient rich and iron depleted media, suggesting that more than one compound was produced that effectively inhibited oomycetes (Wagner, Norris, Chatterjee, et al., 2018).

References

Arsenault-Labrecque, G., Sonah, H., Lebreton, A., et al. (2018). Stable predictive markers for Phytophthora sojae avirulence genes that impair infection of soybean uncovered by whole genome sequencing of 31 isolates. *BMC Biology*, *16*(1), 1–16.

Bernard, R. L., Smith, P. E., Kaufmann, M. J., et al. (1957). Inheritance of resistance to Phytophthora root and stem rot in the soybean. *Agronomy Journal*, *49*(7), 391.

Boevink, P. C., Wang, X., McLellan, H., et al. (2016). A Phytophthora infestans RXLR effector targets plant PP1c isoforms that promote late blight disease. *Nature Communications*, 7(1), 1–14.

Chen, X., Fang, X., Zhang, Y., et al. (2019). Overexpression of a soybean 4-coumaric acid: Coenzyme A ligase (GmPI4L) enhances resistance to Phytophthora sojae in soybean. *Functional Plant Biology*, *46*(4), 304–313.

Cheng, Q., Dong, L., Gao, T., et al. (2018). The bHLH transcription factor GmPIB1 facilitates resistance to Phytophthora sojae in Glycine max. *Journal of Experimental Botany*, *69*(10), 2527–2541.

Cheng, Q., Li, N., Dong, L., et al. (2015). Overexpression of soybean isoflavone reductase (GmIFR) enhances resistance to Phytophthora sojae in soybean. *Frontiers in Plant Science*, *6*, 1024.

Cheng, Y., Ma, Q., Ren, H., et al. (2017). Fine mapping of a Phytophthora-resistance gene RpsWY in soybean (Glycine max L.) by high-throughput genome-wide sequencing. *Theoretical and Applied Genetics*, *130*(5), 1041–1051.

Dong, L., Cheng, Y., Wu, J., et al. (2015). Overexpression of GmERF5, a new member of the soybean EAR motif-containing ERF transcription factor, enhances resistance to Phytophthora sojae in soybean. *Journal of Experimental Botany*, *66*(9), 2635–2647.

Dorrance, A. E., Kurle, J., Robertson, A. E., et al. (2016). Pathotype diversity of Phytophthora sojae in eleven states in the United States. *Plant Disease*, *100*(7), 1429–1437.

Du, Q., Yang, X., Zhang, J., et al. (2018). Over-expression of the Pseudomonas syringae harpin-encoding gene hrpZm confers enhanced tolerance to Phytophthora root and stem rot in transgenic soybean. *Transgenic Research*, *27*(3), 277–288.

Fan, S., Jiang, L., Wu, J., et al. (2015). A novel pathogenesis-related class 10 protein Gly m 4l, increases resistance upon Phytophthora sojae infection in soybean (Glycine max [L.] Merr.). *PLoS One*, *10*(10), e0140364.

Gao, H., & Bhattacharyya, M. K. (2008). The soybean-Phytophthora resistance locus Rps1-k encompasses coiled coil-nucleotide binding-leucine rich repeat-like genes and repetitive sequences. *BMC Plant Biology*, *8*(1), 1–14.

Gao, H., Narayanan, N. N., Ellison, L., et al. (2005). Two classes of highly similar coiled coil-nucleotide binding-leucine rich repeat genes isolated from the Rps1-k locus encode Phytophthora resistance in soybean. *Molecular Plant-Microbe Interactions*, *18*(10), 1035–1045.

Grams, N., & Ospina-Giraldo, M. (2019). Increased expression of Phytophthora sojae genes encoding membrane-degrading enzymes appears to suggest an early onset of necrotrophy during Glycine max infection. *Fungal Genetics and Biology*, *133*, 103268.

Huang, J., Chen, L., Lu, X., et al. (2019). Natural allelic variations provide insights into host adaptation of Phytophthora avirulence effector PsAvr3c. *New Phytologist*, *221*(2), 1010–1022.

Huang, J., Gu, L., Zhang, Y., et al. (2017). An oomycete plant pathogen reprograms host pre-mRNA splicing to subvert immunity. *Nature Communications*, *8*(1), 1–15.

Jackson, T. A., Kirkpatrick, T. L., & Rupe, J. C. (2004). Races of Phytophthora sojae in Arkansas soybean fields and their effects on commonly grown soybean cultivars. *Plant Disease*, *88*(4), 345–351.

Jee, H. J., Kim, W. G., & Cho, W. D. (1998). Occurrence of Phytophthora root rot on soybean (Glycine max) and identification of the causal fungus. *RDA Journal of Crop Protection*, *40*(1), 16–22.

Jiang, B., Cheng, Y., Cai, Z., et al. (2020). Fine mapping of a Phytophthora-resistance locus RpsGZ in soybean using genotyping-by-sequencing. *BMC Genomics*, *21*(1), 1–11.

Jiang, C. J., Sugano, S., Kaga, A., Lee, S. S., Sugimoto, T., Takahashi, M., et al. (2017). Evaluation of resistance to Phytophthora sojae in soybean mini core collections using an improved assay system. *Phytopathology*, *107*, 216–223.

Jing, M., Guo, B., Li, H., et al. (2016). A Phytophthora sojae effector suppresses endoplasmic reticulum stress-mediated immunity by stabilizing plant binding immunoglobulin proteins. *Nature Communications*, 7(1), 1–17.

Kamoun, S., Furzer, O., Jones, J. D. G., et al. (2015). The top 10 oomycete pathogens in molecular plant pathology. *Molecular Plant Pathology*, *16*(4), 413–434.

Kang, I. J., Kang, S., Jang, I. H., et al. (2019). Identification of new isolates of Phytophthora sojae and the reactions of Korean soybean cultivars following hypocotyl inoculation. *The Plant Pathology Journal*, *35*(6), 698.

Kaufmann, M. J., & Gerdemann, J. W. (1958). Root and stem rot of soybean caused by Phytophthora sojae n. sp. [microform]. *Phytopathology*, *48*, 201–208.

Kong, L., Qiu, X., Kang, J., et al. (2017). A Phytophthora effector manipulates host histone acetylation and reprograms defense gene expression to promote infection. *Current Biology*, *27*(7), 981–991.

Li, L., Lin, F., Wang, W., et al. (2016). Fine mapping and candidate gene analysis of two loci conferring resistance to Phytophthora sojae in soybean. *Theoretical and Applied Genetics*, *129*(12), 2379–2386.

Li, Y., Sun, S., Zhong, C., et al. (2017). Genetic mapping and development of co-segregating markers of RpsQ, which provides resistance to Phytophthora sojae in soybean. *Theoretical and Applied Genetics*, *130*(6), 1223–1233.

Li, H., Wang, H., Jing, M., et al. (2018). A Phytophthora effector recruits a host cytoplasmic transacetylase into nuclear speckles to enhance plant susceptibility. *eLife*, 7, e40039.

Ma, Z., Song, T., Zhu, L., et al. (2015). A Phytophthora sojae glycoside hydrolase 12 proteins is a major virulence factor during soybean infection and is recognized as a PAMP. *The Plant Cell*, *27*(7), 2057–2072.

Mumby, M. C., & Walter, G. (1993). Protein serine/threonine phosphatases: Structure, regulation, and functions in cell growth. *Physiological Reviews*, *73*, 673–699.

Niu, J., Guo, N., Sun, J., et al. (2017). Fine mapping of a resistance gene RpsHN that controls Phytophthora sojae using recombinant inbred lines and secondary populations. *Frontiers in Plant Science*, *8*, 1–8.

Niu, L., Yang, J., Zhang, J., et al. (2019). Introduction of the harpin Xooc-encoding gene hrf2 in soybean enhances resistance against the oomycete pathogen Phytophthora sojae. *Transgenic Research*, *28*(2), 257–266.

Qin, J., Song, Q., Shi, A., et al. (2017). Genome-wide association mapping of resistance to Phytophthora sojae in a soybean [Glycine max (L.) Merr.] germplasm panel from maturity groups IV and V. *PLoS One*, *12*(9), e0184613.

Qiu, M., Li, Y. N., Ye, W. W., et al. (2021). A CRISPR/Cas9-mediated in situ complementation method for Phytophthora sojae mutants. *Molecular Plant Pathology*, *22*(3). https://doi.org/10.1111/mpp.13028.

Rizvi, S. S. A., & Yang, X. B. (1996). Fungi associated with soybean seedling disease in Iowa. *Plant Disease*, *80*(1), 57–60.

Sahoo, D. K., Abeysekara, N. S., Cianzio, S. R., et al. (2017). A novel Phytophthora sojae resistance Rps12 gene mapped to a genomic region that contains several Rps genes. *PLoS One*, *12*(1), e0169950.

Schmitthenner, A. F. (1985). Problem sand progress in control of Phytophthora root rot of soybean. *Plant Disease*, *69*(4), 362–368.

Sugimoto, T., Kato, M., Yoshida, S., et al. (2012). Pathogenic diversity of Phytophthora sojae and breeding strategies to develop Phytophthora-resistant soybeans. *Breeding Science*, *61*(5), 511–522.

Tampakaki, A. P., Skandalis, N., Gazi, A. D., Bastaki, M. N., Sarris, P. F., Charova, S. N., et al. (2010). Playing the "harp": Evolution of our understanding of hrp/hrc genes. *Annual Review of Phytopathology*, *48*, 347–370.

Tyler, B. M. (2007). Phytophthora sojae: Root rot pathogen of soybean and model oomycete. *Molecular Plant Pathology*, *8*(1), 1–8.

Wagner, A., Norris, S., Chatterjee, P., et al. (2018). Aquatic pseudomonads inhibit oomycete plant pathogens of Glycine max. *Frontiers in Microbiology*, *9*, 1007. https://doi.org/10.3389/fmicb.2018.01007.

Yang, J., Wang, X., Guo, B., et al. (2019). Polymorphism in natural alleles of the avirulence gene Avr1c is associated with the host adaptation of Phytophthora sojae. *Phytopathology Research*, *1*(1), 1–13.

Yang, J., Ye, W. W., Wang, X. M., et al. (2019). An improved method for the identification of soybean resistance to Phytophthora sojae applied to germplasm resources from the Huanghuaihai and Dongbei regions of China. *Plant Disease*, *104*(2), 408–413.

Yang, J., Zheng, S., Wang, X., et al. (2020). Identification of resistance genes to Phytophthora sojae in domestic soybean cultivars from China using particle bombardment. *Plant Disease*, *104*(7), 1888–1893.

Yao, H. Y., Wang, X. M., Wu, X. F., et al. (2010). Molecular mapping of Phytophthora resistance gene in soybean cultivar Zaoshu18. *Journal of Plant Genetic Resources*, *11*(2), 213–217.

Ye, W. W., Zheng, X. B., & Wang, Y. C. (2020). Research progress on key technologies for monitoring and control of soybean root rot. *Soybean Science*, *39*(5), 804–809.

Zhang, C., Gao, H., Li, R., et al. (2019). GmBTB/POZ, a novel BTB/POZ domain-containing nuclear protein, positively regulates the response of soybean to Phytophthora sojae infection. *Molecular Plant Pathology*, *20*(1), 78–91.

Zhang, Y., Huang, J., Ochola, S. O., et al. (2018). Functional analysis of PsAvr3c effector family from Phytophthora provides probes to dissect SKRP mediated plant susceptibility. *Frontiers in Plant Science*, *9*, 1105.

Zhang, C. Z., Gao, H., Sun, Y., et al. (2021). The BTB/POZ domain protein GmBTB/POZ promotes the ubiquitination and degradation of the soybean AP2/ERF-like transcription factor GmAP2 to regulate the defense response to Phytophthora sojae. *Journal of Experimental Botany*, *72*(22), 7891–7908.

Zhang, J., Sun, S., Wang, G., et al. (2014). Characterization of Phytophthora resistance in soybean cultivars/lines bred in Henan province. *Euphytica*, *196*(3), 375–384.

Zhang, J., Xia, C., Wang, X., et al. (2013). Genetic characterization and fine mapping of the novel Phytophthora resistance gene in a Chinese soybean cultivar. *Theoretical and Applied Genetics*, *126*(6), 1555–1561.

Zhong, C., Sun, S. L., Yao, L. L., et al. (2018a). Next-generation sequencing to identify candidate genes and develop diagnostic markers for a novel Phytophthora resistance gene, RpsHC18, in soybean. *Theoretical and Applied Genetics*, *131*(3), 525–538.

Zhong, C., Sun, S. L., Yao, L. L., et al. (2018b). Fine mapping and identification of a novel Phytophthora root rot resistance locus RpsZS18 on chromosome 2 in soybean. *Frontiers in Plant Science*, *9*, 1–14.

Zhong, C., Yao, L. L., Sun, S. L., et al. (2019). Genetic mapping and molecular characterization of a broad-spectrum Phytophthora sojae resistance gene in Chinese soybean. *International Journal of Molecular Sciences*, *20*(8), 1809.

Zhu, Z. D., Huo, Y. L., Wang, X. M., et al. (2006). Screening for resistance sources to Phytophthora root rot in soybean. *Journal of Plant Genetic Resources*, 7(1), 24–30.

Zhu, L. M., Zhou, Y., Li, X., et al. (2018). Metabolomics analysis of soybean hypocotyls in response to Phytophthora sojae infection. *Frontiers in Plant Science*, *9*, 1530.

CHAPTER FIVE

Insights into soybean with high photosynthetic efficiency

Qinzhen Xu[a], Xiao Liu[a], Chunyang Zhang[a], Weiguang Du[b], Yuefeng Guan[c], and Wenqiang Yang[a,d,e,f,*]

[a]Photosynthesis Research Center, Key Laboratory of Photobiology, Institute of Botany, Chinese Academy of Sciences, Beijing, China
[b]Soybean Research Institute, Heilongjiang Academy of Agricultural Sciences, Harbin, China
[c]FAFU-UCR Joint Center for Horticultural Biology and Metabolomics, Fujian Agriculture and Forestry University, Fuzhou, China
[d]College of Advanced Agricultural Sciences, University of Chinese Academy of Sciences, Beijing, China
[e]Innovative Academy of Seed Design, Chinese Academy of Sciences, Beijing, China
[f]Sino-Danish College, University of Chinese Academy of Sciences, Beijing, China
*Corresponding author: e-mail address: wqyang@ibcas.ac.cn

Contents

Abstract

Photosynthesis is the most important chemical reaction in the world. Improving photosynthetic efficiency has always been the most basic and logical way to increase crop yields. In recent years, many marvelous achievements have been made in rice and tobacco, but research on high photosynthetic efficiency in soybeans has stagnated. This chapter briefly introduces the principles of photosynthesis and the contribution of photosynthesis to soybean yield; reviews the progress of the photosynthates allocation within seeds and nodules, and the further effects of the photosynthates distribution on filling, yield and quality; reviews the contribution of the coordination of two essential elements of carbon and nitrogen between source and sink to soybean yield and nodules; summarizes the critical functions of the various of sugar transporters and key enzymes for carbon related metabolisms; summarizes the contribution of novel tools such as synthetic biology and phenomics to the improvements of soybean photosynthetic efficiency and prospects the potentials of combinations of multiple powerful tools to breed and cultivate soybean materials and varieties with excellent photosynthesis performance. We hope this review to provide new idea on soybean photosynthesis and shed light on the next green revolution of soybean industry.

Advances in Botanical Research, Volume 102
ISSN 0065-2296
https://doi.org/10.1016/bs.abr.2022.02.019

1. Soybean photosynthesis, yield, and seed quality

Since the industrial revolution, the combustion of a large amount of fossil fuel has a great impact on the world climate, mainly including the increase in CO_2 concentration, the destruction of the ozone layer, the increase in fine particulate, and the change in precipitation. In particular, the CO_2 concentration has doubled from 280 ppm before the industrial revolution to current 415 ppm. It was predicted that CO_2 concentration may rise to 900 ppm, which not only increases the temperature and accelerates the melting of glaciers, but also increases the frequency of extreme climate and severe disasters (Myers et al., 2014; Zhao et al., 2017). The continued increase in global atmospheric CO_2 concentration further affects the agricultural productivity, including yield and quality of crops, as well as the diversity and nutritional value of forages (Bhattarai et al., 2017; Bloom et al., 2010; Jones et al., 1984; Myers et al., 2014) and the yield of C_4 plants (maize and sorghum) did not increase as much as C_3 plants (soybean, wheat and rice) do (Long et al., 2005). Higher CO_2 concentrations as carbon fertilization stimulates photosynthesis and increases the ability of soybean leaves to photosynthesize (Campbell et al., 1988; Jancic et al., 2015). Compared to greenhouse-grown plants, field-grown plants maintain high photosynthetic capacity with long-term exposure to high CO_2 treatment (Arp, 1991). Soybean is a main source of edible plant protein (70%) and oil (30%) in the human diet and meets more than 25% of the world's food demand and animal feeding supply (Zhou et al., 2015). Soybean seeds contain roughly 37% protein, 30% carbohydrates (9% of which are fiber), 19% oil (84% unsaturated), 9% moisture, and 5% ash. Soybean is the only legume containing nine essential amino acids for human health. Thus, soybean protein is classified as a complete protein with high-quality. In addition, soybeans are also a good source of nutrients such as phosphorus, potassium, zinc, iron, B vitamins, vitamin A and antioxidant vitamins C and E (Source: USDA Nutrient Database for Standard Reference). With the rapid increase in population and the great potential in biomass energy, it is urgent to cultivate soybean varieties with high yields.

Photosynthesis occurs inside chloroplast, in which light reaction mainly occurs on thylakoid membrane with heterogeneous distribution of various lipids and protein complexes, namely photosystem I (PSI)—light harvesting complex I (LHCI), photosystem II (PSII)—light harvesting complex II (LHCII), cytochrome b_6f, ATP synthase, while the dark reaction is mainly carried out in the stroma (Johnson, 2016; Stirbet et al., 2020). The energy absorbed by the pigments excites electrons from water cleavage accompanied by oxygen production by oxygen evolution complex (OEC), and with protons accumulated in the lumen. By virtue of linear electron transfer flow, electrons are transferred to PSI and as reducing power finally accepted by $NADP^+$ to form NADPH. The differences of proton concentration and pH forms proton motive force (*pmf*), which drives proton back from the lumen side to the stroma side through ATP synthase with ATP from ADP generated as energy for further carbon fixation. When the excess light absorbed and electrons produced by photosystems fail to be consumed in time, photosynthetic organisms not only use a variety of alternative electron outlets (such as Mehler reaction, water-water cycle, and chlororespiration) to consume excess electrons, but also trigger a photoprotective mechanism non-photochemical quenching (NPQ) also known as heat dissipation. The NPQ mainly consists of violaxanthin de-epoxidase (VDE) cycle (qE), state transition (qT) and photoinhibition (qI) (Eberhard et al., 2008; Hohmann-Marriott & Blankenship, 2011). In the dark reaction, one molecule of CO_2 and ribulose 1,5-bisphosphate (RuBP) is combined to produce two molecules of 3-phosphoglycerate (3-PGA) under the catalysis of ribulose-1,5-bisphosphate carboxylase/oxygenase (RuBisCO), and then converted into glyceraldehyde 3-phosphate (GAP) by two molecules of ATP and NADPH. GAP is used to synthesize carbohydrates and other cellular components, such as starch and sucrose, to provide carbon skeleton for biogenesis of diversified carbon related compounds (Giordano et al., 2005; Johnson, 2016). GAP is captured by RuBP with one molecule of ATP as energy and utilized by CBBC (Calvin Benson Bassham cycle) within ten steps reaction, which is divided into three stages: RuBP carboxylation, C_3 product reduction and RuBP regeneration. It is worthy to mention that RuBisCO has the capabilities to react with CO_2 and O_2 (*i.e.*, photorespiration) depending on the concentration of CO_2 and also needs RuBisCO activase to control its enzyme activity. In addition, enzyme activities of RuBisCO can be regulated by light and four enzyme activities in CBBC are regulated via light mediated ferredoxin-thioredoxin system as well as the changes of environmental conditions (*e.g.*, magnesium, and pH).

About 90–95% of the plant's dry weight is the product of photosynthesis. Therefore, the main factors determining crop yield potential include light energy absorption, light energy transfer, light energy conversion efficiency, and the distribution and allocation of photosynthetic products. In the past half-century, the first agricultural "green revolution" has dramatically increased the yield potential of major crops through short stalks and increased harvest index. The harvest index of cereal crops has increased over time, from about 0.3 to 0.4–0.5, and even 0.6 at present. At the same time, through years of unremitting efforts, the light acceptance rate of modern crop varieties reach 0.8–0.9 (Long et al., 2015; Zhu et al., 2010). In theory, the light energy conversion efficiency of C_3 plants can reach 4.6%, and that of C_4 plants can reach 6% (Zhu et al., 2010). Using the energy theory to calculate the relative upper limit of the maximum yield of soybeans, the highest value of the theoretical light energy utilization rate is 1.5–2.4% (Wang & Tian, 2015). Soybean is a typical C_3 crop and large amount of ATP and NADPH are needed for carbon fixation to generate diversified organic compounds including proteins, lipids and carbohydrates. Therefore, soybean photosynthetic efficiency has a lot of room for improvement (Zhu et al., 2008, 2010).

Crop yield is a complex multi-factor trait, which is coordinately determined by multiple quantitative trait loci (QTL) or genes and many environmental cues (Evans, 2013). In general, total number of plants, number of pods, number of seeds in each pod, and the weight of seed are four main factors impacting the final yield, and some other factors like number of branches, the feature of leaf and canopy structure also influence the final yield (Liu et al., 2020). Different cultivation techniques and seed characteristics also indirectly affect the final yield (Liu et al., 2020). In theory, the ultimate source of biomass for plants is the reaction of RuBP and CO_2 catalyzed by RuBisCO, that is, the latter is fixed into carbohydrates through CBBC to form biomass, in which a certain proportion of carbohydrates are allocated to seeds as crop yields. The energy and reducing power sources of these processes are generated by light reactions accomplished by the four protein complexes with lipids and pigments on the thylakoid membrane within chloroplasts. Therefore, as shown in the classic crop genetic yield (Y) model: $Y = 0.487 \times S_t \times \varepsilon_i \times \varepsilon_c \times \varepsilon_p$, where S_t represents the total incident solar radiation obtained by the crop throughout the growing season; ε_i is the intercepted active radiation of crop that can be used for photosynthesis; ε_c represents the ratio of energy intercepted in the canopy into biomass energy, also known as conversion efficiency; ε_p represents the amount of total biomass energy allocated to harvested crops, is not only called distribution efficiency but also named harvest index (Monteith, 1977). The premise of this model is under ideal conditions with optimal cultivation method and lack of stress. It is found that in theory, photosynthesis will lose at least 80% of the intercepted light energy in the light reaction, which is mainly wasted by the narrow absorption spectrum and only 4.6–5% of the total light energy left for photosynthesis. Generally, crop photosynthesis and yield formation are dynamic, especially during the reproductive stage, the photosynthesis efficiency of leaves reaches a peak, and photosynthates at this time are mainly used for yield formation. Therefore, the level of photosynthates decisively depends on leaf area, the duration of photosynthesis and photosynthetic efficiency. Moreover, yield is also affected by the allocation of photosynthetic products and the rates of respiratory consumption. The formation of yield is the result of the interconnections of many internal and external factors affecting photosynthesis efficiency, so it is not surprising that the final yield formation becomes complicated. In fact, a large amount of existing research data showed that the bred varieties with increased yield are accompanied by the increased photosynthetic rate and higher photosynthetic efficiency, indicating the indeed positive correlations exist between photosynthetic efficiency and yield (Curtis et al., 1969; Peng et al., 1991; Zelitch, 1982).

The leaves play a decisive role in the photosynthesis of plants. Soybean leaves are compound, and the leaves will block each other after close planting, which cause poor light transmittance, low light energy interception rate, low light energy utilization rate, and less ventilation. The net photosynthetic rate of leaves is also a factor that affects soybean yield. The higher the net photosynthetic rate of soybean leaves in a day and the longer the duration, the more beneficial to soybean yield. The soybean yield is positively correlated to apparent canopy photosynthesis, especially the period of pod setting and physiological maturity of soybeans (Ashley & Boerma, 1989; Morrison et al., 1999). Improving the distribution of light energy in the canopy of plants, such as reducing the over-distributed light energy in the upper layer and increasing the under-distributed light energy in the lower layer, may increase the light energy utilization efficiency (Ort et al., 2015). At present, the effective distribution and utilization of light energy are mainly studied by changing the angle of different canopies, changing the contents of light-harvesting complexes in different layers, and extending the light absorption spectrum by application of special photosynthetic pigments in the lower leaves (Blankenship & Chen, 2013; Evans, 2013).

The homeostatic regulation of photosynthesis is important for increasing yield. The goal of photosynthesis homeostasis is to maximize the use of available light and minimize the adverse effects of excess light. Photorespiration is essential for oxygenated photosynthesis because it detoxifies and recycles

2-phosphoglycolate (2-PG), which is the main oxidation product catalyzed by RuBisCO, and couples nitrogen metabolism (Levey et al., 2019) (see Fig. 1A). In addition, photorespiration is a necessary alternative electron outlet when the light intensity is over the plants need (Levey et al., 2019). The regulation of photosynthesis is affected by the availability of carbon and nitrogen resources. When there is an imbalance between the source and the sink of the entire plants, the accumulation of carbohydrates in the leaves can lead to a decrease in the expression of photosynthetic genes. Furthermore, the plant hormones and nitrogen supply can also regulate the expression of photosynthesis related genes (Paul & Foyer, 2001). Although increasing plant carbon fixation (*e.g.*, CO_2-concentrating mechanism and carbon fertilization) can increase the potential growth rate and yield, this increase may be offset by a decrease in performance under different stresses (Driever & Kromdijk, 2013).

2. Allocation of photosynthates to major sinks in soybean: Seeds and nodules

The seed yield is determined by both the photosynthesis rate and the transportation of photosynthetic products (Plaut et al., 1987). Photosynthate allocation from source tissues to sink tissues has a huge impact on crop yields. The early stage of soybean grain filling is a critical period for the determination of soybean yield (Shiraiwa et al., 2004). During this period, the production and transportation of soybean photosynthetic products is the key to increase the accumulation of dry matter and the final grain weight of soybean seeds.

Plants transport fixed carbon in the form of sucrose, which is produced in mesophyll cells and uploaded into phloem cells (Chen et al., 2012). In higher plants, sucrose is instantaneously released into apoplasts, and its intercellular transport depends on the activity of the energy-dependent proton-sucrose symporter that is positioned on the plasma membrane (Sauer, 2007). Most plants actively load sucrose into the sieve element-companion cell (SE-CC) complex of the leaves (Regmi et al., 2016) (see Fig. 1C), and three transporter families have been identified as key players in sugar transport: MSTs (monosaccharide transporters), SUTs (sucrose transporters), and SWEETs (hexose and sucrose transporters) (Chen et al., 2015; Saddhe et al., 2021).

According to different substrates and localizations, MSTs can be divided into seven subfamilies, including sugar transporter proteins (STPs),

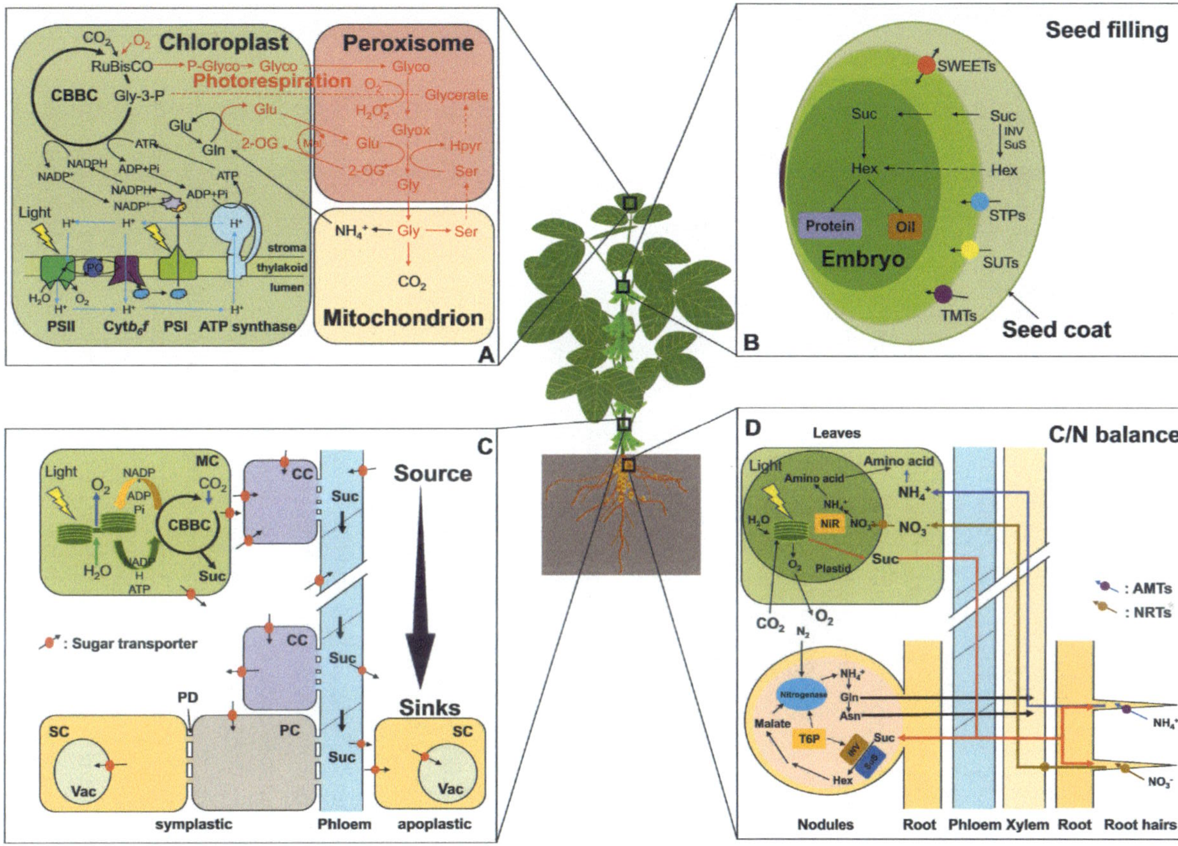

Fig. 1 Effects of photosynthesis on soybean growth and yield. (A) Photorespiration is critical for plant nitrogen utilization and amino acid synthesis. (B) The effect of photosynthetic products on grain filling. (C) The relationship among source, sinks, and flow of photosynthates. (D) The effect of photosynthesis on plant C/N balance. Abbreviations: Suc, Sucrose; Hex, Hexoses; INV, Invertase; SUS, Sucrose Synthase; T6P, Trehalose 6-Phosphate; NiR, Nitrite Reductase; AMT, Ammonium Transporter; NRT, Nitrate Transporter; MC, Mesophyll Cell; Vac, Vacuole; PC, Parenchyma Cell; SC, Sink Cell; PD, Plasmodesmata; CC, Companion Cell. Red arrows represent the transport and distribution of the photosynthates.

polyol/monosaccharide transporters (PMTs), inositol transporters (INTs), vacuolar glucose transporters (VGTs), tonoplast membrane transporters (TMTs), plastidic glucose transporters (pGlcTs), and some unnamed transporters acquired by suppressor screen, including the suppressor of G protein beta 1 (SGB1) and the early-responsive to dehydration six-like (ESLs) (Doidy et al., 2012). VGTs and TMTs act as sugar/H^+ antiporter to load sugar into vacuoles (Cho et al., 2010; Schulz et al., 2011; Wingenter et al., 2010), while ESLs are likely involved in the outflow of sugar from vacuoles (Poschet et al., 2011), and are also responsible for the reuse of stored sugar in response to abiotic or biotic stress (Wormit et al., 2006). pGlcTs contribute to the efflux of sugars from plastids (Doidy et al., 2012). INTs are located in the plasma membrane and vacuolar membrane of the cell and responsible for inositol transportation (Schneider et al., 2008). STPs serve as H^+-hexose symporter to transport glucose, fructose and galactose (Buttner, 2010; Liu et al., 2021; Paulsen et al., 2019), which are located in the plasma membrane of sink and source tissues. STPs also regulate the supply of sugars needed for the development and germination of pollen (Schneidereit et al., 2003; Szenthe et al., 2007; Ylstra et al., 1998). PMTs are broad- spectrum proton symporters involved in transporting polyols, hexoses and/or pentoses (Dusotoit-Coucaud et al., 2010; Klepek et al., 2010). Monosaccharide transporters also provide energy needed for cell division and growth in the early stages of berry development (Vignault et al., 2005; Zhang et al., 2005) and the development of fruit seeds (Wang et al., 2007). Monosaccharide transporter play important roles in both lower and higher plants (Büttner, 2007; Slewinski, 2011), and they functions in the accurate allocation of assimilated carbon between different organs (Büttner & Sauer, 2000) and the effective allocation and utilization of sugars (Komaitis et al., 2020). However, all of the transporters described above have not yet been characterized in soybean.

A series of SWEET transporters (Patil et al., 2015) have been extensively reported functioning in the regulation of sugar distribution between source and sink (Durand et al., 2018). SWEET-mediated sucrose efflux is necessary for soybean seed development (Sosso et al., 2015; Yang et al., 2018). The plasma membrane localized GmSWEET15 mediates the export of sucrose from the endosperm to the embryo essential for further development (Wang et al., 2019a). GmSWEET10a regulates simultaneous increase in size and oil content and a decrease in protein content in soybean seed, while GmSWEET10b further improves the yield and seed quality of modern soybean varieties (Miao et al., 2020; Wang et al., 2020a) (see Fig. 1B).

SUTs are not only involved in the regulation of leaf sugar content and stress signal transduction (Scofield et al., 2007), but also involved in phloem loading, bacterial infection responses, and sugar accumulation (Leach et al., 2017; Marco et al., 2021; Qazi et al., 2012; Xu et al., 2018). The down-regulation and up-regulation of phloem load depends on the transcriptional and post-transcriptional regulation of SUTs, respectively (Scofield et al., 2007; Xu et al., 2018). Other sugar transporters, such as PsSUF1, and PsSUF4 in peas (*Pisum sativum* L.), and PvSUT1and PvSUF1 in beans (*Phaseolus vulgaris* L.), function in the sucrose outflow of seed coats (Zhou et al., 2007). GmSUT1 plays a role in sucrose transport to nodules to promote soybean nodulation (Deng et al., 2021). However, the function of the remaining other SUTs remain to be elucidated.

The utilization of sucrose in sink cells mainly depends on invertase (INV) and sucrose synthase (SuS). Invertase, which hydrolyzes sucrose into glucose and fructose, plays multiple key roles in plant fruit development, regulation of seed quality and development (Jin et al., 2009; Qin et al., 2016; Ruan et al., 2010). INVs occur as distinct isoforms in vacuolar (VIN, vacuolar invertase), cell wall (CWIN, cell wall invertase) and cytosol (CIN, cytosol invertase). INVs are also classified as acid INV (VIN and CWIN) and neutral/alkaline INV (CIN) (Ruan, 2014). Among the three types of invertases, CWIN is responsible for seed filling, and the enhancement of CWIN activity increases crop yield (Li et al., 2013; Tang et al., 2017; Wang et al., 2008; Wang & Ruan, 2012). CWINs co-evolve with the vasculature and may serve as phloem uninstalled functional component (Wan et al., 2018). Overexpression of CIN or VIN can also increase seed oil accumulation and promote fruit ripening, respectively (Qin et al., 2016; Tomlinson et al., 2004). Invertase can regulate many aspects of plant growth and development, from gene expression to long-distance nutrients transportation (Roitsch & González, 2004). For instance, OsVIN2 affects sucrose metabolism, including sugar composition, conversion, and transportation from source to sink (Xu et al., 2019). SuS, which degrades sucrose into UDP-glucose and fructose, is involved in the uploading of sucrose into the phloem (Yao et al., 2020), and closely related to cellulose synthesis and secondary cell wall formation (Coleman et al., 2009). During seed development, SuS is located in the companion cells and may provide energy for phloem transport activities (Fallahi et al., 2008).

Distribution of sugar is regulated by sugar metabolisms and signaling. Trehalose-6-phosphate (T6P), a central sugar signal in plants (Fichtner et al., 2021; Fichtner & Lunn, 2021), shows emerging significance for the utilization and distribution of sucrose, as well as growth and development (Griffiths et al., 2016). Therefore, effective chemical intervention in sugar signals can be used to change the balance of source-sink and potentially increase yields. Throughout the life cycle of a plant, starch plays a dual role in carbon allocation, not only as a carbon source, releasing carbon reserves in leaves for growth and development, but also as a sink in seeds and tubers. Therefore, the presence of starch in tissues and organs has a profound impact on the physiology of growing plants, and starch metabolism affects the yield and quality of crops (MacNeill et al., 2017). Disruption of AGPase enzymes in rice (Ohdan et al., 2005) and potato (Tiessen et al., 2002) causes the stimulation of starch synthesis and the reduction of glycolysis intermediates levels, resulting in the decreased crop yields. In addition, the glycolytic pathway is considered to be essential for oil production during the development of Arabidopsis and oil crop seeds (Andre et al., 2007). Phosphoenolpyruvate, the one of the intermediates in the glycolysis pathway, is the first choice for the synthesis of fatty acids inside the plastids.

As a legume, soybean fixes atmospheric nitrogen gas (N_2) via rhizobia within root nodules. The process requires a large amount of ATP as energy and carbon (C) skeletons generated through photosynthate from the host plants (Libault, 2014; Minchin & Witty, 2005; Prell & Poole, 2006). For instance, *Vigna unguiculata* nodules consume 9% of the net photosynthates produced by plants (Coker & Schubert, 1981). As revealed by ^{14}C quantification, the inoculated roots and nodules gradually have priority of photosynthates distribution and allocation after the emergence of nodules (Vance, 1991) (see Fig. 1D).

Photosynthates are transported to the nodules in the form of sucrose and utilized by SuS and INV enzymes (Flemetakis et al., 2006; Morell & Copeland, 1984; Peiter & Schubert, 2003). Following sucrose catabolism, hexose can be metabolized by glycolysis to produce phosphoenolpyruvate (PEP), which is captured by phosphoenolpyruvate carboxylase (PEPC) to produce oxaloacetate, and further converted into malate by malate dehydrogenase (MDH) (Udvardi & Poole, 2013). It is widely accepted that C_4-dicarboxylic acids like malate, rather than sugars, are the major carbon sources transported to bacteroids (Udvardi & Poole, 2013). The transported malate can be used as carbon source for bacteroid consumption, substrates for the TCA (Tricarboxylic acid) cycle and ammonia assimilation in the Glutamine Synthetase/Glutamine: 2-Oxoglutarate Aminotransferase (the other name is Glutamine oxoglutarate aminotransferase) (GS/GOGAT) cycle (Prell & Poole, 2006; Streeter, 1980). In bacteroids, at the expense of 16 ATP per mole, nitrogenase fixes N_2 to synthetize ammonia. The formula is as following: $\mathbf{N_2 + 8H^+ + 8e^- + 16ATP \rightarrow 2NH_3 + H_2 + 16ADP + 16Pi}$. In the nodules, ammonium is assimilated into glutamine by the combined activity of the enzymes GS/GOGAT pathway (Barsch et al., 2006; Patriarca et al., 2002). Asparagine can be produced from the assimilated glutamine via the concerted activity of glutamine-dependent asparagine synthetase (ASNS) and aspartate aminotransferase (AST). In soybean, a major part of glutamine and asparagine is used for purine synthesis, and then exported to the host plants as ureides in the form of allantoin and allantoate (Prell & Poole, 2006; Streeter, 1980).

Legumes provide the bacteroids with a significant amount of photosynthates, making nodules a vital sink tissue. Cessation of photosynthates supply by decapitation or phloem-girdling can significantly reduce nodule activity. When *Lotus japonicus* plants are exposed to prolonged darkness, photosynthesis deficiency leads to diminished nitrogenase activity and affects many metabolic processes in the nodules (Tsikou et al., 2013). Accordingly, the nodule metabolic genes are down-regulated and the reduced glycolysis result in lower organic acid levels, especially of malate (Tsikou et al., 2013). Since nodule activity is limited by photosynthate supply, it is possible to improve nitrogen fixation by increasing photosynthesis. For instance, nodules of pea grown at increased CO_2 shows a higher sugar accumulation and enhancement of nodule carbon metabolism activities (Cabrerizo et al., 2001). Nevertheless, the specific nitrogen fixation activity is not significantly enhanced, indicating other limiting factors may exist besides carbon source availability (Cabrerizo et al., 2001). Photosynthates and light signals are necessary for nodule development and symbiotic nitrogen fixation of legumes. Recently, the CCaMK-STF-FT module integrates light signals above-ground and symbiotic nitrogen fixation signals underground to regulate the formation of nodules (Wang et al., 2021).

3. Breeding of soybean cultivar with high photosynthetic efficiency

Photosynthetic efficiency improvements have been proved to be critical and feasible in soybean breeding. One potential target of improving photosynthetic efficiency is to optimize light reaction in photosynthesis (Blankenship et al., 2011; Long et al., 2006), mainly to optimize the light capture ability, the electron flow in the electron transport chain and improve the rate of recovery from photoinhibition and the capabilities of photoprotection (Bailey-Serres et al., 2019; Evans, 2013; Johnson, 2016; Stirbet et al., 2020). Light energy is converted into intermediate chemical energy as ATP and NADPH through photosynthetic pigments and four supercomplexes (PSI, PSII, cytochrome b_6f, and ATP synthase) present in chloroplast thylakoid membrane (Allen et al., 2011; Hohmann-Marriott & Blankenship, 2011; Nelson & Ben-Shem, 2004; Silva et al., 2015). This process can be divided into sequential events including light absorption, electron transfer (mainly linear electron transfer and cycle electron transfer), and energy and reducing equivalents formation, which is also the source of energy for survival (Slattery & Ort, 2015; Zhu et al., 2008). Plants generally use chlorophyll *a*, *b* and carotenoids to absorb light, but these two types of pigments can only absorb the red and blue parts of visible light (Eberhard et al., 2008). A large amount of light energy is reflected and lost, which reduces light energy utilization efficiency (Arp et al., 2020). Recently, the regulation and utilization of the yellow green light is reconsidered to increase crop yield and quality (Zhang & Folta, 2012). In addition, other chlorophyll with broader spectrum absorbance (such as chlorophyll *f*; Ho et al., 2016; Nürnberg et al., 2018) and bacterial chlorophylls have great potentials to promote the absorption efficiency of light with broader wavelength (Kramer et al., 2004; Wang et al., 2019b). The chlorophylls are essential for optimizing crop photosynthesis and ensuring leaf development (Shin et al., 2020; Wang et al., 2020b; Wang & Grimm, 2021). Several studies have reported that reducing the biosynthesis and metabolism of chlorophyll and other pigments can increase yield through modulating light harvesting complex (Walker et al., 2018; Wang & Grimm, 2015). Different light such as high light, fluctuating light, and short wavelength light generally brings various of effects to photosynthetic proteins, including D1 (PsbA encoded protein), PetD (Subunit IV of cytochrome b6 f complex), NDHK (NAD(P)H-quinone oxidoreductase subunit K), ZEP (zeaxanthin epoxidase) and PGRL1 (Proton gradient regulation 5-like 1), and eventually causes photoinhibition, photodamage, the failure of energy conversion (Li et al., 2018; Lu, 2016; Shimakawa & Miyake, 2019), and reduced photosynthetic efficiency (Eberhard et al., 2008). Photoprotection ability (Lu, 2016) can be optimized to improve the photosynthetic efficiency, such as the rapid repair of D1 in photosystem II (Ahmad et al., 2020; Chen et al., 2020; Xu et al., 2021), the rapid recovery of NPQ mediated by PsbS (Glowacka et al., 2018; Kromdijk et al., 2016). The more efficient electron transfer and strong VDE activity brought via overexpression of *petC* (*Rieske iron-sulfur subunit of the cytochrome b_6f complex*) will reduce the energy loss caused by photoinhibition (Simkin et al., 2017). As an auxiliary factor of light energy conversion, light harvesting antenna not only assist the transmission of electrons among and within the two photosystems, but also facilitates the protection of the electron chain (Dall'Osto et al., 2015; Shin et al., 2016). Optimizing the size and ratio of light harvesting antenna was proofed to be effective to improve the electron transfer efficiency and photoprotection ability to a certain extent (Beckmann et al., 2009; Jin et al., 2016; Kirst et al., 2017; Song et al., 2017). Plants with engineered photoreceptors such as phytochromes and phototropins (PHOT1 and PHOT2) exhibit better growth and improved biomass accumulation under light-limited conditions (Hart et al., 2019). In addition, cyclic electron transfer, water-water cycle and other pathways are evolved as outlets to consume the excess electrons and maintain the ratio of ATP to NADPH (Endo et al., 2014; Ort et al., 2015). In addition, overexpression of *PsbS* (*Photosystem II Subunit S*), *petC*, and *psbA*, were applied to optimize the efficiency of photosynthesis (Kohzuma et al., 2017).

The dark reaction of photosynthesis (Stirbet et al., 2020) is a series of reactions that use ATP and NADPH produced by the light reaction to fix CO_2 to produce organic matter through three sequential processes (Johnson, 2016). As the material source of the biosphere, photosynthesis has evolved three main carbon fixation pathways including C_3, C_4 and crassulacean acid metabolism (CAM) (Brautigam et al., 2017; Hennacy & Jonikas, 2020; Sage, 2017), and some intermediate types between C_3 and C_4, and even more special carbon fixation patterns. At present, the attempts to improve dark reaction include changing leaf stomata to optimize CO_2 conductivity, remolding and modification of RuBisCO (Evans & Clarke, 2019; Young et al., 2016), increasing RuBP regeneration rate, and rerouting photorespiration (Bailey-Serres et al., 2019; Weber & Bar-Even, 2019). Most of the plants transport CO_2 to mesophyll cells through stomata for carbon fixation. The opening and closure of stomata not only affects CO_2 entry and further photosynthetic efficiency (Hsu et al., 2018), but also affects transpiration, water use efficiency (WUE) (Gersony et al., 2020; Papanatsiou et al., 2019). Therefore, regulation of the stomatal opening and closure and increasing the CO_2 conduction rate inside mesophyll cells will be undoubtedly valid to improve photosynthetic efficiency (Hu et al., 2010; Papanatsiou et al., 2019). During CO_2 carboxylation in the chloroplast, the concentration of CO_2 in most of C_3 plants is lower than that in C_4 and CAM plants because of the special CO_2 concentrating mechanisms in C_4 and CAM plants. Interestingly, almost all of the enzymes required by C_4 and CAM plants to concentrate CO_2 also exist in

C_3 plants with unknown functions (Giordano et al., 2005; Hibberd & Quick, 2002; Schluter & Weber, 2020). These enzymes could be activated to increase the CO_2 concentration and enhance the carbon assimilation rate under some special conditions (Sage, 2017). The most abundant protein in C_3 plant leaf cells is RuBisCO (up to 50%), the most important enzyme in carbon reaction (Evans & Clarke, 2019). Among the four types of RuBisCO found in photosynthetic organisms, mainly type I functions in autotrophs organisms. Remolding and utilization of the excellent feature of other types of RuBisCOs have been attempted to increase photosynthetic efficiency (Lin et al., 2014; Sharwood, 2017). In addition, by analyzing the process of assembly and regulation (Ashida et al., 2019; Ishikawa et al., 2011; Matsumura et al., 2020; Xia et al., 2020), RuBisCO can be modified (Degen et al., 2020; Lin et al., 2014), with optimized efficiency (Orr et al., 2016), which significantly improves the carbon fixation efficiency of plants (Bracher et al., 2017; Laterre et al., 2017; Wilson et al., 2016; Yeates & Wheatley, 2017). However, the carboxylation efficiency of RuBisCO is always affected by the concentration of CO_2 and photorespiration occurs with decreased CO_2 (Eisenhut et al., 2019; Sharwood et al., 2016; Whitney et al., 2011). Although photorespiration consumes the energy and reducing power generated from light reaction (Bauwe et al., 2010; Fernie & Bauwe, 2020; Foyer et al., 2009), the process plays essential functions in primary metabolisms, including nitrogen assimilation, sulfur metabolism and amino acids biogenesis, and connects the metabolites exchange between chloroplast, mitochondria and peroxisomes (Hodges et al., 2016; Maurino & Peterhansel, 2010; Ogren, 1984; Orf et al., 2016). Photorespiration is a fundamental and necessary process for plants to deal with stresses and balance carbon and nitrogen metabolisms (Eisenhut et al., 2019; Foyer et al., 2009; Peterhansel et al., 2013; Rachmilevitch et al., 2004). Recently, through the construction of bypass pathways to generate more CO_2 during photorespiration improved the biomass of tobacco (Simkin et al., 2015), and increase the yield in rice to some extent (Abadie et al., 2016; Betti et al., 2016; Hagemann & Bauwe, 2016). Besides, sedoheptulose-1,7-bisphosphatase (SBPase) as a key and rate-limiting enzyme of CBBC also affects carbon assimilation (Driever et al., 2017; Kohler et al., 2017; Lefebvre et al., 2005), and overexpression of this enzyme in maize was shown to improve the carbon assimilation efficiency and increase the biomass (Evans & Clarke, 2019). Meanwhile, except the CBBC to fix CO_2, multiple active carbon fixation pathways exist in nature, which are also potential targets to improve the efficiency of carbon assimilation (Bar-Even et al., 2010; Ducat & Silver, 2012). Soybeans with relatively high qualities and yields are needed to adapt to the increasing global CO_2 concentration (Fischinger et al., 2010). However, all the attempts mentioned above have not been tested in soybean. It is necessary to cultivate varieties with less CO_2 insensitivity or with stronger CO_2 adaptation capabilities to achieve the idealized balance among light reaction, carbon fixation and photorespiration.

Photosynthesis takes place in the chloroplasts, so metabolic processes occurring inside the chloroplasts will affect photosynthetic efficiency, especially the efficient coordinated coupling of photosynthesis light reaction and dark reaction, and active connections between materials and energy (Kaiser et al., 2019). The ratio of NADPH to ATP is critical to maintain metabolisms for regular plant growth, while the balance of materials and energy will be rescheduled because of the unsound ratio of NADPH to ATP under adverse conditions (Kramer et al., 2004). In addition, the reducing power from light reactions can also be utilized for the essential metabolisms occurring within chloroplasts, such as nitrogen assimilation, sulfur assimilation, fatty acid biosynthesis, phytohormone and their precursor's biosynthesis, and chlorophyll biosynthesis. These processes mainly occur within chloroplasts and are also regulated by the correct chloroplast redox state. Theoretically, soybean yields can be increased through improving chloroplasts performance to better deal with stresses, especially drought and salinity. In addition, the efforts to increase soybean pests and diseases resistance are also potential targets for indirect improvement of photosynthetic efficiency and further yield (Zhou & Zhang, 2020). Many soybean mutants with yellow foliar phenotype and impaired photosynthesis (Sandhu et al., 2018) have been characterized, and the genes have been cloned and identified, such as *PsbP* (*Photosystem II Oxygen Evolution Enhancer Protein*), *Tic110* (*Transposon at the inner envelope membrane of chloroplasts 110*) (Sandhu et al., 2016), *CD-5* (*Chlorophyll-deficient 5*), *CD1* (*Chlorophyll-deficient 1*), and *GmORRM1* (*Organelle RNA Recognition Motif-Containing*) (Zhu et al., 2020). The study on these materials will help for revealing the critical photosynthetic elements and related processes but also open up novel ways to engineer soybean chloroplasts to increase yield. The breakthrough of excellent genetic tools such as chloroplast transformation and genome editing in the near future, will definitely promote better understanding of the regulatory mechanisms during photosynthesis to improve crop quality and yields (Kang et al., 2021).

With the thorough researches on photosynthesis (Bar-On & Milo, 2019), it is found that the distribution of photosynthates and the photosynthesis of organs other than leaves can affect the photosynthetic efficiency (Faralli & Lawson, 2020; Schwander et al., 2016; Simkin et al., 2019; van Bezouw et al., 2019), especially the optimization of the source-sink-flow model play vital roles to increase the coupling efficiency of energy and matter to form the final yield (Araus et al., 2021; Sugiura et al., 2019). However, the detailed regulatory mechanisms of the photosynthates distribution and the contributions of the non-photosynthetic organs is still need to be elucidated (Venegas-Molina et al., 2021). The development of various of omics technologies (Liu et al., 2020; Lloyd & Lister, 2022), especially phenomics can comprehensively display the outcome of traits brought by intricate interactions between genes and environments. The combination of multi-omics will be powerful to dissect the genetic regulatory network of complex traits responsible for plant photosynthetic efficiency and yield (Oellrich et al., 2015; Pieruschka & Schurr, 2019; Topp et al., 2013; van Eeuwijk et al., 2010; Yang et al., 2014, 2020; Zhao et al., 2019) (see Fig. 2A). Multi-omics can be used to study on the phenotypic changes determined by the interaction between genetic regulatory networks and the environments. Then to analyze them from multiple aspects such as genes, proteins and metabolites, and comprehensively analyze the relationship among genotype, phenotype and environments. Genome wide association study (GWAS) based excavation of the critical genes facilitates the analyses the regulatory networks and verifies the function of related elements (Liang et al., 2021). Advanced algorithms such as machine learning and deep learning are used to conduct deeper and more comprehensive research and analyses on the obtained data, and finally are applied to cultivate soybean varieties with high photosynthetic efficiency and yield potentials. Machine learning has been proven to be a powerful tool to excavate the biological mechanisms hidden inside the data and greatly improve the understanding of soybean photosynthesis and other yield related biological processes (Mahood et al., 2020; Walley et al., 2016; Wang, 2020; Wang et al., 2020c; Zhang et al., 2019) (see Fig. 2A). Since the reference genome of soybean variety Williams 82 has been published in 2010 (Schmutz et al., 2010), many reference genomes, such as Zhonghuang 13 and W05, are released. These references genome together with the pan-genome of soybean will accelerate the functional studies of the critical genes involved in the regulatory mechanisms of excellent photosynthesis performance and facilitate the

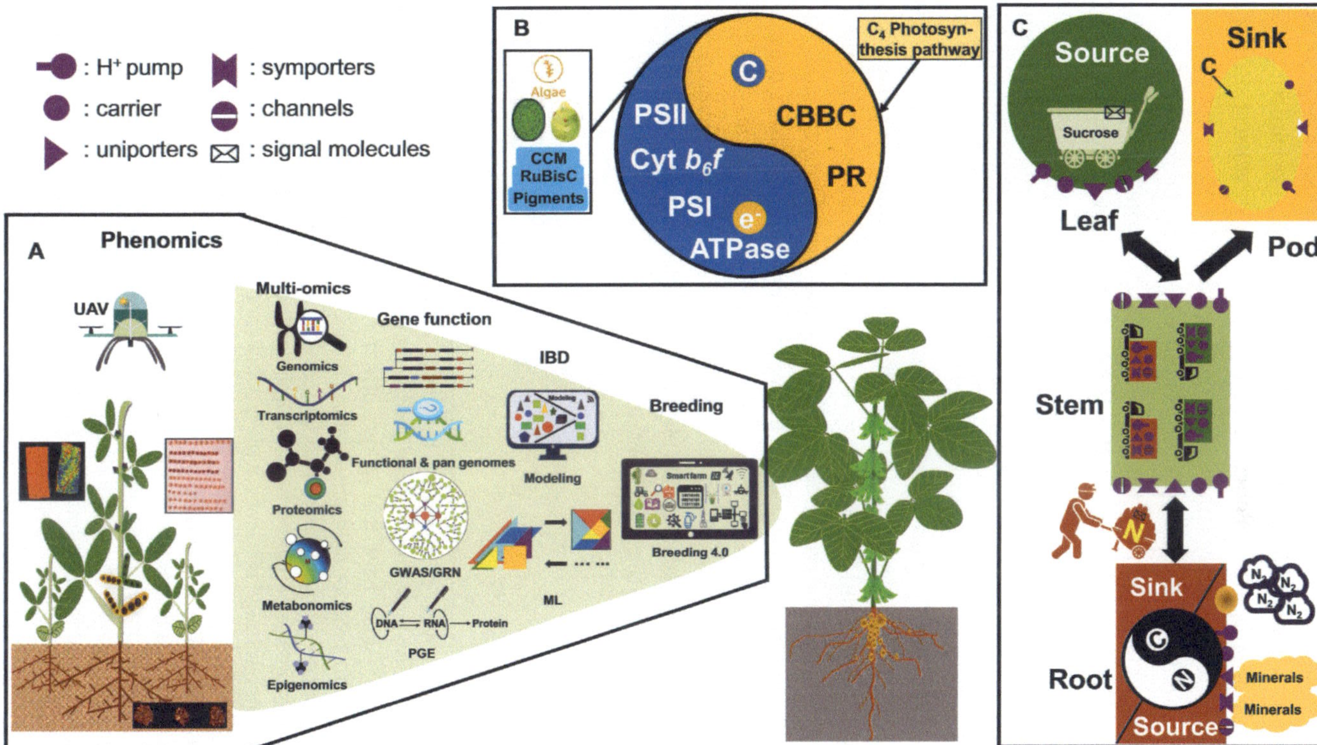

Fig. 2 See figure legend on next page.

future soybean breeding (Bayer et al., 2020; Li et al., 2014; Long et al., 2015; Shen et al., 2018, 2019). Together with multi-omics, bioinformatics and system biology will help to analyze the principles of photosynthetic efficiency. The rapid development of genome editing methods (Shivram et al., 2020) will be applied to inevitably enhance soybean production efficiency (Maurino & Weber, 2013). At present, synthetic biology has achieved to introduce several elementary components of the efficient algal CO_2 concentration mechanism (Cameron et al., 2013; Kubis & Bar-Even, 2019; Lopez-Calcagno et al., 2019; Yu et al., 2018) and build up photorespiration bypasses (Peterhansel et al., 2013; Wang et al., 2020d) to enrich the CO_2 concentration and improve carbon assimilation ability (South et al., 2019) within chloroplast (see Fig. 2B). The combination of multiple methods (Basu et al., 2019; Simkin et al., 2015) will obtain a large amount of data and information (Liu & Yan, 2019; Wallace et al., 2018), which allow us to better and more comprehensively understand the regulatory mechanisms of photosynthesis and thus to cultivate the better environmentally suitable soybean (Song et al., 2020; Yang et al., 2021) (see Fig. 2C). With the widespread and increasing usage of the important technologies (Fernie et al., 2021; Harbinson et al., 2021; Liu & Yan, 2019), soybean breeding with higher photosynthetic efficiency and suitable for environmental changes (Hay et al., 2017), can meet the global demand for soybeans in the future (Liu et al., 2020).

Fig. 2 Perspectives of soybean breeding with high photosynthetic efficiency. (A) Comprehensive tools to facilitate soybean breeding with optimized photosynthetic efficiency. (B) Algae have the various of natural photosynthetic elements such as pigments with broader light absorbance, RuBisCOs with better catalyzing activities, and CO_2 concentrating mechanism (CCM), which can be applied to modify the chloroplast to execute better photosynthesis. Optimization of soybean photosynthesis can be accomplished through efficient coupling of light and dark reactions, characterization of native C_4 pathways in soybean or introduction of new components of C_4 pathways. (C) The reasonable relationship and homeostasis among source, sink and flow is responsible and a guarantee for the distribution, allocation and demands of photosynthates. The photosynthates generated by leaves are transported by different transporters to satisfy the demands from seeds and roots, especially the demands of nodules and nutrients absorption. Abbreviations: UAV, Unmanned Aerial Vehicle; GWAS, Genome Wide Association Study; GRN, Gene Regulation Network; PGE, Precision Genome Editing; IBD, Interactome Big Data; ML, Machine Learning; C, Carbon; PR, Photorespiration; CBBC, Calvin- Benson-Bassham Cycle.

References

Abadie, C., et al. (2016). *In vivo* stoichiometry of photorespiratory metabolism. *Nature Plants*, *2*, 15220–15224. https://doi.org/10.1038/nplants.2015.220.

Ahmad, N., et al. (2020). Alternative routes to improving photosynthesis in field crops. *Trends in Plant Science*, *25*(10), 958–960. https://doi.org/10.1016/j.tplants.2020.07.003.

Allen, J. F., et al. (2011). A structural phylogenetic map for chloroplast photosynthesis. *Trends in Plant Science*, *16*(12), 645–655. https://doi.org/10.1016/j.tplants.2011.10.004.

Andre, C., et al. (2007). A heteromeric plastidic pyruvate kinase complex involved in seed oil biosynthesis in *Arabidopsis*. *Plant Cell*, *19*(6), 2006–2022. https://doi.org/10.1105/tpc.106.048629.

Araus, J. L., et al. (2021). Improving crop yield and resilience through photosynthesis optimisation panacea or pipe dream. *Journal of Experimental Botany*, *72*, 3936–3955. https://doi.org/10.1093/jxb/erab097.

Arp, W. J. (1991). Effects of source-sink relations on photosynthetic acclimation to elevated CO_2. *Plant, Cell & Environment*, *14*, 869–875. https://doi.org/10.1111/j.1365-3040.1991.tb01450.x.

Arp, T. B., et al. (2020). Quieting a noisy antenna reproduces photosynthetic light-harvesting spectra. *Science*, *368*, 1490–1495. https://doi.org/10.1126/science.aba6630.

Ashida, H., et al. (2019). Learning RuBisCO's birth and subsequent environmental adaptation. *Biochemical Society Transactions*, *47*(1), 179–185. https://doi.org/10.1042/BST20180449.

Ashley, D. A., & Boerma, H. R. (1989). Canopy photosynthesis and its association with seed yield in advanced generations of a soybean cross. *Crop Science*, *29*(4), 1042–1045.

Bailey-Serres, J., et al. (2019). Genetic strategies for improving crop yields. *Nature*, *575*, 109–118. https://doi.org/10.1038/s41586-019-1679-0.

Bar-Even, A., et al. (2010). Design and analysis of synthetic carbon fixation pathways. *Proceedings of the National Academy of Sciences of the United States of America*, *107*(19), 8889–8894. https://doi.org/10.1073/pnas.0907176107.

Bar-On, Y. M., & Milo, R. (2019). The biomass composition of the oceans: A blueprint of our blue planet. *Cell*, *179*(7), 1451–1454. https://doi.org/10.1016/j.cell.2019.11.018.

Barsch, A., et al. (2006). GC-MS based metabolite profiling implies three interdependent ways of ammonium assimilation in Medicago truncatula root nodules. *Journal of Biotechnology*, *127*(1), 79–83.

Basu, U., et al. (2019). Genetic dissection of photosynthetic efficiency traits for enhancing seed yield in chickpea. *Plant, Cell & Environment*, *42*(1), 158–173. https://doi.org/10.1111/pce.13319.

Bauwe, H., et al. (2010). Photorespiration: Players, partners and origin. *Trends in Plant Science*, *15*(6), 330–336. https://doi.org/10.1016/j.tplants.2010.03.006.

Bayer, P. E., et al. (2020). Plant pan-genomes are the new reference. *Nature Plants*, *6*(8), 914–920. https://doi.org/10.1038/s41477-020-0733-0.

Beckmann, J., et al. (2009). Improvement of light to biomass conversion by de-regulation of light-harvesting protein translation in *Chlamydomonas reinhardtii*. *Journal of Biotechnology*, *142*(1), 70–77. https://doi.org/10.1016/j.jbiotec.2009.02.015.

Betti, M., et al. (2016). Manipulating photorespiration to increase plant productivity: Recent advances and perspectives for crop improvement. *Journal of Experimental Botany*, *67*(10), 2977–2988. https://doi.org/10.1093/jxb/erw076.

Bhattarai, M. D., et al. (2017). Projecting corn and soybeans yields under climate change in a Corn Belt watershed. *Agricultural Systems*, *152*, 90–99. https://doi.org/10.1016/j.agsy.2016.12.013.

Blankenship, R. E., & Chen, M. (2013). Spectral expansion and antenna reduction can enhance photosynthesis for energy production. *Current Opinion in Chemical Biology*, *17*(3), 457–461. https://doi.org/10.1016/j.cbpa.2013.03.031.

Blankenship, R. E., et al. (2011). Comparing photosynthetic and photovoltaic efficiencies and recognizing the potential for improvement. *Science*, *332*, 805–809. https://doi.org/10.1126/science.1200165.

Bloom, A. J., et al. (2010). Carbon dioxide enrichment inhibits nitrate assimilation in wheat and *arabidopsis*. *Science*, *328*, 899–903. https://doi.org/10.1126/science.1186440.

Bracher, A., et al. (2017). Biogenesis and metabolic maintenance of Rubisco. *Annual Review of Plant Biology*, *68*, 29–60. https://doi.org/10.1146/annurev-arplant-043015-111633.

Brautigam, A., et al. (2017). On the evolutionary origin of CAM photosynthesis. *Plant Physiology*, *174*(2), 473–477. https://doi.org/10.1104/pp.17.00195.

Büttner, M. (2007). The monosaccharide transporter(-like) gene family in *Arabidopsis*. *FEBS Letters*, *581*(12), 2318–2324. https://doi.org/10.1016/j.febslet.2007.03.016.

Buttner, M. (2010). The Arabidopsis sugar transporter (AtSTP) family: An update. *Plant Biology (Stuttgart, Germany)*, *12*(Suppl. 1), 35–41. https://doi.org/10.1111/j.1438-8677.2010.00383.x.

Büttner, M., & Sauer, N. (2000). Monosaccharide transporters in plants: Structure, function and physiology. *Biochimica et Biophysica Acta*, *1465*(2000), 263–274. https://doi.org/10.1016/S0005-2736(00)00143-7.

Cabrerizo, P. M., et al. (2001). Continuous CO_2 enrichment leads to increased nodule biomass, carbon availability to nodules and activity of carbon-metabolising enzymes but does not enhance specific nitrogen fixation in pea. *Physiologia Plantarum*, *113*, 33–40. https://doi.org/10.1034/j.1399-3054.2001.1130105.x.

Cameron, J. C., et al. (2013). Biogenesis of a bacterial organelle: The carboxysome assembly pathway. *Cell*, *155*(5), 1131–1140. https://doi.org/10.1016/j.cell.2013.10.044.

Campbell, W. J., et al. (1988). Effects of CO_2 concentration on rubisco activity, amount, and photosynthesis in soybean leaves. *Plant Physiology*, *88*, 1310–1316. https://doi.org/10.1104/pp.88.4.1310.

Chen, L. Q., et al. (2012). Sucrose efflux mediated by SWEET proteins as a key step for phloem transport. *Science*, *335*(6065), 207–211. https://doi.org/10.1126/science.1213351.

Chen, L. Q., et al. (2015). Transport of sugars. *Annual Review of Biochemistry*, *84*, 865–894. https://doi.org/10.1146/annurev-biochem-060614-033904.

Chen, J. H., et al. (2020). Nuclear-encoded synthesis of the D1 subunit of photosystem II increases photosynthetic efficiency and crop yield. *Nature Plants*, *6*(5), 570–580. https://doi.org/10.1038/s41477-020-0629-z.

Cho, J. I., et al. (2010). Expression analysis and functional characterization of the monosaccharide transporters, *OsTMTs*, involving vacuolar sugar transport in rice (*Oryza sativa*). *New Phytologist*, *186*(3), 657–668. https://doi.org/10.1111/j.1469-8137.2010.03194.x.

Coker, G. T., & Schubert, K. R. (1981). Carbon dioxide fixation in soybean roots and nodules. *Plant Physiology*, *67*(4), 691–696. https://doi.org/10.1104/pp.67.4.691.

Coleman, H. D., et al. (2009). Sucrose synthase affects carbon partitioning to increase cellulose production and altered cell wall ultrastructure. *Proceedings of the National Academy of Sciences of the United States of America*, *106*, 13118–13123. https://doi.org/10.1073/pnas.0900188106.

Curtis, P. E., et al. (1969). Varietal effects in soybean photosynthesis and photorespiration. *Crop Science*, *9*, 323–327. https://doi.org/10.2135/cropsci1969.0011183X000900030021x.

Dall'Osto, L., et al. (2015). Biogenesis of light harvesting proteins. *Biochimica et Biophysica Acta*, *1847*(9), 861–871. https://doi.org/10.1016/j.bbabio.2015.02.009.

Degen, G. E., et al. (2020). An isoleucine residue acts as a thermal and regulatory switch in wheat Rubisco activase. *The Plant Journal*, *103*(2), 742–751. https://doi.org/10.1111/tpj.14766.

Deng, L., et al. (2021). Soybean GmSUT1 transporter participates in sucrose transport to nodules during rhizobial symbiosis. *Plant Growth Regulation*, *96*(1), 119–129. https://doi.org/10.1007/s10725-021-00764-y.

Doidy, J., et al. (2012). Sugar transporters in plants and in their interactions with fungi. *Trends in Plant Science*, *17*(7), 413–422. https://doi.org/10.1016/j.tplants.2012.03.009.

Driever, S. M., & Kromdijk, J. (2013). Will C_3 crops enhanced with the C_4 CO_2-concentrating mechanism live up to their full potential (yield)? *Journal of Experimental Botany*, *64*(13), 3925–3935. https://doi.org/10.1093/jxb/ert103.

Driever, S. M., et al. (2017). Increased SBPase activity improves photosynthesis and grain yield in wheat grown in greenhouse conditions. *Philosophical Transactions Royal Society. Biological Sciences*, *372*(1730). https://doi.org/10.1098/rstb.2016.0384.

Ducat, D.C. & Silver, P.A. (2012). Improving carbon fixation pathways. Current Opinion in Chemical Biology 16(3-4), 337-344. doi:https://doi.org/10.1016/j.cbpa.2012.05.002.

Durand, M., et al. (2018). Carbon source-sink relationship in Arabidopsis thaliana: The role of sucrose transporters. *Planta*, *247*(3), 587–611. https://doi.org/10.1007/s00425-017-2807-4.

Dusotoit-Coucaud, A., et al. (2010). Cloning and characterization of a new polyol transporter (HbPLT2) in Hevea brasiliensis. *Plant & Cell Physiology*, *51*(11), 1878–1888. https://doi.org/10.1093/pcp/pcq151.

Eberhard, S., et al. (2008). The dynamics of photosynthesis. *Annual Review of Genetics*, *42*, 463–515. https://doi.org/10.1146/annurev.genet.42.110807.091452.

Eisenhut, M., et al. (2019). Mechanistic understanding of photorespiration paves the way to a new green revolution. *New Phytologist*, *223*(4), 1762–1769. https://doi.org/10.1111/nph.15872.

Endo, T., et al. (2014). Light energy allocation at PSII under field light conditions: How much energy is lost in NPQ-associated dissipation? *Plant Physiology and Biochemistry*, *81*, 115–120. https://doi.org/10.1016/j.plaphy.2014.03.018.

Evans, J. R. (2013). Improving photosynthesis. *Plant Physiology*, *162*(4), 1780–1793. https://doi.org/10.1104/pp.113.219006.

Evans, J. R., & Clarke, V. C. (2019). The nitrogen cost of photosynthesis. *Journal of Experimental Botany*, *70*(1), 7–15. https://doi.org/10.1093/jxb/ery366.

Fallahi, H., et al. (2008). Localization of sucrose synthase in developing seed and siliques of Arabidopsis thaliana reveals diverse roles for SUS during development. *Journal of Experimental Botany*, *59*(12), 3283–3295. https://doi.org/10.1093/jxb/ern180.

Faralli, M., & Lawson, T. (2020). Natural genetic variation in photosynthesis: An untapped resource to increase crop yield potential? *The Plant Journal*, *101*(3), 518–528. https://doi.org/10.1111/tpj.14568.

Fernie, A. R., & Bauwe, H. (2020). Wasteful, essential, evolutionary stepping stone? The multiple personalities of the photorespiratory pathway. *The Plant Journal*, *102*(4), 666–677. https://doi.org/10.1111/tpj.14669.

Fernie, A. R., et al. (2021). Using precision phenotyping to inform *de novo* domestication. *Plant Physiology*, *186*, 1397–1411. https://doi.org/10.1093/plphys/kiab160.

Fichtner, F., & Lunn, J. E. (2021). The role of trehalose 6-phosphate (Tre6P) in plant metabolism and development. *Annual Review of Plant Biology*, *72*, 737–760. https://doi.org/10.1146/annurev-arplant-050718-095929.

Fichtner, F., et al. (2021). Regulation of shoot branching in *Arabidopsis* by trehalose 6-phosphate. *New Phytologist*, *229*(4), 2135–2151. https://doi.org/10.1111/nph.17006.

Fischinger, S. A., et al. (2010). Elevated CO_2 concentration around alfalfa nodules increases N2 fixation. *Journal of Experimental Botany*, *61*(1), 121–130. https://doi.org/10.1093/jxb/erp287.

Flemetakis, E., et al. (2006). Spatial and temporal organization of sucrose metabolism in Lotus japonicus nitrogen-fixing nodules suggests a role for the elusive alkaline/neutral invertase. *Physiology and Molecular Biology*, *62*(1-2), 53–69. https://doi.org/10.1007/s11103-006-9003-4.

Foyer, C. H., et al. (2009). Photorespiratory metabolism: Genes, mutants, energetics, and redox signaling. *Annual Review of Plant Biology, 60*, 455–484. https://doi.org/10.1146/annurev.arplant.043008.091948.

Gersony, J. T., et al. (2020). Leaf carbon export and nonstructural carbohydrates in relation to diurnal water dynamics in mature oak trees. *Plant Physiology, 183*(4), 1612–1621. https://doi.org/10.1104/pp.20.00426.

Giordano, M., et al. (2005). CO_2 concentrating mechanisms in algae mechanisms, environmental modulation, and evolution. *Annual Review of Plant Biology, 56*, 99–131. https://doi.org/10.1146/annurev.arplant.56.032604.144052.

Glowacka, K., et al. (2018). *Photosystem II Subunit S* overexpression increases the efficiency of water use in a field-grown crop. *Nature Communications, 9*(1), 868–877. https://doi.org/10.1038/s41467-018-03231-x.

Griffiths, C. A., et al. (2016). Chemical intervention in plant sugar signalling increases yield and resilience. *Nature, 540*(7634), 574–578. https://doi.org/10.1038/nature20591.

Hagemann, M., & Bauwe, H. (2016). Photorespiration and the potential to improve photosynthesis. *Current Opinion in Chemical Biology, 35*, 109–116. https://doi.org/10.1016/j.cbpa.2016.09.014.

Harbinson, J., et al. (2021). Designing the crops for the future; the cropbooster program. *Biology, 10*(7), 690–697. https://doi.org/10.3390/biology10070690.

Hart, J. E., et al. (2019). Engineering the phototropin photocycle improves photoreceptor performance and plant biomass production. *Proceedings of the National Academy of Sciences of the United States of America, 116*(25), 12550–12557. https://doi.org/10.1073/pnas.1902915116.

Hay, W. T., et al. (2017). Enhancing soybean photosynthetic CO_2 assimilation using a cyanobacterial membrane protein, ictB. *Journal of Plant Physiology, 212*, 58–68. https://doi.org/10.1016/j.jplph.2017.02.003.

Hennacy, J. H., & Jonikas, M. C. (2020). Prospects for engineering biophysical CO_2 concentrating mechanisms into land plants to enhance yields. *Annual Review of Plant Biology, 71*, 461–485. https://doi.org/10.1146/annurev-arplant-081519-040100.

Hibberd, J. M., & Quick, W. P. (2002). Characteristics of C_4 photosynthesis in stems and petioles of C_3 flowering plants. *Nature, 415*, 451–454. https://doi.org/10.1038/415451a.

Ho, M. Y., et al. (2016). Light-dependent chlorophyll f synthase is a highly divergent paralog of *PsbA* of photosystem II. *Science, 353*(6302). https://doi.org/10.1126/science.aaf9178.aaf9171-9178.

Hodges, M., et al. (2016). Perspectives for a better understanding of the metabolic integration of photorespiration within a complex plant primary metabolism network. *Journal of Experimental Botany, 67*(10), 3015–3026. https://doi.org/10.1093/jxb/erw145.

Hohmann-Marriott, M. F., & Blankenship, R. E. (2011). Evolution of photosynthesis. *Annual Review of Plant Biology, 62*, 515–548. https://doi.org/10.1146/annurev-arplant-042110-103811.

Hsu, P. K., et al. (2018). Abscisic acid-independent stomatal CO_2 signal transduction pathway and convergence of CO_2 and ABA signaling downstream of OST1 kinase. *Proceedings of the National Academy of Sciences of the United States of America, 115*(42), E9971–E9980. https://doi.org/10.1073/pnas.1809204115.

Hu, H., et al. (2010). Carbonic anhydrases are upstream regulators of CO_2-controlled stomatal movements in guard cells. *Nature Cell Biology, 12*(1), 87–93. sup pp 1-18 https://doi.org/10.1038/ncb2009.

Ishikawa, C., et al. (2011). Functional incorporation of sorghum small subunit increases the catalytic turnover rate of Rubisco in transgenic rice. *Plant Physiology, 156*(3), 1603–1611. https://doi.org/10.1104/pp.111.177030.

Jancic, M., et al. (2015). Impact of climate change and carbon dioxide fertilization effect on irrigation water demand and yield of soybean in Serbia. *The Journal of Agricultural Science, 153*(8), 1365–1379. https://doi.org/10.1017/s0021859615000179.

Jin, Y., et al. (2009). Posttranslational elevation of cell wall invertase activity by silencing its inhibitor in tomato delays leaf senescence and increases seed weight and fruit hexose level. *Plant Cell*, *21*(7), 2072–2089. https://doi.org/10.1105/tpc.108.063719.

Jin, H., et al. (2016). Optimization of light-harvesting pigment improves photosynthetic efficiency. *Plant Physiology*, *172*(3), 1720–1731. https://doi.org/10.1104/pp.16.00698.

Johnson, M. P. (2016). Photosynthesis. *Essays in Biochemistry*, *60*(3), 255–273. https://doi.org/10.1042/EBC20160016.

Jones, P. H., et al. (1984). Soybean canopy growth, photosynthesis, and transpiration responses to whole-season carbon dioxide enrichment. *Agronomy Journal*, *76*, 633–637. https://doi.org/10.2134/agronj1984.00021962007600040030x.

Kaiser, E., et al. (2019). Efficient photosynthesis in dynamic light environments: A chloroplast's perspective. *Biochemical Journal*, *476*(19), 2725–2741. https://doi.org/10.1042/BCJ20190134.

Kang, B. C., et al. (2021). Chloroplast and mitochondrial DNA editing in plants. *Nature Plants*, 7(7), 899–905. https://doi.org/10.1038/s41477-021-00943-9.

Kirst, H., et al. (2017). Photosynthetic antenna engineering to improve crop yields. *Planta*, *245*(5), 1009–1020. https://doi.org/10.1007/s00425-017-2659-y.

Klepek, Y. S., et al. (2010). *Arabidopsis thaliana* POLYOL/MONOSACCHARIDE TRANSPORTERS 1 and 2: Fructose and xylitol/H^+ symporters in pollen and young xylem cells. *Journal of Experimental Botany*, *61*(2), 537–550. https://doi.org/10.1093/jxb/erp322.

Kohler, I. H., et al. (2017). Expression of cyanobacterial *FBP/SBPase* in soybean prevents yield depression under future climate conditions. *Journal of Experimental Botany*, *68*(3), 715–726. https://doi.org/10.1093/jxb/erw435.

Kohzuma, K., et al. (2017). The non-mendelian green cotyledon gene in soybean encodes a small subunit of photosystem II. *Plant Physiology*, *173*(4), 2138–2147. https://doi.org/10.1104/pp.16.01589.

Komaitis, F., et al. (2020). Molecular and physiological characterization of the monosaccharide transporters gene family in Medicago truncatula. *Journal of Experimental Botany*, *71*(10), 3110–3125. https://doi.org/10.1093/jxb/eraa055.

Kramer, D. M., et al. (2004). Dynamic flexibility in the light reactions of photosynthesis governed by both electron and proton transfer reactions. *Trends in Plant Science*, *9*(7), 349–357. https://doi.org/10.1016/j.tplants.2004.05.001.

Kromdijk, J., et al. (2016). Improving photosynthesis and crop productivity by accelerating recovery from photoprotection. *Science*, *354*(6314), 851–857. https://doi.org/10.1126/science.aaf9621.

Kubis, A., & Bar-Even, A. (2019). Synthetic biology approaches for improving photosynthesis. *Journal of Experimental Botany*, *70*(5), 1425–1433. https://doi.org/10.1093/jxb/erz029.

Laterre, R., et al. (2017). Photosynthetic trichomes contain a specific Rubisco with a modified pH-dependent activity. *Plant Physiology*, *173*(4), 2110–2120. https://doi.org/10.1104/pp.17.00062.

Leach, K. A., et al. (2017). *Sucrose transporter2* contributes to maize growth, development, and crop yield. *Journal of Integrative Plant Biology*, *59*(6), 390–408. https://doi.org/10.1111/jipb.12527.

Lefebvre, S., et al. (2005). Increased sedoheptulose-1,7-bisphosphatase activity in transgenic tobacco plants stimulates photosynthesis and growth from an early stage in development. *Plant Physiology*, *138*(1), 451–460. https://doi.org/10.1104/pp.104.055046.

Levey, M., et al. (2019). Efficient 2-phosphoglycolate degradation is required to maintain carbon assimilation and allocation in the C_4 plant Flaveria bidentis. *Journal of Experimental Botany*, *70*(2), 575–587. https://doi.org/10.1093/jxb/ery370.

Li, B., et al. (2013). Constitutive expression of cell wall invertase genes increases grain yield and starch content in maize. *Plant Biotechnology Journal*, *11*(9), 1080–1091. https://doi.org/10.1111/pbi.12102.

Li, Y. H., et al. (2014). *De novo* assembly of soybean wild relatives for pan-genome analysis of diversity and agronomic traits. *Nature Biotechnology*, *32*(10), 1045–1052. https://doi.org/10.1038/nbt.2979.

Li, L., et al. (2018). Mechanisms of photodamage and protein turnover in photoinhibition. *Trends in Plant Science*, *23*(8), 667–676. https://doi.org/10.1016/j.tplants.2018.05.004.

Liang, Y., et al. (2021). Natural variation in crops realized understanding, continuing promise. *Annual Review of Plant Biology*, *72*, 7.1–7.29. https://doi.org/10.1146/annurev-arplant-080720-090632.

Libault, M. (2014). The carbon-nitrogen balance of the nodule and its regulation under elevated carbon dioxide concentration. *BioMed Research International*, *2014*, 507946. https://doi.org/10.1155/2014/507946.

Lin, M. T., et al. (2014). A faster Rubisco with potential to increase photosynthesis in crops. *Nature*, *513*(7519), 547–550. https://doi.org/10.1038/nature13776.

Liu, H. J., & Yan, J. (2019). Crop genome-wide association study: A harvest of biological relevance. *The Plant Journal*, *97*(1), 8–18. https://doi.org/10.1111/tpj.14139.

Liu, S., et al. (2020). Toward a "Green Revolution" for soybean. *Molecular Plant*, *13*(5), 688–697. https://doi.org/10.1016/j.molp.2020.03.002.

Liu, J., et al. (2021). AtSTP8, an endoplasmic reticulum-localised monosaccharide transporter from Arabidopsis, is recruited to the extrahaustorial membrane during powdery mildew infection. *New Phytologist*, *230*(6), 2404–2419. https://doi.org/10.1111/nph.17347.

Lloyd, J. P. B., & Lister, R. (2022). Epigenome plasticity in plants. *Nature Reviews Genetics*, *23*, 55–68. https://doi.org/10.1038/s41576-021-00407-y.

Long, S. P., et al. (2005). Global food insecurity. treatment of major food crops with elevated carbon dioxide or ozone under large-scale fully open-air conditions suggests recent models may have overestimated future yields. *Philosophical Transactions of the Royal Society B*, *360*(1463), 2011–2020. https://doi.org/10.1098/rstb.2005.1749.

Long, S. P., et al. (2006). Can improvement in photosynthesis increase crop yields? *Plant, Cell and Environment*, *29*(3), 315–330. https://doi.org/10.1111/j.1365-3040.2005.01493.x.

Long, S. P., et al. (2015). Meeting the global food demand of the future by engineering crop photosynthesis and yield potential. *Cell*, *161*(1), 56–66. https://doi.org/10.1016/j.cell.2015.03.019.

Lopez-Calcagno, P. E., et al. (2019). Overexpressing the H-protein of the glycine cleavage system increases biomass yield in glasshouse and field-grown transgenic tobacco plants. *Plant Biotechnology Journal*, *17*(1), 141–151. https://doi.org/10.1111/pbi.12953.

Lu, Y. (2016). Identification and roles of photosystem II assembly, stability, and repair factors in *Arabidopsis*. *Frontiers in Plant Science*, 7, 168–194. https://doi.org/10.3389/fpls.2016.00168.

MacNeill, G. J., et al. (2017). Starch as a source, starch as a sink: The bifunctional role of starch in carbon allocation. *Journal of Experimental Botany*, *68*(16), 4433–4453. https://doi.org/10.1093/jxb/erx291.

Mahood, E. H., et al. (2020). Machine learning: A powerful tool for gene function prediction in plants. *Applications in Plant Sciences*, *8*(7), e11376–e11387. https://doi.org/10.1002/aps3.11376.

Marco, F. D., et al. (2021). Involvement of SUT1 and SUT2 sugar transporters in the impairment of sugar transport and changes in phloem exudate contents in phytoplasma-infected plants. *International Journal of Molecular Sciences*, *22*(2), 745–768. https://doi.org/10.3390/ijms22020745.

Matsumura, H., et al. (2020). Hybrid Rubisco with complete replacement of rice rubisco small subunits by sorghum counterparts confers C_4 plant-like high catalytic activity. *Molecular Plant*, *13*(11), 1570–1581. https://doi.org/10.1016/j.molp.2020.08.012.

Maurino, V. G., & Peterhansel, C. (2010). Photorespiration: Current status and approaches for metabolic engineering. *Current Opinion in Plant Biology, 13*(3), 249–256. https://doi.org/10.1016/j.pbi.2010.01.006.
Maurino, V. G., & Weber, A. P. (2013). Engineering photosynthesis in plants and synthetic microorganisms. *Journal of Experimental Botany, 64*(3), 743–751. https://doi.org/10.1093/jxb/ers263.
Miao, L., et al. (2020). Natural variation and selection in GmSWEET39 affect soybean seed oil content. *New Phytologist, 225*(4), 1651–1666. https://doi.org/10.1111/nph.16250.
Minchin, F. R., & Witty, J. F. (2005). Respiratory/carbon costs of symbiotic nitrogen fixation in legumes. In H. Lambers, & M. Ribas-Carbo (Eds.), *Plant respiration: From cell to ecosystem* (pp. 195–205). Netherlands, Dordrecht: Springer.
Monteith, J. L. (1977). Climate and the efficiency of crop production in Britain. *Philosophical Transactions Royal Society. Biological sciences, 281*, 277–294. https://doi.org/10.1098/rstb.1977.0140.
Morell, M., & Copeland, L. (1984). Enzymes of Sucrose Breakdown in Soybean Nodules. *Plant Physiology, 74*(4), 1030–1034. https://doi.org/10.1104/pp.74.4.1030.
Morrison, M. J., et al. (1999). Agronomic changes from 58 years of genetic improvement of short-season soybean cultivars in Canada. *Agronomy Journal, 91*(4), 685–689.
Myers, S. S., et al. (2014). Increasing CO_2 threatens human nutrition. *Nature, 510*(7503), 139–142. https://doi.org/10.1038/nature13179.
Nelson, N., & Ben-Shem, A. (2004). The complex architecture of oxygenic photosynthesis. *Nature Reviews. Molecular Cell Biology, 5*(12), 971–982. https://doi.org/10.1038/nrm1525.
Nürnberg, D. J., et al. (2018). Photochemistry beyond the red limit in chlorophyll f–containing photosystems. *Science, 360*, 1210–1213. https://doi.org/10.1126/science.aar8313.
Oellrich, A., et al. (2015). An ontology approach to comparative phenomics in plants. *Plant Methods, 11*, 10–24. https://doi.org/10.1186/s13007-015-0053-y.
Ogren, W. L. (1984). Photorespiration pathways, regulation and modification. *Annual Review of Plant Physiology, 35*, 415–442.
Ohdan, T., et al. (2005). Expression profiling of genes involved in starch synthesis in sink and source organs of rice. *Journal of Experimental Botany, 56*(422), 3229–3244. https://doi.org/10.1093/jxb/eri292.
Orf, I., et al. (2016). Can cyanobacteria serve as a model of plant photorespiration?—A comparative meta-analysis of metabolite profiles. *Journal of Experimental Botany, 67*(10), 2941–2952. https://doi.org/10.1093/jxb/erw068.
Orr, D. J., et al. (2016). Surveying Rubisco diversity and temperature response to improve crop photosynthetic efficiency. *Plant Physiology, 172*(2), 707–717. https://doi.org/10.1104/pp.16.00750.
Ort, D. R., et al. (2015). Redesigning photosynthesis to sustainably meet global food and bioenergy demand. *Proceedings of the National Academy of Sciences of the United States of America, 112*(28), 8236–8529. https://doi.org/10.1073/pnas.1424031112.
Papanatsiou, M., et al. (2019). Optogenetic manipulation of stomatal kinetics improves carbon assimilation, water use, and growth. *Science, 363*(6434), 1456–1459. https://doi.org/10.1126/science.aaw0046. 30923223.
Patil, G., et al. (2015). Soybean (Glycine max) SWEET gene family: Insights through comparative genomics, transcriptome profiling and whole genome re-sequence analysis. *BMC Genomics, 16*, 520–536. https://doi.org/10.1186/s12864-015-1730-y.
Patriarca, E. J., et al. (2002). Key role of bacterial NH_4^+ metabolism in Rhizobium-plant symbiosis. *Microbiology and Molecular Biology Reviews, 66*(2), 203–222.
Paul, M. J., & Foyer, C. H. (2001). Sink regulation of photosynthesis. *Journal of Experimental Botany, 52*, 1383–1400. https://doi.org/10.1093/jexbot/52.360.1383.

Paulsen, P. A., et al. (2019). Crystal structure of the plant symporter STP10 illuminates sugar uptake mechanism in monosaccharide transporter superfamily. *Nature Communications*, *10*(1), 407–415. https://doi.org/10.1038/s41467-018-08176-9.

Peiter, E., & Schubert, S. (2003). Sugar uptake and proton release by protoplasts from the infected zone of *Vicia faba L.* nodules: Evidence against apoplastic sugar supply of infected cells. *Journal of Experimental Botany*, *54*(388), 1691–1700. https://doi.org/10.1093/jxb/erg191.

Peng, S., et al. (1991). Leaf photosynthetic rate is correlated with biomass and grain production in grain sorghum lines. *Photosynthesis Research*, *28*, 1–7.

Peterhansel, C., et al. (2013). Photorespiratory bypasses: How can they work? *Journal of Experimental Botany*, *64*(3), 709–715. https://doi.org/10.1093/jxb/ers247.

Pieruschka, R., & Schurr, U. (2019). Plant phenotyping: Past, present, and future. *Plant Phenomics*, *2019*, 7507131. https://doi.org/10.34133/2019/7507131.

Plaut, Z., et al. (1987). Effect of altered sink: Source ratio on photosynthetic metabolism of source leaves. *Plant Physiology*, *85*(3), 786–791.

Poschet, G., et al. (2011). A novel *Arabidopsis* vacuolar glucose exporter is involved in cellular sugar homeostasis and affects the composition of seed storage compounds. *Plant Physiology*, *157*(4), 1664–1676. https://doi.org/10.1104/pp.111.186825.

Prell, J., & Poole, P. (2006). Metabolic changes of rhizobia in legume nodules. *Trends in Microbiology*, *14*(4), 161–168. https://doi.org/10.1016/j.tim.2006.02.005.

Qazi, H. A., et al. (2012). Stem sugar accumulation in sweet sorghum—Activity and expression of sucrose metabolizing enzymes and sucrose transporters. *Journal of Plant Physiology*, *169*(6), 605–613. https://doi.org/10.1016/j.jplph.2012.01.005.

Qin, G., et al. (2016). A tomato vacuolar invertase inhibitor mediates sucrose metabolism and influences fruit ripening. *Plant Physiology*, *172*(3), 1596–1611. https://doi.org/10.1104/pp.16.01269.

Rachmilevitch, S., et al. (2004). Nitrate assimilation in plant shoots depends on photorespiration. *Proceedings of the National Academy of Sciences of the United States of America*, *101*, 11506–11510. https://doi.org/10.1073/pnas.0404388101.

Regmi, K. C., et al. (2016). Apoplasmic loading in the rice phloem supported by the presence of sucrose synthase and plasma membrane-localized proton pyrophosphatase. *Annals of Botany*, *117*(2), 257–268. https://doi.org/10.1093/aob/mcv174.

Roitsch, T., & González, M.-C. (2004). Function and regulation of plant invertases: Sweet sensations. *Trends in Plant Science*, *9*(12), 606–613. https://doi.org/10.1016/j.tplants.2004.10.009.

Ruan, Y. L. (2014). Sucrose metabolism: Gateway to diverse carbon use and sugar signaling. *Annual Review of Plant Biology*, *65*, 33–67. https://doi.org/10.1146/annurev-arplant-050213-040251.

Ruan, Y. L., et al. (2010). Sugar input, metabolism, and signaling mediated by invertase: Roles in development, yield potential, and response to drought and heat. *Molecular Plant*, *3*(6), 942–955. https://doi.org/10.1093/mp/ssq044.

Saddhe, A. A., et al. (2021). Plant sugars: Homeostasis and transport under abiotic stress in plants. *Physiologia Plantarum*, *171*(4), 739–755. https://doi.org/10.1111/ppl.13283.

Sage, R. F. (2017). A portrait of the C_4 photosynthetic family on the 50th anniversary of its discovery: Species number, evolutionary lineages, and Hall of Fame. *Journal of Experimental Botany*, *68*(2), 4039–4056. https://doi.org/10.1093/jxb/erx005.

Sandhu, D., et al. (2016). Soybean proteins GmTic110 and GmPsbP are crucial for chloroplast development and function. *Plant Science*, *252*, 76–87. https://doi.org/10.1016/j.plantsci.2016.07.006.

Sandhu, D., et al. (2018). Genetics and physiology of the nuclearly inherited yellow foliar mutants in soybean. *Frontiers in Plant Science*, *9*, 471–482. https://doi.org/10.3389/fpls.2018.00471.

Sauer, N. (2007). Molecular physiology of higher plant sucrose transporters. *FEBS Letters*, *581*(12), 2017–2309. https://doi.org/10.1016/j.febslet.2007.03.048.

Schluter, U., & Weber, A. P. M. (2020). Regulation and evolution of C_4 photosynthesis. *Annual Review of Plant Biology*, *71*, 183–215. https://doi.org/10.1146/annurev-arplant-042916-040915.

Schmutz, J., et al. (2010). Genome sequence of the palaeopolyploid soybean. *Nature*, *463*(7278), 178–183. https://doi.org/10.1038/nature08670.

Schneider, S., et al. (2008). Functional and physiological characterization of Arabidopsis INOSITOL TRANSPORTER1, a novel tonoplast-localized transporter for *myo*-inositol. *Plant Cell*, *20*(4), 1073–1087. https://doi.org/10.1105/tpc.107.055632.

Schneidereit, A., et al. (2003). Functional characterization and expression analyses of the glucose-specific AtSTP9 monosaccharide transporter in pollen of *Arabidopsis*. *Plant Physiology*, *133*(1), 182–190. https://doi.org/10.1104/pp.103.026674.

Schulz, A., et al. (2011). Proton-driven sucrose symport and antiport are provided by the vacuolar transporters SUC4 and TMT1/2. *The Plant Journal*, *68*(1), 129–136. https://doi.org/10.1111/j.1365-313X.2011.04672.x.

Schwander, T., et al. (2016). A synthetic pathway for the fixation of carbon dioxide in vitro. *Science*, *354*, 900–904. https://doi.org/10.1126/science.aah5237.

Scofield, G. N., et al. (2007). Involvement of the sucrose transporter, OsSUT1, in the long-distance pathway for assimilate transport in rice. *Journal of Experimental Botany*, *58*(12), 3155–3169. https://doi.org/10.1093/jxb/erm153.

Sharwood, R. E. (2017). Engineering chloroplasts to improve Rubisco catalysis: Prospects for translating improvements into food and fiber crops. *New Phytologist*, *213*(2), 494–510. https://doi.org/10.1111/nph.14351.

Sharwood, R. E., et al. (2016). Temperature responses of Rubisco from Paniceae grasses provide opportunities for improving C_3 photosynthesis. *Nature Plants*, *2*, 16186–16195. https://doi.org/10.1038/nplants.2016.186.

Shen, Y., et al. (2018). *De novo* assembly of a Chinese soybean genome. *Science China. Life Sciences*, *61*(8), 871–884. https://doi.org/10.1007/s11427-018-9360-0.

Shen, Y., et al. (2019). Update soybean Zhonghuang 13 genome to a golden reference. *Science China. Life Sciences*, *62*(9), 1257–1260. https://doi.org/10.1007/s11427-019-9822-2.

Shimakawa, G., & Miyake, C. (2019). What quantity of photosystem I is optimum for safe photosynthesis? *Plant Physiology*, *179*(4), 1479–1485. https://doi.org/10.1104/pp.18.01493.

Shin, W.-S., et al. (2016). Truncated light-harvesting chlorophyll antenna size in *Chlorella vulgaris* improves biomass productivity. *Journal of Applied Phycology*, *28*(6), 3193–3202. https://doi.org/10.1007/s10811-016-0874-8.

Shin, D., et al. (2020). Natural variations at the *Stay-Green* gene promoter control lifespan and yield in rice cultivars. *Nature Communications*, *11*(1), 2819–2829. https://doi.org/10.1038/s41467-020-16573-2.

Shiraiwa, T., et al. (2004). Correlation between yielding ability and dry matter productivity during initial seed filling stage in various soybean genotypes. *Plant Production Science*, 7(2), 138–142.

Shivram, H., et al. (2020). Controlling and enhancing CRISPR systems. *Nature Chemical Biology*, *17*(1), 10–19. https://doi.org/10.1038/s41589-020-00700-7.

Silva, C. S., et al. (2015). Exergy efficiency of plant photosynthesis. *Chemical Engineering Science*, *130*, 151–171. https://doi.org/10.1016/j.ces.2015.02.011.

Simkin, A. J., et al. (2015). Multigene manipulation of photosynthetic carbon assimilation increases CO_2 fixation and biomass yield in tobacco. *Journal of Experimental Botany*, *66*(13), 4075–4090. https://doi.org/10.1093/jxb/erv204.

Simkin, A. J., et al. (2017). Overexpression of the rieskefes protein increases electron transport rates and biomass yield. *Plant Physiology*, *175*(1), 134–145. https://doi.org/10.1104/pp.17.00622.

Simkin, A. J., et al. (2019). Feeding the world: Improving photosynthetic efficiency for sustainable crop production. *Journal of Experimental Botany*, *70*(4), 1119–1140. https://doi.org/10.1093/jxb/ery445.

Slattery, R. A., & Ort, D. R. (2015). Photosynthetic energy conversion efficiency: Setting a baseline for gauging future improvements in important food and biofuel crops. *Plant Physiology*, *168*(2), 383–392. https://doi.org/10.1104/pp.15.00066.

Slewinski, T. L. (2011). Diverse functional roles of monosaccharide transporters and their homologs in vascular plants: A physiological perspective. *Molecular Plant*, *4*(4), 641–662. https://doi.org/10.1093/mp/ssr051.

Song, Q., et al. (2017). The impact of modifying photosystem antenna size on canopy photosynthetic efficiency-development of a new canopy photosynthesis model scaling from metabolism to canopy level processes. *Plant, Cell & Environment*, *40*(12), 2946–2957. https://doi.org/10.1111/pce.13041.

Song, Q., et al. (2020). Decomposition analysis on soybean productivity increase under elevated CO_2 using 3-D canopy model reveals synergestic effects of CO_2 and light in photosynthesis. *Annals of Botany*, *126*(4), 601–614. https://doi.org/10.1093/aob/mcz163.

Sosso, D., et al. (2015). Seed filling in domesticated maize and rice depends on SWEET-mediated hexose transport. *Nature Genetics*, *47*(12), 1489–1493. https://doi.org/10.1038/ng.3422.

South, P. F., et al. (2019). Synthetic glycolate metabolism pathways stimulate crop growth and productivity in the field. *Science*, *363*(6422-30). https://doi.org/10.1126/science.aat9077. eaat9077-9085.

Stirbet, A., et al. (2020). Photosynthesis: Basics, history and modelling. *Annals of Botany*, *126*(4), 511–537. https://doi.org/10.1093/aob/mcz171.

Streeter, J. G. (1980). Carbohydrates in soybean nodules: II. Distribution of compounds in seedlings during the onset of nitrogen fixation. *Plant Physiology*, *66*(3), 471–476.

Sugiura, D., et al. (2019). Interspecific differences in how sink-source imbalance causes photosynthetic downregulation among three legume species. *Annals of Botany*, *123*(4), 715–726. https://doi.org/10.1093/aob/mcy204.

Szenthe, A., et al. (2007). Characterisation and expression of monosaccharide transporters in *lupins, Lupinus polyphyllus* and *L. albus*. *Journal of Plant Research*, *120*(6), 697–705. https://doi.org/10.1007/s10265-007-0112-1.

Tang, X., et al. (2017). Suppression of extracellular invertase inhibitor gene expression improves seed weight in soybean (*Glycine max*). *Journal of Experimental Botany*, *68*(3), 469–482. https://doi.org/10.1093/jxb/erw425.

Tiessen, A., et al. (2002). Starch synthesis in potato tubers is regulated by post-translational redox modification of ADP-glucose pyrophosphorylase: A novel regulatory mechanism linking starch synthesis to the sucrose supply. *Plant Cell*, *14*(9), 2191–2213. https://doi.org/10.1105/tpc.003640.

Tomlinson, K. L., et al. (2004). Evidence that the hexose-to-sucrose ratio does not control the switch to storage product accumulation in oilseeds: Analysis of tobacco seed development and effects of overexpressing apoplastic invertase. *Journal of Experimental Botany*, *55*(406), 2291–2303. https://doi.org/10.1093/jxb/erh251.

Topp, C. N., et al. (2013). 3D phenotyping and quantitative trait locus mapping identify core regions of the rice genome controlling root architecture. *Proceedings of the National Academy of Sciences of the United States of America*, *110*(18), E1695–E1704. https://doi.org/10.1073/pnas.1304354110.

Tsikou, D., et al. (2013). Cessation of photosynthesis in *Lotus japonicus* leaves leads to reprogramming of nodule metabolism. *Journal of Experimental Botany*, *64*(5), 1317–1332. https://doi.org/10.1093/jxb/ert015.

Udvardi, M., & Poole, P. S. (2013). Transport and metabolism in legume-rhizobia symbioses. *Annual Review of Plant Biology*, *64*, 781–805. https://doi.org/10.1146/annurev-arplant-050312-120235.

van Bezouw, R., et al. (2019). Converging phenomics and genomics to study natural variation in plant photosynthetic efficiency. *The Plant Journal*, *97*(1), 112–133. https://doi.org/10.1111/tpj.14190.

van Eeuwijk, F. A., et al. (2010). Detection and use of QTL for complex traits in multiple environments. *Current Opinion in Plant Biology*, *13*(2), 193–205. https://doi.org/10.1016/j.pbi.2010.01.001.

Vance, C. P. (1991). Carbon in N_2 fixation: Limitation or exquisite adaptation. *Annual Review of Plant Biology*, *42*, 373–392. https://doi.org/10.1146/annurev.pp.42.060191.002105.

Venegas-Molina, J., et al. (2021). Why and how to dig into plant metabolite-protein interactions. *Trends in Plant Science*, *26*(5), 472–483. https://doi.org/10.1016/j.tplants.2020.12.008.

Vignault, C., et al. (2005). *VvHT1* encodes a monosaccharide transporter expressed in the conducting complex of the grape berry phloem. *Journal of Experimental Botany*, *56*(415), 1409–1418. https://doi.org/10.1093/jxb/eri142.

Walker, B. J., et al. (2018). Chlorophyll can be reduced in crop canopies with little penalty to photosynthesis. *Plant Physiology*, *176*(2), 1215–1232. https://doi.org/10.1104/pp.17.01401.

Wallace, J. G., et al. (2018). On the road to breeding 4.0: Unraveling the good, the bad, and the boring of crop quantitative genomics. *Annual Review of Genetics*, *52*, 421–444. https://doi.org/10.1146/annurev-genet-120116-024846.

Walley, J. W., et al. (2016). Integration of omic networks in a developmental atlas of maize. *Science*, *353*(6301), 814–818. https://doi.org/10.5061/dryad.v8969).

Wan, H., et al. (2018). Evolution of sucrose metabolism: The dichotomy of invertases and beyond. *Trends in Plant Science*, *23*(2), 163–177. https://doi.org/10.1016/j.tplants.2017.11.001.

Wang, W. (2020). Research of epidemic big data based on improved deep convolutional neural network. *Computational and Mathematical Methods in Medicine*, *2020*. https://doi.org/10.1155/2020/3641745.

Wang, P., & Grimm, B. (2015). Organization of chlorophyll biosynthesis and insertion of chlorophyll into the chlorophyll-binding proteins in chloroplasts. *Photosynthesis Research*, *126*(2-3), 189–202. https://doi.org/10.1007/s11120-015-0154-5.

Wang, P., & Grimm, B. (2021). Connecting chlorophyll metabolism with accumulation of the photosynthetic apparatus. *Trends in Plant Science*, *26*(5), 484–495. https://doi.org/10.1016/j.tplants.2020.12.005.

Wang, L., & Ruan, Y. L. (2012). New insights into roles of cell wall invertase in early seed development revealed by comprehensive spatial and temporal expression patterns of GhCWIN1 in cotton. *Plant Physiology*, *160*(2), 777–787. https://doi.org/10.1104/pp.112.203893.

Wang, Z., & Tian, Z. (2015). Genomics progress will facilitate molecular breeding in soybean. *Science China*, *58*(8), 813–815.

Wang, Y., et al. (2007). Molecular cloning and expression analysis of a monosaccharide transporter gene OsMST4 from rice (*Oryza sativa* L.). *Plant Molecular Biology*, *65*(4), 439–451. https://doi.org/10.1007/s11103-007-9228-x.

Wang, E., et al. (2008). Control of rice grain-filling and yield by a gene with a potential signature of domestication. *Nature Genetics*, *40*(11), 1370–1374. https://doi.org/10.1038/ng.220.

Wang, S., et al. (2019a). The soybean sugar transporter GmSWEET15 mediates sucrose export from endosperm to early embryo. *Plant Physiology*, *180*(4), 2133–2141. https://doi.org/10.1104/pp.19.00641.

Wang, W., et al. (2019b). Structural basis for blue-green light harvesting and energy dissipation in diatoms. *Science*, *363*(6427-6434). https://doi.org/10.1126/science.aav0365.

Wang, S., et al. (2020a). Simultaneous changes in seed size, oil content and protein content driven by selection of SWEET homologues during soybean domestication. *National Science Review*, 7(11), 1776–1786. https://doi.org/10.1093/nsr/nwaa110.

Wang, P., et al. (2020b). Post-translational coordination of chlorophyll biosynthesis and breakdown by BCMs maintains chlorophyll homeostasis during leaf development. *Nature Communications*, *11*(1), 1254–1270. https://doi.org/10.1038/s41467-020-14992-9.

Wang, H., et al. (2020c). Deep learning for plant genomics and crop improvement. *Current Opinion in Plant Biology*, *54*, 34–41. https://doi.org/10.1016/j.pbi.2019.12.010.

Wang, L. M., et al. (2020d). A synthetic photorespiratory shortcut enhances photosynthesis to boost biomass and grain yield in rice. *Molecular Plant*, *13*(12), 1802–1815. https://doi.org/10.1016/j.molp.2020.10.007.

Wang, T., et al. (2021). Light-induced mobile factors from shoots regulate rhizobium-triggered soybean root nodulation. *Science*, *374*, 65–71. https://doi.org/10.1126/science.abh2890.

Weber, A. P. M., & Bar-Even, A. (2019). Update: Improving the efficiency of photosynthetic carbon reactions. *Plant Physiology*, *179*(3), 803–812. https://doi.org/10.1104/pp.18.01521.

Whitney, S. M., et al. (2011). Advancing our understanding and capacity to engineer nature's CO_2-sequestering enzyme, Rubisco. *Plant Physiology*, *155*(1), 27–35. https://doi.org/10.1104/pp.110.164814.

Wilson, R. H., et al. (2016). Evolving *Methanococcoides burtonii* archaeal Rubisco for improved photosynthesis and plant growth. *Scientific Reports*, *6*, 22284–22294. https://doi.org/10.1038/srep22284.

Wingenter, K., et al. (2010). Increased activity of the vacuolar monosaccharide transporter TMT1 alters cellular sugar partitioning, sugar signaling, and seed yield in Arabidopsis. *Plant Physiology*, *154*(2), 665–677. https://doi.org/10.1104/pp.110.162040.

Wormit, A., et al. (2006). Molecular identification and physiological characterization of a novel monosaccharide transporter from *Arabidopsis* involved in vacuolar sugar transport. *Plant Cell*, *18*(12), 3476–3490. https://doi.org/10.1105/tpc.106.047290.

Xia, L. Y., et al. (2020). Molecular basis for the assembly of RuBisCO assisted by the chaperone Raf1. *Nature Plants*, *6*(6), 708–717. https://doi.org/10.1038/s41477-020-0665-8.

Xu, Q., et al. (2018). Regulation of sucrose transporters and phloem loading in response to environmental cues. *Plant Physiology*, *176*(1), 930–945. https://doi.org/10.1104/pp.17.01088.

Xu, X., et al. (2019). *OsVIN2* encodes a vacuolar acid invertase that affects grain size by altering sugar metabolism in rice. *Plant Cell Reports*, *38*(10), 1273–1290. https://doi.org/10.1007/s00299-019-02443-9.

Xu, K., et al. (2021). A critical role of PvFtsH2 in the degradation of photodamaged D1 protein in common bean. *Horticulture Research*, *8*(1), 126–135. https://doi.org/10.1038/s41438-021-00554-7.

Yang, W., et al. (2014). Combining high-throughput phenotyping and genome-wide association studies to reveal natural genetic variation in rice. *Nature Communications*, *5*, 5087–5096. https://doi.org/10.1038/ncomms6087.

Yang, J., et al. (2018). SWEET11 and 15 as key players in seed filling in rice. *New Phytologist*, *218*(2), 604–615. https://doi.org/10.1111/nph.15004.

Yang, W., et al. (2020). Crop phenomics and high-throughput phenotyping: Past decades, current challenges, and future perspectives. *Molecular Plant, 13*(2), 187–214. https://doi.org/10.1016/j.molp.2020.01.008.

Yang, Y., et al. (2021). Extraterrestrial artificial photosynthetic materials for *in-situ* resource utilization. *National Science Review, 8*, 104–129. https://doi.org/10.1093/nsr/nwab104/6297273.

Yao, D., et al. (2020). *Arabidopsis* sucrose synthase localization indicates a primary role in sucrose translocation in phloem. *Journal of Experimental Botany, 71*(6), 1858–1869. https://doi.org/10.1093/jxb/erz539.

Yeates, T. O., & Wheatley, N. M. (2017). Putting the RuBisCO pieces together. *Science, 358*, 1253–1254. https://doi.org/10.1126/science.aar3107.

Ylstra, B., et al. (1998). Hexose transport in growing petunia pollen tubes and characterization of a pollen-specific, putative monosaccharide transporter. *Plant Physiology, 118*, 297–304. https://doi.org/10.1104/pp.118.1.297.

Young, J. N., et al. (2016). Large variation in the Rubisco kinetics of diatoms reveals diversity among their carbon-concentrating mechanisms. *Journal of Experimental Botany, 67*(11), 3445–3456. https://doi.org/10.1093/jxb/erw163.

Yu, H., et al. (2018). Augmenting the Calvin-Benson-Bassham cycle by a synthetic malyl-CoA-glycerate carbon fixation pathway. *Nature Communications, 9*(1), 2008–2017. https://doi.org/10.1038/s41467-018-04417-z.

Zelitch, I. (1982). The close relationship between net photosynthesis and crop yield. *Bioscience, 32*, 796–802. https://doi.org/10.2307/1308973.

Zhang, T., & Folta, K. M. (2012). Green light signaling and adaptive response. *Plant Signaling & Behavior*, 7(1), 75–78. https://doi.org/10.4161/psb.7.1.18635.

Zhang, L., et al. (2005). A monosaccharide transporter is localized to sieve plate and plasmodesmal channel in de-veloping apple fruit. *Chinese Science Bulletin, 50*(5), 430–435. https://doi.org/10.1360/982004-700.

Zhang, H., et al. (2019). QTG-Seq accelerates QTL fine mapping through QTL partitioning and whole-genome sequencing of bulked segregant samples. *Molecular Plant, 12*(3), 426–437. https://doi.org/10.1016/j.molp.2018.12.018.

Zhao, C., et al. (2017). Temperature increase reduces global yields of major crops in four independent estimates. *Proceedings of the National Academy of Sciences of the United States of America, 114*, 9326–9331. https://doi.org/10.1073/pnas.1701762114.

Zhao, C., et al. (2019). Crop phenomics: Current status and perspectives. *Frontiers in Plant Science, 10*, 714–730. https://doi.org/10.3389/fpls.2019.00714.

Zhou, J. M., & Zhang, Y. (2020). Plant immunity: Danger perception and signaling. *Cell, 181*(5), 978–989. https://doi.org/10.1016/j.cell.2020.04.028.

Zhou, Y., et al. (2007). A suite of sucrose transporters expressed in coats of developing legume seeds includes novel pH-independent facilitators. *The Plant Journal, 49*(4), 750–764. https://doi.org/10.1111/j.1365-313X.2006.03000.x.

Zhou, Z., et al. (2015). Resequencing 302 wild and cultivated accessions identifies genes related to domestication and improvement in soybean. *Nature Biotechnology, 33*(4), 408–414. https://doi.org/10.1038/nbt.3096.

Zhu, X., et al. (2008). What is the maximum efficiency with which photosynthesis can convert solar energy into biomass? *Current Opinion in Biotechnology, 19*(2), 153–159. https://doi.org/10.1016/j.copbio.2008.02.004.

Zhu, X., et al. (2010). Improving photosynthetic efficiency for greater yield. *Annual Review of Plant Biology, 61*(1), 235–261. https://doi.org/10.1146/annurev-arplant-042809-112206.

Zhu, X., et al. (2020). Mutation of YL results in a yellow leaf with chloroplast rna editing defect in soybean. *International Journal of Molecular Sciences, 21*(12), 4275–4290. https://doi.org/10.3390/ijms21124275.

CHAPTER SIX

Soybean CLE peptides and their CLAVATA-like signaling pathways

Candice H. Jones, April H. Hastwell, Peter M. Gresshoff, and Brett J. Ferguson*

Integrative Legume Research Group, School of Agriculture and Food Sciences, The University of Queensland, Brisbane, Australia

*Corresponding author: e-mail address: b.ferguson1@uq.edu.au

Contents

Abstract

Plants coordinate development through complex signaling networks involving many different types of molecules, genes, proteins, and hormones. CLAVATA3-Endosperm Surrounding Region-related (CLE) peptides are one such family of signaling molecules that have roles in general plant development as well during interactions with abiotic and biotic factors. Members of the CLE peptide family have been well studied in the model species, *Arabidopsis thaliana*; with extensive CLV3, CLE40 and TDIF signaling pathways having been identified. These CLE peptide pathways tend to have three core elements including a CLE peptide, which is perceived by a Receptor-Kinase that regulates a WUSCHEL-like homeobox transcription factor. Such elements are also present in the Autoregulation of Nodulation pathway that controls nodule number during the legume-rhizobia symbiosis.

Advances in Botanical Research, Volume 102
ISSN 0065-2296
https://doi.org/10.1016/bs.abr.2022.02.006

In the soybean genome, 84 CLE peptide-encoding genes are present, although little research outside of nodulation has been undertaken to understand how they modulate growth in the model legume. Understanding CLE peptides may provide novel targets to improve soybean growth and productivity. Moreover, it also represents a unique opportunity to study CLE peptides in the context of beneficial symbiosis with rhizobia and mycorrhizae of which *A. thaliana* cannot undergo. This chapter reviews the current understanding of CLE peptide signaling in plants, what is currently known about soybean orthologues, and avenues that we can target to improve soybean development using the CLE peptides and their associated molecular networks.

1. Introduction

During their lifetime, a plant will experience many environmental changes and extremes which mean they must maintain plasticity to survive and reproduce. This is especially important for commercial crops like soybean, where their value is directly linked to their ability to yield seed in challenging conditions. Plant development and plasticity is regulated on a molecular scale by complex pathways with numerous signals and crosstalk. With an advance in genomic and proteomic technologies, there has been an increase in the discovery of small signaling molecules, knowns as peptides, that form key components of the pathways that control plasticity and general development. One of the most well studied signal peptide families is the CLAVATA3 (CLV3)/EMBRYO SURROUNDING REGION (ESR)-related (CLE) peptide family.

CLE peptides are small signaling peptides with critical roles in regulating diverse aspects of plant development. The earliest CLE peptides were discovered in the shoot apical meristem (SAM) of *Arabidopsis thaliana* (AtCLV3; Clark, Running, & Meyerowitz, 1995) and the endosperm of *Zea mays* (ZmESR; Opsahl-Ferstad, Deunff, Dumas, & Rogowsky, 1997). Since then, CLE peptides have become the most well-characterized plant peptide family and have been identified in many other plants and plant-symbionts, including early plant species including bryophytes (Cock & McCormick, 2001; Hirakawa et al., 2020; Le Marquer, Bécard, & Frei dit Frey, 2019; Whitewoods et al., 2018). Importantly, CLE peptides have been shown to have significant economic and agricultural potential in fruit size, grain yield and drought response (Liu et al., 2021; Takahashi et al., 2018; Xu et al., 2015). Despite being a well-studied signaling plant peptide family, recent genome-wide identification studies throughout the plant kingdom have demonstrated that there are significantly more family members without a known function (Han, Zhang, Wu, & Wang, 2016; Hastwell, de Bang, Gresshoff, & Ferguson, 2017; Hastwell, Gresshoff, & Ferguson, 2015a; Je et al., 2016; Zhang, Yang, Song, & Wang, 2014).

Genome-wide studies identified 84 CLE peptide encoding genes in soybean (Hastwell et al., 2015a). Other model organisms such as *Arabidopsis thaliana* have 32 CLE peptide encoding genes (Cock & McCormick, 2001) and the closely related *Phaseolus vulgaris* has 46 (Hastwell et al., 2015a). The larger number present in soybean is largely due to a whole genome duplication event which led to most genes having an "a" and "b" copy with extremely high similarity (Schmutz et al., 2010). As a result, there is a high level of redundancy in soybean CLE peptide signal pathways, allowing for neo-functionalization, although this can make the pathways in soybean more difficult to study at the biological level. Fortunately, other model legumes with less genetic redundancy can be used to functionally characterize orthologous CLE peptides and the findings can be carefully extrapolated and confirmed in soybean. The soybean CLE peptides are notably present 2:1 with those of *P. vulgaris* with high homology, with few exceptions, making it the most suitable for comparisons (Hastwell et al., 2015a; Schmutz et al., 2014). It is interesting to note that other model legumes *Lotus japonicus* and *Medicago truncatula* have 52 and 53 CLE peptide encoding genes, respectively, which is much more than *A. thaliana* (Hastwell et al., 2017). One reason that legumes may have an expanded gene family is that CLE peptides are involved in pathways that regulate beneficial root-microbe interactions such as nodulation and mycorrhization that do not occur in *A. thaliana*.

Across the well-studied CLAVATA related pathways, central elements exist including the CLE peptide, Receptor-Like Kinase (RLK), and Homeobox transcription factor (CLE-RLK-WUS). This can be observed in meristematic regions across plant development, including in legume nodulation, and is discussed in further detail below (Corcilius et al., 2017; Depuydt et al., 2013; Kitagawa & Jackson, 2019; Osipova et al., 2012). The consistencies between these molecular pathways in different developmental tissues may be key to identifying molecular components in paralogous mechanisms, as well as new mechanisms that are currently unknown. An evolutionary context could also prove to be a useful tool to unraveling the complexities of CLE peptide signaling, such as that provided in Whitewoods et al. (2018), which examines the expansion of the CLAVATA gene families.

2. CLE peptide structure and function

The structure and activity of CLE peptides has been well reviewed in Gao and Guo (2012) and Hastwell, Gresshoff, and Ferguson (2015b). To summarize, CLE prepropeptides have 3–4 domains; a signal peptide, a variable region, a highly conserved CLE domain and sometimes a C-terminal extension (Hastwell et al., 2015b). The defining feature of CLE peptides is the highly conserved 12–13 amino acid CLE domain (Cock & McCormick, 2001; Oelkers et al., 2008) which is excised and post-translationally modified to form a mature CLE peptide (Ni, Guo, Jin, Hartsell, & Clark, 2011; Ohyama, Shinohara, Ogawa-Ohnishi, & Matsubayashi, 2009; Okamoto, Shinohara, Mori, Matsubayashi, & Kawaguchi, 2013). This then binds directly with receptors to determine pathway specificity and the biological function (Li, Chakraborty, & Xu, 2017; Morita et al., 2016; Ogawa, Shinohara, Sakagami, & Matsubayashi, 2008; Okamoto et al., 2013; Zhang, Lin, Han, Qu, & Chai, 2016). The signal peptide is also essential for biological function and specificity of CLE peptides as it determines the processing and transport (Lim, Lee, & Hwang, 2011; Meng, Ruth, Fletcher, & Feldman, 2010; Ni et al., 2011; Reid, Li, Ferguson, & Gresshoff, 2013). The exact function of the variable region has not yet been determined as in other similar tripartite peptide families, the variable region is inessential (Fernandez et al., 2015; Fiers et al., 2006), however, a subtilase cleavage site was reported within this region of AtCLE40 (Stührwohldt, Ehinger, Thellmann, & Schaller, 2020). Lastly, approximately 23% of soybean CLE peptides have a ~6–9 amino acid C-terminal extension, the role of which is unclear, but it may protect against protease enzymes during systemic transport owing to their presence in long distance transport of nodulation-suppressive CLE peptides and AtCLE25 in drought responses (Hastwell et al., 2015a; Ni et al., 2011; Takahashi et al., 2018).

While there is good understanding of the gene structure-function relationship (Li et al., 2017; Reid et al., 2013; Song, Guo, Ren, Xu, & Liu, 2013), the functional structure of mature CLE peptides has been more difficult to determine. A hydroxyproline *O*-arbinosyltransferase has been found to glycosylate CLE peptides in both the shoot and nodulation pathways (Kassaw, Nowak, Schnabel, & Frugoli, 2017; Li, Kinkema, & Gresshoff, 2009; Novák, 2010; Schnabel et al., 2011; Xu et al., 2015; Yoro et al., 2019; Yoshida, Funayama-Noguchi, & Kawaguchi, 2010). The exact role of this modification is not yet known, but they are thought to potentially impact protein folding and binding with receptors and possibly also aid in peptide stability. Studies with synthetic peptides have shown that arabinosylation increases the biological activity of GmCLE40a in the root apical meristem (Corcilius et al., 2017) and that nodulation suppressive CLE peptides in *Lotus japonicus*, *Pisum sativum* and *Medicago truncatula* require three arabinose sugars to function (Hastwell et al., 2019; Imin, Patel, Corcilius, Payne, & Djordjevic, 2018; Okamoto et al., 2013). Three native CLE peptides have been isolated and structurally characterized; AtCLV3 and LjCLE-RS2 were found to be triarabinosylated, while AtCLE25 was detected only in a dihydroxyproline form (Ohyama et al., 2009; Okamoto et al., 2013; Takahashi et al., 2018). Conversely, AtCLV3 has also been shown to function with and without any post-translational modifications at varying peptide lengths (12 and 13 amino acids; Kim, Wu, Yu, Sheen, & Lee, 2017).

Phylogenetic analysis of soybean CLE peptides identified seven distinct groups based on sequence similarity, particularly within the CLE domain, which may be indicative of function (Table 1; Hastwell et al., 2015a). Of these groups I, III, VI and VII are the most well studied. Group I includes CLV3 and CLE40 which regulate stem cell differentiation in the shoot and root apical meristems, these are notably some of the few CLE peptides that also contain introns. Group III is mostly made up of peptides that share a distinct CLE peptide sequence with TRACHEARY ELEMENT DIFFERENTIATION INHIBITORY FACTOR (TDIF) peptides that regulate xylem development, and one uncharacterised CLE13. These TDIF-related peptides are unique in that they show biological activity without arabinosylation (Zhang et al., 2016). Group VI is the smallest group and contains RHIZOBIA INDUCED CLE (RIC) peptides which regulate nodulation through a well characterized systemic autoregulation pathway (Ferguson et al., 2019; Reid et al., 2011). Group VII is large and has many uncharacterised members, although it contains the NITRATE INDUCED CLE (NIC) peptides which regulate nodulation through a locally acting pathway that is separate to the autoregulation pathway (Reid et al., 2011). Interestingly, the orthologues to several of the uncharacterised peptides in Group VII have been examined in other species and found to have roles in regulating aspects of root development and other root-microbe symbioses (Araya et al., 2014; de Bang et al., 2017; Handa et al., 2015; Karlo et al., 2020). For the remaining Groups II, IV and V, very little is known about the function of these peptides in soybean, or their orthologues in other species. Interestingly, phylogenetic analyses using AtCLE25 which is associated with stomatal control and drought tolerance established that it groups closely

Table 1 The soybean CLE peptide family Groups I to VII (Hastwell et al., 2015a) and their majority consensus sequences, along with functionally characterized orthologous peptides.

Group	Soybean members	Consensus CLE domain	Functional examples	Example References
I	GmCLV3a GmCVL3b GmCLE40a GmCLE40b	RKVPSGPDPLHHN	CLV3 - SAM maintenance CLE40 - RAM maintenance SlCLV3 - regulates fruit size ZmCLE7 (AtCLV3 orthologue) - influences grain yield	Ohyama et al. (2009); Wong, Singh, & Bhalla (2013) Corcilius et al. (2017) Xu et al. (2015) Liu et al. (2021)
II	GmCLE1a GmCLE1b GmCLE2a GmCLE2b GmCLE3a GmCLE3b GmCLE4a GmCLE4b GmCLE5 GmCLE6a GmCLE6b GmCLE7a GmCLE7b GmCLE8a GmCLE8b GmCLE9a GmCLE9b GmCLE10a GmCLE10b GmCLE11a GmCLE11b GmCLE12a GmCLE12b	RRVPSGPDP[I/L] HNR	AtCLE25 (Possible ortholog of GmCLE12a/b) - Stomata control and drought tolerance AtCLE45 (Possible ortholog of GmCLE7–9) - Phloem development	Takahashi et al. (2018); Qian et al., 2018

Table 1 The soybean CLE peptide family Groups I to VII (Hastwell et al., 2015a) and their majority consensus sequences, along with functionally characterized orthologous peptides.—cont'd

Group	Soybean members	Consensus CLE domain	Functional examples	Example References
III	GmCLE13 GmTDIF1a GmTDIF1b GmTDIF2a GmTDIF2b GmTDIF3a GmTDIF3b	HEVPSGPNPISNR	TDIFs - vascular development	Etchells, Smit, Gaudinier, Williams, & Brady (2016)
IV	GmCLE14 GmCLE15a GmCLE15b GmCLE16a GmCLE16b GmCLE17a GmCLE17b	RXVPGGPNPLHN		
V	GmCLE18a GmCLE18b GmCLE19a GmCLE19b GmCLE20a GmCLE20b GmCLE21a GmCLE21b GmCLE22a GmCLE22b GmCLE23a GmCLE23b GmCLE24a GmCLE24b GmCLE25a GmCLE25b GmCLE26 GmCLE27a GmCLE27b	RLVPTGPNPLHNR	AtCLE9 (Possible orthologue of GmCLE24a/b and -25a/b) - ABA-dependent regulation of stomatal opening	Zhang et al. (2019)
VI	GmRIC1a GmRIC1b GmRIC2a GmRIC2b	RLAP[E/G]GPDPX [H/Q]N[F/V]	RIC1/2- autoregulation of nodulation	Reid, Ferguson, and Gresshoff (2011)

Continued

Table 1 The soybean CLE peptide family Groups I to VII (Hastwell et al., 2015a) and their majority consensus sequences, along with functionally characterized orthologous peptides.—cont'd

Group	Soybean members	Consensus CLE domain	Functional examples	Example References
VII	GmCLE28a GmCLE28b GmCLE29a GmCLE29b GmCLE30a GmCLE30b GmCLE31a GmCLE31b GmCLE32 GmCLE33a GmCLE33b GmCLE34a GmCLE34b GmCLE35 GmCLE36a GmCLE36b GmNIC1a GmNIC1b	RLSPGGPDPHHHF	NIC1 - nitrate regulated control of nodulation AtCLE1/3/4/7 - lateral root development, nitrate foraging MtCLE33/E53 - regulate mycorrhizal symbiosis	Reid, et al. (2011) Araya et al. (2014) Karlo et al. (2020)

to GmCLE10a/b and GmCLE12a/b which are orthologous to members of Group II, however this has not been followed up in in any model legumes (Hastwell et al., 2015a; Takahashi et al., 2018). Research into these uncharacterised CLE peptides could unlock new knowledge about plant development and plasticity, with potential applications that could improve adaptation of crops such as soybean in commercial agriculture.

3. CLE peptides in soybean meristems

3.1 CLV3 in SAM

The earliest CLE peptide to be identified was CLAVATA3 (CLV3) which was discovered for its role in regulating stem cell differentiation in the SAM in *A. thaliana* (Clark et al., 1995; Fletcher, Brand, Running, Simon, & Meyerowitz, 1999). Since then, highly conserved CLV3 orthologues have been identified in many dicot species including soybean and other model species (Hastwell et al., 2015a, 2017; Wong et al., 2013; Zhang et al., 2014). Although most of the CLV3 functional characterization has been conducted in Arabidopsis and tomato, it is assumed that the pathway is conserved across dicots. Interestingly in soybean, the CLV3 peptide is encoded by two genes, *GmCLV3a* and *GmCLV3b* (Wong et al., 2013). In the SAM, CLV3 peptides are expressed and triarabinosylated in the central zone, with *GmCLV3b* being expressed at a higher level than *GmCLV3a*, before being secreted to the SAM organizing centre where they are perceived by the leucine rich repeat (LRR)-RLK CLAVATA1 (CLV1) (Fig. 1A; Clark, Williams, & Meyerowitz, 1997; Ohyama et al., 2009; Wong et al., 2013; Xu et al., 2015; Mirzaei et al., 2017). Mutants of *GmCLV1a* have a similar fasciated floral development to that of Arabidopsis (Mirzaei et al., 2017). Other receptor-like proteins are capable of binding to CLV3 either as a CLV1 homodimer or heterodimer with orthologous of GmCLV2, GmCRN or GmRPK2 (Guo et al., 2015; Nimchuk, Tarr, & Meyerowitz, 2011; Nimchuk, Tarr, Ohno, Qu, & Meyerowitz, 2011; Shinohara & Matsubayashi, 2015; Somssich, Je, Simon, & Jackson, 2016; Stone, Trotochaud, Walker, & Clark, 1998). Perception of CLV3 in the organizing centre promotes cell differentiation by triggering a signal cascade to repress WUSCHEL (WUS) transcription factors involved in maintaining stem cell identity (Brand, Fletcher, Hobe, Meyerowitz, & Simon, 2000; Haecker et al., 2004; Schoof et al., 2000; Sun et al., 2019, 2019; Zhou et al., 2015). This is conserved in soybean, although *GmWUS* shows an additional role in floral initial not seen in *A. thaliana* (Wong, Khor, Bhalla, & Singh, 2011). WUS expression promotes CLV3 production, creating a negative CLV3-WUS feedback loop which maintains a perfect balance between SAM stem cell proliferation and differentiation (Brand et al., 2000; Schoof et al., 2000). Interestingly, it was recently shown that during the domestication of tomatoes CLV3 mutations were inadvertently selected for to increase fruit size through fasciation (Xu et al., 2015); and that CLV3 mutations in *Brassica napus* and maize also have the potential to increase yield (Liu et al., 2021; Yang et al., 2018) Furthermore, upstream of the CLV3-WUS feedback loop is miR156b, a repressor of SQUAMOSA PROMOTER BINDING PROTEIN 3 (SBP3) (Gandikota et al., 2007), and the former of which can increase seed weight when overexpressed in soybean, leading to an increase in yield (Sun, Su, et al., 2019; Sun, Zhou, et al., 2019). Considering this, directly modifying the CLV3 pathway could be an ideal target for increasing soybean seed production and yields.

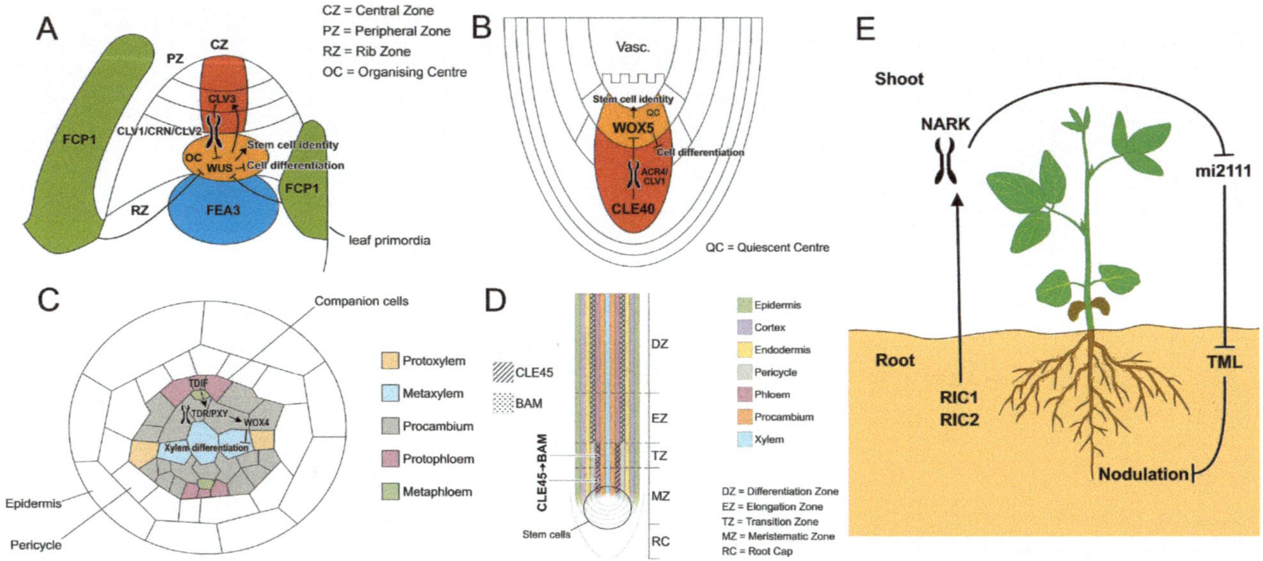

Fig. 1 Conserved CLAVATA-like signaling pathways. (A) CLAVATA3 (CLV3) pathway regulates stem cell maintenance in the shoot apical meristem (SAM). CLV3 is perceived by CLV1 homo- or heterodimers which leads to inhibition of WUSCHEL (WUS). A negative feedback look is created as WUS regulates CLV3 expression. (B) CLE40 pathway regulates stem cell maintenance in the root apical meristem (RAM). CLE40 is perceived by CLV1 and/or ACR4 homo or heterodimers leading to the inhibition of WOX5. (C) TDIF pathway regulates xylem development. TDIF peptides are expressed in the phloem and perceived by TDR/PXY in the procambial stem cells. Perception of TDIFs up-regulates WOX4 and WOX14 which promote proliferation of procambial stem cells. (D) CLE45 pathway regulates phloem development. CLE45 is expressed in the meristematic and transition zones and perceived by BAM3 in the elongation and differentiation zones resulting in reduced protophloem differentiation. (E) Rhizobia Induced CLE peptides (RIC1 and RIC2) regulate nodule development in the Autoregulation of Nodulation (AON) pathway. RIC1 and RIC2 are transported from the roots to the shoot through the xylem to interact with NARK. Perception leads to inhibition of miR2111 which regulates other downstream factors.

3.2 CLE40 in RAM

The root homologue of CLV3 is CLE40 which promotes cell differentiation in the root apical meristem (RAM) (Corcilius et al., 2017; Fiers et al., 2005; Hobe, Müller, Grünewald, Brand, & Simon, 2003; Stahl, Wink, Ingram, & Simon, 2009). Unlike CLV3 which is expressed in stem cells, CLE40 is expressed in differentiated columella cells (Stahl et al., 2009). Once secreted, CLE40 is perceived by a RLK, ARABIDOPSIS CRINKLY4 (ACR4) and CLV1 homo- or heterodimers in the distal root meristem (Fig. 1B; De Smet et al., 2008; Stahl et al., 2009; Guo, Han, Hymes, Denver, & Clark, 2010; Stahl et al., 2013). Perception of CLE40 leads to the repression of a WUS homologue, WOX5, in the quiescent centre which inhibits differentiation of columella daughter cells (Osipova et al., 2012; Sarkar et al., 2007; Stahl et al., 2009). *Medicago truncatula* WOX5 has been further implicated in legume nodulation and genome-wide analyses have been performed to identify orthologues in soybean (Hao, Zhang, Yang, Shan, & Zhou, 2019; Osipova et al., 2012). In soybean, GmCLE40a synthetic peptide application arrests root growth, presumably as the meristematic cells are forced to differentiate (Corcilius et al., 2017). Unlike CLV3 in the SAM, no feedback loop between WOX5 and CLE40 has been identified, but a WOX5 and CLE40 gradient is still required to maintain the quiescent centre at a specified distance from the root tip (Berckmans, Kirschner, Gerlitz, Stadler, & Simon, 2020). Synthetic GmCLE40a peptides require both hydroxylation and triarabinosylation for optimal biological activity (Corcilius et al., 2017) and subtilase processing of the AtCLE40 prepropeptide relies on hydroxylation of the fourth proline (Stührwohldt et al., 2020). While the endogenous AtCLE40 peptide lacks the central seventh proline which is glycosylated in structurally confirmed CLE peptides and is present in the soybean orthologue, the serine in this position is still amenable to *O*-glycosylation (Corcilius et al., 2017; van den Steen, Rudd, Dwek, & Opdenakker, 1998).

3.3 Vascular development

CLE peptides regulate vascular development through unique xylem and phloem pathways (Depuydt et al., 2013; Ito et al., 2006; Wang, 2020). The archetypical peptides such as CLV3 and CLE40 have an arginine at position one of the CLE domain and promote cell differentiation, whereas the slightly atypical peptides in vascular development reduce differentiation. The latter peptide domains begin with histidine and are far fewer in number, these are known as the TRACHEARY ELEMENT DIFFERENTIATION INHIBITORY FACTOR (TDIF) peptides (Ito et al., 2006; Whitford, Fernandez, De Groodt, Ortega, & Hilson, 2008). Seven soybean TDIF-related peptides have been bioinformatically identified and no functional characterization has been conducted, however, it is understood that the molecular mechanisms controlling vascular differentiation are highly conserved in both angiosperms and gymnosperms (Group III, Table 1; Hastwell et al., 2015a; Hirakawa & Bowman, 2015). TDIF peptides are expressed in the phloem and perceived by a LRR RLK called TDIF RECEPTOR (TDR)/PHLOEM INTERCALATED WITH XYLEM (PXY) in the procambial stem cells (Fig. 1C; Fisher & Turner, 2007; Hirakawa et al., 2008; Zhang et al., 2016). Perception of TDIFs by TDR/PXY up-regulates WOX4 and WOX14 which promote proliferation of procambial stem cells unlike their homologues WUS and WOX5 (Etchells, Provost, Mishra, & Turner, 2013; Hirakawa et al., 2008). TDIF-TDR/PXY signaling upregulates Glycogen Synthase Kinase 3 (GSK3), a key regulator of both xylem and phloem development (Kondo et al., 2014; Tamaki et al., 2020). Directly downstream, the BRI1-EMS SUPPRESSOR 1 (BES1) transcription factor is suppressed to reduce procambial stem cell differentiation to xylem (Kondo et al., 2014). Soybean BES1 orthologues have also been identified (Li et al., 2019).

Recently other CLE peptides have been identified to influence xylem development. In poplar, PtrCLE41 and PtrCLE20 are expressed in the phloem and xylem respectively to coordinate secondary growth (Zhu et al., 2020). AtCLE9 and AtCLE10 are required for differentiation of metaxylem, with the former also being implicated in stomata development (Qian et al., 2018). Based on phylogenetic analysis the soybean orthologues of AtCLE9 and AtCLE10 are likely to be GmCLE24a/b or GmCLE25a/b or similar genes located in Group V (Table 1; Hastwell et al., 2015a).

In the phloem development pathway CLE45 is expressed in the meristematic and transition zones and perceived by BARELY ANY MERISTEM3 (BAM3) in the elongation and differentiation zones resulting in reduced protophloem differentiation (Fig. 1D; Depuydt et al., 2013; Hazak et al., 2017). Closely related soybean sequences of CLE45 include GmCLE7-GmCLE9 (Hastwell et al., 2015a). Signals acting downstream of BAM3 have not yet been characterized, although it is possible that WOX transcription factors are involved given the conserved pattern of CLE-WOX signal pathways described here. Identifying orthologues of the vascular development

CLE peptides and their receptors in soybean could benefit the selection of future soybean crop varieties and establish targets for biotechnology approaches that improve growth and yield.

4. CLE peptides regulating symbiosis

4.1 Rhizobia-induced CLE peptides

Nodulation is a species-specific symbiosis that occurs between legumes and nitrogen fixing soil bacteria called rhizobia. It benefits the legume by providing nitrogen but has a high-energy cost so it is tightly regulated by the host plant to maximize resource use. There are two distinct pathways known to regulate nodule development: autoregulation of nodulation (AON) and nitrate regulation of nodulation (Ferguson et al., 2019). In AON, rhizobia infection triggers the production of CLE peptides in the roots; these are known as rhizobia-induced CLE peptide 1 (RIC1) and RIC2 in *G. max* and *P. vulgaris;* CLE-root signal 1 (LjCLE-RS1), LjCLE-RS2 and LjCLE-RS3 in *L. japonicus*, MtCLE12 and MtCLE13 in *M. truncatula* and (Ferguson et al., 2014; Mortier et al., 2010; Nishida, Handa, Tanaka, Suzaki, & Kawaguchi, 2016; Okamoto et al., 2009; Reid et al., 2011). The mature rhizobia-induced CLE peptides are transported from the roots to the shoot through the xylem to interact with NODULE AUTOGREGULATION RECEPTOR KINASE (NARK) in *G. max* and *P. vulgaris*, a CLV1-like LRR RK or HYPERNODULATION ABERRANT ROOT FORMATION (HAR1) in *L. japonicus*, SUPER NUMERIC NODULES (SUNN) in *M. truncatula* (Fig. 1E; Okamoto et al., 2009; Mortier et al., 2010; Reid et al., 2011; Okamoto et al., 2013; Ferguson et al., 2014). Structural analysis of mature LjCLE-RS2 peptides revealed that they are triarabinosylated and that this is essential for interacting with LjHAR1 (Okamoto et al., 2013). The mature structure of other nodulation CLE peptides has not yet been analyzed, but only triarabinosylated synthetic GmRIC1 and GmRIC2 peptides can inhibit nodule formation, indicating that the glycosylation may be critical and conserved across species (Hastwell et al., 2019). However, research in *M. truncatula* revealed that MtCLE12 requires a functional arabinosyltransferase to act through SUNN whereas MtCLE13 does not (Kassaw et al., 2017). Perception of the relevant CLE peptides by the shoot NARK/HAR1/SUNN down-regulates miR2111 production and translocation to the root, where it would have otherwise targeted TOO MUCH LOVE (TML), a negative regulator of nodulation, this complex downstream signal cascade thus inhibits future nodulation events (Fig. 1. Ferguson et al., 2019; Tsikou et al., 2018; Zhang, Su, Gresshoff, & Ferguson, 2021). Similar to SAM and RAM maintenance, *MtWOX5* is expressed in nodules during their development and inhibited downstream of SUNN, supporting the idea that there is a conserved CLE-RLK-WUS pathway involved in nodulation control (Osipova et al., 2012).

4.2 Nitrogen-induced CLE peptides

The primary benefit of nodulation to the host plant is access to nitrogen, therefore if the plant can freely access nitrogen from the rhizophere there is no need to invest energy into nodulation. For this reason, legumes have developed a nitrogen-dependent regulation of nodulation pathway to optimize resource use. When sufficient nitrogen is available in the soil, nitrogen-induced CLE peptides are produced in roots and interact with NARK/HAR1/SUNN to suppress nodulation. The nitrogen-induced CLE peptides currently known in legumes are *GmNIC1a*, *LjCLE-RS2*, *LjCLE-RS3*, *LjCLE40*, MtCLE34 and *MtCLE35* (Mens et al., 2021; Nishida et al., 2016; Okamoto et al., 2009; Reid et al., 2011). GmNIC1a appears to interact with GmNARK in the roots to locally inhibit nodulation and is an orthologue of LjCLE40 and MtCLE34 (Mens et al., 2021; Reid et al., 2011). Whereas LjCLE-RS2 inhibits nodulation locally and systemically and is more closely related to MtCLE35 and GmRIC1a/b (Mens et al., 2021; Okamoto et al., 2009; Reid et al., 2011). The signals downstream of NARK in a locally acting pathway in the root have not yet been identified in the nitrogen-dependent pathway, but it would be worthwhile investigating WOX genes and microRNAs as they have been identified in AON and are conserved across other CLE peptide pathways (Zhang et al., 2021).

4.3 Mycorrhiza symbiosis

There is a high number of common symbiosis genes shared between nodulation and mycorrhization, which is plant-fungi symbiosis that occurs more broadly across land plant species, with a notable exception being Arabidopsis and other *Brassica spp*. (Bonfante & Genre, 2010; Parniske, 2004). Several studies conducted in soybean have identified that some molecular components from AON (including NARK) also regulate mycorrhiza symbiosis (Catford, Staehelin, Lerat, Piché, & Vierheilig, 2003; Meixner et al., 2005; Morandi, Sagan, Prado-Vivant, & Duc, 2000; Schaarschmidt, Gresshoff, & Hause, 2013). Indeed, an AON-like mechanism is also proposed for mycorrhizal regulation as shared receptors have been identified (Meixner et al., 2005; Müller et al., 2019). It is likely that nodulation regulatory pathways were borrowed from existing mycorrhizal regulatory pathways since the mycorrhizal symbiosis evolved >400 million years ago whereas nodulation is ~55 million years old (Remy, Taylor, Hass, & Kerp, 1994; Sprent & James, 2007). Phosphate- and mycorrhiza-responsive CLE peptide-encoding genes LjCLE18, LjCLE19, MtCLE33 and MtCLE53 have been identified in L. *japonicus* and *M. truncatula* (de Bang et al., 2017; Funayama-Noguchi, Noguchi, Yoshida, & Kawaguchi, 2011; Handa et al., 2015; Karlo et al., 2020; Müller et al., 2019). Phylogenetic analysis indicates that the soybean orthologue to these phosphate- and mycorrhiza responsive CLE peptides is likely to be in Group VII (Table 1; Hastwell et al., 2015a). It would be valuable to investigate this group of CLE peptides and further characterize the symbiosis-related pathways in soybean as they could benefit the improvement of crop-microbe relationships in agriculture.

5. CLE peptides of soybean symbionts

From a plant perspective, CLE peptides and the pathways they are involved in give plants the ability to optimize or control their relationship with microorganisms. Interestingly, some organisms outside the plantae kingdom produce CLE peptides presumably to mimic plant signals and hijack existing pathways which enables them to form close associations with host plants. So far CLE peptides have been identified in arbuscular mycorrhiza and sedentary plant-parasitic nematodes, both of which form very close associations with plant hosts (Guo, Ni, Denver, Wang, & Clark, 2011; Le Marquer et al., 2019).

5.1 CLE peptides of mycorrhiza

Arbuscular mycorrhiza species also produce CLE peptides that alter root architecture under symbiotic conditions (Le Marquer et al., 2019). It is interesting to note that the mycorrhizal CLE peptides have a signal peptide and CLE motif but no variable region between them (Le Marquer et al., 2019). Since mycorrhiza and early land plants evolved together so closely it is difficult to know which organism would have developed CLE peptides first. CLE pathways may have been essential for that early symbiosis that enabled plants to colonies land (Whitewoods et al., 2018). Further research into the role of CLE peptides in mycorrhiza-plant interactions may help us to better understand the early evolution of these complex and important pathways.

5.2 CLE peptides of nematodes

Sedentary plant-parasitic nematodes secrete CLE peptides from their dorsal esophageal glands into plant cells (Wang et al., 2001; Olsen & Skriver, 2003; Wang et al., 2005). The nematode CLE peptides are then post-translationally glycosylated and perceived by the plant leading to developmental reprogramming of root tissue to produce active feeding cells (Wang et al., 2010; Chen et al., 2015; Guo et al., 2011; Wang et al., 2011). Interestingly, knocking-down CLE peptide receptors including CLV1, CLV2 and RPK2 in hairy-roots leads to an increase in resistance to *Heterodera glycines* and *Globodera rostochiensis* infection in soybean and potato, respectively (Chen et al., 2015; Guo et al., 2015). This also occurs in Arabidopsis, where CLV2 and CRN is required for successful infection by *Heterodera schachtii*. This demonstrates the potential and agricultural importance of research into these highly complex molecular pathways.

6. Conclusions and future directions

There is great potential in targeting CLE peptides to improve agricultural crop productivity, as highlighted by the recent findings that certain CLV3 alleles increase fruit size in and grain yields (Liu et al., 2021; Xu et al., 2015), the importance of AtCLE25 to drought tolerance (Takahashi et al., 2018) and a greater resistance to nematode infection (Guo et al., 2015). However, further characterization is required in important legume crops such as soybean to understand their diverse and complex roles in regulating legume development and in using this information to improve crop production. CLE peptides involved in SAM, RAM, vascular development, nitrate response and nodulation have been characterized to various extents but there is also a need to further explore the role of CLE peptides in other abiotic and biotic responses and developmental processes. The recent identification of complete CLE peptide families across species along with advances in molecular biology and peptide synthesis create the perfect opportunity to thoroughly investigate the roles of these critical growth-regulating peptides.

Acknowledgments

This work was funded by the Australian Research Council Discovery Project grant (DP190102995) and DECRA grant (DE200100800). The Australian government is thanked for providing an RTP scholarship for CHJ. Vijay Barber is thanked contributing the production of digital illustrations used in this chapter. Dr. Huanan Su is thanked for developing the soybean illustration used in this chapter.

References

Araya, T., Miyamoto, M., Wibowo, J., Suzuki, A., Kojima, S., Tsuchiya, Y. N., et al. (2014). CLE-CLAVATA1 peptide-receptor signaling module regulates the expansion of plant root systems in a nitrogen-dependent manner. *Proceedings of the National Academy of Sciences*, *111*(5), 2029–2034.

Berckmans, B., Kirschner, G., Gerlitz, N., Stadler, R., & Simon, R. (2020). CLE40 signaling regulates root stem cell fate. *Plant Physiology*, *182*(4), 1776–1792.

Bonfante, P., & Genre, A. (2010). Mechanisms underlying beneficial plant–fungus interactions in mycorrhizal symbiosis. *Nature Communications*, *1*(1), 1–11.

Brand, U., Fletcher, J. C., Hobe, M., Meyerowitz, E. M., & Simon, R. (2000). Dependence of stem cell fate in Arabidopsis on a feedback loop regulated by CLV3 activity. *Science*, *289*(5479), 617–619.

Catford, J. G., Staehelin, C., Lerat, S., Piché, Y., & Vierheilig, H. (2003). Suppression of arbuscular mycorrhizal colonization and nodulation in split-root systems of alfalfa after pre-inoculation and treatment with nod factors. *Journal of Experimental Botany*, *54*(386), 1481–1487.

Chen, S., Lang, P., Chronis, D., Zhang, S., De Jong, W. S., Mitchum, M. G., et al. (2015). In planta processing and glycosylation of a nematode CLAVATA3/ENDOSPERM SURROUNDING REGION-like effector and its interaction with a host CLAVATA2-like receptor to promote parasitism. *Plant Physiology*, *167*(1), 262–272.

Clark, S. E., Running, M. P., & Meyerowitz, E. M. (1995). CLAVATA3 is a specific regulator of shoot and floral meristem development affecting the same processes as CLAVATA1. *Development*, *121*(7), 2057–2067.

Clark, S. E., Williams, R. W., & Meyerowitz, E. M. (1997). The *CLAVATA1* gene encodes a putative receptor kinase that controls shoot and floral meristem size in Arabidopsis. *Cell*, *89*(4), 575–585.

Cock, J. M., & McCormick, S. (2001). A large family of genes that share homology with CLAVATA3. *Plant Physiology*, *126*(3), 939–942.

Corcilius, L., Hastwell, A. H., Zhang, M., Williams, J., Mackay, J. P., Gresshoff, P. M., et al. (2017). Arabinosylation modulates the growth-regulating activity of the peptide hormone CLE40a from soybean. *Cell Chemical Biology*, *24*(11), 1347–1355.

de Bang, T. C., Lundquist, P. K., Dai, X., Boschiero, C., Zhuang, Z., Pant, P., et al. (2017). Genome-wide identification of *Medicago* peptides involved in macronutrient responses and nodulation. *Plant Physiology*, *175*(4), 1669–1689.

De Smet, I., Vassileva, V., De Rybel, B., Levesque, M. P., Grunewald, W., Van Damme, D., et al. (2008). Receptor-like kinase ACR4 restricts formative cell divisions in the Arabidopsis root. *Science*, *322*(5901), 594–597.

Depuydt, S., Rodriguez-Villalon, A., Santuari, L., Wyser-Rmili, C., Ragni, L., & Hardtke, C. S. (2013). Suppression of Arabidopsis protophloem differentiation and root meristem growth by CLE45 requires the receptor-like kinase BAM3. *Proceedings of the National Academy of Sciences*, *110*(17), 7074–7079.

Etchells, J. P., Provost, C. M., Mishra, L., & Turner, S. R. (2013). WOX4 and WOX14 act downstream of the PXY receptor kinase to regulate plant vascular proliferation independently of any role in vascular organisation. *Development*, *140*(10), 2224–2234. https://doi.org/10.1242/dev.091314.

Etchells, J. P., Smit, M. E., Gaudinier, A., Williams, C. J., & Brady, S. M. (2016). A brief history of the TDIF-PXY signalling module: Balancing meristem identity and differentiation during vascular development. *New Phytologist*, *209*(2), 474–484.

Ferguson, B. J., Li, D., Hastwell, A. H., Reid, D. E., Li, Y., Jackson, S. A., et al. (2014). The soybean (*Glycine max*) nodulation-suppressive CLE peptide, Gm RIC1, functions interspecifically in common white bean (*Phaseolus vulgaris*), but not in a supernodulating line mutated in the receptor *PvNARK*. *Plant Biotechnology Journal*, *12*(8), 1085–1097.

Ferguson, B. J., Mens, C., Hastwell, A. H., Zhang, M., Su, H., Jones, C. H., et al. (2019). Legume nodulation: The host controls the party. *Plant, Cell & Environment*, *42*(1), 41–51.
Fernandez, A., Drozdzecki, A., Hoogewijs, K., Vassileva, V., Madder, A., Beeckman, T., et al. (2015). The GLV6/RGF8/CLEL2 peptide regulates early pericycle divisions during lateral root initiation. *Journal of Experimental Botany*, *66*(17), 5245–5256.
Fiers, M., Golemiec, E., Van Der Schors, R., Van Der Geest, L., Li, K. W., Stiekema, W. J., et al. (2006). The CLAVATA3/ESR motif of CLAVATA3 is functionally independent from the nonconserved flanking sequences. *Plant Physiology*, *141*(4), 1284–1292.
Fiers, M., Golemiec, E., Xu, J., van der Geest, L., Heidstra, R., Stiekema, W., et al. (2005). The 14-amino acid CLV3, CLE19, and CLE40 peptides trigger consumption of the root meristem in Arabidopsis through a CLAVATA2-dependent pathway. *The Plant Cell*, *17*(9), 2542–2553.
Fisher, K., & Turner, S. (2007). PXY, a receptor-like kinase essential for maintaining polarity during plant vascular-tissue development. *Current Biology*, *17*(12), 1061–1066.
Fletcher, J. C., Brand, U., Running, M. P., Simon, R., & Meyerowitz, E. M. (1999). Signaling of cell fate decisions by CLAVATA3 in Arabidopsis shoot meristems. *Science*, *283*(5409), 1911–1914.
Funayama-Noguchi, S., Noguchi, K., Yoshida, C., & Kawaguchi, M. (2011). Two CLE genes are induced by phosphate in roots of *Lotus japonicus*. *Journal of Plant Research*, *124*(1), 155–163.
Gandikota, M., Birkenbihl, R. P., Höhmann, S., Cardon, G. H., Saedler, H., & Huijser, P. (2007). The miRNA156/157 recognition element in the 3′ UTR of the Arabidopsis SBP box gene SPL3 prevents early flowering by translational inhibition in seedlings. *The Plant Journal*, *49*(4), 683–693.
Gao, X., & Guo, Y. (2012). CLE peptides in plants: Proteolytic processing, structure-activity relationship, and ligand-receptor interaction. *Journal of Integrative Plant Biology*, *54*(10), 738–745.
Guo, X., Chronis, D., De La Torre, C. M., Smeda, J., Wang, X., & Mitchum, M. G. (2015). Enhanced resistance to soybean cyst nematode *Heterodera glycines* in transgenic soybean by silencing putative CLE receptors. *Plant Biotechnology Journal*, *13*(6), 801–810.
Guo, Y., Han, L., Hymes, M., Denver, R., & Clark, S. E. (2010). CLAVATA2 forms a distinct CLE-binding receptor complex regulating Arabidopsis stem cell specification. *The Plant Journal*, *63*(6), 889–900. https://doi.org/10.1111/j.1365-313X.2010.04295.x.
Guo, Y., Ni, J., Denver, R., Wang, X., & Clark, S. E. (2011). Mechanisms of molecular mimicry of plant CLE peptide ligands by the parasitic nematode Globodera rostochiensis. *Plant Physiology*, *157*(1), 476–484.
Haecker, A., Groß-Hardt, R., Geiges, B., Sarkar, A., Breuninger, H., Herrmann, M., et al. (2004). Expression dynamics of *WOX* genes mark cell fate decisions during early embryonic patterning in *Arabidopsis thaliana*. *Development*, *131*(3), 657–668.
Han, H., Zhang, G., Wu, M., & Wang, G. (2016). Identification and characterization of the *Populus trichocarpa* CLE family. *BMC Genomics*, *17*(1), 1–12.
Handa, Y., Nishide, H., Takeda, N., Suzuki, Y., Kawaguchi, M., & Saito, K. (2015). RNA-seq transcriptional profiling of an arbuscular mycorrhiza provides insights into regulated and coordinated gene expression in *Lotus japonicus* and *Rhizophagus irregularis*. *Plant and Cell Physiology*, *56*(8), 1490–1511.
Hao, Q., Zhang, L., Yang, Y., Shan, Z., & Zhou, X. A. (2019). Genome-wide analysis of the *WOX* gene family and function exploration of *GmWOX18* in soybean. *Plants*, *8*(7), 215.
Hastwell, A. H., Corcilius, L., Williams, J. T., Gresshoff, P. M., Payne, R. J., & Ferguson, B. J. (2019). Triarabinosylation is required for nodulation-suppressive CLE peptides to systemically inhibit nodulation in *Pisum sativum*. *Plant, Cell & Environment*, *42*(1), 188–197.

Hastwell, A. H., de Bang, T. C., Gresshoff, P. M., & Ferguson, B. J. (2017). CLE peptide-encoding gene families in *Medicago truncatula* and *Lotus japonicus*, compared with those of soybean, common bean and Arabidopsis. *Scientific Reports*, 7(1), 1–13.

Hastwell, A. H., Gresshoff, P. M., & Ferguson, B. J. (2015a). Genome-wide annotation and characterization of CLAVATA/ESR (CLE) peptide hormones of soybean (*Glycine max*) and common bean (*Phaseolus vulgaris*), and their orthologues of *Arabidopsis thaliana*. *Journal of Experimental Botany*, *66*(17), 5271–5287.

Hastwell, A. H., Gresshoff, P. M., & Ferguson, B. J. (2015b). The structure and activity of nodulation-suppressing CLE peptide hormones of legumes. *Functional Plant Biology*, *42*(3), 229–238.

Hazak, O., Brandt, B., Cattaneo, P., Santiago, J., Rodriguez-Villalon, A., Hothorn, M., et al. (2017). Perception of root-active CLE peptides requires CORYNE function in the phloem vasculature. *EMBO Reports*, *18*(8), 1367–1381.

Hirakawa, Y., & Bowman, J. L. (2015). A role of TDIF peptide signaling in vascular cell differentiation is conserved among euphyllophytes. *Frontiers in Plant Science*, *6*, 1048.

Hirakawa, Y., Fujimoto, T., Ishida, S., Uchida, N., Sawa, S., Kiyosue, T., et al. (2020). Induction of multichotomous branching by CLAVATA peptide in *Marchantia polymorpha*. *Current Biology*, *30*(19), 3833–3840.

Hirakawa, Y., Shinohara, H., Kondo, Y., Inoue, A., Nakanomyo, I., Ogawa, M., et al. (2008). Non-cell-autonomous control of vascular stem cell fate by a CLE peptide/receptor system. *Proceedings of the National Academy of Sciences*, *105*(39), 15208–15213. https://doi.org/10.1073/pnas.0808444105.

Hobe, M., Müller, R., Grünewald, M., Brand, U., & Simon, R. (2003). Loss of CLE40, a protein functionally equivalent to the stem cell restricting signal CLV3, enhances root waving in Arabidopsis. *Development Genes and Evolution*, *213*(8), 371–381.

Imin, N., Patel, N., Corcilius, L., Payne, R. J., & Djordjevic, M. A. (2018). CLE peptide tri-arabinosylation and peptide domain sequence composition are essential for SUNN-dependent autoregulation of nodulation in *Medicago truncatula*. *New Phytologist*, *218*(1), 73–80.

Ito, Y., Nakanomyo, I., Motose, H., Iwamoto, K., Sawa, S., Dohmae, N., et al. (2006). Dodeca-CLE peptides as suppressors of plant stem cell differentiation. *Science*, *313*(5788), 842–845.

Je, B. I., Gruel, J., Lee, Y. K., Bommert, P., Arevalo, E. D., Eveland, A. L., et al. (2016). Signaling from maize organ primordia via FASCIATED EAR3 regulates stem cell proliferation and yield traits. *Nature Genetics*, *48*(7), 785–791.

Karlo, M., Boschiero, C., Landerslev, K. G., Blanco, G. S., Wen, J., Mysore, K. S., et al. (2020). The CLE53-SUNN genetic pathway negatively regulates arbuscular mycorrhiza root colonization in *Medicago truncatula*. *Journal of Experimental Botany*, *71*(16), 4972–4984.

Kassaw, T., Nowak, S., Schnabel, E., & Frugoli, J. (2017). ROOT DETERMINED NODULATION 1 is required for M. truncatula CLE12, but not CLE13, peptide signaling through the SUNN receptor kinase. *Plant Physiology*, *174*(4), 2445–2456.

Kim, H. J., Wu, C. Y., Yu, H. M., Sheen, J., & Lee, H. (2017). Dual CLAVATA3 peptides in Arabidopsis shoot stem cell signaling. *Journal of Plant Biology*, *60*(5), 506–512.

Kitagawa, M., & Jackson, D. (2019). Control of meristem size. *Annual Review of Plant Biology*, *70*, 269–291.

Kondo, Y., Ito, T., Nakagami, H., Hirakawa, Y., Saito, M., Tamaki, T., et al. (2014). Plant GSK3 proteins regulate xylem cell differentiation downstream of TDIF-TDR signalling. *Nature Communications*, *5*(1), 1–11.

Le Marquer, M., Bécard, G., & Frei dit Frey, N. (2019). Arbuscular mycorrhizal fungi possess a CLAVATA3/embryo surrounding region-related gene that positively regulates symbiosis. *New Phytologist*, *222*(2), 1030–1042.

Li, Z., Chakraborty, S., & Xu, G. (2017). Differential CLE peptide perception by plant receptors implicated from structural and functional analyses of TDIF-TDR interactions. *PLoS One*, *12*(4), e0175317.

Li, Q., Guo, L., Wang, H., Zhang, Y., Fan, C., & Shen, Y. (2019). In silico genome-wide identification and comprehensive characterization of the *BES1* gene family in soybean. *Heliyon*, *5*(6), e01868.

Li, D., Kinkema, M., & Gresshoff, P. M. (2009). Autoregulation of nodulation (AON) in *Pisum sativum* (pea) involves signalling events associated with both nodule primordia development and nitrogen fixation. *Journal of Plant Physiology*, *166*(9), 955–967.

Lim, C. W., Lee, Y. W., & Hwang, C. H. (2011). Soybean nodule-enhanced CLE peptides in roots act as signals in GmNARK-mediated nodulation suppression. *Plant and Cell Physiology*, *52*(9), 1613–1627.

Liu, L., Gallagher, J., Arevalo, E. D., Chen, R., Skopelitis, T., Wu, Q., et al. (2021). Enhancing grain-yield-related traits by CRISPR-Cas9 promoter editing of maize CLE genes. *Nature Plants*, 7(3), 287–294.

Meixner, C., Ludwig-Müller, J., Miersch, O., Gresshoff, P., Staehelin, C., & Vierheilig, H. (2005). Lack of mycorrhizal autoregulation and phytohormonal changes in the supernodulating soybean mutant *nts1007*. *Planta*, *222*(4), 709–715.

Meng, L., Ruth, K. C., Fletcher, J. C., & Feldman, L. (2010). The roles of different CLE domains in Arabidopsis CLE polypeptide activity and functional specificity. *Molecular Plant*, *3*(4), 760–772.

Mens, C., Hastwell, A. H., Su, H., Gresshoff, P. M., Mathesius, U., & Ferguson, B. J. (2021). Characterisation of *Medicago truncatula* CLE34 and CLE35 in nitrate and rhizobia regulation of nodulation. *New Phytologist*, *229*(5), 2525–2534.

Mirzaei, S., Batley, J., El-Mellouki, T., Liu, S., Meksem, K., Ferguson, B. J., et al. (2017). Neodiversification of homeologous CLAVATA1-like receptor kinase genes in soybean leads to distinct developmental outcomes. *Scientific Reports*, 7(1), 1–15.

Morandi, D., Sagan, M., Prado-Vivant, E., & Duc, G. (2000). Influence of genes determining supernodulation on root colonization by the mycorrhizal fungus *Glomus mosseae* in *Pisum sativum* and *Medicago truncatula* mutants. *Mycorrhiza*, *10*(1), 37–42.

Morita, J., Kato, K., Nakane, T., Kondo, Y., Fukuda, H., Nishimasu, H., et al. (2016). Crystal structure of the plant receptor-like kinase TDR in complex with the TDIF peptide. *Nature Communications*, 7(1), 1–9.

Mortier, V., Den Herder, G., Whitford, R., Van de Velde, W., Rombauts, S., D'haeseleer, K., et al. (2010). CLE peptides control *Medicago truncatula* nodulation locally and systemically. *Plant Physiology*, *153*(1), 222–237.

Müller, L. M., Flokova, K., Schnabel, E., Sun, X., Fei, Z., Frugoli, J., et al. (2019). A CLE-SUNN module regulates strigolactone content and fungal colonization in arbuscular mycorrhiza. *Nature Plants*, *5*(9), 933–939.

Ni, J., Guo, Y., Jin, H., Hartsell, J., & Clark, S. E. (2011). Characterization of a CLE processing activity. *Plant Molecular Biology*, *75*(1–2), 67–75.

Nimchuk, Z. L., Tarr, P. T., & Meyerowitz, E. M. (2011). An evolutionarily conserved pseudokinase mediates stem cell production in plants. *The Plant Cell*, *23*(3), 851–854.

Nimchuk, Z. L., Tarr, P. T., Ohno, C., Qu, X., & Meyerowitz, E. M. (2011). Plant stem cell signaling involves ligand-dependent trafficking of the CLAVATA1 receptor kinase. *Current Biology*, *21*(5), 345–352.

Nishida, H., Handa, Y., Tanaka, S., Suzaki, T., & Kawaguchi, M. (2016). Expression of the CLE-RS3 gene suppresses root nodulation in *Lotus japonicus*. *Journal of Plant Research*, *129*(5), 909–919.

Novák, K. (2010). Early action of pea symbiotic gene NOD3 is confirmed by adventitious root phenotype. *Plant Science*, *179*(5), 472–478.

Oelkers, K., Goffard, N., Weiller, G. F., Gresshoff, P. M., Mathesius, U., & Frickey, T. (2008). Bioinformatic analysis of the CLE signaling peptide family. *BMC Plant Biology*, *8*(1), 1–15.

Ogawa, M., Shinohara, H., Sakagami, Y., & Matsubayashi, Y. (2008). Arabidopsis CLV3 peptide directly binds CLV1 ectodomain. *Science*, *319*(5861), 294.

Ohyama, K., Shinohara, H., Ogawa-Ohnishi, M., & Matsubayashi, Y. (2009). A glycopeptide regulating stem cell fate in *Arabidopsis thaliana*. *Nature Chemical Biology*, *5*(8), 578–580.

Okamoto, S., Ohnishi, E., Sato, S., Takahashi, H., Nakazono, M., Tabata, S., et al. (2009). Nod factor/nitrate-induced CLE genes that drive HAR1-mediated systemic regulation of nodulation. *Plant and Cell Physiology*, *50*(1), 67–77.

Okamoto, S., Shinohara, H., Mori, T., Matsubayashi, Y., & Kawaguchi, M. (2013). Root-derived CLE glycopeptides control nodulation by direct binding to HAR1 receptor kinase. *Nature Communications*, *4*(1), 1–7.

Olsen, A. N., & Skriver, K. (2003). Ligand mimicry? Plant-parasitic nematode polypeptide with similarity to CLAVATA3. *Trends in Plant Science*, *8*(2), 55–57.

Opsahl-Ferstad, H. G., Deunff, E. L., Dumas, C., & Rogowsky, P. M. (1997). ZmEsr, a novel endosperm-specific gene expressed in a restricted region around the maize embryo. *The Plant Journal*, *12*(1), 235–246.

Osipova, M. A., Mortier, V., Demchenko, K. N., Tsyganov, V. E., Tikhonovich, I. A., Lutova, L. A., et al. (2012). *Wuschel-related homeobox 5* gene expression and interaction of CLE peptides with components of the systemic control add two pieces to the puzzle of autoregulation of nodulation. *Plant Physiology*, *158*(3), 1329–1341.

Parniske, M. (2004). Molecular genetics of the arbuscular mycorrhizal symbiosis. *Current Opinion in Plant Biology*, 7(4), 414–421.

Qian, P., Song, W., Yokoo, T., Minobe, A., Wang, G., Ishida, T., et al. (2018). The CLE9/10 secretory peptide regulates stomatal and vascular development through distinct receptors. *Nature Plants*, *4*(12), 1071–1081.

Reid, D. E., Ferguson, B. J., & Gresshoff, P. M. (2011). Inoculation-and nitrate-induced CLE peptides of soybean control NARK-dependent nodule formation. *Molecular Plant-Microbe Interactions*, *24*(5), 606–618.

Reid, D. E., Li, D., Ferguson, B. J., & Gresshoff, P. M. (2013). Structure-function analysis of the gm RIC1 signal peptide and CLE domain required for nodulation control in soybean. *Journal of Experimental Botany*, *64*(6), 1575–1585.

Remy, W., Taylor, T. N., Hass, H., & Kerp, H. (1994). Four hundred-million-year-old vesicular arbuscular mycorrhizae. *Proceedings of the National Academy of Sciences*, *91*(25), 11841–11843.

Sarkar, A. K., Luijten, M., Miyashima, S., Lenhard, M., Hashimoto, T., Nakajima, K., et al. (2007). Conserved factors regulate signalling in *Arabidopsis thaliana* shoot and root stem cell organizers. *Nature*, *446*(7137), 811–814.

Schaarschmidt, S., Gresshoff, P. M., & Hause, B. (2013). Analyzing the soybean transcriptome during autoregulation of mycorrhization identifies the transcription factors GmNF-YA1a/b as positive regulators of arbuscular mycorrhization. *Genome Biology*, *14*(6), 1–21.

Schmutz, J., Cannon, S. B., Schlueter, J., Ma, J., Mitros, T., Nelson, W., et al. (2010). Genome sequence of the palaeopolyploid soybean. *Nature*, *463*(7278), 178–183.

Schmutz, J., McClean, P. E., Mamidi, S., Wu, G. A., Cannon, S. B., Grimwood, J., et al. (2014). A reference genome for common bean and genome-wide analysis of dual domestications. *Nature Genetics*, *46*(7), 707–713.

Schnabel, E. L., Kassaw, T. K., Smith, L. S., Marsh, J. F., Oldroyd, G. E., Long, S. R., et al. (2011). The ROOT DETERMINED NODULATION1 gene regulates nodule number in roots of *Medicago truncatula* and defines a highly conserved, uncharacterized plant gene family. *Plant Physiology*, *157*(1), 328–340.

Schoof, H., Lenhard, M., Haecker, A., Mayer, K. F., Jürgens, G., & Laux, T. (2000). The stem cell population of Arabidopsis shoot meristems is maintained by a regulatory loop between the *CLAVATA* and *WUSCHEL* genes. *Cell*, *100*(6), 635–644.
Shinohara, H., & Matsubayashi, Y. (2015). Reevaluation of the CLV3-receptor interaction in the shoot apical meristem: Dissection of the CLV3 signaling pathway from a direct ligand-binding point of view. *The Plant Journal*, *82*(2), 328–336.
Somssich, M., Je, B. I., Simon, R., & Jackson, D. (2016). CLAVATA-WUSCHEL signaling in the shoot meristem. *Development*, *143*(18), 3238–3248.
Song, X. F., Guo, P., Ren, S. C., Xu, T. T., & Liu, C. M. (2013). Antagonistic peptide technology for functional dissection of CLV3/ESR genes in Arabidopsis. *Plant Physiology*, *161*(3), 1076–1085.
Sprent, J. I., & James, E. K. (2007). Legume evolution: Where do nodules and mycorrhizas fit in? *Plant Physiology*, *144*(2), 575–581.
Stahl, Y., Grabowski, S., Bleckmann, A., Kühnemuth, R., Weidtkamp-Peters, S., Pinto, K. G., et al. (2013). Moderation of Arabidopsis root stemness by CLAVATA1 and ARABIDOPSIS CRINKLY4 receptor kinase complexes. *Current Biology*, *23*(5), 362–371.
Stahl, Y., Wink, R. H., Ingram, G. C., & Simon, R. (2009). A signaling module controlling the stem cell niche in Arabidopsis root meristems. *Current Biology*, *19*(11), 909–914.
Stone, J. M., Trotochaud, A. E., Walker, J. C., & Clark, S. E. (1998). Control of meristem development by CLAVATA1 receptor kinase and kinase-associated protein phosphatase interactions. *Plant Physiology*, *117*(4), 1217–1225.
Stührwohldt, N., Ehinger, A., Thellmann, K., & Schaller, A. (2020). Processing and formation of bioactive CLE40 peptide are controlled by posttranslational proline hydroxylation. *Plant Physiology*, *184*(3), 1573–1584.
Sun, Z., Su, C., Yun, J., Jiang, Q., Wang, L., Wang, Y., et al. (2019). Genetic improvement of the shoot architecture and yield in soya bean plants via the manipulation of *GmmiR156b*. *Plant Biotechnology Journal*, *17*(1), 50–62.
Sun, B., Zhou, Y., Cai, J., Shang, E., Yamaguchi, N., Xiao, J., et al. (2019). Integration of transcriptional repression and polycomb-mediated silencing of WUSCHEL in floral meristems. *The Plant Cell*, *31*(7), 1488–1505.
Takahashi, F., Suzuki, T., Osakabe, Y., Betsuyaku, S., Kondo, Y., Dohmae, N., et al. (2018). A small peptide modulates stomatal control via abscisic acid in long-distance signalling. *Nature*, *556*(7700), 235–238.
Tamaki, T., Oya, S., Naito, M., Ozawa, Y., Furuya, T., Saito, M., et al. (2020). VISUAL-CC system uncovers the role of GSK3 as an orchestrator of vascular cell type ratio in plants. *Communications Biology*, *3*(1), 1–9.
Tsikou, D., Yan, Z., Holt, D. B., Abel, N. B., Reid, D. E., Madsen, L. H., et al. (2018). Systemic control of legume susceptibility to rhizobial infection by a mobile microRNA. *Science*, *362*(6411), 233–236.
van den Steen, P., Rudd, P. M., Dwek, R. A., & Opdenakker, G. (1998). Concepts and principles of O-linked glycosylation. *Critical Reviews in Biochemistry and Molecular Biology*, *33*(3), 151–208.
Wang, H. (2020). Regulation of vascular cambium activity. *Plant Science*, *291*, 110322.
Wang, J., Replogle, A. M. Y., Hussey, R., Baum, T., Wang, X., Davis, E. L., et al. (2011). Identification of potential host plant mimics of CLAVATA3/ESR (CLE)-like peptides from the plant-parasitic nematode Heterodera schachtii. *Molecular Plant Pathology*, *12*(2), 177–186. https://doi.org/10.1111/j.1364-3703.2010.00660.x.
Wang, X., Allen, R., Ding, X., Goellner, M., Maier, T., de Boer, J. M., et al. (2001). Signal peptide-selection of cDNA cloned directly from the esophageal gland cells of the soybean cyst nematode *Heterodera glycines*. *Molecular Plant-Microbe Interactions*, *14*(4), 536–544.

Wang, J., Lee, C., Replogle, A., Joshi, S., Korkin, D., Hussey, R., et al. (2010). Dual roles for the variable domain in protein trafficking and host-specific recognition of *Heterodera glycines* CLE effector proteins. *New Phytologist, 187*(4), 1003–1017.

Wang, X., Mitchum, M. G., Gao, B., Li, C., Diab, H., Baum, T. J., et al. (2005). A parasitism gene from a plant-parasitic nematode with function similar to CLAVATA3/ESR (CLE) of Arabidopsis thaliana. *Molecular Plant Pathology, 6*(2), 187–191.

Whitewoods, C. D., Cammarata, J., Venza, Z. N., Sang, S., Crook, A. D., Aoyama, T., et al. (2018). CLAVATA was a genetic novelty for the morphological innovation of 3D growth in land plants. *Current Biology, 28*(15), 2365–2376.

Whitford, R., Fernandez, A., De Groodt, R., Ortega, E., & Hilson, P. (2008). Plant CLE peptides from two distinct functional classes synergistically induce division of vascular cells. *Proceedings of the National Academy of Sciences, 105*(47), 18625–18630.

Wong, C. E., Khor, S. Y., Bhalla, P. L., & Singh, M. B. (2011). Novel spatial expression of soybean WUSCHEL in the incipient floral primordia. *Planta, 233*(3), 553–560.

Wong, C. E., Singh, M. B., & Bhalla, P. L. (2013). Spatial expression of *CLAVATA3* in the shoot apical meristem suggests it is not a stem cell marker in soybean. *Journal of Experimental Botany, 64*(18), 5641–5649.

Xu, C., Liberatore, K. L., MacAlister, C. A., Huang, Z., Chu, Y. H., Jiang, K., et al. (2015). A cascade of arabinosyltransferases controls shoot meristem size in tomato. *Nature Genetics, 47*(7), 784–792.

Yang, Y., Zhu, K., Li, H., Han, S., Meng, Q., Khan, S. U., et al. (2018). Precise editing of *CLAVATA* genes in *Brassica napus* L. regulates multilocular silique development. *Plant Biotechnology Journal, 16*(7), 1322–1335.

Yoro, E., Nishida, H., Ogawa-Ohnishi, M., Yoshida, C., Suzaki, T., Matsubayashi, Y., et al. (2019). PLENTY, a hydroxyproline O-arabinosyltransferase, negatively regulates root nodule symbiosis in *Lotus japonicus*. *Journal of Experimental Botany, 70*(2), 507–517.

Yoshida, C., Funayama-Noguchi, S., & Kawaguchi, M. (2010). Plenty, a novel hypernodulation mutant in *Lotus japonicus*. *Plant and Cell Physiology, 51*(9), 1425–1435.

Zhang, H., Lin, X., Han, Z., Qu, L. J., & Chai, J. (2016). Crystal structure of PXY-TDIF complex reveals a conserved recognition mechanism among CLE peptide-receptor pairs. *Cell Research, 26*(5), 543–555.

Zhang, L., Shi, X., Zhang, Y., Wang, J., Yang, J., Ishida, T., et al. (2019). CLE9 peptide-induced stomatal closure is mediated by abscisic acid, hydrogen peroxide, and nitric oxide in *Arabidopsis thaliana*. *Plant, Cell & Environment, 42*(3), 1033–1044.

Zhang, M., Su, H., Gresshoff, P. M., & Ferguson, B. J. (2021). Shoot-derived miR2111 controls legume root and nodule development. *Plant, Cell & Environment, 44*(5), 1627–1641.

Zhang, Y., Yang, S., Song, Y., & Wang, J. (2014). Genome-wide characterization, expression and functional analysis of CLV3/ESR gene family in tomato. *BMC Genomics, 15*(1), 1–12.

Zhou, Y., Liu, X., Engstrom, E. M., Nimchuk, Z. L., Pruneda-Paz, J. L., Tarr, P. T., et al. (2015). Control of plant stem cell function by conserved interacting transcriptional regulators. *Nature, 517*(7534), 377–380.

Zhu, Y., Song, D., Zhang, R., Luo, L., Cao, S., Huang, C., et al. (2020). A xylem-produced peptide PtrCLE20 inhibits vascular cambium activity in Populus. *Plant Biotechnology Journal, 18*(1), 195–206.

CHAPTER SEVEN

Redox metabolism in soybean and its significance in nitrogen-fixing nodules

Karl J. Kunert[a] and Christine H. Foyer[b,*]

[a]Department of Plant and Soil Sciences, FABI, University of Pretoria, Pretoria, South Africa
[b]School of Biosciences, College of Life and Environmental Sciences, University of Birmingham, Edgbaston, United Kingdom
*Corresponding author: e-mail address: c.h.foyer@bham.ac.uk

Contents

Abstract

Reactive oxygen species (ROS) are essential regulators of plant growth, development and defense. The regulated compartment-specific production and processing of ROS are central to the reduction-oxidation (redox) balance of the intracellular and extracellular environments. ROS generation has often been considered to be harmful, because of the potential collateral damage, but their high chemical reactivity is essential to their roles as pro-life signals. Nevertheless, literature evidence demonstrates that the ability to mount a robust antioxidant defense is integral to stress tolerance in soybean as in other plant species. Moreover, the integration of redox signaling with reactive nitrogen species and molecular oxygen signaling is crucial to the functioning of many plant organs, such as seeds and nodules, which are maintained in a state of developmental hypoxia. Here, we describe ROS production, processing and signaling in soybean, particularly in relation to nodule development and protection against environmental stresses. We also consider the interactions between ROS and nitric oxide (NO), highlighting gaps in current knowledge and considering future research directions in exploring redox metabolism in soybean.

Advances in Botanical Research, Volume 102
ISSN 0065-2296
https://doi.org/10.1016/bs.abr.2022.02.020

1. Introduction

1.1 ROS waves

Redox metabolism involves the transfer of electrons, or hydrogen, from one molecule or compound to another as central drives of catabolic and anabolic reactions. In particular, cellular energy metabolism uses reduction and oxidation reactions to drive processes such as photosynthesis and respiration. Molecular oxygen became incorporated into cellular energy metabolism early in evolution. The four-electron reduction of oxygen is catalyzed by the cytochrome C oxidase in respiration, while the four-electron oxidation of water is catalyzed by the water-splitting reaction of photosystem II. This reaction produces oxygen with the release of protons, while electrons are moved through an electron transport system to reduce NADP to NADPH. Other enzymes undertake the two-electron reduction of molecular oxygen to produce hydrogen peroxide (H_2O_2), while other enzymes and processes perform the univalent reduction of oxygen to produce the superoxide radical anion ($O_2^{\bullet-}$). Metal catalyzed reactions between H_2O_2 and superoxide can generate the hydroxyl radical (OH^-) (Das & Roychoudhury, 2014; Halliwell & Gutteridge, 1999). NADPH oxidases and peroxidases (PER) are major sources of superoxide in the apoplast (Das, Nutan, Singla-Pareek, & Pareek, 2015; Demidchik, 2015), while the photosynthetic and respiratory electron transport chains produce superoxide in the chloroplasts and mitochondria, respectively. Photorespiration is a major source of H_2O_2 in the peroxisomes (Foyer & Shigeoka, 2011; Sharma, Jha, Dubey, & Pessarakli, 2012). Together with singlet oxygen, these partially reduced forms of oxygen are collectively called reactive oxygen species (ROS). Any information concerning changes in ROS production in the apoplast and intracellular compartments is finally transmitted to the cytosol and nuclei to allow rapid adjustments in transcription and translation (Dietz, Turkan, & Krieger-Liszkay, 2016).

Each ROS form has distinct biological properties, a short half-life and high chemical reactivity. It is, therefore, not surprising that ROS and redox chemistry is involved in nearly every aspect of plant biology through a network with redox-sensitive proteins including transcription factors and is involved in local and systemic signaling pathways (Noctor, De Paepe, & Foyer, 2007). In particular, ROS can fine-tune signal transduction process associated with plant growth and defense (Fig. 1; Considine & Foyer, 2021a, 2021b). For example, ROS produced within the apoplast are essential for leaf and root growth and differentiation (Chaput, Martin, & Lejay, 2020; Morales & Munné-Bosch, 2016).

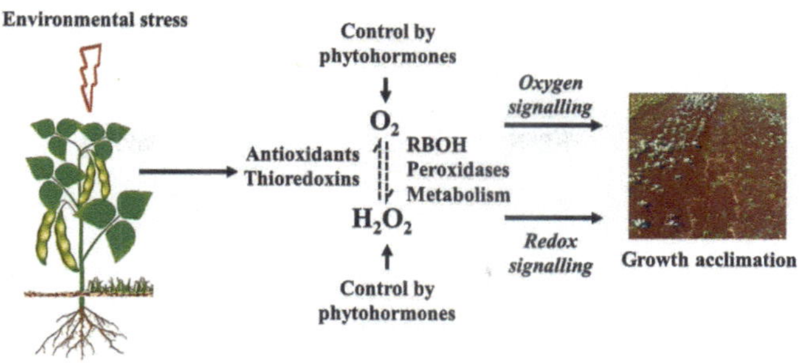

Fig. 1 Simple overview of the key roles of oxygen- and hydrogen peroxide (H_2O_2)-mediated redox signaling in driving plant growth and how environmental stress impacts on this process. Stress-induced ROS production, which is regulated by antioxidants and thioredoxins, is part of a multi-layered reduction/oxidation (redox) response system, in which stress sensing and adaptation are synchronized with energy-producing and energy-consuming pathways of metabolism. ROS production is particularly associated with the photosynthetic and mitochondrial electron transport chains, and also by a large number of enzymes, prominently including NADPH oxidases (also called respiratory burst oxidase homologues, RBOHs) and peroxidases as well as phytohormone pathways that generate ROS as part of the signaling network that drives plant growth and development.

While ROS generation is necessary for growth, development and defense responses of plants, the reactivity of ROS can cause collateral damage to DNA, lipids and proteins, which can ultimately trigger programmed cell death (Karuppanapandian, Kim, Manoharan, & Kim, 2011; Santolini, Wootton, Jackson, & Feelisch, 2019). ROS production that is associated with all biotic and abiotic stress conditions, such as pest and pathogen infestations, temperature extremes, heavy metals, water availability, air pollutants, nutrient deficiency, salt stress and drought, also results in oxidative stress (Czarnocka & Karpiński, 2018; Noctor, Mhamdi, & Foyer, 2014). ROS production in the apoplast is essential to the local and systemic signaling networks that are crucial to the survival of plants exposed to environmental stresses (Fichman, Myers, Grant, & Mittler, 2021). Apoplastic ROS accumulation regulates cell wall composition and properties as well as plasmodesmata (PD) function related to inter- and intracellular signaling (Fichman et al., 2021; Zandalinas et al., 2020). Auto propagating waves of ROS, calcium and electric signals function together to generate rapid systemic cell-to-cell communication (Fig. 2; Wang et al., 2019). The activation of the NADPH oxidase complex and respiratory burst homolog (RBOH) proteins on the plasma membrane generate superoxide radicals in the apoplast, which are converted to hydrogen peroxide either through spontaneous dismutation or the action of superoxide dismutase (SOD) (Wang, Branicky, Noë, & Hekimi, 2018).

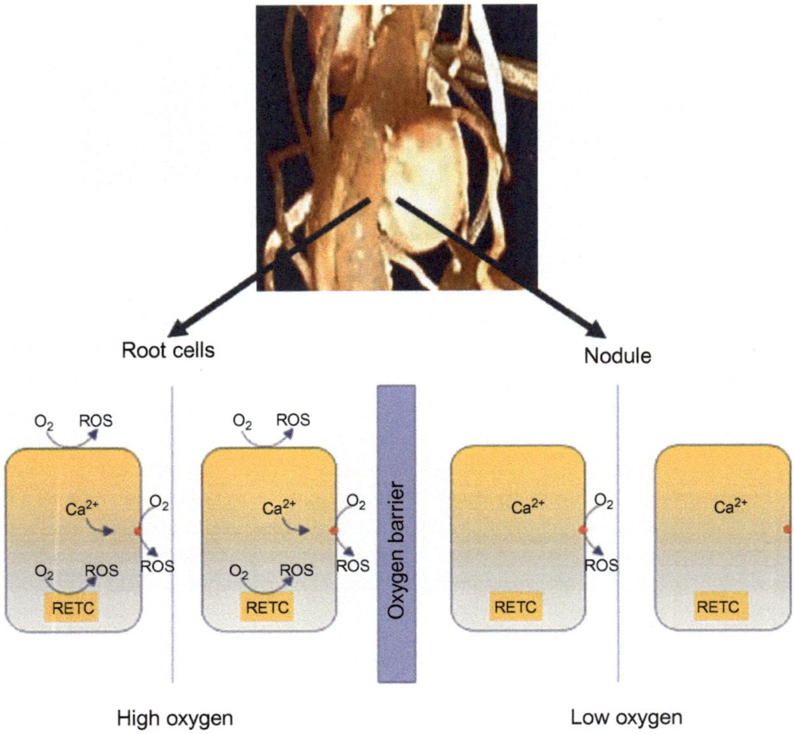

Fig. 2 ROS production in root cells under high and nodules under low oxygen conditions. In root cells under high oxygen conditions ROS are produced by the respiratory electron transport chain (RETC) and by a large number of enzymes including ROS produced by respiratory burst oxidase homologues (RBOH) on the cell surface. The resultant ROS wave passes then from cell to cell, stimulating proliferation followed by cell expansion in the direction of growth. In nodules, an oxygen barrier provided by leghemoglobin limits the availability of molecular oxygen which prevents metabolic ROS production and protects the nitrogen fixation process.

SOD is part of the complex antioxidant defense system that limits ROS accumulation (Dumanović, Nepovimova, Natić, Kuča, & Jaćević, 2021; Foyer & Noctor, 2011). The antioxidant defense system is comprised of enzymatic and non-enzymatic low molecular weight antioxidants that police ROS accumulation in the different compartments of plant cells (Fig. 1; Foyer & Noctor, 2016; Noctor, Mhamdi, & Foyer, 2016). Thioredoxins (Trx), peroxiredoxins (Prx) and glutaredoxins (Grx) are key components of this protective system as are the antioxidant enzymes of the Asada–Halliwell-Foyer cycle that function together with ascorbate, glutathione and NADPH to process ROS (Foyer & Noctor, 2011).

Prx reduce H_2O_2, $ONOO^-$, and lipid peroxides (Hasanuzzaman et al., 2020), while Trx reduce Prx and protein disulfides. Trx reductase catalyzes the NADPH-dependent reduction of oxidized Trx (Dos Santos & Rey, 2006; Jedelská, Luhová, & Petřivalský, 2020). The Asada–Halliwell-Foyer cycle is comprised of ascorbate peroxidase (APX), monodehydroascorbate reductase (MDHAR), dehydroascorbate reductase (DHAR) and glutathione reductase (GR). NADPH provides the reducing power to regenerate reduced glutathione (GSH) from its oxidized form, glutathione disulfide (GSSG). ROS signaling leads to changes in the amounts of GSH and GSSG that are important in redox signal transduction (Foyer & Noctor, 2011; Foyer & Noctor, 2016; Kranner, Birtić, Anderson, & Pritchard, 2006; Noctor & Foyer, 2016). The components that process ROS in the different subcellular compartments relay information that modulates plant growth, development and defense in response to changing environmental conditions (Fig. 1; Foyer & Noctor, 2011).

1.2 ROS signaling

Each ROS form ($^1O_2$, H_2O_2, $O_2^{\bullet-}$) triggers different signaling pathways. The abundance of antioxidants and ROS sensitive targets, such as protein cysteinyl (Cys) thiols, results in very short ROS lifetimes and limits ROS diffusion. H_2O_2 has the longest lifetime of any ROS form, and can be transported by aquaporins to regulate processes such as stomatal closure (Rodrigues et al., 2017). While ROS can react with a diverse range of protein side chains, redox regulation and signal transmission often involves the oxidation and reduction of the Cys groups that are close to ROS production sites. The post-translational modifications (PTMs) of protein Cys residues can occur through either dithiol-disulfide exchange reactions or through reactions in which the Cys residues are oxidized by ROS, reactive nitrogen species (RNS), or by reactive sulfur species (RSS; Zaffagnini et al., 2019). The hydrogen peroxide-dependent two electron oxidation of reactive Cys results in the formation of a sulfenic acid (–SOH), which can react with a proximal thiol from a protein Cys or GSH to form intra/inter-molecular disulfide bonds (–S–S–) or mixed-disulfides (–S–SG, S-glutathionylation). Thus, a wide range of protein Cys oxidation mechanisms mediated by intermediate Cys oxoforms (*i.e.*, sulfenic acids and nitrosothiols) or oxidant molecules (*e.g.*, oxidized glutathione, GSSG) catalyzed by enzymatic systems, are likely to be involved in the redox-mediated cell signaling systems (Zaffagnini et al., 2019).

To date, relatively few proteins that sense hydrogen peroxide by oxidation of protein Cys residues have been characterized. One such protein is the plasma membrane CARD/HPCA receptor kinase that senses hydrogen peroxide through disulfide bond formation between two Cys residues in the extracellular sensing domain. This triggers an influx of calcium ions into the cell, leading to MAP kinase activation and stress acclimation (Wu et al., 2020). Sulfenic acids can act as nucleophiles and can be oxidized to sulfinic ($-SO_2H$) and sulfonic acid ($-SO_3H$). The oxidation of reactive Cys to sulfinic acids is catalyzed by Cys oxidases (PCOs; Weits et al., 2014). PCOs catalyze a key step of the N-degron pathway of regulated proteolysis (White et al., 2017). This pathway is important in the oxygen-mediated regulation of plant growth and development (Graciet et al., 2009; Holman et al., 2009). The pathway regulates the stability of proteins, such as the group VII ethylene response factors (ERF-VIIs), in response to oxygen availability. ERF-VIIs are, however, rapidly degraded in air. Hypoxia limits PCO activity, stabilizing ERF-VIIs and facilitating their functions (Van Dongen & Licausi, 2015). Nitric oxide (.NO) is involved in this regulation. Exposure to hypoxia, or re-oxygenation, results in ROS and RNS accumulation (Gupta et al., 2020; Sasidharan, Schippers, & Schmidt, 2021).

Singlet oxygen ($^1O_2$) signaling can activate different pathways depending whether it is generated within membranes or aqueous environments. In the thylakoid membranes, reactions with carotenoids lead to breakdown products, such as β-CC and dhA that act as signaling molecules increasing stress tolerance (Ramel et al., 2013). β-CC induces the SCARECROW LIKE 14 (SCL14)-controlled xenobiotic detoxification pathway (D'Alessandro, Mizokami, Légeret, & Havaux, 2019) as well as other pathways (Shumbe et al., 2017) that are independent of the EXECUTER (EX)-dependent signaling pathway that is triggered when $^1O_2$ is generated at the membrane surface or in the stroma or cytosol (Ramel et al., 2013; Shumbe et al., 2016).

Soybean, like other legumes, uses the above mechanisms of regulation to control plant growth and defense in response to environmental triggers. Although there are reports of the functions of Trx and Prx in soybeans, particularly in seed and nodule development (Frendo, Matamoros, Alloing, & Becana, 2013; Lee et al., 2005), the literature on redox regulation in soybean is scarce compared to other legumes such as Medicago and Lotus. In Lotus, for example, 7 Prx, 14 Trx, and 3 functional NADPH-thioredoxin reductase (NTRs) genes were identified and characterized (Tovar Mendes, Matamoros, Bustos-Sanmamed, & Becana, 2011). In this review, we will provide an overview of the redox metabolism in soybean, with a particular focus on root nodules, which maintain the hypoxic environment required for symbiotic nitrogen interaction by nitrogen-fixing rhizobium bacteria. We will also consider gaps in knowledge and future research directions to explore the redox metabolism in soybean.

2. Redox defenses against environmental stress

Redox metabolism plays an important role in plant responses to biotic and abiotic stresses. One of the earliest and most important physiological responses to environmental stresses is the oxidative burst. This burst drives signaling pathways that modulate changes in the expression of stress-related genes, resulting in an appropriate defensive response. For example, ROS production by the plant NADPH oxidase RBOHD (Respiratory Burst Oxidase Homolog D) is required for the successful activation of immune responses to pathogen infection. The function of NADPH oxidases is generally associated with developmental with stable functions donating ROS and versatile functions, involved in pollen germination, root development, seed germination, fungal development, symbiosis between plants and microorganisms as well as stress response (Hu et al., 2020). The RBOHproteins not only catalyze the superoxide anion formation and superoxide anion radicals forming hydrogen peroxide (H_2O_2) *via* dismutation (Sagi & Fluhr, 2006), but they also mediate multiple processes to balance cellular homeostasis under stress conditions (Marino, Dunand, Puppo, & Pauly, 2011). Among the RBOHs, RBOHD is, for example, the most highly expressed NADPH oxidase enzyme in pathogen response processes (Suzuki et al., 2011) and RBOHD-dependent ROS being involved in the plant defense reaction to pathogens which is associated with the hypersensitive response (Otulak-Kozieł, Kozieł, & Valverde, 2019; Torres, Dangl, & Jones, 2002). RBOHD is regulated by C-terminal phosphorylation, the PBL13 receptor-like cytoplasmic kinase, and by ubiquitination catalyzed by the E3 ubiquitin ligase called PBL13 interacting RING domain E3 ligase (PIRE). The phosphorylation of RBOHD results in enhanced ubiquitination and decreases the abundance of the protein (Lee et al., 2020).

One presumes that similar regulatory systems operate in legumes, such as soybean, in which 17 RBOHs (GmRBOHs) have been already identified (Ranjan et al., 2018). Of the six identified groups, group VI (*GmRBOH-VI*) RBOHs are induced in response to *Sclerotinia sclerotiorum* challenge. Moreover, silencing of *GmRBOH-VI* proteins leads to enhanced resistance to *S. sclerotiorum* and greater drought tolerance, but also results in a reduced capacity to form nodules (Ranjan et al., 2018). There are several additional reports concerning RBOHD functions in soybean (Lee, Huffaker, Crippen, Robbins, & Goggin, 2018; Wang et al., 2020). For example, soybean has three plant elicitor peptides (Peps) that regulate plant defenses against the root-knot nematode (*Meloidogyne incognita*) and the soybean cyst nematode (*Heterodera glycines*). Of these, GmPep1 induces the expression of RBOHD and the accumulation of ROS in leaves to activate plant defenses through systemic transcriptional reprogramming and ROS signaling (Lee et al., 2018). Similarly, *lesion mimic mutants* (*lmms*), which have enhanced resistance to both bacterial and oomycete pathogen as well as elevated ROS production upon treatment with the bacterial pattern flg22 (a 22-amino-acid peptide, flg22, in the N-terminal fragment of flagellin) identified a malectin-like receptor kinase (RK) (Wang et al., 2020). This kinase, which is upstream of RBOHD, regulates cell death and pattern-triggered immunity in soybean and the Leu-rich repeat transmembrane receptor kinase FLAGELLIN SENSITIVE2 (FLS2) is essential for flagellin perception (Chinchilla, Bauer, Regenass, Boller, & Felix, 2006). The malectin-like receptor kinase GmLMM1 regulates programmed cell death and pattern-triggered immunity in soybean (Wang et al., 2020). This kinase forms part of the pattern recognition receptor (PRR) complex negatively regulating pattern-triggered immunity (PTI) responses, interacting with the flg22 receptor FLS2 and its co-receptor BRASSINOSTEROID-INSENSITIVE 1-ASSOCIATED KINASE (BAK1) before and after immune activation. By suppressing FLS2-BAK1 interactions, GmLMM1 acts as a molecular switch that controls FLS2-BAK1 interaction levels and hence immune activation in soybean (Wang et al., 2020). Soybean has developed polymorphic versions of the flg22 receptors and is hence able to perceive flg22^{Rso} that generally avoid perception in plants (Wei et al., 2020).

As in all plant species, the soybean RBOH proteins function in abiotic stress tolerance as well as in resistance to pathogens and herbivores. The salinity-induced NAM/ATAF1/2/CUC2 (NAC) transcription factor, encoded by *SALT INDUCED NAC1* (*GmSIN1*), promotes root growth and salt tolerance by increasing the expression of *GmRbohBs* and 9-*cis*-epoxycarotenoid dioxygenase coding genes in soybean (Li et al., 2019). The accumulation of ROS and abscisic acid has been suggested to amplify the initial salt stress signal enhancing salinity tolerance in soybean. Silicon (Si) has been also used to alleviate salt tolerance in soybean through the regulation of ROS generation and production of ROS (Chung, Kim, Hamayun, & Kim, 2020). The application of Si decreased the expression of genes associated with redox regulation and decreased the levels of enzymic and non-enzymic antioxidants. The application of salicylic acid (SA) was found to enhance tolerance to aluminium (Al) stress through ROS generation and changes in H_2O_2 accumulation through modulation of the antioxidant enzyme activities in soybean roots (Liu et al., 2017).

As discussed above, ROS are produced in all cells as normal products of metabolism or by the action of RBOHs or other enzymes. ROS production and accumulation is enhanced in response to biotic and abiotic stresses such as drought, salt and cold (Das et al., 2016; Porcel & Ruiz-Lozano, 2004; Tambussi, Bartoli, Guiamet, Beltrano, & Araus, 2004; You & Chan, 2015). In addition, exposure to peroxiding herbicides (Knörzer, Durner, & Böger, 1996) and environmental pollutants, such as ozone (O_3; Langebartels, Wohlgemuth, Kschieschan, Grün, & Sandermann, 2002) leads to increased ROS production and oxidative stress. Studies in Free-Air CO_2 Enrichment (FACE) sites, or experiments in open-top chambers, have revealed that soybean is particularly sensitive to O_3. Elevated CO_2 and O_3 not only induce changes in soybean canopy densities and leaf senescence, but they also modify disease expression. In addition, pod production in field-grown soybean decreases when the plants are exposed to above ambient O_3 levels (Betzelberger et al., 2012). Elevated CO_2 influences O_3 sensitivity in soybean, as well as modifying resistance to important fungal pathogens. Exposure to elevated CO_2 alone, or in combination with O_3 in a 3-year study, significantly reduced the severity of downy mildew infections by up to 66% but significantly increased the severity of brown spot disease (Eastburn, Degennaro, Delucia, Dermody, & Mcelrone, 2010). A recent meta-analysis of FACE studies has shown that there is a strong correlation between yield responses to elevated CO_2 and the genetic yield potential of soybean (Ainsworth & Long, 2021). Ozone enters leaves through the stomata, and permeates the intercellular air spaces, where it reacts with a range of molecules to produce H_2O_2, singlet oxygen, superoxide radicals, hydroxyl radicals and nitric oxide (NO) in the apoplast, triggering apoplastic ROS signaling pathways. While apoplastic ascorbate and other antioxidants are able to scavenge O_3-generated ROS, the extracellular and membrane-bound PRX and RBOH enzymes propagate the ROS signals in the intercellular spaces. Class III PRX family proteins, like the RBOH proteins, also produce ROS and regulate the local concentrations of H_2O_2. However, only between 30% and 50% of the O_3 entering leaves can be detoxified by apoplastic ascorbate. Similarly, the levels of apoplastic ascorbate are also not related to O_3 tolerance and may be, therefore, too low to play a significant role in O_3 detoxification (Ainsworth, 2017). In this regard, an emerging theme is the production of ROS in the apoplast associated with RLK signaling which has, to our knowledge, not been investigated so far in soybean. Redox modulation of apoplastic proteins, including cysteine-rich receptor-like kinases (RLK), can serve as apoplastic ROS sensors (Kimura, Waszczak, Hunter, & Wrzaczek, 2017) triggering local and systemic ROS signaling. In addition, heterotrimeric G-proteins, which are also important in the activation of RBOH under acute O_3 stress, have also not been investigated in greater detail in soybean. Hence, how exposure to O_3 activates redox regulation of proteins and redox signaling leading to altered gene expression and protein abundance (Galant, Koester, Ainsworth, Hicks, & Jez, 2012; Gillespie, Rogers, & Ainsworth, 2011), is also an interesting future research target in soybean

A study of seven soybean genotypes exposed to O_3 throughout the growth season revealed linear increases in antioxidant capacity, together with decreases in photosynthesis, leaf area, seed yield and harvest index (Betzelberger et al., 2012). Proteomic analysis of soybean leaf tissue exposed to high O_3 concentrations lead in the abundance of proteins involved in carbon metabolism, photosynthesis, amino acid synthesis, flavonoid and isoprenoid biosynthesis, as well as antioxidant pathways (Galant et al., 2012). These changes are often accompanied by increases in the ascorbate and glutathione pools but with lower GSH/GSSG and reduced ascorbate to oxidized ascorbate (dehydroascorbate), as is also observed in other abiotic stresses (Knörzer et al., 1996; Wang et al., 2013).

Since exposure to oxidative stress results in enhanced GSH production, GSH and other low molecular weight thiols have been also used as safeners, or bio-stimulants, to ameliorate stress responses. Treatment with folcysteine, or GSH, provided protection of soybean against drought and salt stress (Akram et al., 2017; Schneider et al., 2020). As discussed above, exposure to salinity results in increased ROS production and oxidative stress (Adem, Roy, Zhou, Bowman, & Shabala, 2014; You & Chan, 2015). The amount of total sulfhydryl groups (-SH) and GSH) increases in NaCl-treated soybean cell cultures, together with higher GR activity and a higher affinity (Km) of GR for GSSG (El-Enany, 1997). GSH treatments limit oxidative stress and improves soybean yields (Akram et al., 2017; Nahar, Hasanuzzaman, Alam, & Fujita, 2015). The application of GSH might be, therefore, a suitable approach to enhance stress tolerance in soybean. The increased stress tolerance observed following the addition of GSH is likely to result from effects at multiple levels including the regulation of translation (Cheng et al., 2015; Noctor et al., 2012). It has long been known that overexpression of the enzymes of GSH synthesis, such as glutamylcysteine synthetase, can result in substantial increases in the amount of total glutathione (Noctor et al., 1996; Noctor et al., 2012).

A proteomic analysis of soybean seedlings subjected to drought stress revealed changes in proteins involved in photosynthesis, signaling pathways, and antioxidant metabolism species. Western blot analysis confirmed an increase in the abundance of APX in drought-stressed plants, together with an increase of methionine synthase (Mohammadi, Moieni, Hiraga, & Komatsu, 2012). Similarly, a recent exploration of cross tolerance phenomena, involving either water or heat stress independently or to a combination of these stresses, in two contrasting soybean cultivars revealed that exposure to the stresses together altered the physiological responses of the plants, as well as the proteome and enzymes responses in different ways to the individual stresses (Katam et al., 2020). Some proteins including antioxidant enzymes, such as APX, were more abundant in soybean cultivar PI-471938 (slow wilting and high yielding) than in R95–1705 (high in protein concentration and with moderate yield potential). The observed differences in antioxidant enzymes indicated that PI-471938 is possible better able to restore the cellular redox balance during stress than R95–1705 (Katam et al., 2020).

Transcription factors (TFs), such as the soybean heat shock factor (HSF) GmHSFA1, are activated in response to stress and induce the expression of other heat- and stress-associated TFs to increase or fine-tune the expression of genes leading to thermo-tolerance (Zhu et al., 2006). In general, within the soybean genome, 4342 loci encoding 5035 TF models have been previously already identified which could be grouped into 61 families (Mochida et al., 2009). In particular, plant-specific NAC (NAM, ATAF, and CUC)-domain proteins have been identified playing important roles as transcription factors in soybean development and stress responses (Le et al., 2011). A detailed overview of soybean GmSNAC genes encoding NAC transcription factors has been also recently published (So & Lee, 2019). Eight of 71 GmSNAC genes were differentially induced by abiotic stresses. Based on the structural characteristics and phylogenetic relationships of Arabidopsis, rice, and soybean NACs, these identified soybean GmSNACs could be divided into five subgroups. Also 38 soybean HSFs have been so far identified in soybean which could be grouped into 3 classes and 12 subclasses, located on the 15 soybean chromosomes (Li et al., 2014). The promoter regions of the soybean HSFs contained *cis*-elements that suggest their participation in a wide range of stresses, such as drought, low temperature and other stresses that involve abscisic acid (ABA) responses (Li et al., 2014). *DEHYDRATION-RESPONSIVE ELEMENT BINDING 2A (DREB2A), MULTIPROTEIN BRIDGING FACTOR 1C (MBF1C), HsfA7s, HsfA2,* and *HsfBs* are downstream target genes of HsfA1s in *Arabidopsis* (Ohama, Sato, Shinozaki, & Yamaguchi-Shinozaki, 2017). These TFs modulate the expression of other TFs, chaperones, and ROS-scavenging enzymes and form a regulatory network to regulate stress tolerance (Ohama et al., 2017).

3. Redox metabolism and soybean nodules

3.1 Nodule initiation and development

The symbiotic interaction between legumes and nitrogen-fixing rhizobium bacteria results in the formation of a new organ, the nodule. Most studies reporting the functions of redox metabolism and ROS in nodules have generally focused on legumes such Medicago and common beans (*Phaseolus vulgaris*). The roles of ROS and NO in the control of the early stages of the legume—rhizobium symbiosis have been already reviewed by Damiani, Pauly, Puppo, Brouquisse, and Boscari (2016). Similarly, ROS production and ROS functions in root growth and development have been recently also reviewed (Eljebbawi, Guerrero, Dunand, & Estevez, 2021 Mase & Tsukagoshi, 2021). Root hairs, which are tubular extensions of root epidermal cells, are required for soybean nodule formation as well as nutrition and water absorption. Nodulation of soybean root hairs by the nitrogen-fixing symbiotic bacterium *Bradyrhizobium japonicum* is a complex process coordinated by the mutual exchange of diffusible signal molecules that cause the activation of parallel signaling pathways leading to Ca^{2+} influx and ROS production (Chang, Damiani, Puppo, & Frendo, 2009). The roles of RBOHs proteins is also well characterized in root hair development, which is often divided into the stages of root hair initiation and tip growth. But relatively little information is available on the roles of different RBOH and PRX forms in root hair curling (Montiel, Arthikala, Cardenas, & Quinto, 2016). RBOHC is the main enzyme responsible for ROS-dependent root hair elongation in' Arabidopsis, although RBOHH and RBOHJ also contribute (Mangano et al., 2017). RBOHC actively produces superoxide, and the hydroxyl radicals formed downstream of superoxide are the specific ROS that stimulates root hair elongation (Chapman, Muhlemann, Gayomba, & Muday, 2019). ROS and redox metabolites are important in the initiation of symbiosis and nodule development, as well as the development of the infection thread and nodule meristem formation (Mangano et al., 2017). A recent investigation in common bean has found that the RBOH multigene family has nine members (RbohA–I: Arthikala et al., 2017). RNA interference to impair RBOHB functions limits ROS production and root hair development, impairing rhizobial infection thread progression and also nodule organogenesis (Arthikala et al., 2017). ROS-mediated changes in the expression of auxin- and cyclin-related genes, as well as genes involved in cell growth and division are involved in this regulation.

NO is also an important signaling molecule in the early symbiosis steps (Damiani et al., 2016). Nitrate reductase and the mitochondrial transfer chain are thereby metabolic sources of NO (Horchani et al., 2011; Wany, Foyer, & Gupta, 2018). The micro-symbionts in legumes, such as soybean, provide up to 90% of the NO production due to the action of a nitrite reductase in the denitrification pathway (Calvo-Begueria et al., 2018). In addition, NO synthase activity is important in functional soybean nodule development (Leach, Keyster, Du Plessis, & Ludidi, 2010). The patterns of NO and ROS accumulation are similar during nodule formation and symbiosis. Nodule NO levels are determined by the balance between NO production and scavenging by phytoglobins, which regulate NO levels during the early phase of symbiosis and act as a buffer for oxygen in N-fixing nodules. In particular, *Phytogb1.1* regulates NO concentration during symbiosis establishment and nodule organogenesis in *M. truncatula* (Berger, Guinand, Boscari, Puppo, & Brouquisse, 2020). NO is scavenged by phytoglobins, or non-symbiotic hemoglobins, which have high affinity for both oxygen and NO. Both NR-dependent and mitochondrial NO production are activated by hypoxia. The increased operation of these pathways is correlated with the hypoxic state of functional nodules (Singh, Kumari, Foyer, & Kapuganti, 2020).

An early burst of ROS and NO production occurs in the root hair in response to inoculation with rhizobia, followed by later bursts in the infection thread and during nodule organogenesis, as well as in mature nodules (Hichri et al., 2016; Montiel et al., 2016). ROS production in the apoplast is required for root hair curling, a process that requires a redirection of ROS production away from the root tip to the zone of maximum curvature factors (Cárdenas, Martínez, Sánchez, & Quinto, 2008). ROS production is detected in the tips of actively growing root hairs shortly after addition of

Nod-factors. Phosphoinositide 3-kinases are involved in the growth of Rhizobia in the infection threads within host root hairs (Robert et al., 2018). These kinases are thought to function in micro-symbiont recognition responses that modulate apoplastic ROS production and associated intracellular/endosomal ROS processing in the root hair that control nodulation capacity. Prevention of ROS accumulation impairs root hair curling and infection thread formation (Robert et al., 2018).

ROS and NO are processed by antioxidants, such as TRX, which police ROS amounts during nodule development (Lee et al., 2005). However, the precise roles of antioxidative enzymes, such as catalase (CAT) or glutathione peroxidase (GPX) as well as TRXs and GRXs in ROS and NO processing and signaling still remain to be fully characterized. Transcriptome profiling analysis has provided evidence of the involvement of ROS and NO in the induction of nodulin expression and in the repression of innate defenses that is required to promote the establishment of symbiosis (Damiani et al., 2016).

While the suppression of local immune responses in host legumes, which is induced by the rhizobia, is essential for the establishment of symbiosis, systemic signaling involving ROS occurs in response to perception of the rhizobia. For example, a rapid systemic pathway of redox changes has been reported during the soybean–*Bradyrhizobium japonicum* interaction, with a transient peak of ROS accumulation reported in soybean leaves 30 min after root inoculation with *B. japonicum* (Fernandez-Göbel et al., 2019). The inoculation-dependent accumulation of ROS in soybean leaves is accompanied by changes in the redox state of glutathione and by activation of antioxidant enzymes (Fernandez-Göbel et al., 2019).

3.2 Symbiotic nitrogen fixation

Nodules are rich in reducing compounds, polyunsaturated fatty acids and also O_2-labile proteins, such as the nitrogenase that catalyzes symbiotic nitrogen fixation (SNF). SNF requires a low, but stable, oxygen environment that is controlled by the rate of respiration and the presence of a nodule-localized oxygen diffusion barrier (Fig. 2). Leghemoglobin thereby ensures a continual oxygen flux to support bacteroid respiration through the peribacteroid membrane that separates the microsymbiont from the host cytosol (Fig. 4). The oxygenated form of leghemoglobin (oxyLb) can undergo a slow autoxidation to the ferric form (metLb) producing superoxide and H_2O_2. The ferrous form of Lb can also react with H_2O_2 to form ferrylLb (LbIV), which is, however, unable to transport oxygen (Mathieu et al., 1998). Nodules have also high respiration rates because of the high energy demand of SNF and high respiratory fluxes increase the risk of ROS formation (Dalton, 1995). The regulation of respiration and the efficiency of the oxygen diffusion barrier consequently play a crucial role in the protection of nodule metabolism and this regulation seems to be an active plant-controlled process (Fig. 2; Wei & Layzell, 2006). Stress-induced SNF decreases in soybean are generally associated with modifications in the oxygen concentrations in the nodule, often due to break down of the oxygen diffusion barrier. Stresses, such as drought, salt and chilling, decrease nitrogenase activity and induce structural changes in the oxygen diffusion barrier, altering nodule permeability for O_2 (Del Castillo, Hunt, & Layzell, 1994; Serraj, Fleurat-Lessard, Jaillard, & Drevon, 1995; Van Heerden et al., 2008). Stress-induced changes to nodule O_2 homeostasis can then modulate the oxygenation of leghemoglobin (Kuzma, Topunov, & Layzell, 1995).

3.3 Nodule senescence

Although some legume genes and ROS have been implicated in the later developmental stages of nodule development and/or bacteroid differentiation, the molecular mechanisms that regulate middle or late developmental stages of nodule development are less well characterized. Nodule senescence is a developmentally regulated genetic program in which phytohormones, ROS/RNS antioxidants and proteinases play key roles. Senescence-induced TFs thereby activate proteases and lipases to degrade proteins and lipids, and transporters that export breakdown products (Schippers, Schmidt, Wagstaff, & Jing, 2015). For example, nine cystatins are expressed during nodule development and senescence' of which Glyma18g12240 appears to promote nodulation and delay nodule senescence (Yuan et al., 2017). Although the lifespan of nodules of most legumes is 10–12 weeks, their ability to fix N starts to decline 3–5 weeks post-inoculation (Puppo et al., 2005). Nodules contain high amounts of antioxidants, such as ascorbate and glutathione that are both important in the nodulation process and in nodule senescence, a stage in which the amounts of the metabolites decline (Lee et al., 2005). Nodule senescence is a genetically-controlled process that enables the recycling of nutrients. However, nodule senescence can be triggered by developmental cues and environmental stresses. Any premature nodule senescence severely compromises symbiotic nitrogen fixation and nitrogen availability to the plant.

In addition to the initiation of NOD factor expression, legume-derived flavonoids can induce the pathogenic type III secretion system (T3SS) of rhizobia. This induction injects effector proteins into host cells to modulate nodulation signaling towards nodulation, or to abort the nodulation process, through recognition by the host defense system (Okazaki, Kaneko, Sato, & Saeki, 2013; Nelson & Sadowsky, 2015). Other microbe/pathogen-associated molecular pattern (M/PAMP), such as chitin, can induce the M/PAMP-triggered susceptibility (M/PTS) of host legumes to rhizobia. Similarly, legumes will evolve cell-surface pattern-recognition receptors (PRRs) but little is known about whether nodule development and senescence are directly associated with the re-emergence of plant immunity and defense responses. However, accumulating evidence suggests that ROS and redox signaling and their interactions with phytohormones trigger nodule senescence.

RBOH enzymes play a pivotal role in nodule functioning that correlates with the distribution patterns of ROS within the nodule. It is likely that ROS produced by RBOHs act as signaling molecules to the endosymbiont and also mediating the remodeling of the cell wall. H_2O_2 accumulates in soybean nodules at a late stage of the nodulation process (Puppo, Pauly, Boscari, Mandon, & Brouquisse, 2013). Cell walls and peribacteroid membranes are among the first structures to be degraded. The most visible sign of nodule senescence is, however, a change of the nodule color changing from pink to green (Fig. 3). This change is caused by bilirubin and biliverdin accumulation after disruption of leghemoglobin (Swaraj & Bishnoi, 1996) with the green color partially deriving from leghemoglobin nitration (Sainz et al., 2015).

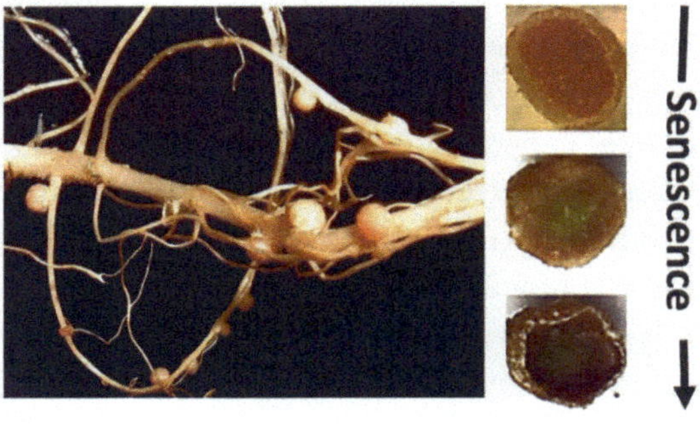

Fig. 3 Root nodules and nodule senescence with progressive degradation of leghemoglobin during nodule senescence. Color change is caused by bilirubin and biliverdin accumulation after disruption of leghemoglobin with the green color partially deriving from leghemoglobin nitration.

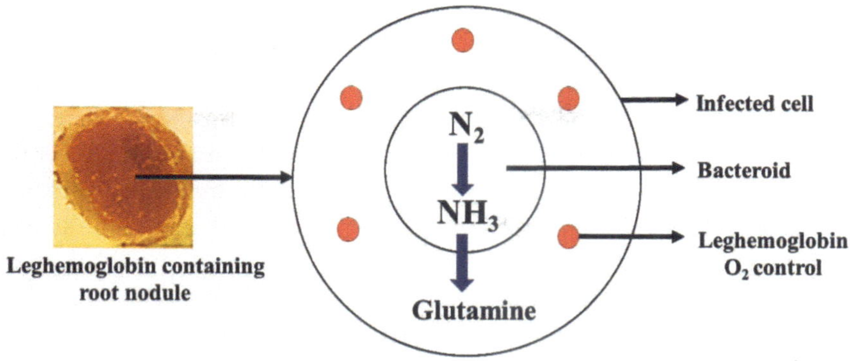

Fig. 4 Leghemoglobin containing root nodule with infected cell containing bacteroid with region of nitrogen fixation and leghemoglobin to buffer the concentration of free oxygen in the cytoplasm of infected plant cells to ensure the proper function of root nodules.

Different RBOH forms probably have different functions at different stages of nodule organogenesis and senescence. For example, *MtRbohB* is expressed in the meristematic region, infection zone, and nitrogen-fixing zone of *M. truncatula* nodules, while *MtRbohE* and *MtRbohF* are mainly expressed in the meristematic region and *MtRbohG* is expressed in the vascular tissue of nodules (Montiel et al., 2016). Senescent soybean and pea nodules have changes in the symbiosome structure that is associated with H_2O_2 and NO accumulation (Bruand & Meilhoc, 2019; Montiel et al., 2016). The addition of a NO donor to mature *M. truncatula* nodules also leads to accelerated senescence (Cam et al., 2012) and NO is also involved in stress-induced senescence of *M. truncatula* and *L. japonicus* nodules (Cam et al., 2012; Fukudome et al., 2019).

Exposure of young and mature nodules to H_2O_2-induced oxidative stress results in changes in leghemoglobin (Mathieu, Davies, et al., 1998; Mathieu, Moreau, Frendo, Puppo, & Davies, 1998). Exposure to salt stress, but not osmotic stress, also alters apoplastic and intracellular ROS accumulation in soybean roots inhibiting root hair curling and leading to cell death (Muñoz, Robert, Melchiorre, Racca, & Lascano, 2012). Exposure of soybean to oxidative stress results in DNA and protein damage at the early stages of nodule development. Increases in peroxides, protein carbonyls and modified DNA base concentrations have been found in older nodules exposed to oxidative stress and the content of catalytic iron increases significantly with nodule age. Iron is thereby liberated from leghemoglobin, especially during nodule senescence. The presence or iron in the peribacteroid space promotes, however, lipid peroxidation, contributing to the degradation of the peribacteroid membrane. Iron storage proteins, such as ferritin multimeric complexes, can scavenge free iron (Theil, 2003). Prevention of leghemoglobin degradation and the release of free iron might, however, delay stress-induced nodule senescence. Silencing of two Medicago ferritin genes, *MtFer2* and *MtFer3*, also limits increases in free iron (Dhanushkodi, Matthew, McManus, & Dijkwel, 2018). Senescence genes are highly expressed in the nodules of a *Mtcas31* (cold acclimation specific31) mutant, that is defective in a dehydrin protein that interacts with the leghemoglobin MtLB120-1. This mutant also shows reduced nitrogenase activity under drought stress (Li et al., 2019). More studies are, however, required to identify the key signaling components and regulators of senescence. Such information is urgently required for developing strategies to improve SNF by delaying nodule senescence.

3.4 Antioxidative systems

Nodules are particularly rich in antioxidant enzymes, such as SOD and CAT, and also proteins including TRX, GRX and PRX, as well as low molecular weight antioxidants such as ascorbate and glutathione. For example, APX represents about 1% of the nodule protein (Dalton, Hanus, Russell, & Evans, 1987). APX is localized in the endodermis of the indeterminate nodules of alfalfa, pea, and clover and in the exterior edge of the parenchyma of determinate nodules in bean and soybean. APX may function in the nodule oxygen barrier removing ROS produced by respiration in the endodermis (indeterminate nodules) and in boundary layers (determinate nodules) (Dalton, Joyner, Becana, Iturbe-Ormaetxe, & Chatfield, 1998). The activities of enzymes involved in the ascorbate–glutathione pathway are higher in N-fixing nodules with high SNF capacity than in nodules with lower SNF capacity (Dalton, Langeberg, & Treneman, 1993). Similar results were obtained in L. *japonicus* lines with leghemoglobin-RNA interference (Gunther, Schlereth, Udvardi, & Ott, 2007; Ott et al., 2005).

GSH and the legume-specific homolog, homoglutathione (hGSH), play important roles in the nodulation process (Frendo et al., 2005; Matamoros, Moran, Iturbe-Ormaetxe, Rubio, & Becana, 1999). The inhibition of GSH and hGSH synthesis in *M. truncatula* decreases root nodule formation (Frendo et al., 2005). GSH and hGSH are localized in the infected zone of bean nodules (Matamoros et al., 1999). High amounts of these thiols are also present in the meristematic zone of the nodule, where they fulfill important roles in symbiosis (Harrison et al., 2005; Muglia, Comai, Spegazzini, Riccillo, & Aguilar, 2008). Strains of *S. meliloti* and *R. tropici*, which are defective in the *gshB* gene that encodes the enzyme catalyzing the second step of GSH synthesis, have delayed nodulation and reduced SNF. GSH also controls glutathione S-transferases (GSTs) activities. These GSTs are involved in defense and catalyze the conjugation of GSH with oxidative degradation products. The amount of the GST9 protein, which is prevalent in soybean nodules, increases during senescence (Dalton et al., 2009). A decrease in SNF in GST9-RNAi silenced soybean nodules is also associated with increases in damaged proteins (Dalton et al., 2009). A detailed analysis of *M. truncatula* TRXs identified two isoforms that are specific to the symbiotic interaction (Alkhalfioui et al., 2008). Soybean plants deficient in Trx have also fewer nodules (Lee et al., 2005).

Bacterial mutants that are deficient in antioxidant defenses, such as catalase or glutathione, not only show a reduced capacity to infect the plant, but also have lower nitrogen-fixation capacities (Harrison et al., 2005; Jamet, Sigaud, Van de Sype, Puppo, & Hérouart, 2003). Similarly, the *rhizobium etli* double mutant, which lacks PRXS and a bifunctional catalase–PER (katG), has a significantly reduced nitrogen fixation capacity (Dombrecht et al., 2005). In addition, *S. meliloti* strains lacking a wide variety of cellular processes, but not directly antioxidant defense, are deficient in oxidative stress protection and symbiosis (Davies & Walker, 2007). Mutations in a *S. meliloti* TRX-like gene, involved in melanin production, impaired responses to paraquat-induced oxidative stress and resulted in lower symbiotic nitrogen fixation rates (Castro-Sowinski, Matan, Bonafede, & Okon, 2007). Finally, *R. leguminosarum* strains deficient in the TRX-like cycY protein are unable to form nitrogen-fixing nodules on pea or vetch, because these mutants could not produce mature c-type cytochromes (Vargas, Wu, Davies, & Downie, 1994).

Drought-induced senescence of soybean nodules has been also linked to oxidation of the ascorbate pool but also to an increased reduction of the glutathione pool (Marquez-Garcia et al., 2015). Decreases in nitrogen-fixation capacity in the senescent zones correlate with diminished antioxidant defenses provided, for example, by GSH/hGSH and/or ascorbate due to higher ROS levels (Evans et al., 1999). Similar results have been obtained in pea (Groten et al., 2006) and common bean (Loscos & Becana, 2008). Lower amounts of low molecular weight antioxidants and antioxidative enzymes have been found in nodules of plants exposed to environmental stress (Moran et al., 1994; Escuredo et al., 1996;

Gogorcena, Iturbe-Ormaetxe, Escuredo, & Becana, 1995; Matamoros et al., 1999; Loscos & Becana, 2008). Salicylic acid-induced oxidation and programmed cell death are finally the last stages of the senescence process (Gechev, Van Breusegem, Stone, Denev, & Laloi, 2006; Noctor et al., 2007; Shao, Chu, Shao, Jaleel, & Mi, 2008). Sequences encoding putative proteins involved in redox-related processes and derived from the soybean database have been also analyzed in relation to the effects of drought stress on nodules (Noctor et al., 2014). The abundance of relatively few redox-related transcripts changes in nodules in response to drought when compared to the response found in leaves. The abundance of transcripts encoding PRX1 (Glyma07g09240) and PRX2 lyma09g32540 (also called TPX1, because it encodes a TRX-dependent peroxidase) were significantly higher in plants exposed to drought than in a well-watered control. PRX1 has homology to PRXIIC, which is expressed in response to oxidants, such as H_2O_2 and lipid peroxides, and also by the addition of ascorbate (Horling et al., 2003).

4. Future perspectives

Given that soybean is a commercially important crop plant, relatively little research on the redox proteins and enzymes has been so far reported in the literature compared to other legumes. While Medicago and *Lotus japonicus* are good model legumes, much more effort is, therefore, required to understand the nuances of redox regulation in soybean, particularly with regard to the regulation of SNF and stress responses. Soybean, as a subtropical/tropical legume, is more difficult to grow in the Northern Hemisphere, where growth conditions are not optimal for soybean growth. Optimum soybean growing conditions require high light intensities and daytime air temperatures over 30 °C. Similarly, soybean is highly sensitive to dark chilling and only shows poor recovery from nighttime temperatures below 10 °C. Scientists in the Northern hemisphere are, therefore, less attracted to soybean as a legume of choice for research. However, recent research has already shown that a climate change will have a great impact on soybean growing areas, with early maturing varieties predicted to mature as far north as Scotland (Coleman et al., 2021). With future increase in atmospheric CO_2 levels, increases in yields of about $0.5\,t\,ha^{-1}$ are already predicted at some sites. Such data suggest that soybean is likely to be a viable crop in Northern Europe and for a similar climate (Coleman et al., 2021). Many European counties are already growing more soybeans in a bid to increase protein security. For example, German soybean production has increased more than tenfold since 2012 because of a steady expansion in production area. According to the German Federal Statistical Office record levels (94,100 t) of soybeans were harvested in Germany in 2020. Hence, varieties must be chosen to align with the growing season and more intensive studies are required to determine the physiological and genetic responses underpinning acclimation to changing climates and the factors determining sustainable yields (Kulkarni et al., 2018).

Until recently soybean was predominantly grown in the US (https://www.pioneer.com/us/agronomy/nitrogen_fertilizer_soybean.html) and South America, applying N-fertilization. Therefore, understanding the constraints on soybean nodule formation and SNF was not a priority. Hence, less research was undertaken on soybean nodules, which are highly sensitive to oxidative stress. Understanding nodule redox metabolism and the prevention of oxidative stress is, therefore, urgently required. A marked exception is Brazil, where researchers have investigated the use of soybean inoculants for many years, with a view to increasing yields without N-fertilization because of high fertilizer costs (Hungria et al., 2006). In contrast, an N-fertilizer is generally not required for alfalfa production (https://ucanr.edu/blogs/blogcore/postdetail.cfm?postnum=10478). Soybean has determinate nodules, which have a meristem at the periphery that is only active at early stages of development. Such activity results in a spherical shape. In contrast, *M. truncatula* has indeterminate nodules with a persistent meristem at the apices, which is scientifically interesting in terms of regrowth of nodules.

Model systems such Lotus and *M. truncatula* have been extensively studied and hence there is a large repository of genomic information that has already been produced, together with a range of useful molecular tools to assist researchers. These model species have relatively small genomes, and their genomes have been extensively sequenced and characterized. The soybean genome has been, however, completely sequenced and is readily available (http://www.phytozome.net/soybean). A recent review catalogs and contextualizes the plant genes that are required for SNF in two model legume species, and also in the crop species soybean and common bean (Roy et al., 2020). Moreover, a reference-grade genome of the wild soybean accession W05, has been reported (Xie et al., 2019) with a final assembled genome size of 1013.2 Mb and a contig N50 of 3.3 Mb. The analytical power of the W05 genome has wide applications in comparative genomic and evolutionary studies, as well as in crop breeding and improvement programs. The relatively large size of the soybean genome might have advantages in terms of analysis, particularly in relation to the analysis of transcriptomic, metabolomic and proteomic data (Ferguson & Gresshoff, 2009). Mutant collections are becoming increasingly more available for elucidation of the molecular mechanisms involved in nodule development and functioning, particularly those involving redox metabolism (Ren, 2018).

There are many reasons for studying redox metabolism in soybean, the most economically significant legume globally, not least that the functions of so few redox proteins, particularly TRX, GRX and PRXs have been characterized. Together with pea, the study of soybean has provided in the past substantial insights into legume development. However, limitations at the molecular genetic level have meant that soybean research has been neglected, and replaced by other legume models (Ferguson & Gresshoff, 2009), but one should always remember that many well-studied model legumes, such as Medicago and Lotus, are temperate, and not sub-tropical, species.

Gene discovery and confirmation of findings in Medicago, Lotus and common beans is now technically relatively simple. Genome editing (GE) might also become an efficient tool in soybean for the manipulation of useful traits such as protection against oxidative stress (Bao et al., 2020). Although Agrobacterium-mediated transformation might still the preferred choice for characterizing gene functions, stable soybean genetic transformation is far from routine. Soybean remains a recalcitrant crop for transformation and regeneration. In addition, Agrobacterium-mediated transformation efficiencies are affected by different tissue types, cultivars and/or species (Homrich, Wiebke-Strohm, Weber, & Bodanese-Zanettini, 2012). Hence, few soybean varieties can be directly improved through genetic manipulation. The application of GE might also effectively shorten the time required to develop stress-resilient soybean varieties. A major advantage of the GE technologies over GM technologies is that the edited genome contains no foreign DNA, which is a major concern for GM (Du et al., 2016; Turnbull, Lillemo, & Hvoslef-Eide, 2021). GE technology includes, for example, the Zn finger nuclease approach (ZFN), transcription activator-like effector-based nucleases (TALEN) or the clustered regulatory interspaced short palindromic repeats editing (CRISPR/Cas) (Tiwari, Trivedi, & Pandey, 2020). CRISPR/Cas has already been used to edit target genes in legumes (Bhowmik et al., 2021; Ji et al., 2019; Liu et al., 2019). Such technologies can be applied to make more rapid progress in our understanding of redox metabolism in soybean, particularly, for example, in understanding the role of redox PTMs in the control of metabolism, signaling and gene expression, as well as the structure/function relationships of different components. Soybean genome sequences are available and annotation is also much improved. The study of soybean mutants produced using the GE technology will, therefore, greatly aid future studies in soybean redox metabolism. Mutants in genes involved in redox metabolism are essential to understand the roles of different components in redox regulation and oxidative stress tolerance, which are likely to play important roles in soybean responses to a changing climate. It is equally important to study the roles of ROS and redox proteins in long distance signaling between above-ground and below-ground organs (roots and particularly nodules). It is important to identify bottlenecks in signaling and antioxidant production. Other possibilities would be to increase production and accumulation of antioxidants, such as glutathione, by modulation of GSH biosynthesis genes. This strategy would not only increase stress tolerance but also soybean yields and improving SNF. Soybean breeding programs have rarely, if ever, directly selected for the effectiveness of SNF in crop or forage legumes. GE would be also useful in investigations of specific gene functions, particularly in multigene families, whose roles in SNF have thus far been masked by genetic redundancy, as well as other genes that are less frequently mutated by conventional approaches. Much more information is, however, required to fully understand the redox regulation and signaling networks that operate in soybean. The application of GE technologies will greatly help resolve these issues.

Acknowledgment

KJK was funded by NRF SA incentive funding.

References

Adem, G. D., Roy, S. J., Zhou, M., Bowman, J. P., & Shabala, S. (2014). Evaluating contribution of ionic, osmotic and oxidative stress components towards salinity tolerance in barley. *BMC Plant Biology*, *14*, 113.

Ainsworth, E. A. (2017). Understanding and improving global crop response to ozone pollution. *Plant Journal*, *90*, 886–897.

Ainsworth, E. A., & Long, S. L. (2021). 30 years of free-air carbon dioxide enrichment (FACE): What have we learned about future crop productivity and its potential for adaptation? *Global Change Biology*, *27*, 27–49.

Akram, S., Siddiqui, M. N., Hussain, B. M. N., Al Bari, A., Mostofa, M. G., Hossain, M. A., et al. (2017). Exogenous glutathione modulates salinity tolerance of soybean [Glycine max (L.) Merrill] at reproductive stage. *Journal of Plant Growth Regulation*, *36*, 877–888.

Alkhalfioui, F., Renard, M., Frendo, P., Keichinger, C., Myer, Y., Gelhaye, E., et al. (2008). A novel type of thioredoxin dedicated to symbiosis in legumes. *Plant Physiology*, *148*, 424–435.

Arthikala, M.-K., Montiel, J., Sánchez-López, R., Nava, N., Cárdenas, L., & Quinto, C. (2017). Respiratory burst oxidase homolog gene a is crucial for rhizobium infection and nodule maturation and function in common bean. *Frontiers in Plant Science, 8*, 2003.

Bao, A., Zhang, C., Huang, Y., Chen, H., Zhou, X., & Cao, D. (2020). Genome editing technology and application in soybean improvement. *Oil Crop Science, 5*, 31–40.

Berger, A., Guinand, S., Boscari, A., Puppo, A., & Brouquisse, R. (2020). *Medicago truncatula* phytoglobin 1.1 controls symbiotic nodulation and nitrogen fixation via the regulation of nitric oxide concentration. *New Phytologist, 227*, 84–98.

Betzelberger, A. M., Yendrek, C. R., Sun, J., Leisner, C. P., Nelson, R. L., Ort, D. R., et al. (2012). Ozone exposure response for U.S. soybean cultivars: Linear reductions in photosynthetic potential, biomass, and yield. *Plant Physiology, 160*, 1827–1839.

Bhowmik, P., Konkin, D., Polowick, P., Hodgins, C. L., Subedi, M., Xiang, D., et al. (2021). CRISPR/Cas9 gene editing in legume crops: Opportunities and challenges. *Legume Science*, 1–16.

Bruand, C., & Meilhoc, E. (2019). Nitric oxide in plants: Pro- or anti-senescence. *Journal of Experimental Biology, 70*, 4419–4427.

Calvo-Begueria, L., Rubio, M. C., Martinez, J. I., Perez-Rontome, C., Delgado, M. J., Bedmar, E. J., et al. (2018). Redefining nitric oxide production in legume nodules through complementary insights from electron paramagnetic resonance spectroscopy and specific fluorescent probes. *Journal of Experimental Botany, 69*, 3703–3714.

Cam, Y., Pierre, O., Boncompagni, E., Hérouart, D., Meilhoc, E., & Bruand, C. (2012). Nitric oxide (NO): A key player in the senescence of Medicago truncatula root nodules. *New Phytologist, 196*, 548–560.

Cárdenas, L., Martínez, A., Sánchez, F., & Quinto, C. (2008). Fast, transient and specific intracellular ROS changes in living root hair cells responding to Nod factors (NFs). *Plant Journal, 56*, 802–813.

Castro-Sowinski, S., Matan, O., Bonafede, P., & Okon, Y. (2007). A thioredoxin of *Sinorhizobium meliloti* CE52G is required for melanin production and symbiotic nitrogen fixation. *Molecular Plant-Microbe Interaction, 20*, 986–993.

Chang, C., Damiani, I., Puppo, A., & Frendo, P. (2009). Redox changes during the legume–rhizobium symbiosis. *Molecular Plant, 2*, 370–377.

Chapman, J. M., Muhlemann, J. K., Gayomba, S. R., & Muday, G. (2019). RBOH-dependent ROS synthesis and ROS scavenging by plant specialized metabolites to modulate plant development and stress responses. *Chemical Research in Toxicology, 32*, 370–396.

Chaput, V., Martin, A., & Lejay, L. (2020). Redox metabolism: The hidden player in carbon and nitrogen signaling? *Journal of Experimental Botany, 71*, 3816–3826.

Cheng, M.-C., Ko, K., Chnag, W.-L., Kuo, W.-C., Chen, G.-H., & Lin, T.-P. (2015). Increased glutathione contributes to stress tolerance and global translational changes in Arabidopsis. *Plant Journal, 83*, 926–939.

Chinchilla, D., Bauer, Z., Regenass, M., Boller, T., & Felix, G. (2006). The Arabidopsis receptor kinase FLS2 binds flg22 and determines the specificity of flagellin perception. *Plant Cell, 8*, 465–476.

Chung, Y. S., Kim, K.-S., Hamayun, M., & Kim, Y. (2020). Silicon confers soybean resistance to salinity stress through regulation of reactive oxygen and reactive nitrogen species. *Frontiers in Plant Science, 10*, 1725.

Coleman, K., Whitmore, A. P., Hassall, K. L., Shield, I., Semenov, M. A., Dobermannd, A., et al. (2021). The potential for soybean to diversify the production of plant-based protein in the UK. *Science of the Total Environment, 767*, 144903.

Considine, M. J., & Foyer, C. H. (2021a). Oxygen and reactive oxygen species-dependent regulation of plant growth and development. *Plant Physiology, 186*(1), 79–92. https://doi.org/10.1093/plphys/kiaa077.

Considine, M. J., & Foyer, C. H. (2021b). Stress effects on the reactive oxygen species-dependent regulation of plant growth and development. *Journal of Experimental Botany*, *72*, 5795–5806.

Czarnocka, W., & Karpiński, S. (2018). Friend or foe? Reactive oxygen species production, scavenging and signaling in plant response to environmental stresses. *Free Radical Biology and Medicine*, *122*, 4–20.

D'Alessandro, S., Mizokami, Y., Légeret, B., & Havaux, M. (2019). The apocarotenoid β-cyclocitric acid elicits drought tolerance in plants. *iScience*, *19*, 461–473.

Dalton, D. A. (1995). Antioxidant defenses of plants and fungi. In S. Ahmad (Ed.), *Oxidative stress and antioxidant defenses in biology* (pp. 298–355). New York, NY: Chapman and Hall.

Dalton, D. A., Boniface, C., Turner, Z., Lindahl, A., Kim, H. J., Jelinek, L., et al. (2009). Physiological roles of glutathione S-transferases in soybean root nodules. *Plant Physiology*, *150*, 521–530.

Dalton, D. A., Hanus, F. J., Russell, S. A., & Evans, H. J. (1987). Purification, properties, and distribution of ascorbate peroxidase in legume root nodules. *Plant Physiology*, *83*, 789–794.

Dalton, D. A., Joyner, S. L., Becana, M., Iturbe-Ormaetxe, I., & Chatfield, J. M. (1998). Antioxidant defenses in the peripheral cell layers of legume root nodules. *Plant Physiology*, *116*, 37–43.

Dalton, D. A., Langeberg, L., & Treneman, N. (1993). Correlations between the ascorbate-glutathione pathway and effectiveness in legume root nodules. *Physiologia Plantarum*, *87*, 365–370.

Damiani, I., Pauly, N., Puppo, A., Brouquisse, R., & Boscari, A. (2016). Reactive oxygen species and nitric oxide control early steps of the legume—rhizobium symbiotic interaction. *Frontiers in Plant Science*, *7*, 454.

Das, A., Eldakak, M., Paudel, B., Kim, D.-W., Hemmati, H., Basu, C., et al. (2016). Leaf proteome analysis reveals prospective drought and heat stress response mechanisms in soybean. *Biomed Research International*, *2016*. Article ID 6021047.

Das, P., Nutan, K. K., Singla-Pareek, S. L., & Pareek, A. (2015). Oxidative environment and redox homeostasis in plants: Dissecting out significant contribution of major cellular organelles. *Frontiers in Environmental Science*, *2*, 70.

Das, K., & Roychoudhury, A. (2014). Reactive oxygen species (ROS) and response of antioxidants as ROS-scavengers during environmental stress in plants. *Frontiers in Environmental Science*, *2*, 53.

Davies, B. W., & Walker, G. C. (2007). Identification of novel *Sinorhizobium meliloti* mutants compromised for oxidative stress protection and symbiosis. *Journal of Bacteriology*, *189*, 2110–2113.

Del Castillo, L. D., Hunt, S., & Layzell, D. B. (1994). The role of oxygen in the regulation of nitrogenase activity in drought stressed soybean nodules. *Plant Physiology*, *106*, 949–955.

Demidchik, V. (2015). Mechanisms of oxidative stress in plants: From classical chemistry to cell biology. *Environmental and Experimental Botany*, *109*, 212–228.

Dhanushkodi, R., Matthew, C., McManus, M. T., & Dijkwel, P. P. (2018). Drought-induced senescence of *Medicago truncatula* nodules involves serpin and ferritin to control proteolytic activity and iron levels. *New Phytologist*, *220*, 196–208.

Dietz, K. J., Turkan, I., & Krieger-Liszkay, A. (2016). Redox- and reactive oxygen species-dependent signaling into and out of the photosynthesizing chloroplast. *Plant Physiology*, *171*, 1541–1550.

Dombrecht, B., Heusdens, C., Beullens, S., Verreth, C., Mulkers, E., Proost, P., et al. (2005). Defence of *rhizobium etli* bacteroids against oxidative stress involves a complexly regulated atypical 2-Cys peroxiredoxin. *Molecular Microbiology*, *55*, 1207–1221.

Dos Santos, C. V., & Rey, R. (2006). Plant thioredoxins are key actors in the oxidative stress response. *Trends in Plant Science*, *11*, 329–334.

Du, H., Zeng, X., Zhao, M., Cui, X., Wang, Q., Yang, H., et al. (2016). Efficient targeted mutagenesis in soybean by TALENs and CRISPR/Cas9. *Journal of Biotechnology*, *217*, 90–97.

Dumanović, J., Nepovimova, E., Natić, M., Kuča, K., & Jaćević, V. (2021). The significance of reactive oxygen species and antioxidant defense system in plants: A concise overview. *Frontiers in Plant Science*, *11*, 552969.

Eastburn, D. M., Degennaro, M. M., Delucia, E. H., Dermody, O., & Mcelrone, A. J. (2010). Elevated atmospheric carbon dioxide and ozone alter soybean diseases at SoyFACE. *Global Change Biology*, *16*(16), 320–330.

El-Enany, A. (1997). Glutathione metabolism in soybean callus-cultures as affected by salinity. *Biologia Plantarum*, *39*, 35–39.

Eljebbawi, A. H., Guerrero, Y. D. C. R., Dunand, C., & Estevez, J. M. (2021). Highlighting reactive oxygen species (ROS) as multitaskers in root development. *iScience*, *24*, 101978.

Escuredo, P. R., Minchin, F. R., Gogorcena, Y., Iturbe-Ormaetxe, I., Klucas, R. V., & Becana, M. (1996). Involvement of activated oxygen in nitrate-induced senescence of pea root nodules. *Plant Physiology*, *110*, 1187–1195.

Evans, P., Gallesi, D., Mathieu, C., Puppo, A., Hernandez, M. J., & De Felipe, M. (1999). Oxidative stress occurs during soybean nodule senescence. *Planta*, *208*, 73–79.

Ferguson, B. J., & Gresshoff, P. M. (2009). *Soybean as a model legume*. Available from: https://www.researchgate.net/publication/255909192_Soybean_as_a_model_legume (accessed July 17, 2021).

Fernandez-Göbel, T. F., Deanna, R., Muñoz, N. B., Robert, G., Asurmendi, S., & Lascano, R. (2019). Redox systemic signaling and induced tolerance responses during soybean–*Bradyrhizobium japonicum* interaction: Involvement of Nod factor receptor and autoregulation of nodulation. *Frontiers in Plant Science*, *10*, 141.

Fichman, Y., Myers, R. J., Grant, D. G., & Mittler, R. (2021). Plasmodesmata-localized proteins and ROS orchestrate light-induced rapid systemic signaling in *Arabidopsis*. *Science Signaling*, *14*. eabf0322.

Foyer, C. H., & Noctor, G. (2011). Ascorbate and glutathione: The heart of the redox hub. *Plant Physiology*, *155*, 2–18.

Foyer, C. H., & Noctor, G. (2016). Stress-triggered redox signalling: What's in pROSpect? *Plant Cell & Environment*, *39*, 951–964.

Foyer, C. H., & Shigeoka, S. (2011). Understanding oxidative stress and antioxidant functions to enhance photosynthesis. *Plant Physiology*, *155*, 93–100.

Frendo, P., Harrison, J., Norman, C., Hernández Jiménez, M. J. H., Van de Sype, G., Gilabert, A., et al. (2005). Glutathione and homoglutathione play a critical role in the nodulation process of *Medicago truncatula*. *Molecular Plant-Microbe Interactions*, *18*, 254–259.

Frendo, P., Matamoros, M. A., Alloing, G., & Becana, M. (2013). Thiol-based redox signaling in the nitrogen-fixing symbiosis. *Frontiers in Plant Science*, *4*, 376.

Fukudome, M., Watanabe, E., Osuki, K. I., Imaizumi, R., Aoki, T., Becana, M., et al. (2019). Stably-transformed Lotus japonicus plants overexpressing phytoglobin LjGlb1-1 show decreased nitric oxide levels in roots and nodules as well as delayed nodule senescence. *Plant and Cell Physiology*, *60*, 816–825.

Galant, A., Koester, R., Ainsworth, E. A., Hicks, L. M., & Jez, J. M. (2012). From climate change to molecular response: Redox proteomics of ozone-induced responses in soybean. *New Phytologist*, *194*, 220–229.

Gechev, T. S., Van Breusegem, F., Stone, J. M., Denev, I., & Laloi, C. (2006). Reactive oxygen species as signals that modulate plant stress responses and programmed cell death. *BioEssays*, *28*, 1091–1101.

Gillespie, K. M., Rogers, A., & Ainsworth, E. A. (2011). Growth at elevated ozone or elevated carbon dioxide concentration alters antioxidant capacity and response to acute oxidative stress in soybean (Glycine max). *Journal of Experimental Botany*, *62*, 2667–2678.

Gogorcena, Y., Iturbe-Ormaetxe, I., Escuredo, P. R., & Becana, M. (1995). Antioxidant defenses against activated oxygen in pea nodules subjected to water stress. *Plant Physiology*, *108*, 753–759.

Graciet, E., Walter, F., O'Maoileidigh, D., Pollmann, S., Meyerowitz, E. M., Varshavsky, A., et al. (2009). The N-end rule pathway controls multiple functions during Arabidopsis shoot and leaf development. *Proceedings of the National Academy of Sciences of the United States of America*, *106*, 13618–13623.

Groten, K., Dutilleul, C., van Heerden, P. D. R., Vanacker, H., Bernard, S., Finkemeier, I., et al. (2006). Redox regulation of peroxiredoxin and proteinases by ascorbate and thiols during pea root nodule senescence. *FEBS Letters*, *580*, 1269–1276.

Gunther, C., Schlereth, A., Udvardi, M., & Ott, T. (2007). Metabolism of reactive oxygen species is attenuated in leghemoglobin-deficient nodules of *Lotus japonicas*. *Molecular Plant-Microbe Interaction*, *20*, 1596–1603.

Gupta, K. J., Mur, L. A. J., Wany, A., Kumari, A., Fernie, A. R., & Ratcliffe, R. G. (2020). The role of nitrite and nitric oxide under low oxygen conditions in plants. *New Phytologist*, *225*, 1143–1151.

Halliwell, B., & Gutteridge, J. M. C. (1999). *The chemistry of free radicals and related 'reactive species'* (pp. 36–104). New York: Oxford University Press.

Harrison, J., Jamet, A., Muglia, C. I., Van de Sype, G., Aguilar, O. M., Puppo, A., et al. (2005). Glutathione plays a fundamental role in growth and symbiotic capacity of *Sinorhizobium meliloti*. *Journal of Bacteriology*, *187*, 168–174.

Hasanuzzaman, M., Bhuyan, M. H. M. B., Zulfiqar, F., Raza, A., Mohsin, S. M., Mahmud, J. A., et al. (2020). Reactive oxygen species and antioxidant defense in plants under abiotic stress: Revisiting the crucial role of a universal defense regulator. *Antioxidants*, *9*, 681.

Hichri, I., Meilhoc, E., Boscari, A., Bruand, C., Frendo, P., & Brouquisse, R. (2016). Nitric oxide: Jack-of-all-trades of the nitrogen- fixing symbiosis? *Advances in Botanical Research*, *77*, 193–218.

Holman, T. J., Jones, P. D., Russell, L., Medhurst, A., Tomas, S. U., Talloji, P., et al. (2009). The N- end rule pathway promotes seed germination and establishment through removal of ABA sensitivity in Arabidopsis. *Proceedings of the National Academy of Sciences of the United States of America*, *106*, 4549–4554.

Homrich, M. S., Wiebke-Strohm, B., Weber, R. L., & Bodanese-Zanettini, M. H. (2012). Soybean genetic transformation: A valuable tool for the functional study of genes and the production of agronomically improved plants. *Genetics and Molecular Biology*, *35*, 998–1010.

Horchani, F., Prevot, M., Boscari, A., Evangelisti, E., Meilhoc, E., Bruand, C., et al. (2011). Both plant and bacterial nitrate reductases contribute to nitric oxide production in *Medicago truncatula* nitrogen- fixing nodules. *Plant Physiology*, *155*, 1023–1036.

Horling, F., Lamkemeyer, P., König, J., Finkemeier, I., Kandlbinder, A., Baier, M., et al. (2003). Divergent light-, ascorbate-, and oxidative stress-dependent regulation of expression of the peroxiredoxin gene family in Arabidopsis. *Plant Physiology*, *131*, 317–325.

Hu, C. H., Wang, P. Q., Zhang, P. P., Nie, X.-M., Li, B.-B., Li, T., et al. (2020). NADPH oxidases: The vital performers and center hubs during plant growth and signaling. *Cell*, *9*, 437.

Hungria, M., Franchini, J. C., Campo, R. J., Crispino, C. C., Moraes, J. Z., Sibaldelli, R. N. R., et al. (2006). Nitrogen nutrition of soybean in Brazil: Contributions of biological N2 fixation and N fertilizer to grain yield. *Canadian Journal of Plant Science*, *86*, 927–939.

Jamet, A., Sigaud, S., Van de Sype, G., Puppo, A., & Hérouart, D. (2003). Expression of the bacterial catalase genes during *Sinorhizobium meliloti-Medicago sativa* symbiosis and their crucial role during the infection process. *Molecular Plant-Microbe Interactions*, *16*, 217–225.

Jedelská, T., Luhová, L., & Petřivalský, M. (2020). Thioredoxins: Emerging players in the regulation of protein S-nitrosation in plants. *Plants*, *9*, 1426.

Ji, J., Zhang, C., Sun, Z., Wang, L., Duanmu, D., & Fan, Q. (2019). Genome editing in cowpea *Vigna unguiculata* using CRISPR-Cas9. *International Journal of Molecular Sciences*, *20*, 2471.

Karuppanapandian, T., Moon, J.-C., Kim, C., Manoharan, K., & Kim, W. (2011). Reactive oxygen species in plants: Their generation, signal transduction, and scavenging mechanisms. *Australian Journal of Crop Science*, *5*, 709–725.

Katam, R., Shokri, S., Murthy, N., Singh, S. K., Suravajhala, P., Khan, M. N., et al. (2020). Proteomics, physiological, and biochemical analysis of cross tolerance mechanisms in response to heat and water stresses in soybean. *PLoS One*, *15*, e0233905.

Kimura, S., Waszczak, C., Hunter, K., & Wrzaczek, M. (2017). Bound by fate: The role of reactive oxygen species in receptor-like kinase signalling. *The Plant Cell*, *29*, 638–654.

Knörzer, O. C., Durner, J., & Böger, P. (1996). Alterations in the antioxidative system of suspension-cultured soybean cells (Glycine max) induced by oxidative stress. *Physiologia Plantarum*, *97*, 388–396.

Kranner, I., Birtić, S., Anderson, K. M., & Pritchard, H. W. (2006). Glutathione half-cell reduction potential: A universal stress marker and modulator of programmed cell death? *Free Radical Biology and Medicine*, *40*, 2155–2165.

Kulkarni, K. P., Tayade, R., Asekova, S., Song, J. T., Shannon, J. G., & Lee, J.-D. (2018). Harnessing the potential of forage legumes, alfalfa, soybean, and cowpea for sustainable agriculture and global food security. *Frontiers in Plant Science*, *9*, 1314.

Kuzma, M. M., Topunov, A. F., & Layzell, D. B. (1995). Effects of temperature on infected cell O2 concentration and adenylate levels in attached soybean nodules. *Plant Physiology*, *107*, 1209–1216.

Langebartels, C., Wohlgemuth, H., Kschieschan, S., Grün, S., & Sandermann, H. (2002). Oxidative burst and cell death in ozone-exposed plants. *Plant Physiology and Biochemistry*, *40*, 567–575.

Le, D. T., Nishiyama, R., Watanabe, Y., Mochida, K., Yamaguchi-Shinozaki, K., Shinozaki, K., et al. (2011). Genome-wide survey and expression analysis of the plant-specific NAC transcription factor family in soybean during development and dehydration stress. *DNA Research*, *18*, 263–296.

Leach, J., Keyster, M., Du Plessis, M., & Ludidi, N. (2010). Nitric oxide synthase activity is required for development of functional nodules in soybean. *Journal of Plant Physiology*, *167*, 1584–1591.

Lee, M. W., Huffaker, A., Crippen, D., Robbins, R. T., & Goggin, F. L. (2018). Plant elicitor peptides promote plant defences against nematodes in soybean. *Molecular Plant Pathology*, *19*, 858–869.

Lee, D. H., Lal, N. K., Lin, Z.-J. D., Ma, S., Liu, J., Castro, B., et al. (2020). Regulation of reactive oxygen species during plant immunity through phosphorylation and ubiquitination of RBOHD. *Nature Communications*, *11*, 1838.

Lee, M. Y., Shin, K. H., Kim, Y. K., Suh, J. Y., Gu, Y. Y., Kim, M. R., et al. (2005). Induction of thioredoxin is required for nodule development to reduce reactive oxygen species levels in soybean roots. *Plant Physiology*, *139*, 1881–1889.

Li, S., Wang, N., Ji, D., Zhang, W., Wang, Y., Yu, Y., et al. (2019). A GmSIN1/GmNCED3s/GmRbohBs feed-forward loop acts as a signal amplifier that regulates root growth in soybean exposed to salt stress. *The Plant Cell*, *31*, 2107–2130.

Li, P.-S., Yu, T.-F., He, G.-H., Chen, M., Zhou, Y.-B., Chai, S.-C., et al. (2014). Genome-wide analysis of the Hsf family in soybean and functional identification of GmHsf-34 involvement in drought and heat stresses l. *BMC Genomics*, *15*, 1009.

Liu, J., Gunapati, S., Mihelich, N. T., Stec, A. O., Michno, J. M., & Stupar, R. M. (2019). Genome editing in soybean with CRISPR/Cas9. *Methods in Molecular Biology*, *1917*, 217–234.

Liu, N., Song, F., Zhu, X., You, J., Yang, Z., & Li, X. (2017). Salicylic acid alleviates aluminum toxicity in soybean roots through modulation of reactive oxygen species metabolism. *Frontiers in Chemistry*, *5*, 96.

Loscos, J., Matamoros, M. A., & Becana, M. (2008). Ascorbate and homoglutathione metabolism in common bean nodules under stress conditions and during natural senescence. *Plant Physiology*, *146*, 1282–1292.

Mangano, S., Denita-Juarez, S. P., Choi, H.-S., Marzol, E., Hwang, Y., Ranocha, P., et al. (2017). Molecular link between auxin and ROS-mediated polar growth. *Proceedings of the National Academy of Sciences of the United States of America*, *114*, 5289–5294.

Marino, D., Dunand, C., Puppo, A., & Pauly, N. (2011). A burst of plant NADPH oxidases. *Trends in Plant Science*, *17*, 9–15.

Marquez-Garcia, B., Shaw, D., Cooper, J. W., Karpinska, B., Quain, M. D., Matome Makgopa, E. M., et al. (2015). Redox markers for drought-induced nodule senescence, a process occurring after drought-induced senescence of the lowest leaves in soybean (Glycine max). *Annals of Botany*, *116*, 497–510.

Mase, K., & Tsukagoshi, H. (2021). Reactive oxygen species link gene regulatory networks during Arabidopsis root development. *Frontiers in Plant Science*, *12*, 660274.

Matamoros, M. A., Moran, J. F., Iturbe-Ormaetxe, I., Rubio, M. C., & Becana, M. (1999). Glutathione and homoglutathione synthesis in legume root nodules. *Plant Physiology*, *121*, 879–888.

Mathieu, C., Davies, M. J., Moreau, S., Hérouart, D., Frendo, P., Puppo, A., et al. (1998). Radical formation during soybean nodule senescence. In C. Elmerich (Ed.), *Vol 31. Biological nitrogen fixation for the 21st century. Current plant science and biotechnology in agriculture*. Dordrecht: Springer. https://doi.org/10.1007/978-94-011-5159-7_313.

Mathieu, C., Moreau, S., Frendo, P., Puppo, A., & Davies, M. J. (1998). Direct detection of radicals in intact soybean nodules: Presence of nitric oxide leghemoglobin complexes. *Free Radical Biology and Medicine*, *24*, 1242–1249.

Mochida, K., Yoshida, T., Sakurai, T., Yamaguchi-Shinozaki, K., Shinozaki, K., & Phan Tran, L.-S. (2009). *In silico* analysis of transcription factor repertoire and prediction of stress responsive transcription factors in soybean. *DNA Research*, *16*, 353–369.

Mohammadi, P. P., Moieni, A., Hiraga, S., & Komatsu, S. (2012). Organ-specific proteomic analysis of drought-stressed soybean seedlings. *Journal of Proteomics*, *75*, 1906–1923.

Montiel, J., Arthikala, M. K., Cardenas, L., & Quinto, C. (2016). Legume NADPH oxidases have crucial roles at different stages of nodulation. *International Journal of Molecular Sciences*, *17*, 680.

Morales, M., & Munné-Bosch, S. (2016). Oxidative stress: A master regulator of plant trade-offs? *Trends in Plant Science*, *21*, 996–999.

Moran, J. F., Becana, M., Iturbe-Ormaetxe, I., Frechilla, S., Klucas, R. V., & Aparicio-Tejo, P. (1994). Drought induces oxidative stress in pea plants. *Planta*, *194*, 346–352.

Muglia, C., Comai, G., Spegazzini, E., Riccillo, P. M., & Aguilar, O. M. (2008). Glutathione produced by *rhizobium tropici* is important to prevent early senescence in common bean nodules. *FEMS Microbiology Letters*, *23*, 191–198.

Muñoz, N., Robert, G., Melchiorre, M., Racca, R., & Lascano, R. (2012). Saline and osmotic stress differentially affects apoplastic and intracellular reactive oxygen species production, curling and death of root hair during *Glycine max L. – Bradyrhizobium japonicum* interaction. *Environmental and Experimental Botany*, *78*, 76–83.

Nahar, K., Hasanuzzaman, M., Alam, M. M., & Fujita, M. (2015). Glutathione-induced drought stress tolerance in mung bean: Coordinated roles of the antioxidant defence and methylglyoxal detoxification systems. *AoB Plants*, *7*, plv069.

Nelson, M. S., & Sadowsky, M. J. (2015). Secretion systems and signal exchange between nitrogen-fixing rhizobia and legumes. *Frontiers in Plant Science*, *6*, 491.

Noctor, G., De Paepe, R., & Foyer, C. H. (2007). Mitochondrial redox biology and homeostasis in plants. *Trends in Plant Science*, *12*, 125–134.

Noctor, G., & Foyer, C. H. (2016). Intracellular redox compartmentation and ROS-related communication in regulation and signalling. *Plant Physiology, 171*, 1582–1592.
Noctor, G., Mhamdi, A., & Foyer, C. H. (2014). The roles of reactive oxygen metabolism in drought: Not so cut and dried. *Plant Physiology, 164*, 1636–1648.
Noctor, G., Mhamdi, A., & Foyer, C. H. (2016). Oxidative stress and antioxidative systems: Recipes for successful data collection and interpretation. *Plant Cell & Environment, 39*, 1140–1160.
Noctor, G., Mhamdi, A., Queval, G., Chaouch, S., Han, Y., Neukermans, J., et al. (2012). Glutathione functions in plants: An integrated overview. *Plant, Cell & Environment, 35*, 454–484.
Noctor, G., Strohm, M., Jouanin, L., Kunert, K. J., Foyer, C. H., & Rennenberg, H. (1996). Synthesis of glutathione in leaves of transgenic poplar overexpressing [gamma]-glutamylcysteine synthetase. *Plant Physiology, 112*, 1071–1078.
Ohama, N., Sato, H., Shinozaki, K., & Yamaguchi-Shinozaki, K. (2017). Transcriptional regulatory network of plant heat stress response. *Trends in Plant Science, 22*, 53–65.
Okazaki, S., Kaneko, T., Sato, S., & Saeki, K. (2013). Hijacking of leguminous nodulation signaling by the rhizobial type III secretion system. *Proceedings of the National Academy of Sciences of the United States of America, 110*, 17131–17136.
Ott, T.v., Dongen, J. T., Gunther, C., Krusell, L., Desbrosses, G., Vigeolas, H., et al. (2005). Symbiotic leghemoglobins are crucial for nitrogen fixation in legume root nodules but not for general plant growth and development. *Current Biology, 15*, 531–535.
Otulak-Kozieł, K., Kozieł, E., & Valverde, R. A. (2019). The respiratory burst oxidase homolog D (RbohD). Cell and tissue distribution in potato-potato virus Y (PVY^{NTN}) hypersensitive and susceptible reactions. *International Journal of Molecular Sciences, 20*, 2741.
Porcel, R., & Ruiz-Lozano, J. M. (2004). Arbuscular mycorrhizal influence on leaf water potential, solute accumulation, and oxidative stress in soybean plants subjected to drought stress. *Journal of Experimental Botany, 55*(403), 1743–1750.
Puppo, A., Groten, K., Bastian, F., Carzaniga, R., Soussi, M., Lucas, M. M., et al. (2005). Legume nodule senescence: Roles for redox and hormone signalling in the 994 orchestration of the natural aging process. *New Phytologist, 165*, 683–701.
Puppo, A., Pauly, N., Boscari, A., Mandon, K., & Brouquisse, R. (2013). Hydrogen peroxide and nitric oxide: Key regulators of the legume-rhizobium and mycorrhizal symbioses. *Antioxidants and Redox Signaling, 18*, 2202–2219.
Ramel, F., Ksas, B., Akkari, E., Mialoundama, A. S., Monnet, F., Krieger-Liszkay, A., et al. (2013). Light-induced acclimation of the Arabidopsis chlorina1 mutant to singlet oxygen. *The Plant Cell, 25*, 1445–1462.
Ranjan, A., Jayaraman, D., Grau, C., Hill, J. H., Whitham, S. A., Ané, J.-M., et al. (2018). The pathogenic development of *sclerotinia sclerotiorum* in soybean requires specific host NADPH oxidases. *Molecular Plant Pathology, 19*, 700–714.
Ren, G. (2018). The evolution of determinate and indeterminate nodules within the Papilionoideae subfamily. Wageningen University. doi.org/10.18174/429101.
Robert, G., Muñoz, N., Alvarado-Affantranger, X., Saavedra, L., Davidenco, V., Rodríguez-Kessler, M., et al. (2018). Phosphatidylinositol 3-kinase function at very early symbiont perception: A local nodulation control under stress conditions? *Journal of Experimental Botany, 69*, 2037–2048.
Rodrigues, O., Reshetnyak, G., Grondin, A., Saijo, Y., Leonhardt, N., Maurel, C., et al. (2017). A signaling role of aquaporins in guard cells. *Proceedings of the National Academy of Sciences, 114*, 9200–9205.
Roy, S., Liu, W., Nandety, R. S., Crook, A., Mysore, K. S., Pislariu, C. I., et al. (2020). Celebrating 20 years of genetic discoveries in legume nodulation and symbiotic nitrogen fixation. *The Plant Cell, 32*, 15–41.
Sagi, M., & Fluhr, R. (2006). Production of reactive oxygen species by plant NADPH oxidases. *Plant Physiology, 141*, 336–340.

Sainz, M., Calvo-Begueria, L., Pérez-Rontomé, C., Wienkoop, S., Abián, J., Staudinger, C., et al. (2015). Leghemoglobin is nitrated in functional legume nodules in a tyrosine residue within the heme cavity by a nitrite/peroxide-dependent mechanism. *The Plant Journal, 81*, 723–735.

Santolini, J., Wootton, S. A., Jackson, A. A., & Feelisch, M. (2019). The redox architecture of physiological function. *Current Opinion in Physiology, 9*, 34–47.

Sasidharan, R., Schippers, J. H. M., & Schmidt, R. R. (2021). Redox and low-oxygen stress: Signal integration and interplay. *Plant Physiology, 186*, 66–78.

Schippers, J. H., Schmidt, R., Wagstaff, C., & Jing, H.-C. (2015). Living to die and dying to live: The survival strategy behind leaf senescence. *Plant Physiology, 169*, 914–930.

Schneider, J. R., Müller, M., Klein, V. A., Rossato-Grando, L. G., Barcelos, R. P., Dalmago, G. A., et al. (2020). Soybean plant metabolism under water deficit and xenobiotic and antioxidant agent application. *Biology, 9*, 266.

Serraj, R., Fleurat-Lessard, P., Jaillard, B., & Drevon, J. J. (1995). Structural changes in the inner-cortex cells of soybean root nodules are induced by short-term exposure to high salt or oxygen concentrations. *Plant Cell & Environment, 18*, 455–462.

Shao, H. B., Chu, L. Y., Shao, M. A., Jaleel, C. A., & Mi, H. M. (2008). Higher plant antioxidants and redox signaling under environmental stresses. *Comptes Rendus Biologies, 33*, 433–441.

Sharma, P., Jha, A. B., Dubey, R. S., & Pessarakli, M. (2012). Reactive oxygen species, oxidative damage, and antioxidant defence mechanism in plants under stressful conditions. *Journal of Botany, 2012*. Article ID 217037.

Shumbe, L., Chevalier, A., Legeret, B., Taconnat, L., Monnet, F., & Havaux, M. (2016). Singlet oxygen-induced cell death in Arabidopsis under high-light stress is controlled by OXI1 kinase. *Plant Physiology, 170*, 1757–1771.

Shumbe, L., D'Alessandro, S., Shao, N., Chevalier, A., Ksas, B., Bock, R., et al. (2017). METHYLENE BLUE SENSITIVITY 1 (MBS1) is required for acclimation of Arabidopsis to singlet oxygen and acts downstream of β-cyclocitral. *Plant Cell & Environment, 40*, 216–226.

Singh, P., Kumari, A., Foyer, C. H., & Kapuganti, J. G. (2020). The power of the phytoglobin–NO cycle in the regulation of nodulation and symbiotic nitrogen fixation. *New Phytologist, 227*, 84–98.

So, H. A., & Lee, J. H. (2019). NAC transcription factors from soybean (*Glycine max* L.) differentially regulated by abiotic stress. *Journal of Plant Biology, 62*, 147–160.

Suzuki, N., Miller, G., Morales, J., Shulaev, V., Torres, M. A., & Mittler, R. (2011). Respiratory burst oxidases: The engines of ROS signaling. *Current Opinion in Plant Biology, 14*, 691–699.

Swaraj, K., & Bishnoi, N. R. (1996). Physiological and biochemical basis of nodule senescence in legumes: A review. *Plant Physiology and Biochemistry, 23*, 105–116.

Tambussi, E. A., Bartoli, C. G., Guiamet, J. J., Beltrano, J., & Araus, J. L. (2004). Oxidative stress and photodamage at low temperatures in soybean (Glycine max L. Merr.) leaves. *Plant Science, 167*, 19–26.

Theil, E. C. (2003). Ferritin: At the crossroads of iron and oxygen metabolism. *ChemInform, 34*, no-no.

Tiwari, M., Trivedi, P. K., & Pandey, A. (2020). Emerging tools and paradigm shift of gene editing in cereals, fruits, and horticultural crops for enhancing nutritional value and food security. *Food and Energy Security, 10*, e258.

Torres, M. A., Dangl, J. L., & Jones, J. D. G. (2002). Arabidopsis gp91(phox) homologues AtrbohD and AtrbohF are required for accumulation of reactive oxygen intermediates in the plant defense response. *Proceedings of the National Academy of Sciences USA, 99*, 517–522.

Tovar Mendes, A., Matamoros, M. A., Bustos-Sanmamed, P. B., & Becana, M. (2011). Peroxiredoxins and NADPH-dependent thioredoxin systems in the model legume *Lotus japonicus*. *Plant Physiology, 156*, 1535–1547.

Turnbull, C., Lillemo, M., & Hvoslef-Eide, T. A. K. (2021). Global regulation of genetically modified crops amid the gene edited crop boom – A review. *Frontiers in Plant Science, 12*, 630396.

Van Dongen, J. T., & Licausi, F. (2015). Oxygen sensing and signaling. *Annual Review of Plant Biology, 66*, 345–367.

Van Heerden, P. D., Kiddle, G., Pellny, T. K., Mokwala, P. W., Jordaan, A., Strauss, A. J., et al. (2008). Roles for the regulation of respiration and the oxygen diffusion barrier in soybean in the protection of symbiotic nitrogen fixation from chilling-induced inhibition and shoots from premature senescence. *Plant Physiology, 148*, 316–327.

Vargas, C., Wu, G., Davies, A. E., & Downie, J. A. (1994). Identification of a gene encoding a thioredoxin-like product necessary for cytochrome c biosynthesis and symbiotic nitrogen fixation in *rhizobium leguminosarum*. *Journal of Bacteriology, 176*, 4117–4123.

Wang, Y., Branicky, R., Noë, A., & Hekimi, S. (2018). Superoxide dismutases: Dual roles in controlling ROS damage and regulating ROS signaling. *Journal Cell Biology, 217*, 1915–1928.

Wang, G., Hu, C., Zhou, J., Liu, Y., Cai, J., Pan, C., et al. (2019). Systemic root-shoot signaling drives jasmonate-based root defense against nematodes. *Current Biology, 29*, 3430–3437.

Wang, D., Liang, X., Bao, Y., Yang, S., Zhang, X., Yu, H., et al. (2020). A malectin-like receptor kinase regulates cell death and pattern-triggered immunity in soybean. *EMBO Reports, 21*, e50442.

Wang, Y., Zhao, T., Wang, J., Zhao, Y., Guo, W., & Cao, M. (2013). Reactive oxygen species, ascorbic acid and glutathione to ozone stress in soybean (Glycine max) leave. In *Third International Conference on Intelligent System Design and Engineering Applications, Hong Kong, China* (pp. 1084–1087).

Wany, A., Foyer, C. H., & Gupta, K. J. (2018). Nitrate, NO and ROS signaling in stem cell homeostasis. *Trends in Plant Science, 12*, 1041–1044.

Wei, Y., Balaceanu, A., Rufian, J. S., Segonzac, C., Zhao, A., Morcillo, R. J. L., et al. (2020). An immune receptor complex evolved in soybean to perceive a polymorphic bacterial flagellin. *Nature Communications, 11*, 3763.

Wei, H., & Layzell, D. B. (2006). Adenylate-coupled ion movement. A mechanism for the control of nodule permeability to O2 diffusion. *Plant Physiology, 141*, 280–287.

Weits, D., Giuntoli, B., Kosmacz, M., Parlanti, S., Hubberten, H.-M., Riegler, H., et al. (2014). Plant cysteine oxidases control the oxygen-dependent branch of the N-end-rule pathway. *Nature Communications, 5*, 3425.

White, M. D., Klecker, M., Hopkinson, R. J., Weits, D. A., Mueller, C., Naumann, C., et al. (2017). Plant cysteine oxidases are dioxygenases that directly enable arginyl transferase-catalysed arginylation of N-end rule targets. *Nature Communications, 8*, 14690.

Wu, F., Chi, Y., Jiang, Z., Xu, Y., Xie, L., Huang, F., et al. (2020). Hydrogen peroxide sensor HPCA1 is an LRR receptor kinase in Arabidopsis. *Nature, 578*, 577–581.

Xie, M., Chung, C. Y.-L., Li, M.-W., Wong, F.-L., Wang, X., Liu, A., et al. (2019). A reference-grade wild soybean genome. *Nature Communications, 10*, 1216.

You, J., & Chan, Z. (2015). ROS regulation during abiotic stress responses in crop plants. *Frontiers in Plant Science, 6*, 1092.

Yuan, S. L., Li, R., Chen, H. F., Zhang, C. J., Chen, L. M., Hao, Q. N., et al. (2017). RNA-seq analysis of nodule development at five different developmental stages of soybean (Glycine max) inoculated with *Bradyrhizobium japonicum* strain 113-2. *Scientific Reports*, 7, 42248.

Zaffagnini, M., Fermani, S., Marchand, C. H., Costa, A., Sparla, F., Rouhier, N., et al. (2019). Redox homeostasis in photosynthetic organisms: Novel and established thiol-based molecular mechanisms. *Antioxidants & Redox Signaling, 31*, 155–210.

Zandalinas, S. I., Fichman, Y., Devireddy, A. R., Sengupta, S., Azad, R. K., & Mittler, R. (2020). Systemic signaling during abiotic stress combination in plants. *Proceedings of the National Academy of Sciences of the United States of America*, *117*, 13810–13820.

Zhu, B., Ye, C., Lü, H., Chen, X., Chai, G., Chen, J., et al. (2006). Identification and characterization of a novel heat shock transcription factor gene, GmHsfA1, in soybeans (Glycine max). *Journal of Plant Research*, *119*, 247–256.

CHAPTER EIGHT

Pursuing greener farming by clarifying legume-insect pest interactions and developing marker-assisted molecular breeding

Kejing Fan[a,†], Ho Yin Yip[a,b], Satyanarayana Taddi[a,b,†], Yi Ying Huang[a,b,†], Fuk Ling Wong[a], Jerome H.L. Hui[a,b,*], and Hon-Ming Lam[a,*]

[a]Centre for Soybean Research of the State Key Laboratory of Agrobiotechnology and School of Life Sciences, The Chinese University of Hong Kong, Shatin, Hong Kong Special Administrative Region, China
[b]Simon F.S. Li Marine Science Laboratory, School of Life Sciences, The Chinese University of Hong Kong, Shatin, Hong Kong Special Administrative Region, China
*Corresponding authors: e-mail address: jeromehui@cuhk.edu.hk; honming@cuhk.edu.hk

Contents

† These authors contributed equally.

Advances in Botanical Research, Volume 102
ISSN 0065-2296
https://doi.org/10.1016/bs.abr.2022.02.018

Abstract

Legumes are important for both food security and sustainable agriculture. Given their relatively high protein content, more than 700 million people worldwide rely on them as food or poultry feeds. They are also important in intercropping systems for improving soil quality, due to their nitrogen-fixing ability. One major challenge to the yield of legumes is infestation by insect pests and pathogens both in the field and during storage. Most farmers have responded by planting resistant strains of legume crops and spraying them with insecticide and fungicide. Nevertheless, the continuous use of biocides, despite their cost-effectiveness, results in resistance development by the pests and pathogens, and raises environmental safety concerns for both humans and off-target beneficial species of insects and microbes. In this review, we discuss the most up-to-date thinking on the interactions between legumes and their insect pests and current farming practices, explain the latest techniques used in identifying molecular markers to aid in the breeding of insect-resistant cultivars, and highlight areas that require further development for effective and ecofriendly integrated pest management.

1. Introduction

Grain legumes (a category which includes soybeans, chickpeas, pigeon peas, cowpeas, field peas, lentils, greengrams, blackgrams, faba beans, and grasspeas) are the main dietary protein sources among vegetarians and form an integral part of the daily diet of human beings all over the world. The cultivation of legumes can contribute to the reduction of malnutrition and improvement of soil fertility (Abate, 2012; Schreinemachers et al., 2014). The total area under soybean, groundnut and pulses cultivation worldwide is estimated at 249 million hectares (FAOSTAT, 2021). Despite the importance of legumes, it has been a struggle to increase their production relative to cereal crops (Foyer et al., 2016).

One major reason is the damage to such crops caused by insect pests. For example, the African bollworm (*Helicoverpa armigera*) alone caused an annual loss in pigeon pea and chickpea production worth US$645 million (ICRISAT, 1992). Another example is the soybean aphid, *Aphis glycines*, which could cause up to 58% of soybean crop yield loss, worth an estimated US$2.4 billion annually (Song, Swinton, Difonzo, O'Neal, & Ragsdale, 2006; Tilmon, Hodgson, O'Neal, & Ragsdale, 2011; Wang, Fang, Lin, Zhang, & Wang, 1994). Table 1 summarizes some

Table 1 Summary of the distribution, infestation area and damage caused by different insect pests.

Legume crop	Insect pest	Scientific name	Distribution[a]	Attack area(s)	Infestation rate (IR)/ yield loss (YL)	Reference(s)
Common bean	Bean weevil	*Acanthoscelides obtectus*	Worldwide	Seed	YL: 13.7%	Allotey, Segwabe, and Randome (2016)
Pea and lentil	Pea aphid	*Acyrthosiphon pisum*	Worldwide	Leaf, stem and pod	YL: Up to 35.7%	Sandhi and Reddy (2020)
Cowpea	Striped bean weevil	*Alcidodes leucogrammus*	Africa	Root and stem	IR: 13.8–45.8%	Amoako-Atta (1983)
Cowpea	Coreid pod-bug	*Anoplocnemis curvipes*	Africa and Iran	Seed	IR: 15.0–17.2%	Koona, Osisanya, Jackai, and Tonye (2004)
Soybean	Velvetbean caterpillar	*Anticarsia gemmatalis*	North and South America	Leaf	—	Anazonwu and Johnson (1986)
Leguminosae family[b]	Cowpea aphid	*Aphis craccivora*	Worldwide	Leaf, stem and pod	IR: 32.6% (Cowpea)	Mehrparvar, Madjdzadeh, Mahdavi Arab, Esmaeilbeygi, and Ebrahimpour (2012); Ofuya (1991)
Broad bean	Black bean aphid	*Aphis fabae*	Worldwide except Oceania	Leaf, stem and pod	—	Völkl and Stechmann (1998)
Soybean	Soybean aphid	*Aphis glycines*	Asia, Russia, USA, Canada and Australia	Whole plant	IR: Up to 80%	Venette and Ragsdale (2004)
Chickpea	Semi loopers	*Autographa nigrisigna*	Asia and Russia	Leaf and pod	—	Sharma et al. (2007)
Soybean	Bean leaf beetle	*Ceratoma trifurcata*	North America	Leaf, stem and pod	YL: 3.06 kg/ha	Smelser and Pedigo (1992)
Pigeon pea	Pod bug	*Clavigralla scutellaris*	Africa and Asia	Pod and seed	YL: 35–65%	Shanower, Romeis, and Minja (1999)
Cowpea	African pod bug	*Clavigralla tomentosicollis*	Africa	Pod and seed	IR: 57.0–60.0%	Koona et al. (2004)

Continued

Table 1 Summary of the distribution, infestation area and damage caused by different insect pests.—cont'd

Legume crop	Insect pest	Scientific name	Distribution[a]	Attack area(s)	Infestation rate (IR)/ yield loss (YL)	Reference(s)
Soybean	Leafhopper	*Empoasca terminalis*	India and Indonesia	Pod and seed	YL: 26%	Nasruddin, Fattah, Baco, and Said (2015)
Pigeon pea	Leafhopper	*Empoasca kerri*	India, Bangladesh and USA	Leaf	YL: 10%	Rachappa, Shivayogiyappa, Harischandra, and Yelshetty (2018)
Common bean	Bean ladybeetle	*Epilachna varivestis*	North America, Japan and Colombia	Leaf	YL: 2–54.9%	Kabissa and Fronk (1986)
Soybean	Limabean pod borer	*Etiella zinckenella*	Worldwide	Pod and seed	YL: Up to 80%	Kuswantoro, Bayu, Baliadi, and Tengkano (2017)
Pigeon pea	Leaf webber	*Grapholita critica*	Asia	Leaf	YL: 1.3–50.56 kg/ha	Kumar et al. (2014)
Leguminosae family[b]	American bollworm	*Helicoverpa armigera*	Worldwide	Pod, seed and flower	IR: 20.7–37.0% (Pigeon pea) YL: 12.5–17.8% (Common bean) IR: 16.4–26.3% (Cowpea)	Ali et al. (2009); Karel (1985); Ofuya (1991)
Chickpea	Leaf miner	*Liriomyza cicerina*	Africa, India, Middle East Asia and Europe	Leaf	IR: Up to 40%	Toker, Erler, Ceylan, and Canci (2010)
Leguminosae family[b]	Legume pod borer	*Maruca vitrata*	Worldwide	Flower bud and pod	YL: 23.3% (Pigeon pea)	Sharma, Saxena, and Bhagwat (1999)
Cowpea	Flower bud thrips	*Megalurothrips sjostedti*	Africa and Saudi Arabia	Pod	YL: 20–100%	Singh (1987)
Leguminosae family[b]	Bean flower thrips	*Megalurothrips usitatus*	Asia and Oceania	Flower	—	Tang et al. (2015)
Pigeon pea	Pod fly	*Melanagromyza obtusa*	Kenya, Asia, North America, Australia, Papua New Guinea and Colombia	Pod and seed	IR: 29.7–52.0%	Sharma, Bhosle, Kamble, Bhede, and Seeras (2011)

Leguminosae family[b]	Soybean stem fly	*Melanagromyza sojae*	Egypt, South Africa, Asia, Spain, Australia, Solomon Islands and South America	Stem	YL: 2–51% (Soybean)	Vitorio et al. (2019)
Leguminosae family[b]	Green stink bug	*Nezara viridula*	Worldwide	Pod	—	Ntonifor and Jackai (1994)
Cowpea	Foliage beetle	*Ootheca mutabilis*	Africa	Flower and pods	—	Egho (2010a)
Soybean	Stemfly	*Ophiomyia centrosematis*	Africa, Asia and Australia	Stem and seed	—	Talekar and Lee (1988)
Common bean	Bean stem maggot	*Ophiomyia phaseoli*	Africa, Asia, USA and Oceania	Stem, leaf and seed	YL: 8–100%	Ochilo and Nyamasyo (2010)
Soybean	Stink bug	*Piezodorus guildinii*	USA, Cuba and South America	Seed	—	Panizzi and Slansky (1985)
Leguminosae family[b]	Soybean looper	*Pseudoplusia includens*	Australia, North and South America	Leaf	IR: Up to 91%	Kogan and Cope (1974)
Soybean	Bean bug	*Riptortus clavatus*	Japan, South Korea and Taiwan	Seed	IR: 43.6–52%	Lee et al. (2004)
Soybean and cowpea	Pod-sucking bug	*Riptortus dentipes*	Africa	Pod	—	Ntonifor and Jackai (1994)
Soybean	Stink bug	*Riptortus pedestris*	Asia	Pod and seed	IR: 46.7–56.4%	Li et al. (2021)
Leguminosae family[b]	Fall armyworm	*Spodoptera frugiperda*	Worldwide	Pod	—	Andrews (1980)
Leguminosae family[b]	Two-spotted spider mite	*Tetranychus urticae*	Worldwide	Seed	—	Razmjou, Tavakkoli, and Nemati (2009)

[a]CAB Internation (https://www.cabi.org/) or EPPO Global Database (https://gd.eppo.int/).
[b]Polyphagous pest.

of the major global insect pests specific to legumes (Abate, 2012; Abate & Ampofo, 1996; Clement et al., 2000).

2. Mechanisms of host plant resistance against insects

The interaction between the host plant and its insect pests is a continuously evolving form of one-upmanship. Plants have evolved numerous defense mechanisms against insect attacks, including chemical and physical barriers such as volatile compounds that attract the predators of herbivorous insects, secondary metabolites that impart an unpleasant taste or are toxic to herbivores, and increased trichome density as both a physical barrier and a reservoir of antiherbivory compounds. The main mechanisms of host plant resistance are: antixenosis (adverse effects on the behavior of pests), antibiosis (adverse effects on the life history of pests), and tolerance (ease of recovery from insect damage).

2.1 Antixenosis

Antixenosis is also referred to as a nonpreference, in which a plant characteristic poses a chemical or biophysical barrier that deters or repels the arthropod without causing it harm. It may be affected by morphological factors (such as trichomes, waxiness, or pigmentation) or chemical factors (such as phenolic compounds, volatile compounds, enzymes, or other secondary metabolites) that adversely affect the actions of insects, prompting them to choose an alternative host plant. Although this mode of repelling insect pests does not harm them directly, it can increase the pests' exposure to parasitoids or predators by increasing their search time for a more palatable host (Belmain, Haggar, Holt, & Stevenson, 2013). This approach could therefore be part of an integrated pest management (IPM) strategy that incorporates the use of predators or parasitoids of the target insect pest species.

Trichomes have been found in alfalfa/lucerne to offer effective resistance to sapsucking insects such as the potato leafhopper (Shade, Doskocil, & Maxon, 1979). These glandular hairs are likely to provide some level of resistance against other sapsucking pests as well, including aphids (Shade & Kitch, 1983). Trichomes seem also to be able to prevent heavy aphid infestations on chickpeas, as aphids feed more effectively on the areas of these vegetables where trichomes are absent (Edwards, 2001). Trichome secretions often contain the immobilized bodies of young nymphs and larvae of a number of insect herbivores (Boethel, 1999). Resistance seems to be related to the orientation and size of trichomes in leafhopper-resistant soybeans, rather than their density (Turnipseed, 1977).

Soybean trichome length and density have also been implicated in a plant's resistance to beanflies, whiteflies, pod borers, and bean leaf beetles (Chiang & Norris, 1983; Lam & Pedigo, 2001; Lambert, McPherson, & Espelie, 1995; Talekar & Lin, 1994). Pod trichomes have been found to be involved in Coreid bug resistance (Koona et al., 2002). The morphology of pods and seeds may also lead to resistance against insect herbivory. For cowpeas, resistance to weevils is strongly correlated with seed coat thickness (Kitch, Shade, & Murdock, 1991), while their resistance to hemipteran and lepidopteran pod pests is correlated with a variety of morphological characteristics, including pod toughness, hull thickness, peduncle length, and seed position in the pod (Koona et al., 2002; Tayo, 1989).

2.2 Antibiosis

Antibiosis is defined as a mechanism of pest resistance in which the plant produces an adverse biological effect on the arthropod herbivore, either by killing it or making it too sick to do further damage and cause yield loss. Usually, this resistance mechanism relies on the deployment of plant biochemicals, such as free amino acids, fatty acids, terpenoids, flavonoids and phenolic compounds (Tables 2 and 3), which may weaken an insect attempting to colonize or feed on the plant and harm its biological output (Smith, 2005). Antibiosis is employed in the resistance of pigeon peas and chickpeas to *H. armigera*, resulting in decreased larval survival, decreased fertility, lower weight gain and prolonged larval growth of the herbivore (Kumari, Sharma, & Reddy, 2010). Accessions of wild *Cajanus* spp. (relatives of the pigeon pea) have elevated manifestations of pod borer antibiosis (Sharma, 2005). Antibiosis in the forms of decreased larval and pupal weights, increased larval mortality, extended larval and pupal periods, failure to pupate, reduced fecundity and egg viability contributes to the resistance to *H. armigera* in chickpeas (Narayanamma, Sharma, Gowda, & Sriramulu, 2007). The cowpea cultivar MNC 99-541 F21 demonstrated antibiosis against whitefly (*Bemisia tabaci* biotype B), prolonging the insect's life cycle, and the cowpea genotypes BRS-Urubuquara, Canapu, and TE97-304 G-4 also showed antibiosis toward whitefly, causing elevated nymphal mortality (Cruz, Baldin, de Jesus, & de Castro, 2014).

Table 2 Biochemical substances and mechanisms involved in host plant resistance against insects.

Biochemical substances	Functions and mechanisms	References
Terpenoids (monoterpenoids, sesquiterpenoids, diterpenoids, triterpenoids)	Antibiotic, feeding deterrents, toxicants, oviposition deterrents	Nishida (2002)
Flavonoids (flavonols, flavones, isoflavonoids, anthocyanins)	Phytoalexins, antifeedant, inhibition of mitochondrial oxidation	Ananthakrishnan (1999)
Tannins and lignins	Unpalatability, mechanical barrier postabsorptive inhibition	De Bruxelles and Roberts (2001)
Aldehydes, ketones, alkanes, waxes	Protective layer	Panda and Khush (1995)
Phenolics	Ovicidal, toxicants, photosensitizing insects	Nishida (2002)
Glucosinolates	Antibiotic, toxicant, repellent, irritant	Kliebenstein et al. (2001)
Alkaloids	Digestive enzyme inhibitor, toxicants, interfere in the nerve system, antifeedant, glucosidase inhibitor	Panda and Khush (1995)

Table 3 Protein-derived molecules involved in host plant resistance against insects.

Protein-derived molecules	Function	References
Chitinases	Damage the insect midgut	Falco, Marbach, Pompermayer, Lopes, and Silva-Filho (2001)
Proteinase inhibitors	Digestive enzyme inhibitors	Pompermayer et al. (2001)
Lectins	Increase the absorption of toxic substances, interfere in the absorption of nutrients	Falco et al. (2001)
Polyphenol oxidase	Reduce the nutritive value of the proteins	Falco et al. (2001)
Peroxidase	Hypersensitive response, lignin biosynthesis	Arimura et al. (2000)

Plant secondary metabolites involved in environmental adaptations and stress tolerance can be broadly classified into phenolics, flavonoids, alkaloids, and terpenoids. Terpenoids are the most diverse among these classes of compounds, and play a significant role in the area of chemical ecology (Gershenzon & Dudareva, 2007).

Terpenoids are known to mediate interactions between plants and insects (e.g., pollinators, predators, parasitoids, and herbivores). These interactions generally work to the advantage of the plants and to the detriment of the herbivores. The emission of volatile organic compounds (VOCs), including terpenoids, is believed to mediate these interactions. It is believed that the characteristics and effects of the VOC mixture depend on the identity and proportion of the individual components, and that each species of attacking herbivore is susceptible to a specific blend of VOCs (Boncan et al., 2020). The emitted VOCs may include one or a few compounds with toxic, deterrent/repellent, or attractive properties.

Terpenoids are classified (based on the number of carbon atoms they possess) into hemiterpenoids (C5), monoterpenoids (C10), homoterpenoids (C11, C16), sesquiterpenoids (C15), diterpenoids (C20), sesterpenoids (C25), triterpenoids (C30), tetraterpenoids (C40), and polyterpenoids ($C>40$, higher-order terpenoids) (Boncan et al., 2020). Volatile terpenes (VTs) are terpenoids that are released to interact with the environment, and include hemi-, homo-, mono-, sesqui-, and some diterpenoids. Certain chemical properties, viz. vapor pressure and hydrophobicity, influence the storage and volatility of these terpenoids, while their emission rates and patterns depend on abiotic and biotic factors, including temperature, humidity, seasonality, irradiance, and interactions with other plants and organisms (Yazaki, Arimura, & Ohnishi, 2017).

Terpenoids play essential roles in various biological processes, not only for plant defenses but also for the growth of insects. The well-known methyl farnesoate (MF), farnesoic acid (FA), and juvenile hormone (JH) contain insect sesquiterpenoids (Beran, Kollner, Gershenzon, & Tholl, 2019; Cheong, Huang, Bendena, Tobe, & Hui, 2015; Qu, Bendena, Tobe, & Hui, 2018).

Plant tissues with specialized structures, such as secretory cavities, resin canals, latex canals, and glandular trichomes (Holopainen, Himanen, Yuan, Chen, & Stewart, 2013), are involved in the preservation and storage of terpenoids. In the presence of a high concentration of terpinolene, less damage than usual was caused by *Paropsisterna tigrina* adults in *Melalecua alternifolia* (Bustos-Segura, Külheim, & Foley, 2015). Besides directly attacking insect pests, terpenoids may also serve as elicitors to activate defense mechanisms in plants against pests. On the other hand, terpenoids that attract herbivores can also serve as bioherbicides that facilitate weed control in rice paddies and other cultivated areas (Mitra, Karmakar, Mukherjee, & Barik, 2017). In addition to promoting crop yield by weed control, pest management, and pollination, terpenoids are also used to improve or change the scents and flavors of food products to suit consumer preferences, through engineering the terpenoid biosynthesis pathways.

2.3 Tolerance

Tolerance is a polygenic trait that helps a plant to tolerate or recover from an insect infestation. It also denotes the degree to which an insect population can be sustained by a plant without significant loss of vigor or yield. However, due to environmental factors, tolerance is more variable than antixenosis and antibiosis. The degree of tolerance is also markedly affected by the age or size and the general vigor of the plant, as well as by the size of the insect-resistant plant population vs the magnitude of the infestation (Belmain et al., 2013).

The degree of reduction in grain yield is also a good indicator of the agronomic efficiency and genotypic ability to withstand damage to insect pests. In chickpeas, better recovery and a lower reduction in grain yield were obtained with the *H. armigera*-tolerant cultivars ICC 12476, ICC 12479 ICC 506 EB, and ICCV 2, than with the vulnerable control ICCC 37 (Narayanamma et al., 2007). Pigeon pea cultivars ICPL 98008, ICPL 187-1, ICP 7203-1, T 21, ICPL 332, and ICP 7035 exhibited moderate levels of resistance to *H. armigera* over the growing season. Under unprotected conditions, ICP 7203-1, ICPL 87119, ICPL 187-1, ICPL 84060, and ICPL 332 exhibited lower grain yield losses than the controls, ICPL 87091 and ICPL 87 (Kumari et al., 2010).

To fully utilize resistant cultivars as effective tools in modern pest control schemes, a keen understanding of the types and mechanisms of resistance is required. Some of the pest-resistance mechanisms in legumes are shown in Table 4.

Table 4 Legume-insect interactions and host resistance mechanisms.

Legume crop	Insect pests	Resistance mechanisms	Genetics	References
Soybean (*Glycine max*)	Stink bugs, *Nezara viridula*	Antibiosis and antixenosis	Quantitative	Lopes, Destro, Montalvan, Ventura, and Guerra (1997)
	Soybean aphid, *Aphis glycines*	Antibiosis	*Rag1*	Li, Zhao, et al. (2004), Li, Hill, and Hartman (2004)
	Potato leafhopper, *E. fabae*	Antixenosis (trichomes)	Quantitative	Elden and Elgin (1992)
	Various caterpillars: *Spodoptera exigua, Pseudoplusia includens, H. zea, H. virescens*	Antibiosis, some antixenosis	Recessive or partial dominance, quantitative	Lambert and Kilen (1984); Kilen and Lambert (1986); Kilen and Lambert (1998)
Chickpea (*Cicer arietinum*)	Pod borers, *Helicoverpa armigera* and *H. punctigera*	Chemical antibiosis (trichome exudates)	Quantitative	Yoshida, Cowgill, and Wightman (1997)
Alfalfa (*Medicago sativa*)	Pea aphid, *Acyrthosiphon pisum*	Antibiosis	Polygenic	Julier, Bournoville, Landre, Ecalle, and Carre (2004)
	Potato leafhopper, *E. fabae*	Tolerance	Quantitative	Sorensen and Horber (1974)
Pea (*Pisum sativum*)	Pea aphid, *Acyrthosiphon kondoi*	Indirect via leaf	*st*	Kareiva and Sahakian (1990)
		Morphology	*Af*	Soroka and MacKay (1990)
	Pea weevil, *Bruchus pisorum*	Hypersensitivity	*Np*	Doss et al. (2000)

Continued

Table 4 Legume-insect interactions and host resistance mechanisms.—cont'd

Legume crop	Insect pests	Resistance mechanisms	Genetics	References
Common bean (*Phaseolus vulgaris*)	Leafhoppers, *Empoasca* spp.	Tolerance, antixenosis	Quantitative	Galwey and Evans (1982); Kornegay and Temple (1986)
	Mexican bean beetle, *Epilachna varivestis*	Antibiosis	Recessive, quantitative	Rufener, St. Martin, Cooper, and Hammond (1989)
	Mexican bean weevil, *Zabrotes subfasciatus*	Antibiosis	*Arc*	Osborn, Blake, Gepts, and Bliss (1986); Romero-Andreas, Yandell, and Bliss (1986)
	Bean weevil, *Acanthoscelides obtectus*	Antibiosis	Recessive, quantitative	Kornegay and Cardona (1991)
	Bean pod weevil, *Apion godmani*	Antibiosis	*Agr, Agm*	Garza, Cardona, and Singh (1996)

Rag1, Resistance to Aphis glycines gene 1;*st, Reduced stipule* gene; *Af, Afila* gene; *Np, Neoplastic pod* gene; *Arc*, Arcelin; *Agr, A. godmani* resistance; *Agm: A. godmani* resistance modifier.

3. Insect-borne plant pathogens

The interactions among insect pests, microorganisms and plants are complex (Franco, Moura, Vivanco, & Silva-Filho, 2017). In these interactions, insect pests are not merely herbivores, but can also act as vectors for the transmission of plant viruses (Whitfield, Falk, & Rotenberg, 2015). Viral diseases are the main biotic stress factors on legume yields in the tropics and subtropics (Sastry & Zitter, 2014). Around 80% of the 1480 known plant viruses rely on insect vectors for transmission (Eigenbrode, Bosque-Pérez, & Davis, 2018). In this section we discuss four major insect pests responsible for transmitting legume viruses, viz. aphids, whiteflies, beetles and thrips (Table 5).

Table 5 Major pest vectors for the transmission of legume crop viruses.

Pest	Crop	Disease	Virus
Aphids	Soybean	Mosaic	Soybean mosaic virus (SMV)
		Dwarf	Soybean dwarf virus (SDV)
	Groundnut	Rosette	Groundnut rosette assistor virus (GRAV)
			Satellite RNA
			Groundnut rosette umbravirus (GRV)
		Mottle	Peanut mottle virus (PeMoV)
		Stripe	Peanut stripe virus (PStV)
		Yellow mosaic	Cucumber mosaic virus (CMV)
	Common bean	Common mosaic and black root	Bean common mosaic virus (BCMV), Bean common mosaic necrosis virus (BCMNV)
		Mosaic due to CMV	CMV
	Cowpea	Mosaic due to potyviruses	Cowpea aphid-borne mosaic virus (CABMV), Bean common mosaic virus-blackeye cowpea mosaic strain (BCMV-BICM)
		Stunt	CMV
			BCMv-BICM

Continued

Table 5 Major pest vectors for the transmission of legume crop viruses.—cont'd

Pest	Crop	Disease	Virus
	Chickpea	Stunt	Bean leaf roll virus (BLRV), SDV, Beet western yellows virus (BWYV), Legume yellows virus (LYV) and Chickpea luteovirus (CpLV)
		Chlorotic stunt	Chickpea chlorotic stunt virus (CpCSV)
	Pea	Mosaic due to potyviruses	Pea seed-borne mosaic virus (PSbMV)
			Bean yellow mosaic virus (BYMV)
		Enation mosaic	Pea enation mosaic virus-1 (PEMV-1)
			Pea enation mosaic virus-2 (PEMV-2)
		Top yellows	BLRV
	Faba bean	Necrotic yellows	Faba bean necrotic yellows virus (FBNYV), Faba bean necrotic stunt virus (FBNSV)
		Leaf roll	BLRV
		Mosaic and necrosis	BYMV
	Lentil	Yellows and stunt	BLRV
			FBNYV
		Mosaic and mottle	PSbMV
			CMV
Beetles	Soybean	Bud light	Cowpea serve mosaic virus (CPSMV)
		Leaf yellow and roll	Bean pod mottle virus (BPMV)
	Cowpea	Mosaic due to comoviruses	Cowpea mosaic virus (CPMV) and CPSMV
		Mosaic due to sobemovirus	Southern bean mosaic virus (SBMV)
		Chlorotic mottle	Cowpea chlorotic mottle virus (CCMV)
		Mottle	Cowpea mottle virus (CPMov)
	Faba bean	Mottle	Broad bean mottle virus (BBMV)
	Lentil	Mosaic and mottle	Broad bean stain virus (BBSV)

Table 5 Major pest vectors for the transmission of legume crop viruses.—cont'd

Pest	Crop	Disease	Virus
Thrips	Soybean	Bud blight	Peanut bud necrosis virus (PBNV) and Soybean vein necrosis virus (SVNV)
		Brazilian bud blight	Tobacco streak virus (TSV)
	Groundnut	Stem necrosis	TSV
		Spotted wilt	Tomato spotted wilt virus (TSWV)
		Bud necrosis	PBNV
	Mungbean and Urdbean	Leaf curl	PBNV
Whiteflies	Soybean	Yellow mosaic	Begomoviruses (e.g., Bean golden mosaic virus (BGMV))
	Common Bean	Golden mosaic	BGMV
		Golden yellow mosaic	Bean golden yellow mosaic virus (BGYMV)
		Dwarf mosaic	Bean dwarf mosaic virus (BDMV)
	Cowpea	Golden mosaic and yellow mosaic	Cowpea golden mosaic virus (CGMV)
			Mungbean yellow mosaic India virus (MYMIV)
			Dolichos yellow mosaic virus (DoYMV)
		Mild mottle	Cowpea mild mottle virus (CMMV)
	Pigeon pea	Yellow mosaic	MYMV, Rhynchosia mosaic virus
	Mungbean and Urdbean	Yellow mosaic	MYMV, MYMIV and Horsegram yellow mosaic virus (HgYMV)

3.1 Aphids

Aphids transmit viruses that infect legume crops, which could either kill the plant or significantly reduce the crop yield (Ng & Perry, 2004). For instance, the aphids *A. glycines* can transmit the soybean mosaic virus (SMV) (Rupe & Luttrell, 2008), and the occurrence rate of SMV in a field is highly correlated with the flights of aphids, resulting in huge yield losses (Widyasari, Alazem, & Kim, 2020; Yang et al., 2008).

In Japan, Indonesia, Australia, New Zealand and some parts of Africa, soybean dwarf virus (SDV) is an important soybean disease. This can also be transmitted by aphids such as *A. glycines* (Damsteegt et al., 2011).

Groundnut rosette disease (GRD) is a severe disease of groundnuts in Uganda, and can sometimes result in a complete loss of yield (Okello et al., 2014). GRD is caused by three causal agents: groundnut rosette assistor virus (GRAV), groundnut rosette umbravirus (GRV) and satellite RNA. The transmission of these three agents requires aphids as vectors (Okello et al., 2014). As viral infections in groundnuts cannot be directly controlled once they have gained hold, the control of aphids is the only way of preventing them from spreading further (Alegbejo & Abo, 2002).

Peanut mottle disease, caused by peanut mottle virus (PeMoV), is also spread by aphids, notably *Aphis craccivora* Koch and *Myzus persicae* (Bock, 1973). Another aphid-transmitted virus, peanut stripe virus (PStV), was first reported in China and causes a mild mottle in the seed.

3.2 Whiteflies

Begomoviruses can infect legumes and cause the legume yellow mosaic disease (YMD) (Alabi, Kumar, Mgbechi-Ezeri, & Naidu, 2010). Begomovirus transmission is highly dependent on the whitefly, *B. tabaci* (Hema, Sreenivasulu, Patil, Kumar, & Reddy, 2014; Mar et al., 2017). In India, the yield loss caused by begomoviruses has exceeded $300 million in mungbean, soybean and black gram (Varma & Malathi, 2005). Whiteflies feed on plant phloem sap through injecting their stylets into the vascular tissue (Morgan et al., 2013). During the ingestion process, the whitefly ingests begomoviruses from the phloem sap, which can then infect other legume crops such as soybeans, common beans, cowpeas, pigeon peas, mungbeans and urdbeans, as the insect moves on to feed on the next plant (Czosnek, Ghanim, & Ghanim, 2005; Hema et al., 2014; Morales & Anderson, 2001).

3.3 Beetles

Beetles can cause mosaic and mottle diseases in several legume crops through the transmission of mosaic and mottle viruses (Bradshaw, Rice, & Hill, 2008). The bean leaf beetle is the main cause of the spread of bean pod mottle virus (BPMV), which causes the young leaves of soybeans to yellow and pucker (Yadav et al., 2015). The transmission of cowpea mosaic virus

(CPMV) is highly dependent on beetles. This virus reduces the plant's fresh weight and reduces the number and weight of pods (Lomonossoff & Shanks, 1999).

3.4 Thrips

Thrips palmi Karny has been reported to be the vector for peanut bud necrosis virus (PBNV), and can spread the virus among several legume crops including soybeans, groundnuts, mungbeans and urdbeans (Hema et al., 2014; Pensuk, Jogloy, Wongkaew, & Patanothai, 2008; Thekke-Veetil et al., 2020). Recently, the nucleic acid sequences of 12 arthropod viruses and 12 plant viruses were found in the soybean thrips, *Neohydatothrips variabilis*, suggesting that thrips could acquire viruses from various host plants and transmit them to soybeans (Thekke-Veetil et al., 2020).

4. Pest adaptations to plant defenses

The interactions between insects and plants are bidirectional. Plants have evolved different mechanisms to defend themselves against herbivores through physical or chemical barriers. At the same time, insects have also developed counteradaptations by secreting effectors to suppress plant defenses triggered by the initial insect-produced elicitors (Chen & Mao, 2020). For instance, the salivary protein C002 produced by pea aphids (*Acyrthosiphon pisum*) shortens the feeding span of some insects on faba beans (Mutti et al., 2008). Another effector, Armet, has been detected in the phloem sap of infested faba beans, and its knockdown through RNAi shortens aphid feeding and life spans (Wang et al., 2015). Furthermore, the aphid symbiont *Serratia symbiotica* promotes the expression of a histidine-rich Ca^{2+}-binding protein-like gene in the aphid salivary gland to prolong the feeding by its host on *Medicago truncatula*. It does this by suppressing the accumulation of Ca^{2+} and reactive oxygen species (ROS), the initial plant defense response to insect herbivory (Wang et al., 2020). In the soybean aphid *A. glycines*, downregulation of effectors and upregulation of transposable elements have been proposed as putative mechanisms adopted by the insect to counteract soybean aphid resistance (Yates-Stewart et al., 2020). Similarly, nontransient microbiota isolated in the stink bug, *Nezara viridula*, help the herbivore to adapt to soybean defenses (Medina et al., 2018).

5. Worldwide insecticide use in agriculture

Pest management technology has undergone a range of advances in recent decades. The widespread indiscriminate killing of insects, including agricultural pests, through the use of synthetic insecticides in the latter half of the 20th century has given way to more agroecologically sensitive strategies aimed at controlling pest populations by interfering with their reproduction, attracting predators, and using insect pheromone traps. Although the application of synthetic biocides to protect crops against pests and diseases has protected crop yields, the use of these chemicals has had harmful effects both on the environment and on human health. In recent years, attempts have been made to use molecular approaches to develop cultivars or varieties that are resistant to insect pests in grain legumes, in order to reduce reliance on pesticide use and make agriculture more ecofriendly (Sharma, 2016). These molecular approaches include genetic or quantitative trait loci (QTL)-related introgression/pyramiding, marker-assisted selections and wide hybridization for the generation of insect pest-resistant cultivars (Khera et al., 2013).

Total insecticide usage worldwide in agriculture rose from a low in 1994, at 569,465 t, to a peak in 2010, at 720,538 t (Fig. 1A). Average total insecticide application by continent between 1990 and 2019 was highest in Asia, which accounted for 60.3% of the world's total, followed by the Americas (26.2%), Europe (8.3%), Africa (3.8%), and Oceania (1.4%) (Fig. 1B). The top 10 countries with the highest average insecticide usage from 1990 to 2019 were China (272,792.9 t), the USA (79,817.13 t), Brazil (44,011.97 t), India (29,641.23 t), Russia (23,018 t), Japan (12,561.03 t), Turkey (12,600.1 t), Italy (11,305.63 t), Vietnam (11,268.43 t), and Spain at 9799.97 t (Fig. 1C) (FAOSTAT, 2021).

6. Integrated pest management (IPM) and current farming practices

The oldest strategies for controlling insect species were related to modifications in agricultural practices. However, these strategies were soon abandoned or deemphasized with the development of synthetic pesticides, and research on them was largely discontinued. They are reliant on long-term planning because agricultural controls are often proactive rather

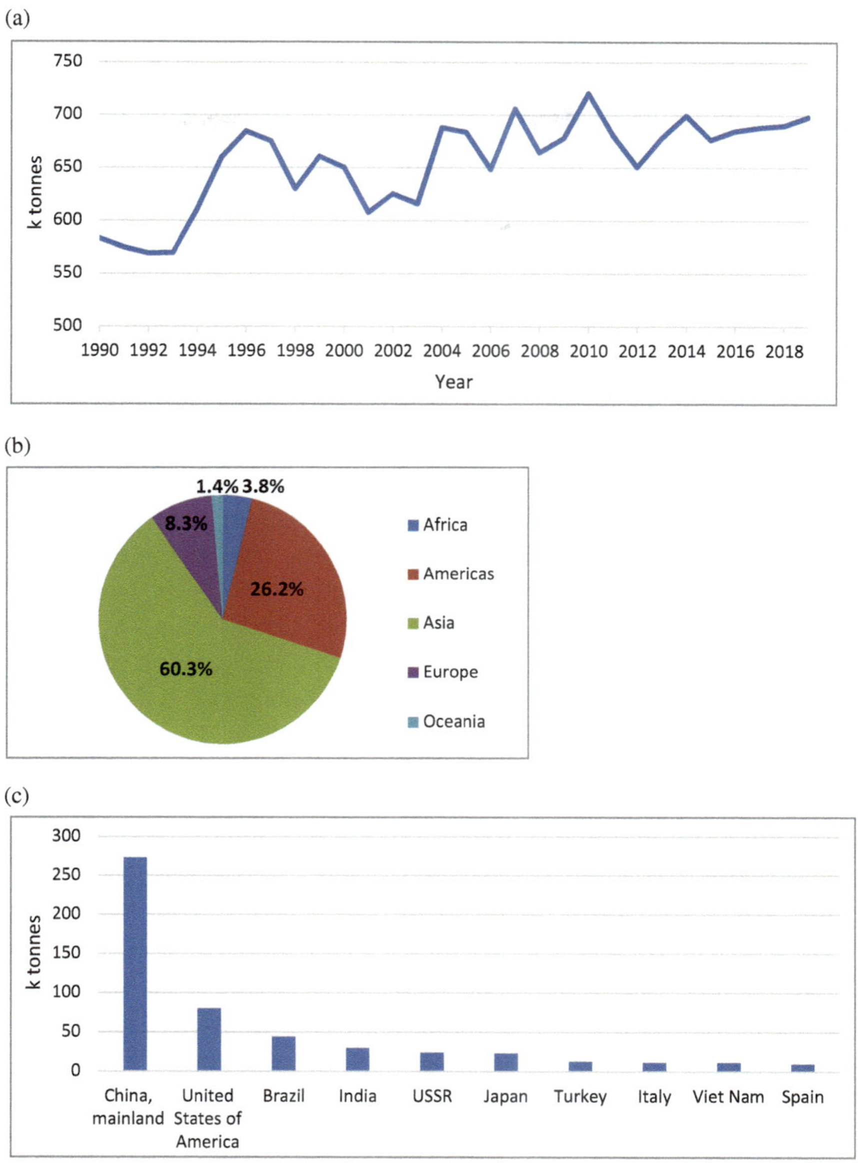

Fig. 1 Insecticide usage database on the quantity (in tonnes) of insecticides used or sold for crops in the agricultural sector worldwide between 1990 and 2019 (FAOSTAT, 2021). There was no information on the amount applied to single crops. (A) Total insecticide usage worldwide in tons by year. (B) Insecticide usage (% of world total) by continents based on the annual average between 1990 and 2019. (C) Top 10 average insecticide usage in tonnes by countries.

than reactive. Also, since their effectiveness relies on a thorough knowledge of the ecology of the many different pest species, most of which have been poorly understood in the past, the outcomes have been highly variable and their effectiveness often difficult to determine. Understandably, most farmers were gradually won over to the more direct and less experience- and skill-dependent solutions to pest problems by the use of toxic chemicals (Hill, 1989).

The situation today is very different. We now have a much better understanding of the ecological interactions within crop systems. We have access to predictive computer models for certain pests, and governments are now expected to rely less on toxic chemicals to solve their problems. As a result of multiple tragedies related to pesticide misuses or spills, the discovery of pesticide residues in recreational areas, drinking water, food and human tissues, and the rise in the occurrence of pesticide-related allergies and petrochemical sensitivities, public interest in the relationship between food quality and human health has grown, accompanied by a rise in the demand for organic foods (Hill, 1987). Integrated pest management (IPM) has thus been developed for pest control. The Food and Agriculture Organization (FAO) defines IPM as "the careful consideration of all available pest control techniques and subsequent integration of appropriate measures that discourage the development of pest populations and keep pesticides and other interventions to levels that are economically justified and reduce or minimise risks to human health and the environment." IPM emphasizes the cultivation of a healthy crop with the least possible disruption to agroecosystems, and encourages natural pest control mechanisms (FAO, 2021).

It has been shown that the stink bug-resistant soybean cultivars (PI229358, PI171451, PI171444) experienced a lower pentatomid infestation than other varieties (Jones & Sullivan, 1982). Some experiments have shown that certain breeding lines of soybean may provide possible genetic materials for breeding stink bug resistance into cultivated soybeans (Campos, Knutson, Heitholt, & Campos, 2010). These resistant lines will reduce the damage caused by stink bugs, and hence reduce the need for insecticide applications.

The resistance of soybean plants to whitefly infestation can be used in an integrated program for pest management. Although the types and mechanisms of host plant resistance against insect attacks have been identified (Wiseman, 1985), few researchers have investigated the actual mechanisms of soybean resistance to whiteflies. A combination of factors is believed to impart resistance to certain soybean genotypes and not to others (Bueno, Raetano, Junior, & Carvalho, 2017). These observations are based on research findings from a number of other plant species on their resistance to whiteflies and other insect pests. An example is the presence of hooked trichomes on field beans as a form of physical barrier (Pillemer & Tingey, 1976). Soybean resistance to whiteflies may also be attributable to variations in trichome density, length, and erectness (Helm, Kogan, & Hill, 1980; Turnipseed, 1977).

Since information regarding the mechanisms of host plant selection by whiteflies is lacking, scientists have begun to test selected whitefly-resistant soybean germplasms. Possible physical mechanisms such as the trichome density and erectness of soybean leaf pubescence, plant height, and canopy closure index, have been examined, as well as possible chemical resistance mechanisms such as soybean epicuticular lipids.

The cowpea aphid, *A. craccivora* Koch, is a polyphagous pest that attacks crops by feeding on all plant parts, resulting in substantial losses of yield (Keatinge et al., 2015). Among the management strategies for *A. craccivora*, the use of chemical insecticides is ranked first by farmers because there are several chemical insecticides available for use against aphids (Egho, 2010b). Cowpea aphids cause damage to the crop plant through sucking and extracting plant sap, thus reducing the amount of nutrients and water available to the crop and enabling the transmission of plant viruses. The intercropping of cowpeas with cereals such as maize, sorghum, and millet has been used as a pest management technique against *A. craccivora*, but intercropping alone cannot entirely repel the insects (Hassan, 2013).

For pod borer control, farmers rely primarily on insecticides. Insecticides are expensive and have caused pesticide resistance and environmental degradation through their indiscriminate use (Singh & Mandal, 2013). That being the case, more environmentally friendly and sustainable pest management strategies are needed, such as integrating the use of tolerant/resistant crop cultivars with adaptive agronomic methods and the use of biological controls. Some of these legume farming practices for managing insect pests are summarized in Table 6.

Successfully controlling pests is one of the greatest challenges for farmers and ranchers who want to operate sustainably. They need management options that are effective and profitable to decrease their use of off-farm chemical inputs. Farmers across the United States are gradually finding that ecological methods for controlling pests meet these needs. For example, practices that promote the natural enemies of pests on the farm provide

Table 6 Insect pests of legume crops and current pest-control farming practices.

Crop	Insect pests	Farming practices	References
Soybean	Bean fly, *Melanagromyza sojae*	Late-maturing or long-duration soybean varieties can reduce the yield losses due to *M. sojae*	Srinivasan (2014); Talekar (1990); Sariah and Makundi (2007)
		Early sowing of soybean crop in Tanzania helps reduce bean fly infestations	
	Beet army worm, *Spodoptera exigua*	Botanical pesticides such as neem that are safer for the parasitoids can be used to encourage the natural enemies of the pests	Packiam and Ignacimuthu (2012)
		Sex pheromones of *S. exigua* can be used for monitoring as well as mass-trapping	
	Bean bugs, *Riptortus pedestris*	Delayed sowing of soybean helps avoid damage from bean bugs due to lower pest densities	Wada, Endo, and Takahashi (2006)
		Intercropping of soybean with trap crops (e.g., sesame and corn) can enhance the performance of parasitoids and thus reduce the infestation of *Riptortus* bugs	Youn and Jung (2008)
		Entomopathogenic fungi such as *B. bassiana* could be used to manage *Riptortus* bugs	Srinivasan (2014)
Pigeon pea	Gram pod borer, *Helicoverpa armigera*	Placing a high concentration of sex pheromone in a slow-release formulation	Srinivasan (2014)
		H. armigera sex pheromone traps can be used to monitor the pest population status	
		Crop rotation	
	Bean fly, *Melanagromyza obtuse*	Used resistant or moderately resistant cultivars. Varieties having purple stems, thinner stems and smaller pith diameter are resistant to bean flies	Srinivasan (2014); Talekar (1990)

Cowpea	Aphids, *Aphis craccivora*	Avoid monoculture and apply crop rotation. The selected field should be located away from other legume crops	Srinivasan (2014)
		Use neem oil, either alone or in combination with the EPF biopesticides	El-Hawary and Abd El-Salam (2008); Halder, Rai, and Kodandaram (2013)
Chickpea	*Helicoverpa. armigera*	Early planting of crop can help avoid periods of peak abundance of *H. armigera*	Weigand et al. (1994)
Mungbean	Leafhoppers, *Empoasca kerri*	Okra, sorghum and pearl millet can be grown as intercrops in mungbean fields	Chakravorty and Yadav (2013)
		This can reduce the leafhopper damage significantly	
		Use neem-based biopesticides at recommended doses	
	Thrips, *Megalurothrips distalis*	Use blue sticky traps to monitor thrips at regular intervals and determine when other pest management controls are required. Entomopathogenic fungi could reduce thrips damage	Ekesi and Maniania (2000)
Lentil	Pod borer, *Helicoverpa armigera*	Insecticide application necessary, using methidathion (SupracideR @ 0.5 kg a.i./ha), and/or deltamethrin (Decis @ 38 g a.i./ha)	Beniwal, Bayaa, Weigand, Makkouk, and Saxena (1993)
		Use endosulfan (Thiodan 35 @ 3 mL/L) at the time of flowering/early pod-setting	Stevenson, Dhillon, Sharma, and Bouhssini (2007)
	Armyworm, *Spodoptera exigua*	Collect and destroy the caterpillar. Spray 500 mL of Dichlorvos 85 SL or 3 kg of carbaryl 50 WP or 1 L of Quinalphos 25 EC in 500 L of water per ha	https://www.digokrishi.com/lentil-insect-pest
	Cutworm, *Agrotis ipsilon*	Destroy the holes where they hide. Deep plowing of the field in between the crops	

an estimated $4.5 billion worth of pest suppression each year in the United States. Other traditional practices include rotating in new crops that interrupt the pests' life cycles, improving the ability of a crop to compete against weeds, and building up soils that provide unfavorable conditions for disease-causing microorganisms naturally. The ecological approach uses awareness, sustainability and resource management to grow crops safely while keeping insect pests, diseases and weeds at bay (Hinds, 2020).

7. Strategies for effective and ecofriendly IPM

Now that researchers are armed with reliable information on the life cycles and ecology of the various insect pests, the next stage in the development of IPM will be to look for more innocuous pesticide substitutes such as hormones, pheromones, traps, biological control agents, bioengineered pathogens, and the release of sterile males. While useful as less invasive curative methods, such approaches still avoid addressing the root causes of the issue. They also keep the farmer reliant on experts and product suppliers, and perpetuate the image of pests as enemies to be eradicated. We are now on the threshold of a third step that recognizes pests not as rivals, but as indicators of problems in agricultural system design and management. It is also becoming apparent that most of the keys to suitable environmental designs and management methods capable of keeping pest numbers at sustainable levels are held by ecological expertise. With this strategy, through the incorporation of a variety of agricultural and bioecological controls, potential infestations are prevented from becoming serious problems, while the more familiar invasive controls, such as pesticides, are reserved for emergencies only (Hill, 1989).

Effective IPM needs a comprehensive knowledge of the interactions among crop plants, pests, and the surrounding environment, as well as an understanding of farmers' knowledge and expectations. Therefore, to develop sustainable pest control strategies for grain legumes, it is critical to: (1) survey the awareness and attitudes of farmers toward grain legume pests, (2) investigate existing farming practices on the management of grain legume pests, and (3) recognize other limiting factors on the productivity of grain legumes (Adati et al., 2007; Nwilene, Nwanze, & Youdeowei, 2008).

Sometimes biocides (including pesticides and herbicides) are another cause of disruption in the field. By using herbicides to remove noncrop vegetation which provides a major overwintering environment for beneficial species, local biodiversity is reduced. The use of pesticides can also destroy nontarget animal species that are important natural enemies of the insect pests, either as competitors or predators. Instead, alternative tactics such as perimeter trap cropping, where another more enticing noncrop plant is grown around the perimeter of a crop field to lure pests away from the cash crop, should be considered. Simultaneously, a selective pesticide can be strategically applied at the trap crop border to destroy the target pests and reduce the total amount of pesticide required. While insecticides are effective in the short term against sucking pests and the pod borer complex, the development of insecticide resistance among insect pests, the presence of insecticide residues on crops, and off-target environmental hazards require alternative and ecologically safe long-term management methods, such as planting pest-resistant cultivars (Soundararajan, Narayanasami, & Geetha, 2013).

One of the best strategies for effective and ecofriendly insect pest management is the development of pest-resistant cultivars. To achieve this goal, legume breeding, the use of wild legumes as genetic resources, and genome-wide association studies (GWAS) analyses all play important roles.

7.1 Legume breeding for enhanced insect resistance

Grain legume crops around the world are damaged by many pest species. The most destructive insect pests are: pod borer and leaf miner in chickpeas, stink bug and white flies in soybeans, *Sitona* weevil, a seed weevil, in lentils, pod fly and pod borers in pigeon peas, bruchid in mungbeans, and leaf miner in groundnuts (Weigand et al., 1994). Genetic sources of pest resistance can be harnessed as a part of the integrated pest control strategies in conjunction with cropping systems and agricultural practices that help minimize the release of pesticides into the environment, even with a small number of pest-resistant cultivars. If no resistant cultivar can be found (as is the case with *Aphis fabae* Scopoli, which causes serious damage to faba beans in temperate regions), plant characteristics such as height, sowing date and density may still help identify ideotypes that can reduce the input of insecticides (Stoddard, Nicholas, Rubiales, Thomas, & Villegas-Fernandez, 2010).

In common beans, the resistance to storage bean weevils (*Acanthoscelides obtectus* and *Zabrotes subfasciatus*) is a target for breeders, and some progress has been made with the use of inhibitors in the development of some cultivars with antibiosis properties. With various modes of action, the arcelin-phytohemagglutinin-α-amylase inhibitor (Arc-Pha-αAI) locus of the common bean, *Phaseolus vulgaris* L., has been found to encode plant defense proteins. In some cases, introgression of this locus into cultivated varieties from resistant wild bean accessions was capable of conferring resistance against the Mexican bean weevil, *Z. subfasciatus* (Boheman), but not against the bean weevil, *A. obtectus* (Say). Resistance can be the product of one or a combination of insect inhibitory proteins in this locus with respect to the Mexican bean weevil (Fory et al., 1996). Recently, wild accessions and conventional landraces have been established as major sources of resistance against bean weevils (Zaugg et al., 2012). Introgressed lines and segregating populations may be used to conduct QTL analyses to fine-tune QTL mapping and to discover genes that regulate agronomic characteristics relevant to the resistance against these pests.

Mungbean (*Vigna radiata* L.) is one of the key beans consumed in China, while bruchid (*Callosobruchus chinensis* L.) is a storage pest that causes major losses of mungbean seeds. Breeding for bruchid-resistant varieties is an important target for crop improvement in mungbean and other legumes (Cheng, Wang, & Wu, 2005; Sun, Cheng, & Wang, 2007). The bruchid-immune line V2709 (introduced from AVRDC-Thailand) was identified following the screening of a mungbean germplasm set (Sun et al., 2007). A single dominant *Br2* gene anchored by molecular markers, which can be used in marker-assisted selection, appeared to be associated with the resistance against bruchid (Sun et al., 2007). The bruchid-resistant variety "Zhong Lv No.4" was developed with V2709 as a progenitor. It was released in 2004 in China, and removed the need for chemical fumigation to prevent in-storage bruchid infestation.

The pod borer, *H. armigera* (Hubner), is a major chickpea pest in the Indian subcontinent. Several germplasm accessions/breeding lines/cultivars with mild resistance, as well as some annual and perennial wild *Cicer* accessions, have been reported to offer some degree of resistance against *H. armigera* (Dar, Reddy, Gowda, & Ramesh, 2006; Sharma et al., 2007). Genetic transformants of chickpea expressing a sequence-modified *Cry2Aa* gene showed differential resistance to pod borer larvae in proportion to the level of expression of the Cry2Aa protein. A high-expressing line has been found to confer near-complete pod borer protection (Sumita et al., 2010). The spreading types of chickpea appear to be more susceptible to damage by *H. armigera* than the erect types (Yadav et al., 2006). Therefore, the erect plant habit could be the distinguishing feature of a resistant ideotype (Table 7).

Table 7 Identification and exploitation of host plant resistance to insect pests in grain legumes.

Crop	Genotypes	Insect pest	Reference
Soybean	Nimsoy	Leaf miner	Dhaliwal, Singh, and Jindal (2004)
	PI 227687 PI 171444	Stink bugs	Talekar (1987)
Cowpea	VITA5, VITA 4, Banswara, TVu946	Pod borer, *Maruca vitrata*	Singh (1978); Lal (1987); Chanchal and Singh (2014)
	IT82E-16, P 1473, MS 9369, P 1476	Aphid, *Aphis craccivora*	Benchasri, Nualsri, Santipracha, and Ngampongsai (2007)
	IT99K-429-2, IT89KD-288	Bruchids, *Callosobruchus maculatus*	Obadofin (2014)
Green gram	V2817, V1128	Bruchids	Somta et al. (2008)
	PDM-84-143, TAM-20	Stem fly	Devasthali and Joshi (1994)
	P 526, J1, LM 11	Pod borer	Lakshminarayan, Singh, and Mishra (2008)
Black gram	PLU 54, CBG 08-011	Pod borer, *Helicoverpa armigera*	Lal (1987)
Field pea	Bonville, EC 33860; PS 410, T 6113	Pod borer, *Etiella zinckenella*	Lal (1987), Teshome et al. (2014)
	T 6113, PS 41-6; KMPR 9, PS 40	Leaf miner, *Chromatomyia horticola*	Teshome et al. (2014)
Pigeon pea	BDN 2, ICPL 84060, ICPL 4, Bori	Pod borer, *Helicoverpa armigera*	Lateef and Pimbert (1990)
	MPG 679, ICPL 88034	Legume pod borer, *Maruca vitrata*	Saxena, Lateef, Fonseka, Ariyaratne, and Dharamsena (1996)
	ICP 7941E1, ICP 10531-E1	Pod fly, *Melanagromyza obtusa*	Lateef and Pimbert (1990); Moudgal, Lakra, Dahiya, and Dhillon (2008)
Chickpea	ICC 09314, ICC 506, ICCV 10, ICCV 7	Pod borer, *Helicoverpa armigera*	Bhagwat, Aherker, Satpute, and Thakre (1995); Sharma et al. (2017)
	ICC 12475	Beet armyworm *Spodoptera exigua*	Sharma et al. (2017)
	JG 315, DCP 923, BG 372,BG 1003	Bruchid	Sharma et al. (2017)
	ILC 5901, ILC 380, ILC 7738	Leaf miner, *Liriomyza cicerina*	Girija, Sallmath, Patll, Gowda, and Sharma (2008); Shankar, Sharma, Ramesh Babu, and Sridevi (2013)

7.2 Wild relatives of legume crops as genetic resources for insect resistance

One of the possible crop improvement options is the transfer of genes from wild legume relatives to the cultivars to confer resistance against legume pests. For example, wild *Cajanus* spp. are the repository of many significant trait-specific genes and can be used to enhance crop cultivars, enrich variability and diversity, and increase the genetic base and prebreeding populations of wild *Cajanus* spp. In the cultivated crops, the genes responsible for resistance to insect pests are often sacrificed in the process of domestication to secure other desirable agronomic traits, but they are still retained in many wild accessions or relatives of the crop plant (Table 8). High levels of pest resistance have been reported in a few cases in beans, field peas (Clement, Hardie, & Elberson, 2002), cowpeas (Redden, Dobie, & Gatehouse, 1983), and black grams (Dongre, Pawar, Thakare, & Harwalkar, 1996). The wild relatives or accessions of pigeon peas and chickpeas are important sources of genes for insect pest resistance (Dhillon & Sharma, 2012).

7.3 Developing molecular markers for insect resistance in legumes by GWAS

The terminology for phenotypic characterizations across different GWAS studies on plants needs to be standardized in order for their findings to be integrated successfully (Zhao et al., 2019). It is therefore important to use uniform, popular descriptors or ontologies (Shrestha et al., 2010). Although the use of ontology by research groups has increased over the last decade, more improvement is needed (Ćwiek-Kupczyńska et al., 2016; Krajewski et al., 2015; Walls et al., 2019). Ontologies are relatively standardized for a variety of legume crops (e.g., chickpeas, faba beans, lentils, and soybeans) and can be accessed at http://www.cropontology.org.

To facilitate data integration and usage, enhanced data management and sharing of crop genotypes, phenotypes, and GWAS datasets are also required. To identify single-nucleotide polymorphism (SNP)-trait associations in legume crops, there is a need to build open-source databases and repositories for genotypic and phenotypic datasets coupled with easy-to-use GWAS pipelines with embedded standardized genotypic data.

One of the major limiting factors for soybean production is insect damage. Hitherto, insect control to protect soybean yields has relied primarily on insecticide applications. Some soybean plants have been shown to be

Table 8 Wild species of legume crops identified as genetic resources for breeding resistance to different insect pests.

Crop	Wild species/accessions	Insect pests	References
Soybean	PI 171444 (MG VI)	*Anticarsia gemmatalis* (Hubner)	Kester, Smith, and Gilman (1984)
	PI 229358, PI 227687		Hoffmann-Campo, Ramos Neto, Oliveira, and Oliveira (2006)
	PI 274454		Ortega, All, Boerma, and Parrott (2016)
	PI 227687	*Trichoplusia ni* caterpillars	Liu, Norris, and Lyne (1989)
	PI 567598B	*Megacopta punctatissima* Montandon	Bray, McDaniel, All, Li, and Parrott (2016)
Groundnut	*Arachis cardenasii*	Leaf miner	Sharma, Sujana, and Rao (2009)
	A. duranensis	*Helicoverpa armigera*	
	A. kempff-mercadoi	*Spodoptera litura*	
Mungbean	*V. radiata* var. *sublobata*	Bruchid	Fujii and Miyazaki (1987); Sarkar and Bhattacharyya (2015)
Cowpea	*Vigna vexillata* (L.) Benth	Pod-sucking bug	Jackai and Oghiakhe (1989)
		Clavigralla tomentosicollis	
		Stal, spotted pod borer, *M. vitrata*	

Continued

Table 8 Wild species of legume crops identified as genetic resources for breeding resistance to different insect pests.—cont'd

Crop	Wild species/accessions	Insect pests	References
Pigeon pea	ICPW 214 (*Cajanus bracteata*)	Pod fly, *M. obtuse*	Sharma, Pampapathy, and Reddy (2003)
	ICPW 278, ICPW 280 (*C. scarabaeoides*)	Pod borer, *H. armigera*	Sujana, Sharma, and Rao (2008)
	ICWP 016 (*Cajanus albicans*)	*Meloidogyne javanica*	Sharma, Jaba, and Vashisth (2017)
	ICWP 062 (*C. platycarpus*)		
Chickpea	*Cicer bijugum*, *C. reticulatum*	*H. armigera*	Sharma, Pampapathy, Lanka, and Ridsdill-Smith (2005)
	C. pinnatifidum, *C. bijugum*, *C. echinospermum*	Bruchid, *C. chinensis*	Singh, Ocampo, and Robertson (1998)
Lentil	ILWL 245 (*L. culinaris*)	*Sitona* weevil (*Sitona crinitus*)	El Bouhssini, Sarker, Erskine, and Joubi (2008)
	L. nigricans, and *L. lamottei*	Seed weevil (*Bruchus* spp.)	
	L. culinaris Medikus subsp	Bruchids (*Bruchus* spp.)	Laserna-Ruiz, De-Los-Mozos-Pascual, Santana-Méridas, Sánchez-Vioque, and Rodríguez-Conde (2012)

resistant against certain insect pests. Such plants, e.g. those resistant to soybean aphid (SBA), have been used to develop resistant soybean cultivars. Phenotypic data for resistance against beet soybean looper (SBL), Mexican bean beetle (MBB), armyworm, potato leafhopper (PLH), SBA, velvet bean caterpillar (VBC) and chewing damage caused by unspecified insects can be integrated with the data from the genotyping array, SoySNP50K, and the statistical capacity of GWAS for a detailed understanding of insect resistance among the United States Department of Agriculture Soybean Germplasm Collection. Important SNP markers for SBL, MBB, PLH and VBC have been identified. The PLH resistance locus is located close to, but distinct from, a locus for soybean pubescence density on chromosome 12. These findings provide genetic evidence that the density of pubescence and PLH resistance may be two distinct traits. This research provides a new insight into the QTL mapping of the resistance of soybeans to four major insect pests of soybean (Chang & Hartman, 2017).

By correlating phenotypic data with SNPs across a genome, genome-wide association mapping can help identify resistant accessions. From published studies screening cultivated soybeans (*G. max*) and wild soybeans (*Glycine soja* Siebold & Zucc.) using aphids exhibiting Biotype 1, 2 or 3 characteristics, aphid population counts were collected from 2366 soybean accessions. The SoySNP50 K high-density genotyping array used earlier to genotype the USDA Soybean Germplasm Collection was used to obtain genotypic results. Important associations have been identified on 18 of the 20 soybean chromosomes between SNPs and soybean aphid counts. On chromosomes 7, 8, 13, and 16 with known *Rag* genes, a substantial number of associated SNPs have been found. On chromosomes 1, 2, 4 to 6, 9 to 12, 14, and 17 to 20 where *Rag* genes have not yet been mapped, there were also prominent associations with SNPs, indicating that many *Rag* genes remain to be discovered. These SNPs can be used to identify promising accessions for breeding programs with new aphid-resistance characteristics (Hanson et al., 2018).

In a recent genome-wide association analysis, two SNP markers (C35011941-894 and Scaffold30061-3363) were closely correlated with aphid resistance across three models with a log of odds (LOD) value greater than 2.5. These findings will provide valuable information on the genetic availability of aphid resistance for the molecular breeding of resistant types of cowpea (Qin et al., 2017).

Molecular markers may play an important role in accelerating the introgression of specific genes that confer resistance against target insects on high-yield cultivars, by recognizing the essence of gene actions while reducing the deleterious effects of introducing unwanted genes from wild species through linkage drag. Molecular breeding also provides the ability to pyramid numerous resistance sources that, due to similar phenotypes, could not be effectively selected by traditional breeding methods, thus enhancing the resistance level and/or potentially producing more stable resistant cultivars.

Considerable progress has been made in the development of genetic linkage maps of chickpeas, cowpeas, and soybeans, though much remains to be done regarding pigeon peas, common beans, lentils, and field peas. The preliminary identification of molecular markers for insect resistance has been done in soybeans, chickpeas, mungbeans, field peas, and cowpeas. However, no distinct benefit over the traditional method could be found by using marker-supported selection for the resistance to insect pests, and in most instances, epistatic resistance was also very high. For breeding insect-resistant varieties, the identification of the genes responsible for resistance and the determination of their locations on the chromosomes are important.

7.4 Genetic methods to improve the legume resistance to insect pests

With the advancement of genomics and genetics methods, a number of insect resistant genes have been identified in different plants. These discoveries offer new opportunities to improve legume yield by making them resistant to insect pests. One such method is the transfer of exogenous genes into the host plant genomes, such as the vegetative insecticidal proteins (VIP) and secretable proteins Bt or Cry isolated from the soil bacterium *Bacillus thuringiensis* (Estruch et al., 1996; Gu et al., 2021). In 1997 Peggy Ozias-Akins and Chong Singsit were the first to develop transgenic alfalfa and peanut carrying *Cry* to improve the legume resistance against insects (Singsit et al., 1997). Since then, other transgenic legumes have also been developed, including the cowpea, pigeon pea and chickpea (Alok, Annapragada, Singh, Murugesan, & Singh, 2020; Kumar, Jaiwal, Sreevathsa, Chaudhary, & Jaiwal, 2021; Singh et al., 2018). Instead of introducing exogenous genes, several transgenic soybean lines were created through the expression of their own genes. For instance, the terpene synthase gene *GmAFS* was overexpressed in soybean and enhanced its resistance to cyst nematodes (Lin et al., 2017).

Another method of improving legume crop breeding lies with the rapid development of gene editing technologies such as CRISPR/Cas9.

This include engineering legume agronomic traits which include enhancing the formation of defense hormones, knocking out the legume susceptibility genes, changing the pathogen effector-target interaction, modifying the insect genome to reduce their resistance to insecticides, and inducing sterility in insect pests (Bisht, Bhatia, & Bhattacharya, 2019).

In addition to increasing the legume crop yield, other benefits of obtaining genetically modified legumes, such as the soybeans, could significantly improve farmers' incomes, reduce the use of insecticides, fuels and tillage uses, resulting in a lower carbon footprint, and contribute to the development of a more sustainable environment (Brookes & Barfoot, 2017). Despite these potential advantages, permission for their commercial release has been delayed by public suspicion of new biotechnologies and concerns about the safety of genetically modified crops (Tyczewska, Woź niak, Gracz, Kuczyński, & Twardowski, 2018).

8. Effects of global warming and climate change on insect pests

Global warming and climate change will trigger significant changes in the diversity and abundance of arthropods, the spatial distribution of insect pests, the dynamics of insect populations, the biotypes of insects, the interactions between herbivores and plants, the abundance of natural enemies of these pests and the possible extinction of some of these species, and the effectiveness of crop protection technologies. The distribution of insect pests will also be affected by the changes in cropping patterns caused by climate change. Alterations in the geographical range and abundance of the insect pests may exacerbate the severity of crop losses, and will therefore have a significant impact on crop production and food security.

Major insect pests may move into more temperate regions from the tropics, such as pod borers (genera *Spodoptera*, *Helicoverpa* and *Maruca*), whiteflies and aphids, resulting in increased damage to grain legumes. Host-plant resistance, natural enemies, and biopesticides are some of the possible tools for IPM. However, as a consequence of global warming, the relative effectiveness of these pest control tactics is likely to change. Climate change will also result in the increased spread of insect-transmitted diseases. Altogether, these phenomena will have significant implications for food security and crop protection, especially in developing countries where the need to increase and maintain food production is most urgent (Sharma, 2014).

Insects have evolved various host plant-association patterns, life histories and feeding strategies tailored to exploiting their hosts (Panda & Khush, 1995). One foreseeable development is the evolution of minor but lasting improvements in the assortment of insect-derived enzymes, that allow insect pests to also feed on other plant species closely related to their original hosts, as they expand the range of their habitats as a result of global warming.

9. Conclusion

For hundreds of millions of people in Southeast Asia, Africa, and Latin America, tropical grain legumes are a major protein source. They are also important sources of income generated from exporting these grains to other parts of the world. Among the main biotic constraints hindering the productivity of these crops are insect pests, especially those insects that damage the legume pods, as they are the key cause of yield losses in tropical grain legumes.

For most insect pests, chemical control measures are available but are not easily affordable or applicable to the conditions of smallholder farms in which most tropical grain legumes are grown. The only viable solution is an IPM approach that relies more on planting strategies. Diversification of crops is an important agricultural strategy and can be accomplished by intercropping, strip-cropping or trap-cropping, which also provides a natural biological control of insect pests. Legume varieties that are resistant or tolerant to many of the major insect pests have been developed through conventional and molecular breeding. In addition to finding new means of pest management, researchers must also pay careful attention to the scale of available technologies in developing future insect pest controls for tropical grain legumes.

Secondary metabolites (such as terpenoids, flavonoids, tannins, phenolics, and alkaloids) are involved in abiotic and biotic stress responses by plants, and many of them are released in the form of volatile organic compounds. In particular, terpenoids play important roles in reducing herbivory and attracting beneficial species, such as those involved in tritrophic and multitrophic interactions. Interactions between plants and insects can be characterized as a perpetual arms race. In both plants and insects, these interactions have resulted in an unparalleled diversity of terpenes and terpenoids. During the next few years scientists must obtain a better understanding of their complex interactions and to make effective use of them in biotechnology. Achieving adequate precision and accuracy in phenotyping insect pest resistance remains a major bottleneck. In order to develop insect pest-resistant cultivars, improved higher-precision phenotyping systems would directly benefit both marker-assisted selection and traditional breeding, along with further strategic research that feeds into these endeavors.

The coevolution of plants and insects advances in lockstep. In order to protect themselves from herbivory, plants have established defense mechanisms, while insects have found numerous ways to evade the harmful effects of the host plants' defenses. Although researchers have attempted to dissect plant-insect interactions, our knowledge on this subject is still limited. A better understanding of this process will enable us to design more effective methods for the biological control of insect pests using natural products and by developing new plant varieties with enhanced chemical defenses. Future research must concentrate on environmentally sustainable methods for pest control that are consistent with the needs of legume farmers and the constraints that act upon them.

Acknowledgments

The authors thank Ms. Jee Yan Chu and Dr. David Wilmshurst for copyediting the manuscript, Ms. Yauling for preparing some of the figures, and Mr. Li Chade for formatting.

Conflicts of interest

The authors declare no conflict of interest.

Author contributions

S.T. wrote the first draft of this manuscript. K.F., F.L.W. and H.Y.Y. provided scientific inputs and contributed to the writing. J.H.L.H. and H.-M.L. coordinated the writing effort and prepared the final manuscript.

Funding

This work was funded by the Hong Kong Research Grants Council Area of Excellence Scheme (AoE/M403/16) and the Lo Kwee-Seong Biomedical Research Fund. S.T. was supported by IPDF provided by The Chinese University of Hong Kong.

References

Abate, T. (2012). A knowledge-based strategy towards tropical grain legumes research and development. *Journal of SAT Agricultural Research*, *10*, 1–12.

Abate, T., & Ampofo, J. K. O. (1996). Insect pests of beans in Africa: Their ecology and management. *Annual Review of Entomology*, *41*, 45–73.

Adati, T., Tamò, M., Yusuf, S. R., Downham, M. C. A., Singh, B. B., & Hammond, W. (2007). Integrated pest management for cowpea-cereal cropping systems in the West African savannah. *International Journal of Tropical Insect Science*, *27*, 123–137.

Alabi, O. J., Kumar, P. L., Mgbechi-Ezeri, J. U., & Naidu, R. A. (2010). Two new 'legumoviruses' (genus *Begomovirus*) naturally infecting soybean in Nigeria. *Archives of Virology, 155*(5), 643–656.

Alegbejo, M. D., & Abo, M. E. (2002). Etiology, ecology, epidemiology and control of groundnut rosette disease in Africa. *Journal of Sustainable Agriculture, 20*(2), 17–29.

Ali, A., Choudhury, R. A., Ahmad, Z., Rahman, F., Khan, F. R., & Ahmad, S. K. (2009). Some biological characteristics of *Helicoverpa armigera* on chickpea. *Tunisian Journal of Plant Protection, 4*(1), 99–106.

Allotey, J., Segwabe, M., & Randome, I. (2016). Damage caused, loss assessment and emergence pattern of *Acanthoscelides obtectus* (Say) on the bean *Phaseolus vulgaris* Linn. in Gaborone. *Journal of Applied Zoological Research, 27*, 159–163.

Alok, D., Annapragada, H., Singh, S., Murugesan, S., & Singh, N. P. (2020). Symbiotic nitrogen fixation and endophytic bacterial community structure in Bt-transgenic chickpea (Cicer arietinum L). *Scientific Reports, 10*(1), 1–12.

Amoako-Atta, B. (1983). Observation on the pest status of the striped bean weevil *Alcidodes leucogrammus* Erichs. on cowpea under intercropping systems in Kenya. *International Journal of Tropical Insect Science, 4*(4), 351–356.

Ananthakrishnan, T. N. (1999). Induced responses, signal diversity and plant defense: Implications in insect phytophagy. *Current Science, 76*, 285–290.

Anazonwu, D. L., & Johnson, S. J. (1986). Effects of host and density on larval color, size, and development of the Velvetbean caterpillar, *Anticarsia gemmatalis* (Lepidoptera: Noctuidae). *Environmental Entomology, 15*(4), 779–783.

Andrews, K. L. (1980). The whorlworm, *Spodoptera frugiperda* in Central America and neighbouring areas. *Florida Entomologist, 63*(4), 456–467.

Arimura, G., Ozawa, R., Shimoda, T., Nishioka, T., Boland, W., & Takabayashi, J. (2000). Herbivory-induced volatiles elicit defence genes in lima bean leaves. *Nature, 406*(6795), 512–515.

Belmain, S. R., Haggar, J., Holt, J., & Stevenson, P. C. (2013). *Managing legume pests in sub-Saharan Africa: Challenges and prospects for improving food security and nutrition through agroecological intensification*. Chatham Maritime (United Kingdom): Natural Resources Institute, University of Greenwich.

Benchasri, S., Nualsri, C., Santipracha, Q., & Ngampongsai, A. (2007). Evaluation of aphid (*Aphis craccivora* Koch) resistance in 24 accessions of yardlong bean and cowpea. In *Proceeding of the 1st joint PSU–UNS international conference on bioscience: Food, agriculture, and the environment, Songkhla, Thailand* (pp. 215–222).

Beniwal, S. P. S., Bayaa, B., Weigand, S., Makkouk, K. H., & Saxena, M. C. (1993). *Field guide to lentil diseases and insect pests*. Aleppo: International Center for Agriculture Research in Dry Areas (ICARDA).

Beran, F., Kollner, T. G., Gershenzon, J., & Tholl, D. (2019). Chemical convergence between plants and insects: Biosynthetic origins and functions of common secondary metabolites. *New Phytologist, 223*, 52–67.

Bhagwat, V. R., Aherker, S. K., Satpute, V. S., & Thakre, H. S. (1995). Screening of chickpea (*Cicer arietinum L.*) genotypes for resistance to *Helicoverpa armigera* (Hb.) and its relationship with malic acid in leaf exudates. *Journal of the Entomological Research Society, 19*, 249–253.

Bisht, D. S., Bhatia, V., & Bhattacharya, R. (2019). Improving plant-resistance to insect-pests and pathogens: The new opportunities through targeted genome editing. *Seminars in Cell & Developmental Biology, 96*, 65–76.

Bock, K. R. (1973). Peanut mottle virus in East Africa. *Annals of Applied Biology, 74*(2), 171–179.

Boethel, D. J. (1999). Assessment of soybean germplasm for multiple insect resistance. In L. Clement, & S. S. Quisenberry (Eds.), *Global plant genetic resources for insect-resistant crops* (pp. 101–130). Boca Raton: CRC Press.

Boncan, D. A. T., Tsang, S. S. K., Li, C., Lee, I. H. T., Lam, H. M., Chan, T. F., et al. (2020). Terpenes and terpenoids in plants: Interactions with environment and insects. *International Journal of Molecular Sciences*, *21*(19), 7382.

Bradshaw, J. D., Rice, M. E., & Hill, J. H. (2008). Evaluation of management strategies for bean leaf beetles (Coleoptera: Chrysomelidae) and bean pod mottle virus (Comoviridae) in soybean. *Journal of Economic Entomology*, *101*(4), 1211–1227.

Bray, A., McDaniel, L., All, J., Li, Z., & Parrott, W. (2016). Phenotyping techniques and identification of soybean resistance to the kudzu bug. *Crop Science*, *56*, 1807–1816.

Brookes, G., & Barfoot, P. (2017). Environmental impacts of genetically modified (GM) crop use 1996–2015: Impacts on pesticide use and carbon emissions. *GM Crops and Food*, *8*, 117–147. https://doi.org/10.1080/21645698.2017.1309490.

Bueno, R. C., Raetano, C. G., Junior, J. D., & Carvalho, F. K. (2017). Integrated management of soybean pests: The example of Brazil. *Outlooks on Pest Management*, *28*(4), 149–153.

Bustos-Segura, C., Külheim, C., & Foley, W. (2015). Effects of terpene chemotypes of *Melaleuca alternifolia* on two specialist leaf beetles and susceptibility to myrtle rust. *Journal of Chemical Ecology*, *41*, 937–947.

Campos, M., Knutson, A., Heitholt, J., & Campos, C. (2010). Resistance to seed feeding by southern green stink bug, *Nezara viridula* (Linnaeus), in soybean, *Glycine max* (L.) Merrill. *Southwestern Entomologist*, *35*(3), 233–239.

Chakravorty, S., & Yadav, D. K. (2013). Effect of intercrops on jassid (*Empoasca kerri Pruthi*) infesting green gram, *Vigna radiata* (L.) Wilczek. *Annals of Plant Protection Sciences*, *21*(2), 229–232.

Chanchal, S., & Singh, N. N. (2014). Evaluation of responses of cowpea (*Vigna unguiculata* Walpers) genotypes to infestation of legume pod borer (*Maruca vitrata* Fabricius). *Journal of Food Legumes*, *27*(4), 334–339.

Chang, H. X., & Hartman, G. L. (2017). Characterization of insect resistance loci in the USDA soybean germplasm collection using genome-wide association studies. *Frontiers in Plant Science*, *8*, 670.

Chen, C. Y., & Mao, Y. B. (2020). Research advances in plant-insect molecular interaction. *F1000Research*, *9*. F1000 Faculty Rev-198.

Cheng, X., Wang, S., & Wu, S. (2005). Tagging and utilization of bruchid resistance gene using PCR markers in mungbean. *Scientia Agricultura Sinica*, *38*, 1534–1539.

Cheong, S. P., Huang, J., Bendena, W. G., Tobe, S. S., & Hui, J. H. L. (2015). Evolution of ecdysis and metamorphosis in arthropods: The rise of regulation of juvenile hormone. *Integrative and Comparative Biology*, *55*(5), 878–890.

Chiang, H. S., & Norris, D. M. (1983). Morphological and physiological parameters of soybean resistance to agromyzid beanflies. *Environmental Entomology*, *12*, 260–265.

Clement, S. L., Hardie, D. C., & Elberson, L. R. (2002). Variation among accessions of Pisum fulvum for resistance to pea weevil. *Crop Science*, *42*, 2167–2173.

Clement, S. L., Wightman, J. A., Hardie, D. C., Bailey, P., Baker, G., & McDonald, G. (2000). Opportunities for integrated management of insect pests of grain legumes. In R. Knight (Ed.), *Linking research and marketing opportunities for pulses in the 21st century*. Dordrecht: Springer.

Cruz, P. L., Baldin, E. L. L., de Jesus, P., & de Castro, M. (2014). Characterization of antibiosis to the silverleaf whitefly *Bemisia tabaci* biotype B (Hemiptera: Aleyrodidae) in cowpea entries. *Journal of Pest Science*, *87*, 639–645.

Ćwiek-Kupczyńska, H., Altmann, T., Arend, D., Arnaud, E., Chen, D., Cornut, G., et al. (2016). Measures for interoperability of phenotypic data: Minimum information requirements and formatting. *Plant Methods*, *12*, 44.

Czosnek, H., Ghanim, M., & Ghanim, M. (2005). The circulative pathway of begomoviruses in the whitefly vector *Bemisia tabaci*—Insights from studies with *tomato yellow leaf curl virus*. *Annals of Applied Biology*, *140*(3), 215–231.

Damsteegt, V. D., Stone, A. L., Kuhlmann, M., Gildow, F. E., Domier, L. L., Sherman, D. J., et al. (2011). Acquisition and transmissibility of U.S. soybean dwarf virus isolates by the soybean aphid, *Aphis glycines*. *Plant Disease*, *95*(8), 945–950.

Dar, W. D., Reddy, B. V. S., Gowda, C. L. L., & Ramesh, S. (2006). Genetic resources enhancement of ICRISAT-mandate crops. *Current Science*, *91*(7), 880–884.

De Bruxelles, G. L., & Roberts, M. R. (2001). Signals regulating multiple responses to wounding and herbivores. *Critical Reviews in Plant Sciences*, *20*(5), 487–521.

Devasthali, S., & Joshi, M. (1994). Infestation and varietal preference of insect pests in green gram. *Indian Agriculturist*, *38*, 263–272.

Dhaliwal, G. S., Singh, R., & Jindal, V. (2004). Host plant resistance and insect pest management: Progress and potential. In G. S. Dhaliwal, & R. Singh (Eds.), *Host plant resistance to insects: Concepts and applications* (pp. 517–558). New Delhi, India: Panima Publishing Corporation.

Dhillon, M., & Sharma, H. (2012). Paradigm shifts in research on host plant resistance to insect pests. *Indian Journal of Plant Protection*, *40*, 1–11.

Dongre, T. K., Pawar, S. E., Thakare, R. G., & Harwalkar, M. R. (1996). Identification of resistant sources to cowpea weevil (*Callosobruchus maculatus* (F.)) in Vigna sp. and inheritance of their resistance in black gram (*Vigna mungo var. mungo*). *Journal of Stored Products Research*, *32*, 201–204.

Doss, R. P., Oliver, J. E., Proebsting, W. M., Potter, S. W., Kuy, S., Clement, S. L., et al. (2000). Bruchins: Insect-derived plant regulators that stimulate neoplasm formation. *Proceedings of the National Academy of Sciences of the United States of America*, *97*(11), 6218–6223.

Edwards, O. (2001). Interspecific and intraspecific variation in the performance of three aphid species on five-grain legume hosts. *Entomologia Experimentalis et Applicata*, *100*, 21–30.

Egho, E. O. (2010a). Comparative studies on insect species of cowpea (Vigna unguiculata) (L) walp in two agro-ecological zones during the early cropping season, in Delta State, southern Nigeria. *Agriculture and Biology Journal of North America*, *1*(5), 946–949.

Egho, E. O. (2010b). Management of major field insect pests and yield of cowpea (*Vigna unguiculata* (L) walp) under calendar and monitored application of synthetic chemicals in Asaba, southern Nigeria. *African Journal of General Agriculture*, *6*, 263–273.

Eigenbrode, S. D., Bosque-Pérez, N. A., & Davis, T. S. (2018). Insect-borne plant pathogens and their vectors: Ecology, evolution, and complex interactions. *Annual Review of Entomology*, *63*, 169–191.

Ekesi, S., & Maniania, N. K. (2000). Susceptibility of *Megalurothrips sjostedti* developmental stages to *Metarhizium anisopliae* and the effects of infection on feeding, adult fecundity, egg fertility and longevity. *Entomologia Experimentalis et Applicata*, *94*(3), 229–236.

El Bouhssini, M., Sarker, A., Erskine, W., & Joubi, A. (2008). First sources of resistance to Sitona weevil (*Sitona crinitus* Herbst) in wild Lens species. *Genetic Resources and Crop Evolution*, *55*, 1–4.

Elden, T. C., & Elgin, J. H., Jr. (1992). Mechanisms of resistance to the potato leafhopper (Homoptera: Cicadellidae) in selected alfalfa clones. *Journal of Economic Entomology*, *85*(2), 576–582.

El-Hawary, F. M., & Abd El-Salam, A. M. E. (2008). Effect of neem and antitranspirant products against *Aphis craccivora* Koch and its biology. *Egyptian Academic Journal of Biological Sciences*, *1*(2), 189–196.

Estruch, J. J., Warren, G. W., Mullins, M. A., Nye, G. J., Craig, J. A., & Koziel, M. G. (1996). Vip3A, a novel Bacillus thuringiensis vegetative insecticidal protein with a wide spectrum of activities against lepidopteran insects. *Proceedings of the National Academy of Sciences of the United States of America*, *93*, 5389–5394.

Falco, M. C., Marbach, P. A. S., Pompermayer, P., Lopes, F. C. C., & Silva-Filho, M. C. (2001). Mechanisms of sugarcane response to herbivory. *Genetics and Molecular Biology, 24*, 113–122.

FAO. (2021). *NSP-integrated pest management*. http://www.fao.org/agriculture/crops/core-themes/theme/pests/ipm/en/.

FAOSTAT. (2021). *Food and agriculture data*. http://www.fao.org/faostat/en/#data.

Fory, L. F., Finardi-Filho, F., Quintero, C. M., Osborn, T. C., Cardona, C., Chrispeels, M. J., et al. (1996). α-Amylase inhibitors in resistance of common beans to the Mexican bean weevil and the bean weevil (Coleoptera: Bruchidae). *Journal of Economic Entomology, 89*, 204–210.

Foyer, C. H., Lam, H. M., Nguyen, H. T., Siddique, K. H., Varshney, R. K., Colmer, T. D., et al. (2016). Neglecting legumes has compromised human health and sustainable food production. *Nature Plants, 2*, 16112.

Franco, F. P., Moura, D. S., Vivanco, J. M., & Silva-Filho, M. C. (2017). Plant-insect-pathogen interactions: A naturally complex ménage à trois. *Current Opinion in Microbiology, 37*, 54–60.

Fujii, K., & Miyazaki, S. (1987). Infestation resistance of wild legumes (*Vigna sublobata*) to azuki bean weevil, *Callosobruchus chinensis* and its relationship with cytogenetic classification. *Applied Entomology and Zoology, 22*, 229–230.

Galwey, N. W., & Evans, A. M. (1982). The inheritance of resistance to *Empoasca kraemeri* Ross and Moore in the common bean, *Phaseolus vulgaris* L. *Euphytica, 31*, 933–952.

Garza, R., Cardona, C., & Singh, S. P. (1996). Inheritance of resistance to the bean-pod weevil (*Apion godmani* Wagner) in common beans from Mexico. *Theoretical and Applied Genetics, 92*, 357–362.

Gershenzon, J., & Dudareva, N. (2007). The function of terpene natural products in the natural world. *Nature Chemical Biology, 3*(7), 408–414.

Girija, Sallmath, P. M., Patll, S. A., Gowda, C. L. L., & Sharma, H. C. (2008). Biophysical and biochemical basis of host plant resistance to pod borer *Helicoverpa armigera* (Hubner) in chickpea (*Cicer arietinum* L.). *Indian Journal of Genetics and Plant Breeding, 68*(3), 320–323.

Gu, J., Ye, R., Xu, Y., Yin, Y., Li, S., & Chen, H. (2021). A historical overview of analysis systems for Bacillus thuringiensis (Bt) Cry proteins. *Microchemical Journal, 165*, 106137.

Halder, J., Rai, A., & Kodandaram, M. H. (2013). Compatibility of neem oil and different entomopathogens for the management of major vegetable sucking pests. *National Academy Science Letters, 36*, 19–25.

Hanson, A. A., Lorenz, A. J., Hesler, L. S., Bhusal, S. J., Bansal, R., Michel, A. P., et al. (2018). Genome-wide association mapping of host-plant resistance to soybean aphid. *The Plant Genome, 11*(3).

Hassan, S. (2013). Effect of variety and intercropping on two major cowpea (*Vigna unguiculata* L. Walp) field pests in Mubi, Adamawa State, Nigeria. *International Journal of Agricultural Research and Development, 1*(5), 108–109.

Helm, C. G., Kogan, M., & Hill, B. G. (1980). Sampling leafhoppers on soybean. In M. Kogan, & D. C. Herzog (Eds.), *Sampling methods in soybean entomology* (pp. 260–282). New York: Springer.

Hema, M., Sreenivasulu, P., Patil, B. L., Kumar, P. L., & Reddy, D. V. (2014). Tropical food legumes: Virus diseases of economic importance and their control. *Advances in Virus Research, 90*, 431–505.

Hill, S. (1987). Cultural pest control. *American Journal of Alternative Agriculture, 2*(4), 191.

Hill, S. (1989). Cultural methods of pest, primarily insect, control. In *Ecological agriculture projects*. http://www. https://eap.mcgill.ca/publications/eap58.htm.

Hinds, J. (2020). A whole farm approach to managing pests. In *SARE outreach*. https://sare.org/wp-content/uploads/A_Whole-Farm_Approach_to_Managing_Pests.pdf.

Hoffmann-Campo, C. B., Ramos Neto, J. A., Oliveira, M. C. N. D., & Oliveira, L. J. (2006). Detrimental effect of rutin on *Anticarsia gemmatalis*. *Pesquisa Agropecuária Brasileira, 41*, 1453–1459.

Holopainen, J. K., Himanen, S. J., Yuan, J. S., Chen, F., & Stewart, C. N. (2013). Ecological functions of terpenoids in changing climates. In *Natural products* (pp. 2913–2940). Springer Science and Business Media LLC: Berlin/Heidelberg, Germany.

ICRISAT. (1992). *International Crops Research Institute for the semi-arid tropics. The medium-term plan*. Patancheru.

Jackai, L. E. N., & Oghiakhe, S. (1989). Pod wall trichomes and resistance of two wild cowpea, *Vigna vexillata*, accessions to *Maruca testualis* (Geyer) (Lepidoptera: Pyralidae) and *Clavigralla tomentosicollis* Stal. (Hemiptera: Coreidae). *Bulletin of Entomological Research, 79*, 595–605.

Jones, W., & Sullivan, M. (1982). Role of host plants in population dynamics of stink bug pests of soybean in South Carolina. *Environmental Entomology, 11*, 867–875.

Julier, B., Bournoville, R., Landre, B., Ecalle, C., & Carre, S. (2004). Genetic analysis of lucerne (*Medicago sativa* L.) seedling resistance to pea aphid (Acyrthosiphon pisum Harris). *Euphytica, 138*, 133–139.

Kabissa, J., & Fronk, W. D. (1986). Bean foliage consumption by Mexican bean beetle (Coleoptera: Coccinellidae) and its effect on yield. *Journal of the Kansas Entomological Society, 52*(2), 275–279.

Kareiva, P., & Sahakian, R. (1990). Tritrophic effects of a simple architectural mutation in pea plants. *Nature, 345*, 433–434.

Karel, A. K. (1985). Yield losses from and control of bean pod borers, *Maruca testulalis* (Lepidoptera: Pyralidae) and *Heliothis armigera* (Lepidoptera: Noctuidae). *Journal of Economic Entomology, 78*(6), 1523–1526.

Keatinge, J. D. H., Wang, J.-F., Dinssa, F. F., Ebert, A. W., Hughes, J. D. A., Stoilova, T., et al. (2015). Indigenous vegetables worldwide: Their importance and future development. *Acta Horticulturae, 1102*, 1–20.

Kester, K., Smith, C., & Gilman, D. (1984). Mechanisms of resistance in soybean (*Glycine max* [L.] Merrill) genotype PI 171444 to the southern green stink bug., *Nezara viridula* (L.) (Hemiptera: Pentatomidae). *Environmental Entomology, 13*, 1208–1215.

Khera, P., Upadhyaya, H. D., Pandey, M. K., Roorkiwal, M., Sriswathi, M., Pasupuleti, J., et al. (2013). Single nucleotide polymorphism–based genetic diversity in the reference set of peanut (*Arachis* spp.) by developing and applying cost-effective kompetitive allele specific polymerase chain reaction genotyping assays. *The Plant Genome, 6*(3), 1–11.

Kilen, T. C., & Lambert, L. (1986). Evidence for different genes controlling insect resistance in three soybean genotypes. *Crop Science, 26*, 869–871.

Kilen, T. C., & Lambert, L. (1998). Genetic control of insect resistance in soybean germplasm PI 417061. *Crop Science, 38*, 652–654.

Kitch, L. W., Shade, R. E., & Murdock, L. L. (1991). Resistance to the cowpea weevil (*Callosobruchus maculatus*) larva in pods of cowpea (*Vigna unguiculata*). *Entomologia Experimentalis et Applicata, 60*, 183–192.

Kliebenstein, D. J., Kroymann, J., Brown, P., Figuth, A., Pedersen, D., Gershenzon, J., et al. (2001). Genetic control of natural variation in *Arabidopsis glucosinolate* accumulation. *Plant Physiology, 126*(2), 811–825.

Kogan, M., & Cope, D. (1974). Feeding and nutrition of insects associated with soybeans. 3. Food intake, utilization, and growth in the soybean looper, *Pseudoplusia includens*. *Annals of the Entomological Society of America, 67*(1), 66–72.

Koona, P., Osisanya, E. O., Jackai, L. E. N., Tamo, M., Reeves, J., & Hughes, J. D. A. (2002). Pod surface characteristics in wild and cultivated Vigna species and resistance to the coreid bug *Clavigralla tomentosicollis* Stal. (Hemiptera: Coreidae). *International Journal of Tropical Insect Science, 22*(1), 1–7.

Koona, P., Osisanya, E. O., Jackai, L. E. N., & Tonye, J. (2004). Infestation and damage by *Clavigralla tomentosicollis* and *Anoplocnemis curvipes* (Hemiptera: Coreidae) in cowpea plants with modified leaf structure and pods in different positions relative to the canopy. *Environmental Entomology, 33*(3), 471–476.

Kornegay, J. L., & Cardona, C. (1991). Inheritance of resistance to *Acanthoscelides obtectus* in a wild common bean accession crossed to commercial bean cultivars. *Euphytica, 52,* 103–111.

Kornegay, J. L., & Temple, S. R. (1986). Inheritance and combining ability of leafhopper defense mechanisms in common bean 1. *Crop Science, 26,* 1153–1158.

Krajewski, P., Chen, D., Ćwiek, H., van Dijk, A. D., Fiorani, F., Kersey, P., et al. (2015). Towards recommendations for metadata and data handling in plant phenotyping. *Journal of Experimental Botany, 66*(18), 5417–5427.

Kumar, A., Jaiwal, R., Sreevathsa, R., Chaudhary, D., & Jaiwal, P. K. (2021). Transgenic cowpea plants expressing Bacillus thuringiensis Cry2Aa insecticidal protein imparts resistance to Maruca vitrata legume pod borer. *Plant Cell Reports, 40,* 583–594.

Kumar, B. N., Yelshetty, S., Rachappa, V., Naganagoud, A., Katti, P., & Amaresh, Y. S. (2014). Assessment of crop loss due to leaf webber, *Grapholita critica* Meyr. in pigeonpea. *Journal of Experimental Zoology, 17*(1), 235–236.

Kumari, D. A., Sharma, H. C., & Reddy, D. J. (2010). Incorporation of lyophilized leaves and pods into artificial diet to assess antibiosis component of resistance to pod borer in pigeonpea. *Journal of Food Legumes, 23*(1), 57–65.

Kuswantoro, H., Bayu, M. S. Y. I., Baliadi, Y., & Tengkano, W. (2017). Resistance of advanced soybean lines to pod borrer (*Etiella zinckenella*). *Biosaintifika Journal of Biology & Biology Education, 9*(2), 317–324.

Lakshminarayan, S., Singh, P. S., & Mishra, D. S. (2008). Relationship between whitefly population, YMV disease and morphological parameters of green gram germplasm. *Environment and Ecology, 26,* 978–982.

Lal, S. S. (1987). Insect pests of mung, urd, cowpea and pea and their management. In M. V. Rao, & S. Sithanantham (Eds.), *Plant protection in field crops* (pp. 185–201). Hyderabad: Plant Protection Association of India.

Lam, W. K., & Pedigo, L. P. (2001). Effect of trichome density on soybean pod feeding by adult bean leaf beetles (Coleoptera: Chrysomelidae). *Journal of Economic Entomology, 94*(6), 1459–1463.

Lambert, L., & Kilen, T. C. (1984). Influence of three soybean plant genotypes and their F1 intercrosses on the development of five insect species. *Journal of Economic Entomology,* 77(3), 622–625.

Lambert, A. L., McPherson, R. M., & Espelie, K. E. (1995). Soybean host plant resistance mechanisms that alter abundance of whiteflies (Homoptera: Aleyrodidae). *Environmental Entomology, 24,* 1381–1386.

Laserna-Ruiz, I., De-Los-Mozos-Pascual, M., Santana-Méridas, O., Sánchez-Vioque, R., & Rodríguez-Conde, M. F. (2012). Screening and selection of lentil (*Lens* Miller) germplasm resistant to seed bruchids (*Bruchus* spp.). *Euphytica, 188*(2), 153–162.

Lateef, S. S., & Pimbert, M. P. (1990). The search for host plant resistance of *Helicoverpa armigera* in chickpea and pigeonpea at ICRISAT. In *Proceedings of the consultative group meeting on the host selection behavior of Helicoverpa armigera, ICRISAT, Hyderabad, March 5–7* (pp. 185–192).

Lee, G. H., Paik, C. H., Choi, M. Y., Oh, Y. J., Kim, D. H., & Na, S. Y. (2004). Seasonal occurrence, soybean damage and control efficacy of bean bug, *Riptortus clavatus* Thunberg (Hemiptera: Alydidae) at soybean field in Honam Province. *Korean Society of Applied Entomology, 43*(3), 249–255.

Li, W., Guo, Y., Hu, Y., Chen, J., Zhang, J., & Shi, S. (2021). Field cage assessment of feeding damage by *Riptortus pedestris* on soybeans in China. *Insects, 12*(3), 255.

Li, Y., Hill, C. B., & Hartman, G. L. (2004). Effect of three resistant soybean genotypes on the fecundity, mortality, and maturation of soybean aphid (Homoptera: Aphididae). *Journal of Economic Entomology*, *97*(3), 1106–1111.

Li, L., Zhao, Y., McCaig, B. C., Wingerd, B. A., Wang, J., Whalon, M. E., et al. (2004). The tomato homolog of CORONATINE-INSENSITIVE1 is required for the maternal control of seed maturation, jasmonate-signaled defense responses, and glandular trichome development. *The Plant Cell*, *16*(1), 126–143.

Lin, J., Wang, D., Chen, X., Köllner, T. G., Mazarei, M., Guo, H., et al. (2017). An (E,E)-α-farnesene synthase gene of soybean has a role in defence against nematodes and is involved in synthesizing insect-induced volatiles. *Plant Biotechnology Journal*, *15*, 510.

Liu, S. H., Norris, D. M., & Lyne, P. (1989). Volatiles from the foliage of soybean, *Glycine max*, and lima bean, *Phaseolus lunatus*: Their behavioral effects on the insects *Trichoplusia ni* and *Epilachna varivestis*. *Journal of Agricultural and Food Chemistry*, *37*, 496–501.

Lomonossoff, G. P., & Shanks, M. (1999). Comoviruses (Comoviridae). In A. Granoff, & R. G. Webster (Eds.), *Encyclopedia of virology* (pp. 285–291). United States: Elsevier.

Lopes, E. C. A., Destro, D., Montalvan, R., Ventura, M. U., & Guerra, E. P. (1997). Genetic gain and correlations among traits for stink bug resistance in soybeans. *Euphytica*, *97*, 161–166.

Mar, T. B., Mendes, I. R., Lau, D., Fiallo-Olivé, E., Navas-Castillo, J., Alves, M. S., et al. (2017). Interaction between the New World begomovirus Euphorbia yellow mosaic virus and its associated alphasatellite: Effects on infection and transmission by the whitefly *Bemisia tabaci*. *The Journal of General Virology*, *98*(6), 1552–1562.

Medina, V., Sardoy, P. M., Soria, M., Vay, C. A., Gutkind, G. O., & Zavala, J. A. (2018). Characterized non-transient microbiota from stinkbug (*Nezara viridula*) midgut deactivates soybean chemical defenses. *PLoS One*, *13*(7).

Mehrparvar, M., Madjdzadeh, S. M., Mahdavi Arab, N., Esmaeilbeygi, M., & Ebrahimpour, E. (2012). Morphometric discrimination of Black Legume Aphid, *Aphis craccivora* Koch (Hemiptera: Aphididae), populations associated with different host plants. *North-Western Journal of Zoology*, *8*(1), 172–180.

Mitra, S., Karmakar, A., Mukherjee, A., & Barik, A. (2017). The role of leaf volatiles of *Ludwigia octovalvis* (Jacq.) Raven in the attraction of *Altica cyanea* (Weber) (Coleoptera: Chrysomelidae). *Journal of Chemical Ecology*, *43*(7), 679–692.

Morales, F. J., & Anderson, P. K. (2001). The emergence and dissemination of whitefly-transmitted geminiviruses in Latin America. *Archives of Virology*, *146*(3), 415–441.

Morgan, J. K., Luzio, G. A., Ammar, E.-D., Hunter, W. B., Hall, D. G., & Shatters, R. G., Jr. (2013). Formation of Stylet Sheaths *in āere* (in air) from eight species of phytophagous hemipterans from six families (suborders: Auchenorrhyncha and Sternorrhyncha). *PLoS One*, *8*(4), e62444.

Moudgal, R. K., Lakra, R., Dahiya, B., & Dhillon, M. (2008). Physico-chemical traits of *Cajanus cajan* (L.) Millsp. pod wall affecting *Melanagromyza obtusa* (Malloch) damage. *Euphytica*, *161*, 429–436.

Mutti, N. S., Louis, J., Pappan, L. K., Pappan, K., Begum, K., Chen, M. S., et al. (2008). A protein from the salivary glands of the pea aphid, *Acyrthosiphon pisum*, is essential in feeding on a host plant. *Proceedings of the National Academy of Sciences of the United States of America*, *105*(29), 9965–9969.

Narayanamma, V. L., Sharma, H. C., Gowda, C. L. L., & Sriramulu, M. (2007). Mechanisms of resistance to *Helicoverpa armigera* and introgression of resistance genes into F1 hybrids in chickpea. *Arthropod-Plant Interactions*, *1*(4), 263.

Nasruddin, A., Fattah, A., Baco, M. S., & Said, A. E. (2015). Potential damages, seasonal abundance and distribution of Empoasca terminalis Distant (Homoptera: Cicadellidae) on soybean in South Sulawesi. *Jurnal Entomologi Indonesia*, *11*(2), 93–102.

Ng, J. C. K., & Perry, K. L. (2004). Transmission of plant viruses by aphid vectors. *Molecular Plant Pathology, 5*, 505–511.

Nishida, R. (2002). Sequestration of defensive substances from plants by Lepidoptera. *Annual Review of Entomology, 47*, 57–92.

Ntonifor, N. N., & Jackai, L. E. N. (1994). Comparative growth and development of *Riptortus dentipes* and *Nezara viridula* on cowpea and soybean pods. *International Journal of Tropical Insect Science, 15*(4–5), 453–459.

Nwilene, F., Nwanze, K., & Youdeowei, A. (2008). Impact of integrated pest management on food and horticultural crops in Africa. *Entomologia Experimentalis et Applicata, 128*, 355–363.

Obadofin, A. A. (2014). Screening of some cowpea varieties for resistance to *Callosobruchus maculatus*. *International Journal of Pure and Applied Sciences and Technology, 22*, 9–17.

Ochilo, W. N., & Nyamasyo, G. H. (2010). Pest status of bean stem maggot (*Ophiomyia* spp.) and black bean aphid (*Aphis fabae*) in Taita District, Kenya. *Tropical and Subtropical Agroecosystems, 13*(1), 91–97.

Ofuya, T. I. (1991). Observations on insect infestation and damage in cowpea (*Vigna unguiculata*) intercropped with tomato (*Lycopersicon esculentum*) in a rain forest area of Nigeria. *Experimental Agriculture, 27*(4), 407–412.

Okello, D. K., Akello, L. B., Tukamuhabwa, P., Odong, T. L., Ochwo-Ssemakula, M., Adriko, J., et al. (2014). Groundnut rosette disease symptoms types distribution and management of the disease in Uganda. *African Journal of Plant Science, 8*(3), 153–163.

Ortega, M. A., All, J. N., Boerma, H. R., & Parrott, W. A. (2016). Pyramids of QTLs enhance host-plant resistance and Bt-mediated resistance to leaf-chewing insects in soybean. *Theoretical and Applied Genetics, 129*(4), 703–715.

Osborn, T. C., Blake, T., Gepts, P., & Bliss, F. A. (1986). Bean arcelin: 2. Genetic variation, inheritance and linkage relationships of a novel seed protein of *Phaseolus vulgaris* L. *Theoretical and Applied Genetics, 71*(6), 847–855.

Packiam, M., & Ignacimuthu, S. (2012). Effect of PONNEEM on *Spodoptera litura* (Fab.) and its compatibility with *Trichogramma chilonis* Ishii. *Brazilian Archives of Biology and Technology, 55*, 291–298.

Panda, N., & Khush, G. S. (1995). *Host plant resistance to insects*. Wallingford: CAB International.

Panizzi, A. R., & Slansky, F., Jr. (1985). Review of phytophagous pentatomids (Hemiptera: Pentatomidae) associated with soybean in the Americas. *Florida Entomologist, 68*(1), 184–214.

Pensuk, V., Jogloy, S., Wongkaew, S., & Patanothai, A. (2008). Generation means analysis of resistance to peanut bud necrosis caused by peanut bud necrosis tospovirus in peanut. *Plant Breeding, 123*(1), 90–92.

Pillemer, E. A., & Tingey, W. M. (1976). Hooked trichomes: A physical plant barrier to a major agricultural pest. *Science (New York, N.Y.), 193*(4252), 482–484.

Pompermayer, P., Lopes, A., Terra, W., Parra, J., Falco, M., & Silva-Filho, M. (2001). Effects of soybean proteinase inhibitor on development, survival and reproductive potential of the sugarcane borer, *Diatraea saccharalis*. *Entomologia Experimentalis et Applicata, 99*, 79–85.

Qin, J., Shi, A., Mou, B., Bhattarai, G., Yang, W., Weng, Y., et al. (2017). Association mapping of aphid resistance in USDA cowpea (*Vigna unguiculata* L. Walp.) core collection using SNPs. *Euphytica, 213*.

Qu, Z., Bendena, W. G., Tobe, S. S., & Hui, J. (2018). Juvenile hormone and sesquiterpenoids in arthropods: Biosynthesis, signaling, and role of MicroRNA. *The Journal of Steroid Biochemistry and Molecular Biology, 184*, 69–76.

Rachappa, V., Shivayogiyappa, Harischandra, N., & Yelshetty, S. (2018). Assessment of crop loss due to leafhopper, (*Empoasca kerri Pruthi*) in pigeonpea. *Legume Research, 41*, 155–158.

Razmjou, J., Tavakkoli, H., & Nemati, M. (2009). Life history traits of *Tetranychus urticae* Koch on three legumes (Acari: Tetranychidae). *Munis Entomology & Zoology*, *4*(1), 204–211.

Redden, R. J., Dobie, P., & Gatehouse, A. M. R. (1983). The inheritance of seed resistance to *Callosobruchus maculatus* F. in cowpea (*Vigna unguiculata* L. Walp.). I. Analyses of parental, F1, F2, F3 and backcross seed generations. *Australian Journal of Agricultural Research*, *34*, 681–695.

Romero-Andreas, J., Yandell, B. S., & Bliss, F. A. (1986). Bean arcelin: 1. Inheritance of a novel seed protein of *Phaseolus vulgaris* L. and its effect on seed composition. *Theoretical and Applied Genetics*, *72*, 123–128.

Rufener, G. K., II, St. Martin, S. K., Cooper, R. L., & Hammond, R. B. (1989). Genetics of antibiosis resistance to Mexican bean beetle in soybean. *Crop Science*, *29*, 618–622.

Rupe, J., & Luttrell, R. G. (2008). Effect of pests and diseases on soybean quality. In L. A. Johnson, P. J. White, & R. Galloway (Eds.), *Soybeans: Chemistry, production, processing, and utilization* (pp. 93–116). Urbana, USA: American Oil Chemists' Society.

Sandhi, R. K., & Reddy, G. V. (2020). Biology, ecology, and management strategies for pea aphid (Hemiptera: Aphididae) in pulse crops. *Journal of Integrated Pest Management*, *11*(1), 1–20.

Sariah, J. B., & Makundi, R. H. (2007). Effect of sowing time on infestation of beans (*Phaseolus vulgaris* L.) by two species of the Bean Stem Maggot, Ophiomyia spencerella and Ophiomyia phaseoli (Diptera: Agromyzidae). *Archives of Phytopathology and Plant Protection*, *40*(1), 45–51.

Sarkar, S., & Bhattacharyya, S. (2015). Screening of greengram genotypes for bruchid (*Callosobruchus chinensis* L.) resistance and selection of parental lines for hybridization programme. *Legume Research*, *38*(5), 704–706.

Sastry, K. S., & Zitter, T. A. (2014). Management of virus and viroid diseases of crops in the tropics. In K. S. Sastry, & T. A. Zitter (Eds.), *Plant virus and viroid diseases in the tropics* (pp. 149–480). Dordrecht: Springer.

Saxena, K. B., Lateef, S. S., Fonseka, H. H. D., Ariyaratne, H. P., & Dharamsena, C. M. D. (1996). *Maruca testulalis* damage in determinate and indeterminate lines of pigeonpea. *International Chickpea Newsletter*, *3*, 91–93.

Schreinemachers, P., Ramasamy, S., Wu, M. H., Bhattarai, M., Patricio, R., Yule, S., et al. (2014). Safe and sustainable management of legume pests and diseases in Thailand and Vietnam: A situational analysis. *International Journal of Tropical Insect Science*, *34*, 88–97.

Shade, R. E., Doskocil, M. J., & Maxon, N. P. (1979). Potato leafhopper resistance in glandular-haired alfalfa species. *Crop Science*, *19*, 287–289.

Shade, R. E., & Kitch, L. W. (1983). Pea aphid (Homoptera: Aphididae) biology on glandular-haired medicago species. *Environmental Entomology*, *12*(1), 237–240.

Shankar, M., Sharma, H. C., Ramesh Babu, T., & Sridevi, D. (2013). Evaluation of chickpea genotypes for resistance to beet armyworm, Spodoptera exigua. *Indian Journal of Plant Protection*, *41*(4), 275–281.

Shanower, T. G., Romeis, J., & Minja, E. M. (1999). Insect pests of pigeonpea and their management. *Annual Review of Entomology*, *44*, 77–96.

Sharma, H. C. (2005). Heliothis/Helicoverpa management "emerging trends & strategies for future research". In H. C. Sharma (Ed.), *Heliothis/Helicoverpa management: Emerging trends and strategies for future research* (pp. 453–463). New Delhi, India: Oxford & IBH Publishing Co. Pvt. Ltd.

Sharma, H. C. (2014). Climate change effects on insects: Implications for crop protection and food security. *Journal of Crop Improvement*, *28*(2), 229–259.

Sharma, H. C. (2016). Host plant resistance to insect pests in pigeonpea: Potential and limitations. *Legume Perspectives*, *11*, 24–29.

Sharma, O. P., Bhosle, B. B., Kamble, K. R., Bhede, B. V., & Seeras, N. R. (2011). Management of pigeonpea pod borers with special reference to pod fly (*Melanagromyza obtusa*). *Indian Journal of Agricultural Sciences*, *81*(5), 539–543.

Sharma, H. C., Gowda, C. L., Stevenson, P. C., Ridsdill-Smith, T. J., Clement, S. L., Rao, G. V. R., et al. (2007). Host plant resistance and insect pest management in chickpea. In S. S. Yadav, R. J. Redden, W. Chen, & B. Sharma (Eds.), *Chickpea breeding and management* (pp. 520–537). Wallingford, UK: CAB International.

Sharma, H. C., Jaba, J., & Vashisth, S. (2017). Distinguishing proof and utilization of resistance of insect pests in grain legumes: Progress and limitations. In R. Arora, & S. Sandhu (Eds.), *Breeding insect resistant crops for sustainable agriculture*. Singapore: Springer.

Sharma, H. C., Pampapathy, G., Lanka, S. K., & Ridsdill-Smith, T. J. (2005). Exploitation of wild Cicer reticulatum germplasm for resistance to *Helicoverpa armigera*. *Journal of Economic Entomology*, *98*(6), 2246–2253.

Sharma, H. C., Pampapathy, G., & Reddy, L. J. (2003). Wild relatives of pigeonpea as a source of resistance to the pod fly (*Melanagromyza obtusa* Malloch) and pod wasp (*Tanaostigmodes cajaninae* La Salle). *Genetic Resources and Crop Evolution*, *50*, 817–824.

Sharma, H. C., Saxena, K. B., & Bhagwat, V. R. (1999). *The legume pod borer, Maruca vitrata: Bionomics and management. Information bulletin no. 55*. Patancheru, Andhra Pradesh, India: International Crops Research Institute for the Semi-Arid Tropics.

Sharma, H. C., Sujana, G., & Rao, D. M. (2009). Morphological and chemical components of resistance to pod borer, *Helicoverpa armigera* in wild relatives of pigeonpea. *Arthropod-Plant Interactions*, *3*(3), 151–161.

Shrestha, R., Arnaud, E., Mauleon, R., Senger, M., Davenport, G. F., Hancock, D., et al. (2010). Multifunctional crop trait ontology for breeders' data: Field book, annotation, data discovery and semantic enrichment of the literature. *AoB Plants*, *2010*.

Singh, S. R. (1978). Resistance to pests of cowpea in Nigeria. In S. R. Singh, H. F. Van Emden, & T. A. Taylor (Eds.), *Pest of grain legumes: Ecology and control* (pp. 267–279). New York, USA: Academic Press.

Singh, S. R. (1987). Host plant resistance for cowpea insect pest management. *International Journal of Tropical Insect Science*, *8*(4–6), 756–769.

Singh, S., Kumar, N. R., Maniraj, R., Lakshmikanth, R., Rao, K. Y. S., Muralimohan, N., et al. (2018). Expression of Cry2Aa, a Bacillus thuringiensis insecticidal protein in transgenic pigeon pea confers resistance to gram pod borer, Helicoverpa armigera. *Scientific Reports*, *8*(1), 1–12.

Singh, B., & Mandal, K. (2013). Environmental impact of pesticides belonging to newer chemistry. In A. K. Dhawan, B. Singh, M. Brar-Bhullar, & R. Arora (Eds.), *Integrated pest management* (pp. 152–190). Jodhpur: Scientific Publishers.

Singh, K., Ocampo, B., & Robertson, L. (1998). Diversity for abiotic and biotic stress resistance in the wild annual Cicer species. *Genetic Resources and Crop Evolution*, *45*, 9–17.

Singsit, C., Adang, M. J., Lynch, R. E., Anderson, W. F., Wang, A., Cardineau, G., et al. (1997). Expression of a Bacillus thuringiensis cryIA(c) gene in transgenic peanut plants and its efficacy against lesser cornstalk borer. *Transgenic Research*, *6*(2), 169–176.

Smelser, R. B., & Pedigo, L. P. (1992). Bean leaf beetle (Coleoptera: Chrysomelidae) herbivory on leaf, stem, and pod components of soybean. *Journal of Economic Entomology*, *85*(6), 2408–2412.

Smith, C. M. (2005). *Plant resistance to arthropods—Molecular and conventional approaches*. Dordrecht: Springer Verlag.

Somta, C., Somta, P., Tomooka, N., Ooi, P., Vaughan, D. A., & Srinives, P. (2008). Characterization of new sources of mungbean (*Vigna radiata* (L.) Wilczek) resistance to bruchids, *Callosobruchus* spp. (Coleoptera: Bruchidae). *Journal of Stored Products Research*, *44*, 316–321.

Song, F., Swinton, S., Difonzo, C., O'Neal, M., & Ragsdale, D. (2006). Profitability analysis of soybean aphid control treatments in three north-central states. In *Michigan State University, Department of Agricultural Economics staff paper no. 2006-24.*

Sorensen, E. L., & Horber, E. K. (1974). Selecting alfalfa seedlings to resist the potato leafhopper. *Crop Science, 14*, 85–86.

Soroka, J. J., & MacKay, P. A. (1990). Seasonal occurrence of the pea aphid, *Acyrthosiphon pisum* (Harris) (Homoptera: Aphididae), on cultivars of field peas in Manitoba and its effect on pea growth and yield. *The Canadian Entomologist, 122*, 503–513.

Soundararajan, R. P., Narayanasami, C., & Geetha, S. (2013). Host plant resistance to insect pests of grain legumes—A review. *Agricultural Reviews, 34*, 176.

Srinivasan, R. (2014). *Insect and mite pests on vegetable legumes: A field guide for identification and management*. AVRDC—The World Vegetable Center.

Stevenson, P. C., Dhillon, M. K., Sharma, H. C., & Bouhssini, M. E. (2007). Insect pests of lentil and their management. In S. S. Yadav, D. L. McNeil, & P. C. Stevenson (Eds.), *Lentil*. Dordrecht: Springer.

Stoddard, F. L., Nicholas, A. H., Rubiales, D., Thomas, J., & Villegas-Fernandez, A. M. (2010). Integrated pest management in faba bean. *Field Crops Research, 115*, 308–318.

Sujana, G., Sharma, H. C., & Rao, D. M. (2008). Antixenosis and antibiosis components of resistance to pod borer *Helicoverpa armigera* in wild relatives of pigeonpea. *International Journal of Tropical Insect Science, 28*(4), 191–200.

Sumita, A., Bidyut, K. S., Ananda, P., Karen, O., Rod, M., William, J. M., et al. (2010). Transgenic chickpeas (*Cicer arietinum* L.) expressing a sequence-modified cry2Aa gene. *Plant Science, 178*, 333–339.

Sun, L., Cheng, X., & Wang, L. (2007). Current progress in research of bruchid resistance in mungbean. *Journal of Plant Genetic Resources, 8*, 113–117.

Talekar, N. S. (1987). Host plant resistance to insects attacking soybean and mungbean in the tropics. *International Journal of Tropical Insect Science, 8*, 777–782.

Talekar, N. S. (1990). *Agromyzid flies of food legumes in the tropics* (pp. 267–282). Delhi, India: Wiley Eastern Limited.

Talekar, N. S., & Lee, Y. H. (1988). Biology of *Ophiomyia centrosematis* (Diptera: Agromyzidae), a pest of soybean. *Annals of the Entomological Society of America, 81*(6), 938–942.

Talekar, N. S., & Lin, C. P. (1994). Characterization of resistance to limabean pod borer (Lepidoptera: Pyralidae) in soybean. *Journal of Economic Entomology, 87*, 821–825.

Tang, L. D., Yan, K. L., Fu, B. L., Wu, J. H., Liu, K., & Lu, Y. Y. (2015). The life table parameters of (Thysanoptera: Thripidae) on four leguminous crops. *Florida Entomologist, 98*(2), 620–625.

Tayo, T. O. (1989). Anatomical basis of cowpea resistance to the pod borer, *Maruca testulalis* (Geyer). *International Journal of Tropical Insect Science, 10*, 631–638.

Teshome, A., Mendesil, E., Geleta, M., Andargie, D., Anderson, P., Rämert, B., et al. (2014). Screening the primary gene pool of field pea (*Pisum sativum* L. subsp. sativum) in Ethiopia for resistance against pea weevil (*Bruchus pisorum* L.). *Genetic Resources and Crop Evolution, 62*, 525–538.

Thekke-Veetil, T., Lagos-Kutz, D., McCoppin, N. K., Hartman, G. L., Ju, H. K., Lim, H. S., et al. (2020). Soybean thrips (Thysanoptera: Thripidae) harbor highly diverse populations of arthropod, fungal and plant viruses. *Viruses, 12*(12), 1376.

Tilmon, K. J., Hodgson, E. W., O'Neal, M. E., & Ragsdale, D. W. (2011). Biology of the soybean aphid, *Aphis glycines* (Hemiptera: Aphididae) in the United States. *Journal of Integrated Pest Management, 2*, A1–A7.

Toker, C., Erler, F., Ceylan, F. O., & Canci, H. (2010). Severity of leaf miner [*Liriomyza cicerina* (Rondani, 1875) (Diptera: Agromyzidae)] damage in relation to leaf type in chickpea. *Turkish Journal of Entomology, 34*(2), 211–225.

Turnipseed, S. G. (1977). Influence of trichome variations on populations of small phytophagous insects in soybean. *Environmental Entomology*, *6*(6), 815–817.

Tyczewska, A., Woźniak, E., Gracz, J., Kuczyński, J., & Twardowski, T. (2018). Towards food security: Current state and future prospects of agrobiotechnology. *Trends in Biotechnology*, *36*, 1219–1229.

Varma, A., & Malathi, V. G. (2005). Emerging geminivirus problems: A serious threat to crop production. *Annals of Applied Biology*, *142*(2), 145–164.

Venette, R. C., & Ragsdale, D. W. (2004). Assessing the invasion by soybean aphid (Homoptera: Aphididae): Where will it end? *Annals of the Entomological Society of America*, *97*(2), 219–226.

Vitorio, L., Bazan, A. F. C., Arnemann, J. A., Pozebon, H., Ugalde, G. A., Marques, R. P., et al. (2019). First record of the soybean stem fly *Melanagromyza sojae* (Diptera: Agromyzidae) in Bolivia. *Genetics and Molecular Research*, *18*(1), 245–252.

Völkl, W., & Stechmann, D. H. (1998). Parasitism of the black bean aphid (Aphis fabae) by *Lysiphlebus fabarum* (Hym., Aphidiidae): The influence of host plant and habitat. *Journal of Applied Entomology*, *122*(1–5), 201–206.

Wada, T., Endo, N., & Takahashi, M. (2006). Reducing seed damage by soybean bugs by growing small-seeded soybeans and delaying sowing time. *Crop Protection*, *25*, 726–731.

Walls, R. L., Cooper, L., Elser, J., Gandolfo, M. A., Mungall, C. J., Smith, B., et al. (2019). The plant ontology facilitates comparisons of plant development stages across species. *Frontiers in Plant Science*, *10*, 631.

Wang, W., Dai, H., Zhang, Y., Chandrasekar, R., Luo, L., Hiromasa, Y., et al. (2015). Armet is an effector protein mediating aphid-plant interactions. *FASEB Journal*, *29*(5), 2032–2045.

Wang, X., Fang, Y., Lin, Z., Zhang, L., & Wang, H. (1994). A study on the damage and economic threshold of the soybean aphid at the seedling stage. *Plant Protection (Institute of Plant Protection, CAAS, China)*, *20*, 12–13.

Wang, Q., Yuan, E., Ling, X., Zhu-Salzman, K., Guo, H., Ge, F., et al. (2020). An aphid facultative symbiont suppresses plant defence by manipulating aphid gene expression in salivary glands. *Plant, Cell & Environment*, *43*(9), 2311–2322.

Weigand, S., Lateef, S., Sharaf El-Din, N., Mahmoud, S. F., Ahmed, K., & Ali, K. (1994). Integrated control of insect pests of cool season food legumes. In F. J. Muehlbauer, & W. J. Kaiser (Eds.), *Expanding the production and use of cool season food legumes* (pp. 679–694). Dordrecht: Kluwer Academic.

Whitfield, A. E., Falk, B. W., & Rotenberg, D. (2015). Insect vector-mediated transmission of plant viruses. *Virology*, *479–480*, 278–289.

Widyasari, K., Alazem, M., & Kim, K. H. (2020). Soybean resistance to soybean mosaic virus. *Plants (Basel, Switzerland)*, *9*(2), 219.

Wiseman, B. R. (1985). Types and mechanisms of host plant resistance to insect attack. *International Journal of Tropical Insect Science*, *6*, 239–242.

Yadav, M. K., Aravindan, S., Mukherjee, A. K., Kumar Bag, M., Lenka, S., & Ghritlahre, S. K. (2015). Viral diseases of soybean. *Popular Kheti*, *3*(3), 124–127.

Yadav, S., Kumar, J., Yadav, S., Singh, S., Yadav, V., Turner, N. C., et al. (2006). Evaluation of *Helicoverpa* and drought resistance in desi and kabuli chickpea. *Plant Genetic Resources*, *4*(3), 198–203.

Yang, X., Thannhauser, T. W., Burrows, M., Cox-Foster, D., Gildow, F. E., & Gray, S. M. (2008). Coupling genetics and proteomics to identify aphid proteins associated with vector-specific transmission of polerovirus (*luteoviridae*). *Journal of Virology*, *82*(1), 291–299.

Yates-Stewart, A. D., Daron, J., Wijeratne, S., Shahid, S., Edgington, H. A., Slotkin, R. K., et al. (2020). Soybean aphids adapted to host-plant resistance by down regulating putative effectors and up regulating transposable elements. *Insect Biochemistry and Molecular Biology*, *121*, 103363.

Yazaki, K., Arimura, G. I., & Ohnishi, T. (2017). 'Hidden' terpenoids in plants: Their biosynthesis, localization and ecological roles. *Plant and Cell Physiology*, *58*, 1615–1621.

Yoshida, M., Cowgill, S. E., & Wightman, J. A. (1997). Roles of oxalic acid and malic acids in chickpea trichome exudates in hostplant resistance to *Helicoverpa armigera*. *Journal of Chemical Ecology*, *23*, 1195–1210.

Youn, H. S., & Jung, C. (2008). Effect of trap cropping on the egg parasitism of *Riptortus clavatus* (Thunberg) in soybean fields. *Journal of Asia-Pacific Entomology*, *11*, 73–76.

Zaugg, I., Magni, C., Panzeri, D., Daminati, M., Bollini, R., Benrey, B., et al. (2012). QUES, a new *Phaseolus vulgaris* genotype resistant to common bean weevils, contains the Arcelin-8 allele coding for new lectin-related variants. *Theoretical and Applied Genetics*, *126*, 647–661.

Zhao, J., Sauvage, C., Zhao, J., Bitton, F., Bauchet, G., Liu, D., et al. (2019). Meta-analysis of genome-wide association studies provides insights into genetic control of tomato flavor. *Nature Communications*, *10*(1), 1534.

CHAPTER NINE

Genetic regulations of the oil and protein contents in soybean seeds and strategies for improvement

Ailin Liu[a], Sau-Shan Cheng[a], Wai-Shing Yung[a], Man-Wah Li[a,b], and Hon-Ming Lam[a,b,*]

[a]Centre for Soybean Research of the State Key Laboratory of Agrobiotechnology and School of Life Sciences, The Chinese University of Hong Kong, Shatin, Hong Kong Special Administrative Region, China
[b]Shenzhen Research Institute, The Chinese University of Hong Kong, Shenzhen, China
*Corresponding author: e-mail address: honming@cuhk.edu.hk

Contents

Abstract

Soybean [*Glycine max* (L.) Merr.] is an important legume crop that provides high-quality vegetable protein and oil. In general, oil makes up around 19% of the dry weight of a soybean seed, with the five most abundant fatty acids being palmitic, stearic, oleic, linoleic, and linolenic acids. Storage proteins make up roughly 40% of the dry seed weight,

Advances in Botanical Research, Volume 102
ISSN 0065-2296
https://doi.org/10.1016/bs.abr.2022.03.002

mainly consisting of β-conglycinin (7S globulin) and glycinin (11S globulin). The content and composition of soybean seeds are valuable traits controlled by a complex genetic background and various growth conditions. Environmental stresses, including drought, temperature, and salinity, could severely reduce seed productivity and alter the chemical compositions. Understanding the genetic regulations of oil and protein biosynthesis and how they are affected by abiotic stresses is imperative to enhancing the seed quality. This chapter reviews how oil and protein contents and compositions are affected by environmental conditions, summarizes the genes involved in protein and oil biosynthesis/metabolism, emphasizes the genetic mechanisms of oil/protein accumulation, and offers strategies for improving soybean production both quantitatively and qualitatively.

1. Introduction

Protein and oil are the two principal seed constituents that make soybean (*Glycine max*) an important crop. Due to population growth and strong demand for oil and feed protein, the soybean acreage has increased globally. Soybean plantation areas in the United States was around 87.6 million acres (USDA, 2022) in 2021. Between 2000 and 2020, soybean production more than doubled, from 161 to nearly 353 million tons, and soybean oil production increased from 25.6 to 59.9 million tons (FAOSTAT, 2020). The United States Department of Agriculture (USDA) food composition data show that soybean seeds contain 36.49% protein, 19.94% oil, of which more than half are unsaturated fatty acids (11.2%), 30.16% carbohydrates, and 9.3% fiber (FoodData [USDA], 2022). The protein concentration in soybean ranges from 341 to 568 g/kg of total seed weight, and oil content ranges from 83 to 279 g/kg (Wilson, 2004). Increasing the oil accumulation and protein content in soybean seeds has been a major objective of soybean breeding programs for decades (Van & McHale, 2017). However, the negative correlations in soybean seeds between protein content and oil content as well as yield render it challenging to increase soybean protein content while maintaining the desired seed oil level and seed yield (Patil et al., 2017; Rincker et al., 2014).

The wide range in protein and oil concentrations is not only genetically controlled but is also affected by multiple conditions like agronomic practices, soil conditions, disease pressure and environmental factors including climate change, increasing average temperature and decreasing water availability (Alsajri et al., 2020; Miransari, 2016a). Genetic mapping studies utilizing biparental populations have been carried out for the identification of quantitative trait loci (QTLs) for soybean seed composition. Nowadays, over 700 QTLs controlling soybean seed oil and protein contents have been identified using accessions with different genetic backgrounds, and 57 QTLs have been confirmed (Van & McHale, 2017). The biosynthesis of oil has been well studied, and the fatty acid (FA) composition is found to be highly regulated via the spatial separation of biosynthetic steps within different subcellular compartments.

A better understanding of the biosynthetic genes and regulatory mechanisms of soybean protein and oil biosynthesis could facilitate the breeding of varieties with high protein or oil content via genetic engineering approaches. Thus, this chapter describes the soybean seed compositions of protein and oil, how seed contents are altered under different environmental conditions, and summarizes the biosynthetic genes and genetic regulations determining protein and FA compositions. Also, we highlight the strategies for improving protein and oil contents in soybean from multiple approaches.

2. Oil and protein contents in soybean seeds

2.1 Oil content and fatty acid composition of soybean seeds

As a dominant oilseed, the oil content in dry soybean seeds averages around 19% and varies from 6.5% to 28.7% depending on the soybean varieties and growth conditions (Greenberg & Hartung, 1998). The yield of soybean oil accounted for nearly 60% of the world's oilseed production in 2019/2020 (SoyStats, 2022). Soybean oil mainly consists of high amounts of fatty acids (FAs), tocopherols and triacylglycerols (TAGs) (Jokić et al., 2013). The fiver major FAs found in soybean seeds are palmitic acid (C16:0), stearic acid (C18:0), oleic acid (C18:1), linoleic acid (C18:2), and linolenic acid (C18:3), with concentrations accounting for 10, 4, 18, 55, and 13%, respectively (Clemente & Cahoon, 2009; Lee, Bilyeu, & Shannon, 2007).

The soybean oil quality is determined by the FA composition. Modification of the ratios between saturated FAs (C16:0 and C18:0), oleic acid, and linolenic acid is an important goal in improving the oil quality. Nowadays, soybeans are bred to have higher amounts of monounsaturated FAs and lower amounts of polyunsaturated FAs (PUFA). Unsaturated FAs play an important role in immune system regulation, blood clotting, cholesterol metabolism, and the structure of membrane phospholipids in the brain and retina (Djuricic & Calder, 2021). Foods high in saturated FAs are associated with high cholesterol and high risk for heart disease. Lowering the content of saturated FAs makes soybean oil more acceptable for consumers and food manufacturers (Wilson, 2004). Moreover, PUFAs such as linoleic (C18:2) and linolenic acids (C18:3) are susceptible to oxidation due to a high number of double bonds, which reduce the oxidative stability and shelf life of the final product. The common way to improve the stability of soybean oil is to increase the saturation by hydrogenation. Yet, there is also a major concern about the trans-fats produced during the process. Thus, increasing the oleic acid (C18:1) content is more favorable for both health and product stability concerns (Yao et al., 2020). It is found that the linolenic acid concentration is the most stable across environments. Soybean lines with stable oleic acid and linolenic acid across environments could thus be used as a source of germplasms with a desirable FA composition (Oliva et al., 2006). Studies on soybean seed fatty acid profiles of various germplasms under different growth conditions have been conducted (Abdelghany et al., 2020; Anwar, Kamal, Nadeem, & Shabir, 2016; Ciabotti et al., 2019; Do Wee, Hashiguchi, Anai, Suzuki, & Akashi, 2017; La et al., 2019; Song et al., 2016). Breeding programs that can make use of this information to alter the oil composition is a priority in improving soybean oil for both food and industrial uses.

2.2 Protein content and composition of soybean seeds

Soybean is grown for its production of seeds with a rich protein content and substantial nutritional value for human and animal consumption worldwide. On average, the protein content accounts for around 40% of the dry weight of soybean seeds, consisting mainly of storage proteins (Krishnan, Natarajan, Mahmoud, & Nelson, 2007; Medic, Atkinson, & Hurburgh, 2014). Glycinin (11S globulin) and conglycinin (7S globulin) are two major components, constituting around 70% of the total storage proteins (Liu et al., 2007; Medic et al., 2014; Paek, Imsande, Shoemaker, & Shibles, 1997; Panthee et al., 2004). Soybean seeds also contain a considerable amount of Bowman-Birk inhibitors (BBI) and Kunitz trypsin inhibitors (KTI). During food processing, inactivation by cleavage or aggregation of these inhibitors is needed due to their inhibitory effects on protein digestion by trypsin and chymotrypsin (Gillman, Kim, & Krishnan, 2015). In spite of the adverse effects on the nutritional value, these protease inhibitors were reported to possess anticarcinogenic activities (Kennedy, 1998; Kobayashi, 2013).

Soybean seeds also comprise about 1% lectin, which is a class of glycoproteins that specifically bind to carbohydrates (Sharon & Lis, 2002). Similar to the protease inhibitors, lectins have been shown to adversely impact nutrient uptake in animals (Friedman & Brandon, 2001). Although lectins cause damaging effects when bound to the intestinal epithelium, they have also been reported to have potential anticarcinogenic activity and therapeutic uses (de Mejia, Bradford, & Hasler, 2003; George, Bhide, Thengane, Hosseini, & Manjaya, 2008; Kik et al., 1989). Besides, the immunodominant allergen, P34, constitutes around 1% of the soybean seed proteins. P34 was identified as a 34-kDa protein localized in the protein storage vacuoles (Kalinski, Melroy, Dwivedi, & Herman, 1992). Around 2% of the soybean seed proteome is contributed by oleosins, which play an essential role in maintaining the stability of oleosomes and thus affect lipid storage (Schmidt et al., 2011). The presence of active ureases also reduces the nutritional value of soybean meal especially when mixed with other food sources in animal feed (Real-Guerra, Staniscuaski, & Carlini, 2013).

The soybean seed protein content is tightly regulated, as numerous early attempts have failed to utilize the soybean system for the significant accumulation of recombinant proteins, such as vaccine antigens, milk protein and growth hormone (Herman, 2014; Lau & Sun, 2009; Oakes, Bost, & Piller, 2009; Philip, Darnowski, Maughan, & Vodkin, 2001; Russell et al., 2005). Also, the innate program in soybean to rebalance the seed protein composition was functional even the synthesis of major storage proteins was disrupted. In transgenic plants with the α and α' subunits of conglycinin genes silenced, glycinin, proglycinin and a precursor of P34 were significantly accumulated in seeds, resulting in a similar seed protein content to the wild type. The accumulated proteins were also stored in protein bodies in addition to the normal storage vacuoles (Kinney, Jung, & Herman, 2001). Moreover, knocking down both conglycinin and glycinin synthesis did not lead to a significant change in the total protein content. A detailed investigation in the protein composition identified that the lack of conglycinin and glycinin was compensated by the accumulation of other storage proteins such as lectin, P34 and KTI, while increases in the abundance of other proteins may also have participated in the rebalancing. Further analysis also indicated that the amino acid profile of the mutant seeds was similar to that of the wild type (Schmidt et al., 2011).

3. Alterations in oil and protein contents and compositions under diverse environmental conditions

Crop production is hampered by environmental stresses such as drought, flooding, salinity and heat. Stresses adversely affect plant physiology and metabolism, inhibiting plant growth and eventually leading to deterioration in seed quality and yield. Indeed, environment has been reported as the major factor influencing protein concentration in soybean (Assefa et al., 2019). It is well documented that a negative correlation exists between oil and protein contents, and when oil concentration decreases, the protein content increases (Bandillo et al., 2015; Chung et al., 2003; Cober, Voldeng, & H., 2000; Wilcox, 1998). These environmental factors are also likely the cause of variations between seed protein and oil concentrations from different geographic locations. Numerous studies have reported the impacts of different stress factors on the oil and protein production of soybean seeds and they are summarized in this section.

3.1 Changes in oil and protein contents due to drought stress

Drought stress is a severe abiotic stress factor limiting soybean production worldwide. It is reported that soybean yield is most sensitive to drought stress during the pod filling stage of development (Rodrigues et al., 2012). The drought effects at the pod filling stage were first assessed in the soybean cultivars "Gnome" and "Hodgson 78." Both protein and oil contents were affected. Drought stress increased seed protein content by around 10% and the protein contents in both cultivars increased linearly with the increase in drought stress intensity (Dornbos & Mullen, 1992). On the other hand, the oil content of "Gnome" decreased up to 12.4% mainly due to a reduction in oleic acid content (Dornbos & Mullen, 1992).

Lipids are a major form of carbon and energy storage in the seed for germination. Drought causes oil content reduction owing to the inhibition on the production of digestible carbohydrates such as glucose, fructose, and sucrose. Drought also affects the FA composition due to reduced unloading of sugars from the stem to developing seeds (Bellaloui, Hu, Mengistu, Kassem, & Abel, 2013). A study on the effects of drought treatment on the slow-wilting genotypes, NTCPR94-5157 and N04-9646, showed that seed protein, oleic acid and sugar contents increased whereas seed oil, linoleic and linolenic acids decreased, and concluded that the accumulation of oleic acid could be an adaptive mechanism to maintain cell turgor under drought stress (Bellaloui, Gillen, et al., 2013). Seeds of the soybean cultivar, Hutcheson, grown under drought conditions (−90 to −100 kPa soil water potential) had higher total oil and oleic acid concentrations and lower linoleic and linolenic concentrations compared to irrigated soybeans (−15 to −20 kPa soil water potential) (Bellaloui, Hu, et al., 2013). A study on the irrigation effect on seven soybean germplasms found that it was not significant on the PUFA concentration but it changes the composition depending on the genotypes (Lee, Oliva, Sleper, & Shannon, 2008).

The effect of drought stress on the reduction of seed oil is more significant than on the seed protein content (Miransari, 2016a). The explanation was that drought stress could trigger the remobilization of nitrogen sources from the leaf (Turner, Davies, Plummer, & Siddique, 2005). However, most of the later studies reported a negative impact of drought stress on seed protein content (Rotundo & Westgate, 2009). It was found that seeds from the cultivars M7, M9 and Hobit stressed by drought at the seed filling stage contained significantly lower protein contents than the control, while the same stress applied at earlier stages such as the flowering and podding stages had no significant effects on seed protein content (Maleki et al., 2013). In line with this finding, Nakagawa et al. (2018) reported that the seed protein content was >50% lower in drought-stressed soybean cultivar "Fukuyutaka" compared to that in control. A similar reduction was reported in another study on "Sinara" and "Sigalia," although the drought treatment was applied at the flowering stage (Anda, Simon, Soos, da Silva, & Menyhart, 2021). The seed protein contents of "Asgrow AG5332" and "Progeny 5333RY" were also reported to correlate positively with the degree of drought stress in terms of soil moisture (Wijewardana, Reddy, & Bellaloui, 2019).

In a study investigating the different effects drought had on the components of seed storage proteins between a drought-tolerant (BOSA) and a drought-sensitive (L 121) genotype under drought stress (Blanusa, Stikic, Vucelic-Radovic, Barac, & Velickovic, 2000), it was found that drought affected the stage at which the accumulation of the β subunit of conglycinin took place in each genotype, although the differences were not observed at consistent stages in two experimental years (Blanusa et al., 2000). The alterations in the accumulation of the β subunit of conglycinin were also reported in another study on the cultivated soybean Harosoy, in which the accumulation was observed at a later developmental stage when under drought stress (Samarah, Mullen, Cianzio, & Scott, 2006). Also, an earlier accumulation of heat-stable dehydrin proteins was observed under greenhouse conditions but not in the field (Samarah et al., 2006).

3.2 Changes in oil and protein contents under salinity stress

Seed yield, oil, FA composition and protein production are all affected by salinity (Miransari, 2016b). Under salt stress, the percentage of oil in soybean seeds declined while the total oil content per seed increased during the seed filling stage (Ghassemi-Golezani & Farhangi-Abriz, 2018). Irrigation with 6000 mg/L saline solution on the soybean cultivar Giza 111 led to a significant decrease in seed yield (the product of pod weight and seed number/plant) and oil content compared to the control plants irrigated with tap water (Sadak et al., 2020). It was found that different salt concentrations might have opposite effects on the seed protein content (Phang, Shao, & Lam, 2008; Wan, Shao, Chen, & Yan, 2002).

A life-long exposure to mild salinity at 4 dS/m had no significant effect on the proportion of protein in the seeds of cultivar M7, while higher salinity conditions at 7 and 10 dS/m reduced the proportion of protein in seeds (Farhangi-Abriz & Ghassemi-Golezani, 2016). The seedlings grown under all the above salinity conditions had lower overall yield and total protein production per plant. Although most of the amino acids had higher concentrations per gram dry weight, those of leucine and valine were reduced under higher salinity (Farhangi-Abriz & Ghassemi-Golezani, 2016). Alterations in the metabolic pathways of branched-chain amino acids were reported as one of the key features of salt stress adaptation in soybean and therefore may affect the resultant accumulation of these amino acids in seeds (Liu et al., 2019; Yung et al., 2022).

3.3 Effects of heat stress on oil and protein contents in soybean seeds

Heat stress harms almost all the stages of soybean development from germination to maturity, and results in a substantial decrease in crop yield. Heat stress inhibits photosynthetic activities and hastens senescence prematurely, which decreases the synthesis and distribution of assimilates to seeds (McDonald & Paulsen, 1997). Increasing temperature can lead to a reduced number of cells per cotyledon, later onset of seed fill and maturity, and lower seed weight and size (Tacarindua, Shiraiwa, Homma, Kumagai, & Sameshima, 2012). The soybean oil content is determined by both soybean varieties and temperature conditions. A study on the temperature effect on

20 soybean cultivars across 60 locations found that seed oil concentration increased with temperature (14–28°C) (Piper & Boote, 1999). In an earlier study, a slight increase in seed protein concentration was observed in "Hodgson 78" grown at 27, 29, 33, and 35°C (Dornbos & Mullen, 1992). At 29°C the percentage of protein increased by 4% and oil content decreased by 2.6%, compared to those at 35°C. High temperature decreased the proportion of unsaturated acid including linoleic and linolenic acids in seeds but increased oleic acid concentration (Dornbos & Mullen, 1992). A study on different soybean varieties including Hutcheson, DT97-4290, AG 4403 and AG 4903 showed that growing at higher temperatures (36/28°C, day/night) resulted in higher total oil and oleic acid concentrations and lower linoleic and linolenic concentrations than those grown at normal temperatures (25/20°C, day/night) (Bellaloui, Reddy, & Mengistu, 2015).

The effect of elevated temperatures on the protein concentration in soybean seeds was variable among cultivars (Ortiz, De Smet, Sozzani, & Locke, 2022; Piper & Boote, 1999). Generally, an elevated temperature led to an increase in protein concentration and a decrease in total protein content in seeds (Rotundo & Westgate, 2009). Similar to drought stress, heat stress applied from the seed filling stage onwards led to a lower percentage of protein in the seeds of cultivar "Fukuyutaka" (Nakagawa et al., 2020). A severe heat regime of 42°C/26°C (day/night) cycles starting from the flower initiation stage was reported to impair the accumulation of glycinin and conglycinin in both heat-tolerant (DS25-1) and sensitive cultivars (DT97-4290) (Chebrolu et al., 2016). When these cultivars were exposed to moderate heat stress (36°C/24°C day/night cycles), the levels of proline and leucine in seeds were higher, while the levels of aspartate were lower, than those grown in the optimal condition (Chebrolu et al., 2016). Similarly, a lower accumulation of the β subunit of conglycinin, lipoxygenase1, sucrose binding proteins, P34 and Bowman-Birk inhibitor was observed in the seeds of "DS25-1" and "DT97-4290" exposed to the same severe heat stress condition in a later study (Krishnan, Kim, Oehrle, Smith, & Gillman, 2020). As expected, the seeds from both cultivars contained higher levels of heat shock proteins HSP70 and HSP17.6 under severe heat stress (Krishnan et al., 2020). As shown in the transmission electron microscopy images, the cotyledon cells were plasmolyzed and the protein storage vacuoles were destroyed in the sensitive cultivar (DT97-4290) under severe heat, whereas the ultrastructure in the tolerant cultivar (DS25-1) was much less affected, although the plant storage vacuoles were only partially filled (Krishnan et al., 2020).

3.4 Other abiotic stresses and factors

The seed protein and oil contents of soybean are also impacted by cold and flooding stresses. Under low temperature conditions, various studies have shown a decrease in the oil content in soybean seeds (Alsajri et al., 2020; Khan et al., 2011; Rotundo, Miller-Garvin, & Naeve, 2016). However, cold stress was found to have no significant effect on the seed protein content, although the exposure to long-term cold stress (12°C/6°C day/night cycles for 9 days) after seed sowing resulted in higher seed yields in 12 out of 16 selected cultivars (Staniak, Stepien-Warda, Czopek, Kocira, & Baca, 2021). Similarly, flooding stress was reported to cause significant changes in other seed components such as fatty acids and isoflavones but not in the protein content (VanToai et al., 2012). In addition, phosphate deficiency was found to reduce the protein yield per plant while an enhanced carbon dioxide concentration at 800 μmol mol^{-1} exerted an opposite effect (Singh, Barnaby, Reddy, & Sicher, 2016).

4. Genetic strategies for improving oil and protein contents

4.1 Genetic mapping and identification of genes associated with oil and protein contents

Seed oil content is a quantitative trait that is governed by multiple genes. In the soybean genome, there are more than 1100 annotated acyl lipid genes (Schmutz et al., 2010), about double that in the model plant Arabidopsis. This is probably due to the recent whole-genome duplication of the soybean genome and could potentially explain the high oil content in the soybean seed. Thus far, QTL mapping and genome-wide association study (GWAS) have identified a large number of genomic regions controlling the soybean seed oil content (Cao et al., 2017; Di et al., 2022; Eskandari, Cober, & Rajcan, 2013; Huang et al., 2020; Hwang et al., 2014; Karikari et al., 2019; Li, Xu, Yang, & Zhao, 2019; Liu et al., 2020; Priolli, Campos, Stabellini, Pinheiro, & Vello, 2015; Silva et al., 2021; Zhang et al., 2018; Zhao et al., 2019). One of the main goals of soybean breeding is to increase the seed oil content. It has been found that a large portion of a selected region in the soybean genome is occupied by overlapping QTL regions related to oil content (Zhou et al., 2015), some of which contain known biosynthetic genes for fatty acids.

Soybean seed protein and oil contents often share overlapping QTL regions. As previously reviewed, improvement in the soybean protein content is challenging as soybean oil content and yield are negatively correlated with its protein content (Bandillo et al., 2015; Chaudhary et al., 2015; Pandurangan et al., 2012; Patil et al., 2017). Since the selection for a higher and better oil content is a higher priority than for the protein content, any improvements in the protein content often takes a backseat in breeding programs. Analyses on the breeding efforts over the past 60 years revealed that efforts were biased toward selecting for higher oil content rather than protein content in the hope of improving nutritional values (Mahmoud et al., 2006). Wild soybean or its recombinant inbred lines serve as a good reservoir of the genetic variations for different traits. Based on restriction fragment length polymorphism (RFLP) analyses, *G. max* contains alleles that are associated with higher oil contents than in *G. soja*, and vice versa for the alleles associated with protein contents (Diers, Keim, Fehr, & Shoemaker, 1992). Wild soybean often contains specific QTLs that are associated with higher protein contents and higher heterogeneity of the proteins (Natarajan, Xu, Bae, Caperna, & Garrett, 2006; Sebolt, Shoemaker, & Diers, 2000). The wild soybean usually contains a higher 11S:7S ratio than the cultivated germplasms, leading to the suggestion that some regulatory genes or elements in the cultivars might have been altered during the domestication of soybean (Kwanyuen, Pantalone, Burton, & Wilson, 1997). These qualities of the wild soybean enable it to be a good genetic resource for improved protein levels.

Undoubtedly, cultivated soybean might also possess certain desirable traits after years of breeding and selection. It is reported that the wild soybean genotype contains more abundant allergenic and antinutritional proteins such as the subunits of β-conglycinin, glycinin, and KTI than the cultivated genotypes (Natarajan et al., 2006). Some *G. max* accessions such as PI603570A and PI567476 were reported to have a lower level of the allergen P34 (Bilyeu, Ren, Nguyen, Herman, & Sleper, 2009; Joseph, Hymowitz, Schmidt, & Herman, 2006; Koo et al., 2011). Some of the cultivars carry four amino acid substitutions in the P34 sequence, and some only express a truncated P34 protein (Bilyeu et al., 2009; Joseph et al., 2006; Koo et al., 2011). Some of the cultivated germplasms might show high protein contents upon selection in some geographical regions. PI407788A and "Enrei" contain the (CC+) *GmSWEET39* allele at chromosome 15 for higher protein content in South Korea and Japan respectively (Brzostowski & Diers, 2017; Shimomura et al., 2015; Zhang et al., 2020). Other successful breeding of the high-protein elites, such as R95-1705, R05-1415, and R05-1772, were reported to obtain stable high-protein-content germplasms by crossing (Chen, Ishibashi, Dombek,

& Rupe, 2011; Chen, Sneller, Ishibashi, & Cornelious, 2008). Thus, the breeding program for developing protein-rich soybean should include these different lines as the parents.

4.2 Gene regulations on the content, nutrition and stability of soybean oil

4.2.1 The FA biosynthesis pathway in soybean seeds

The biosynthesis of oils in the seed has been well described (Bates, Stymne, & Ohlrogge, 2013). In brief, FAs are assembled in the plastid. Acetyl-CoA is first converted to malonyl-CoA by acetyl-CoA carboxylase (ACCase). The acyl chain is elongated on the acyl carrier protein (ACP) by fatty acid synthase (FAS) by adding two carbon atoms every time until 16 carbon acyl-ACP (C16:0) is produced. The 16:0-ACP may be converted further to 18:0-ACP by ketoacyl-ACP synthase II (KASII) and may subsequently be desaturated by stearoyl-ACP desaturase (SAD). Desaturation of the fatty acids may also take place through the action of fatty acid desaturase (FAD) to form polyunsaturated FAs (PUFAs). The free fatty acids (C16:0, C18:0, C18:1, etc.) are then released from ACP by acyl-ACP thioesterase (FAT) and exported to the endoplasmic reticulum. There, in the de novo TAG synthesis pathway (Kennedy Pathway), free fatty acids will be incorporated into glycerol-3-phosphate (G3P) by acyl-CoA:G3P acyltransferase (GPAT) to form lyso-phosphatidic acid (LPA). Further additions of FA to LPA by acyl-CoA:LPA acyltransferase (LPAAT) will yield phosphatidic acid (PA), which will be converted to diacylglycerol (DAG) by PA phosphatase (PAP). At the final step, acyl-CoA:DAG acyltransferase (DGAT) catalyzes the formation of triacylglycerol (TAG) from DAG through the addition of the last FA. Alternatively, TAG can also be produced through the phosphatidylcholine (PC) pathway, although this is not the major source of TAG in soybean seeds. The TAG in the seed is mostly stored in oleosomes (oil bodies), which are wrapped in a single layer of membrane compositing of phospholipids and proteins such as oleosins (Nikiforidis, 2019).

4.2.2 Regulatory genes that can improve the FA composition of soybean seeds

There are two major goals in improving the soybean seed oil content. The first is to increase the total oil content of the seed, ideally without sacrificing the overall yield. The second is to "improve" the oil quality by increasing the beneficial fatty acids while reducing those detrimental ones. Thus far, a number of genes, including not only the biosynthetic ones, but also a number of transcription factors from soybean have been characterized.

The most common genetic strategy to reduce the PUFA content in seeds is to impair the function of fatty acid desaturase (FAD) to create a high-oleic acid low-linoleic acid soybean. In the past few decades, different molecular techniques have been used to manipulate *FAD* genes in soybean, including induced mutagenesis (Anai et al., 2008), RNA interference (RNAi) (Yang et al., 2018), transcription activator-like effector nucleases (TALENs) genome editing (Demorest et al., 2016; Haun et al., 2014), CRISPR-cas9-mediated genome editing (al Amin et al., 2019; Wu et al., 2020). FAD2 and FAD3 desaturate oleic acid and linoleic acid, respectively. The mutation of either copy of *FAD2* genes (*GmFAD2-1A* and *GmFAD2-1B*) can increase the seed oleic acid content to 50% of the total and reduce linoleic acid down to 30% (Anai et al., 2008). The *Gmfad2-1A Gmfad2-1B* double mutant can further increase oleic acid to 80% and reduce linoleic acid to less than 5% (Pham, Shannon, & Bilyeu, 2012). There are three *FAD3* genes in the soybean genome, namely *GmFAD3A*, *GmFAD3B*, and *GmFAD3C* (Bilyeu, Palavalli, Sleper, & Beuselinck, 2005). The *Gmfad3A Gmfad3C* double mutant has a 3% linolenic acid content (Bilyeu et al., 2005), while the triple mutant has the linolenic acid content as low as 1% (Pham et al., 2012). Stacking the mutations of *fad2* and *fad3* together generated lines with high oleic acid, low linoleic acid and low linolenic acid (Demorest et al., 2016; Pham et al., 2012). A high-oleic low-linolenic soybean produced by TALEN genome editing has been deemed non-regulated by USDA in 2020, making it the first commercially available genome-edited crop.

Acyl-ACP thioesterase (FAT) is responsible for the release of saturated FAs from ACPs. TILLING-by-Sequencing+ identified a number of soybean mutants with amino acid substitutions in different *FAT* genes (Zhou et al., 2021). These mutations cause a reduction in saturated FAs and an increase in oleic acid (Zhou et al., 2021). The loss of function of either *GmFAT1a* or *GmFAT1b* can reduce ~50% palmitic and ~30% stearic acid contents without altering the unsaturated FA content (Ma, Sun, Whelan, & Shou, 2021). Although the *Gmfat1a Gmfat1b* double mutant has significantly reduced saturated FA contents in the leaf, it also suffers from growth retardation and male sterility, which is not practical for crop production (Ma et al., 2021).

Through genetic mapping, a mutation disrupting the donor splice site of *GmKASIIIA* (*Glyma09g41380*) was found to be associated with reduced palmitic acid content (Cardinal et al., 2014). Given that KASIII participates in the reaction between acetyl-CoA and malonyl-ACP, the reason why the mutation of *GmKASIIIA* should lead to a reduction in palmitic acid is a mystery. However, both the mutation of *KASIIIA* (Cardinal et al., 2014) and the ectopic expression of *KASIIIA* caused a reduction in the total oil content (Dehesh, Tai, Edwards, Byrne, & Jaworski, 2001). This suggests that the ratio of KASIII to malonyl-ACP may be crucial for efficient FA biosynthesis.

In some food applications, oils with a higher saturated FA content are needed. To fulfill this demand, soybean lines with high stearic acid are also being generated. A germplasm with as high as 35% stearic acid has been found (Pantalone, Wilson, Novitzky, & Burton, 2002). Yet, the cause for the high stearic acid content remains unknown. Through mutant screening, a mutation in a delta-9-stearoyl-acyl-carrier protein desaturase (SACPD-C)-encoding gene was found to increase the stearic acid content in soybean seeds (Carrero-Colon, Abshire, Sweeney, Gaskin, & Hudson, 2014). SPACD-C is responsible for converting stearic acid to oleic acid. Thus, impairing the function of SPACD-C can increase stearic acid at the expense of oleic acid (Carrero-Colon et al., 2014). This is also true in high-oleic acid lines. Stacking the *sacpd* mutation with the *fad2-1a fad2-1b* double mutation caused a 10% increase in stearic acid but also a 10% decrease in oleic acid in the triple mutant compared to the *fad2-1a fad2-1b* double mutant (Bilyeu et al., 2018). An increase in stearic acid content can also be achieved by the ectopic expression of the stearoyl-ACP thioesterase-encoding gene (Park, Graef, Xu, Tenopir, & Clemente, 2014). The overexpression of a codon-optimized stearoyl-ACP thioesterase from mangosteen in soybean led to a significant enhancement in the stearic acid content in the seed from 3% in the wild type to around 10% in the transgenic lines (Park et al., 2014). Nevertheless, a slight reduction in other FAs was also observed in these transgenic lines (Park et al., 2014).

4.2.3 Transcription regulation of oil accumulation in soybean seeds

Oil biosynthesis in the soybean seed is governed by a complex transcription regulation network. WRINKLED1 (WRI) is a transcription factor in the APETALA2/ethylene responsive element-binding protein (AP2/EREBP) family and is responsible for the biosynthesis of storage compounds in Arabidopsis. Its homologs in soybean appear to play a central role in oil accumulation in the seed. There are two WRI1-encoding genes in the soybean genome, *GmWRI1a* (*Glyma15g34770/Glyma.15G221600*) and *GmWRI1b* (*Glyma08g24420/Glyma.08G227700*). *GmWRI1a* is highly expressed in maturing soybean seeds (Chen et al., 2018). Overexpression of *GmWRI1a* with a seed-specific promoter resulted in significantly higher total oil and fatty acid contents in the seed, accompanied by a bigger seed size and higher total seed weight per plant and per hundred seeds (Chen et al., 2018). The most significant changes in fatty acid contents were the significant increases in oleic acid (18:1) and linoleic acid (C18:2). Results from gene expression study and electrophoretic mobility shift assay (EMSA) showed that GmWRI1a is able to act on the AW-box on the promoter of genes encoding ketoacyl-ACP synthase (KASII and KASIII), enoyl-ACP reductase, stearoyl-ACP desaturase (SAD), acyl-ACP thioesterase, and long-chain acyl-CoA synthetase (Chen et al., 2018).

Another seed-specific DREB-type AP2 transcription factor, GmDREBL, also governs the oil accumulation in soybean seeds (Zhang et al., 2016). The expression of *GmDREBL* is positively regulated by GmABI3 and GmABI5 (Zhang et al., 2016). Overexpression of *GmDREBL* in soybean hairy roots increased the expression of *GmWRI1a*, but not that of *GmWRI1b* (Chen et al., 2018; Zhang et al., 2016). The ectopic expression of *GmDREBL* in Arabidopsis enhanced the expressions of *KAS1*, *FATA*, *LPD1* and *SAD* (*At5g16240*) through the activation of *WRI1* (Zhang et al., 2016). The activation of these genes resulted in larger seeds and higher FA contents (C18:1, C18:2, C18:3, and C20:1) in the seed (Zhang et al., 2016). Evidence also showed that *GmDREBL* may be subjected to selection during domestication, as cultivated soybeans, which have high oil contents, tend to have a higher expression of *GmDREBL* (Zhang et al., 2016).

A nuclear transcription factor Y subunit A-encoding gene (*GmNYFA*, *Glyma02g47380*) was differentially expressed between the developing seed of cultivated soybeans and wild soybeans (Lu et al., 2016). This gene is also located within a QTL for seed oil content (Lu et al., 2016). Similar to *GmDREBL*, the overexpression of *GmNYFA* in Arabidopsis also enhanced the expressions of *WRI1*, *KAS1*, *LPD1*, *FATA*, *SAD*, and a gene encoding thioesterase, resulting in an elevation in seed FAs (C16:0, C18:1, C18:2, C18:3, and C20:1) and oil content (Lu et al., 2016). It is probable that GmNYFA acts directly on the CCAAT box on the *WRI1* promoter to boost the expression of *WRI1* and indirectly increase the expression of the other oil biosynthetic genes (Lu et al., 2016). As evidence of a potential target of soybean domestication, the promoter of *GmNYFA* in cultivated soybeans and some wild soybeans harbor a 1500-bp insertion. This insertion is associated with the higher expression of *GmNYFA* and oil content in cultivated soybeans (Lu et al., 2016).

The expression of *GmWRI1a* was found to be regulated by GmLEC2a, a B-box-containing transcription factor expressed mainly in the flower, pod and seed (Manan et al., 2017). The ectopic overexpression of *GmLEC2a* can enhance the total FA content in transgenic Arabidopsis seeds and soybean hairy roots (Manan et al., 2017). Comparisons between the transcriptomes of the wild type and transgenic hairy roots expressing *GmLEC2a* revealed that GmLEC2a is a positive regulator for a group of transcription factors involved in TAG biosynthesis, including *GmWRI1a*, *GmWRI1b*, *GmDof11*, *GmFUS3*, *GmABI3* and *GmLEC1*. The expressions of TAG biosynthetic genes, including *KAS*, *GPAT*, *DGAT*, *PGAT*, *PAP* and so on, were also regulated by the overexpression of *GmLEC2a*. In contrast, sucrose synthase genes were largely suppressed to shut off the biosynthesis of starch in order to redirect the materials for lipid biosynthesis (Manan et al., 2017).

GmZF351 encodes a tandem CCCH zinc finger protein in soybean, functioning as a transcription activator that governs oil accumulation in seed (Li et al., 2017). Ectopic overexpression of *GmZF351* can increase FA and TAG accumulation in the seed of both transgenic Arabidopsis and soybean (Li et al., 2017). GmZF351 can directly activate the expression of *WRI1*, *Biotin Carboxyl Carrier Protein 2* (*BCCP2*), *KASIII*, *DGAT1*, *OLE2* in soybean (Li et al., 2017). Through the activation of *WRI1*, lipid biosynthetic genes downstream of *WRI1* are also regulated by GmZF351 (Li et al., 2017). In general, the expression of *GmZF351* is higher in cultivated soybean than in wild soybean and it is positively correlated with the lipid content in seeds (Li et al., 2017). An analysis of the promoter of *GmZF351* identified four major haplotypes, one for cultivated soybean and three for wild soybean (Li et al., 2017). This suggests that domestication may have selected a haplotype of *GmZF351* with a higher expression that can in turn increase seed oil content.

Two soybean DNA binding with one finger (Dof) transcription factors, GmDof4 and GmDof11, were also found to be involved in lipid accumulation (Wang et al., 2007). The expression of *GmDof4* is predominantly in flower and pod while that of *GmDof11* is constitutively expressed in leaf, flower and pod (Wang et al., 2007). As mentioned in a previous paragraph, *GmDof11*, among other transcription factors, is also a target of GmLEC2a (Manan et al., 2017). The ectopic expressions of *GmDof4* and *GmDof11* can elevate the total FA contents in Arabidopsis (Wang et al., 2007), *Brassica napus* L. (Sun et al., 2018) and *Chlorella ellipsoidea* (Zhang et al., 2014). Since the gene expression study was done in a heterologous transgenic system, it may not truly reflect the functions of GmDof4 and GmDof11 in soybean (Sun et al., 2018; Wang et al., 2007; Zhang et al., 2014). However, the available evidence suggests that GmDof4 and GmDof11 may target different sets of genes related to TAG biosynthesis and accumulation. While only GmDof4 was shown to interact with *accD* that encodes the β-subunit of ACCase in Arabidopsis (Wang et al., 2007), the ectopic expressions of both *GmDof4* and *GmDof11* could enhance the expression of *accD* in *B. napus* L. (Sun et al., 2018). Furthermore, GmDof11 was shown to positively regulate a long-chain-acyl CoA synthetase-encoding gene in Arabidopsis while GmDof4 was shown to act on *FAB2*, *FAD3* and *FAD8* in *B. napus* L. (Sun et al., 2018; Wang et al., 2007).

GmbZIP123 is expressed highly in multiple organs, especially root and developing seed (Song, Li, et al., 2013). The translation of the Arabidopsis homologs of *GmbZIP123* is regulated by sucrose due to the presence of upstream open reading frames (uORFs) in their transcripts (Song, Li, et al., 2013). Two uORFs are also found in the transcript of *GmbZIP123*, implying that the translation of *GmbZIP123* may also be regulated by sucrose. The ectopic overexpression of *GmbZIP123* in Arabidopsis significantly increased the FA content in seeds without altering the 1000-seed weight (Song, Li, et al., 2013). Unlike *GmWRI1a* which altered only oleic acid and linoleic acid contents, the overexpression of *GmbZIP123* altered all major FA contents (C16:0, C18:1, C18:2, C18:3 and C20:1) (Song, Li, et al., 2013). RNA-seq analyses showed that instead of acting directly on FA biosynthetic genes, GmbZIP123 enhanced the expression of sucrose transporter-encoding genes including *SUC1*, *SUC5*, *cell wall invertase 1* (*cwINV1*), *cwINV3*, and *cwINV6* in transgenic Arabidopsis (Song, Li, et al., 2013). Consistent with the expression study, the glucose, fructose, and sucrose levels were all higher in the developing seeds of the transgenic line than those in the wild-type plant (Song, Li, et al., 2013). These results suggested that GmbZIP123 can enhance FA accumulation by allocating carbon resources to the developing seeds.

GmMYB73 is highly expressed in the flower, root, and early developing seed in soybean (Liu et al., 2014). The ectopic overexpression of *GmMYB73* in Arabidopsis caused multiple phenotypic changes including the repression of trichome formation, increase in seed size, and enhanced oil accumulation (Liu et al., 2014). The overexpression can also improve the FA accumulation in *Lotus japonica* seed and soybean hairy roots (Liu et al., 2014), suggesting the function of GmMYB73 is conserved in these species. The overexpression of *GmMYB73* functionally resembled its Arabidopsis homolog,

CAPRICE-like proteins (CPC and TRY), by derepressing the expression of *phospholipase D alpha 1* (*PLDα1*) and other genes through the repression of *GLABRA 2* (*GL2*) by forming a repressor complex with GL3 and Enhance of GLABRA (EGL3) in Arabidopsis (Liu et al., 2014).

4.2.4 Altering soybean oil accumulation by modifying biosynthetic genes

DGAT is responsible for the acylation of the sn-3 position of DAG to form TAG. DGAT is considered to be the rate-limiting enzyme in TAG biosynthesis and the expression of *DGAT* is correlated with TAG accumulation during seed development (Settlage, Kwanyuen, & Wilson, 1998). There are three major DGATs, namely DGAT1, DGAT2, and DGAT3, each with different affinities and activities toward different acyl-CoAs. At least three GmDGAT1-, five GmDGAT2-, and two GmDGAT3-encoding genes have been identified in the soybean genome (Xu et al., 2021; Zhao et al., 2019). Ectopic expressions of the three *GmDGAT1s* in Arabidopsis led to a significant increase in oil content (Xu et al., 2021; Zhao, Bi, et al., 2019; Zhao, Chang, et al., 2019), while the expressions of *GmDGAT2* and *GmDGAT3* mainly altered the FA composition, with less prominent effects on total oil content (Xu et al., 2021; Zhao, Bi, et al., 2019; Zhao, Chang, et al., 2019). This makes *DGAT1* an attractive target for genetic manipulation to increase soybean oil production. In a later experiment, a 3% increase in total FA was observed in transgenic soybean seeds overexpressing *GmDGAT1B* (*Glyma17g06120*), accompanied by a significant increase in oleic acid but a decrease in linoleic acid (Xu et al., 2021). Although the manipulation of soybean *DGAT* yielded seeds with an improved oil content, it also came with penalties in the protein content. However, a study has demonstrated that the expression of *DGAT2A* from a soil fungus, *Umbelopsis ramanniana*, can significantly increase soybean seed oil content with a minimal trade-off in the protein content (Lardizabal et al., 2008). Furthermore, expressing a modified GmDGAT with high oleic acid affinity (based on the analysis of the mutant library of DGAT from *Corylus americana*, which is a species with high oleic acid and high oil content) added an extra 3% of oil to the transgenic soybean seed (Roesler et al., 2016). This implies protein engineering using base-editing technology to modify the endogenous *GmDGAT* genes may be a possible way to improve soybean oil content in the future.

G3P is the essential building block of TAG. Therefore improving the G3P supply could also increase TAG accumulation. The overexpression of a plastidic NAD^+-dependent glycerol-3-phosphate dehydrogenase gene, *GmGPDHp1*, produced transgenic soybean seeds with a higher G3P content (Zhao et al., 2021). The increase in TAG also brought about a 20% increase in seed oil content, but potentially at the expense of soluble sugars in mature seeds (Zhao et al., 2021). The transgenic seeds have lower free FAs but higher PC, PE, DAG, and TAG, suggesting that the overexpression of *GmGPDHp1* has enhanced the incorporation of FAs with G3P.

Impairing the catabolism of TAG is also an effective way to improve soybean seed oil content. TAG lipases hydrolyze TAG into DAG and a carboxylate. Four TAG lipase-encoding genes, namely *Sugar Dependent1-1* (*GmSDP1-1*), *GmSDP1-2*, *GmSDP1-3*, and *GmSDP1-4*, are identified in soybean (Kanai, Yamada, Hayashi, Mano, & Nishimura, 2019). Knocking down the expressions of all four *GmSDPs* by RNAi improved the FA content in the seed and the overall oil yield per plant (Kanai et al., 2019). Moreover, without altering the seed number per plant, the seed yield was also improved due to the increase in seed weight (Kanai et al., 2019).

4.2.5 Improving oil accumulation by altering the sink and the source

Apart from directly improving the biosynthesis of oil and FAs, enhancing the sink can also improve the oil content in soybean seed. A GWAS of 219 diverse soybean accessions identified an oleosin protein-encoding gene (*GmOLEO1*) that affects oil content in soybean seeds, which is a potential target of artificial selection (Zhang et al., 2019). Oleosin is an important protein constituent of the oil body membrane. The constitutive overexpression of *GmOLEO1* in soybean led to a 10% increase in oil contents in the seed, which was mainly attributable to the increase in linoleic acid (C18:2) and linolenic acid (C18:3) (Zhang et al., 2019). Although the seed size was reduced in the transgenic soybean, the overall seed yield in total seed weight per plant showed a significant increase (Zhang et al., 2019). The higher *GmOLEO1* expression in the transgenic line facilitated the formation of smaller but more numerous oil bodies in the seed (Zhang et al., 2019), which may be the reason why more oil could be packed. At the same time, the enlargement in the sink also provided a positive feedback to increase the expressions of oil biosynthesis-related genes and other oleosin-encoding genes (Zhang et al., 2019). Tajima's D analysis demonstrated that the promoter, but not the coding region, of *GmOLEO1* was under positive selection (Zhang et al., 2019).

The biosynthesis of FAs in the seed largely relies on the carbon source from photosynthetic tissues. The efficiency of sucrose reallocation can thus determine the rate of oil accumulation in the seed. Through population studies, two Sugar Will Eventually be Exported Transporter-encoding genes (*GmSWEET39/GmSWEET10a/Glyma.15G049200* and *GmSWEET10b/Glyma.08G183500*) were identified to be involved in the domestication-related enhancement of oil accumulation in soybean seed (Miao et al., 2020; Wang et al., 2020; Zhang et al., 2020). They are mainly expressed in the thin-walled parenchyma of the seed coat, and the resulting proteins are localized on the plasma membrane (Miao et al., 2020; Wang et al., 2020; Zhang et al., 2020). Both GmSWEET10a and GmSWEET10b are able to transport sucrose and hexoses in a heterologous cell culture system and Xenopus oocytes (Wang et al., 2020). Swapping promoters and coding sequences of different alleles of *GmSWEET10a* in transgenic analyses demonstrated that both the gene expression level and the protein sequence of GmSWEET10a contributed to the oil accumulation (Wang et al., 2020). The overexpression of *GmSWEET10a* and *GmSWEET10b* led to higher oil content, larger seed size, but lower protein content in the transgenic soybean seeds (Wang et al., 2020). Consistent with the results from overexpression, knockout mutants of these two genes had lower oil content, smaller seed size, and higher protein content (Wang et al., 2020). Together with the fact that knocking out both genes caused a significant reduction in sucrose and hexoses in the developing embryo but not in the seed coat (Wang et al., 2020), it is suggested that GmSWEET10a and GmSWEET10b function to upload sugar into the embryo from the seed coat.

4.3 Gene regulation and strategy for improving protein contents in soybean seeds

4.3.1 Regulatory genes involved in soybean protein contents

There are over 200 protein-related quantitative trait loci (QTLs) in the soybean genome as recorded by Soybase (Grant, Nelson, Cannon, & Shoemaker, 2009). Unlike the seed oil content, there is no clear story on how soybean seed protein content is regulated. QTLs controlling soybean proteins are scattered among all the soybean chromosomes except chromosome 16 (Huang et al., 2019). Previous studies showed that the region mapped to the soybean linkage group I (LG-I) on chromosome 20 showed a strong correlation with seed protein content (Bandillo et al., 2015; Bolon et al., 2010; Diers et al., 1992; Pandurangan et al., 2012; Sebolt et al., 2000). The seed protein-associated QTLs in this region on chromosome 20 were further narrowed down to an 8.4-Mb region (Bandillo et al., 2015), a 3-Mb region (Hwang et al., 2014) and a 1-Mb region (Vaughn, Nelson, Song, Cregan, & Li, 2014), respectively. Yet these regions were still too large for the selection of protein content-controlling genes for characterization. Moreover, the soybean protein content is thought to be regulated by multiple genes, while the phenotype is highly subjected to environmental influences such as temperature, latitude, and humidity (Bolon et al., 2010; Huang et al., 2019; Qin et al., 2014). Moreover, the soybean genome duplication event has led to the rearrangement of homologous chromosomes and resulted in the duplication of different QTL regions, eventually diluting the efforts of studying protein content regulation (Lestari, Van, Lee, Kang, & Lee, 2013). These factors increase the difficulty in studying protein content regulating mechanisms. Recent researches have coupled QTL analyses with microarrays or RNA-sequencing to screen for potential gene candidates that regulate soybean seed protein contents (Bolon et al., 2010; Wang et al., 2021). This approach was able to help resolve single nucleotide polymorphisms (SNPs) and insertions/deletions (indels) between the high- and low-protein genotypes (Hwang et al., 2014; Wang et al., 2021). Furthermore, QTL analyses related to the soybean seed content could then focus more on the non-shared gene contents within the duplicated genomic regions to consolidate research efforts (Lestari et al., 2013). Although a relationship between the protein content and some genes was proposed, not many of them were validated for their corresponding gene functions.

4.3.2 Rebalancing storage proteins in soybean seeds by the molecular manipulation of globulin genes

There are two principal components of soybean seed storage proteins as categorized by their sedimentation coefficients, the 7S β-conglycinin, and the 11S glycinin. These proteins constitute about 70% of the total storage proteins in the seed (Liu et al., 2007). The 7S β-conglycinin is a trimer composed of three subunits: α′, α, and β subunits (Koshiyama, 1968). The 11S glycinin is a hexamer encoded by the *Gy1*, *Gy2*, and *Gy3* (group I) and *Gy4* and *Gy5* (group II) glycinin-encoding genes (Fischer & Goldberg, 1982; Nielsen et al., 1989). The β-conglycinin and glycinin-encoding genes consist of multiple gene members, and each of them is regulated transcriptionally, translationally, or post-translationally after flowering and during embryogenesis (Dickinson, Hussein, & Nielsen, 1989; Harada, Barker, & Goldberg, 1989). The methionine and cysteine concentrations showed a strong correlation with the 11S:7S globulin ratio, while threonine and tyrosine showed a moderate correlation (Kwanyuen et al., 1997). The 11S glycinin contains more cysteine, methionine, threonine, and tyrosine. It is believed that a higher 11S:7S ratio could lead to beneficial changes in the nutritional quality of soybean meal.

Soybean protein accumulation is subjected to delicate adjustments during seed filling to maintain a relatively stable total protein content in the seed. Repression of the storage proteins would result in other proteins being accumulated as compensation (Kwanyuen et al., 1997; Ogawa, Tayama, Kitamura, & Kaizuma, 1989; Schmidt et al., 2011; Yang, Yu, Zheng, & James, 2016). The low-7S β-conglycinin soybean varieties, Keburi and Mo-shi-dou Gong 503, were found not to express the α′ subunit of β-conglycinin and, as a result, had a lowered level of the α and β subunits of β-conglycinin, without showing any abnormality in the agronomic traits (Ogawa et al., 1989). The co-suppression of β-conglycinin α′ and α subunits resulted in similar seed storage protein rebalancing as observed in the low-7S β-conglycinin soybean varieties, Keburi and Mo-shi-dou Gong 503 (Kinney et al., 2001; Ogawa et al., 1989). As shown in the electron microscopy-immunogold assays, P34 and 11S glycinin concentrations were boosted in the protein bodies in these two soybean varieties (Kinney et al., 2001). It is believed that the accumulation of P34 and 11S glycinin in the endoplasmic reticulum (ER)–derived protein bodies was enhanced to compensate for the loss of seed storage proteins (Kinney et al., 2001). Concomitant soybean protein rebalancing was also observed in the transgenic soybean with glycinin and β-conglycinin-encoding genes suppressed by RNA interference (RNAi) (Schmidt et al., 2011). The concentrations of P34 and KTI surged in the soybean seed to compensate for the missing storage proteins, eventually attaining a comparable seed protein content with that of untransformed lines (Schmidt et al., 2011). However, the introgression of exogenous green fluorescence protein (GFP) into the co-repressed background did not alter the proteome of the soybean seed, suggesting the presence of an intrinsic selection mechanism for specific compensating proteins (Schmidt et al., 2011). The soybean seed globulin rebalancing was also reported in studies conducted with cultivars expressing different levels of a subunit of the 11S glycinin, 11SA4 (Yang et al., 2016). It was demonstrated that the 11SA4 deficiency in the soybean seed would induce a compensatory accumulation of 7S globulins and result in minimal changes in the total protein content (Yang et al., 2016).

The compensatory nature of soybean seed storage proteins suggests an approach for improving the nutritional value of soybean seeds, such as higher cysteine and methionine contents (Kwanyuen et al., 1997). The regulatory pathway for glycinin and β-conglycinin rebalancing and compensation might provide good study models for understanding protein loading to seeds for storage in soybean (Kinney et al., 2001). Undoubtedly, manipulating the ratio of the 11S and 7S globulins could affect the nutritional value of the soybean seed. It was reported that the ratio of globulin members might affect the texture and processing method required in food processing, such as the ability to form aggregates and precipitates, as 11S and 7S globulins have different association and dissociation properties upon thermal treatments (Saio & Watanabe, 1978; Utsumi, Damodaran, & Kinsella, 1984). Hence the resulting changes in physical properties should be considered in making soybean seed protein content improvements (Saio & Watanabe, 1978).

4.3.3 Improving the nutritional value of soybean seed by increasing protein content and changing amino acid composition

A lot of effort has been made on manipulating the original soybean seed composition for better nutritional values, but manipulating proteins other than glycinin and β-conglycinin might be an instant solution before we can achieve a better understanding of the protein content regulation mechanisms in soybean seeds. The ectopic expression of *β-zein*, a methionine-rich storage protein from maize, could increase the soybean seed methionine level but the impact was relatively small (Guo et al., 2020). GmSWEET39 is a sucrose transporter that facilitates sugar transport and the supply of intermediates for protein and oil metabolism in soybean seeds (Zhai, Liu, Xu, & Shanklin, 2017). It was reported that a 2-bp (CC) deletion in *GmSWEET39* encoded a truncated transporter that resulted in a high-oil low-protein content in the seed (Zhang et al., 2020). The high level of asparagine synthetase (AS) in soybean leaves was positively correlated with high seed protein concentration among different cultivars with varying degrees of AS expression (Wan, Shao, Shan, Zeng, & Lam, 2006). Furthermore, the asparagine synthetase (AS) and asparaginase (ASPG) levels were positively correlated with the protein contents from a study of 73 recombinant inbred lines (Pandurangan et al., 2012). It is suggested that the transcript and protein (enzyme) levels, and enzyme activities in the seed coat might explain the differences in protein contents among these lines (Pandurangan et al., 2012). There are not many well-known transcriptional and translational controls on soybean seed protein composition. One of them is the regulation by C4 of nuclear factor Y (NF-YC4) and Qua-Quine Starch (QQS). *Qua-Quine Starch* (QQS) is an orphan gene from *Arabidopsis thaliana*, and the ectopic expression of *QQS* modulated protein compositions in the leaf and seed by affecting the carbon and nitrogen partitioning in soybean (Li et al., 2015; Li & Wurtele, 2015). QQS also interacts with the conserved subunit C4 of nuclear factor Y (NF-YC4) to mediate the transcription of a variety of CCAAT box-containing genes (Li et al., 2015). The ectopic expression of *QQS* could lead to an overall increase in protein content by around 20% with a reduction in oil content by around 10% (Li & Wurtele, 2015). Other than the ectopic expression of *QQS* in soybean, overexpression of its interacting partner *GmNF-YC4-2* (*Glyma.04g196200*) also resulted in a higher protein level in the soybean seed along with increased yield (O'conner et al., 2021). Besides, *GmNF-YC4-2*-overexpressing lines showed increased resistance to biotic stress such as bacterial, viral, and fungal challenges (O'conner et al., 2021). As mentioned in a previous section on oil accumulation regulation in the soybean seed by transcription factors, GmLEC2 also acts as a master regulator of soybean seed protein content. The ectopic expression of *GmLEC2a* in Arabidopsis resulted in an increase in protein, while the *Atlec2* loss-of-function mutant seeds contained approximately 40% less protein than wild type seeds (Manan et al., 2017).

Altering the amino acid content in the soybean seed might be a feasible way to improve its nitrogen-based nutritional value (Ishimoto et al., 2010; Pandurangan et al., 2012; Wan et al., 2006). As already mentioned, selective breeding in the past 60 years has minimally affected the overall amino acid composition of cultivated soybean seeds, while those from the wild soybean *G. soja* contain a significantly higher level of various amino acids (Mahmoud et al., 2006). Several studies utilized feedback-insensitive enzymes in amino acid metabolism for boosting the respective amino acid contents in seeds (Falco et al., 1995; Ishimoto et al., 2010; Kita et al., 2010; Song, Hou, et al., 2013; Yu et al., 2018). The ectopic expression of a rice tryptophan-feedback-resistant subunit of anthranilate synthase (OASA1D) increased the tryptophan level in the seed, using either a 35S promoter or a seed-specific promoter (Ishimoto et al., 2010; Kita et al., 2010). OASA1D is a regulatory enzyme in the tryptophan (Trp) biosynthesis pathway. The overexpression of *OASA1D* bypassed the negative tryptophan feedback mechanism, which led to the accumulation and deposition of tryptophan in the soybean seed without detrimental effects on its growth and development (Ishimoto et al., 2010). The ectopic expression of *OASA1D* in both low- and high-free amino acid breeding lines of soybean also resulted in a similar boost in the level of tryptophan in the seed (Kita et al., 2010). The ectopic expression of an Arabidopsis methionine-feedback-resistant form of cystathionine γ-synthase (CGS) in soybean using either a seed-specific or a constitutive promoter was shown to boost the methionine storage in seed (Song, Hou, et al., 2013; Yu et al., 2018). Cystathionine γ-synthase (CGS) is the first committed enzyme bridging the aspartate family biosynthesis pathway with methionine metabolism. Similar to the bypass of the negative tryptophan feedback mechanism, the transgenic soybean seeds had an elevated methionine content without affecting their agronomic performance (Song, Hou, et al., 2013; Yu et al., 2018). Such uses of feedback-insensitive enzymes in amino acid metabolism also worked on soybean seed lysine content. The ectopic expression of lysine-feedback-insensitive *dihydrodipicolinate synthase* (*DHDPS*) from *E. coli* and *aspartokinase* (*AK*) from *Corynebacterium* using a seed-specific promoter resulted in a five-fold increase in the seed lysine level (Falco et al., 1995).

5. Conclusions

Seed protein and oil contents are two valuable traits determined by genotypes and planting conditions. To produce soybean seeds with higher protein/oil contents, efforts should be made to screen the wild soybeans and germplasms from various geographical locations and under different stress environments for tolerant genotypes and incorporate them into breeding programs. It is also vital to identify all the genes, regulators, proteins, and metabolites involved in the protein and oil metabolism pathways to decipher the molecular mechanisms that determine seed composition. Understanding the development, growth, seed productivity, seed quality and sink-source relations in the soybean seed will give us better insights into the physiological responses and underlying genetics of stress tolerance, with the ultimate goal of enhancing soybean seed yield and quality.

Acknowledgments

This work was supported by Hong Kong Research Grants Council Area of Excellence Scheme (AoE/M-403/16), Lo Kwee-Seong Biomedical Research Fund and Guangdong Provincial Department of Science and Technology 2020 Key Areas Research and Development Programs: Breeding for High Yield and High Quality New Soybean Cultivars for Tropics and Subtropics (2020B020220008). J.Y. Chu copy-edited the text. Any opinions, findings, conclusions or recommendations expressed in this publication do not reflect the views of the Government of Hong Kong Special Administrative Region or the Innovation and Technology Commission.

References

Abdelghany, A. M., Zhang, S., Azam, M., Shaibu, A. S., Feng, Y., Qi, J., et al. (2020). Natural variation in fatty acid composition of diverse world soybean germplasms grown in China. *Agronomy*, *10*, 24.

al Amin, N., Ahmad, N., Wu, N., Pu, X. M., Ma, T., Du, Y. Y., et al. (2019). CRISPR-Cas9 mediated targeted disruption of FAD2-2 microsomal omega-6 desaturase in soybean (*Glycine max*. L). *BMC Biotechnology*, *19*.

Alsajri, F. A., Wijewardana, C., Irby, J. T., Bellaloui, N., Krutz, L. J., Golden, B., et al. (2020). Developing functional relationships between temperature and soybean yield and seed quality. *Agronomy Journal*, *112*, 194–204.

Anai, T., Yamada, T., Hideshima, R., Kinoshita, T., Rahman, S. M., & Takagi, Y. (2008). Two high-oleic-acid soybean mutants, M23 and KK21, have disrupted microsomal omega-6 fatty acid desaturase, encoded by GmFAD2-1a. *Breeding Science*, *58*, 447–452.

Anda, A., Simon, B., Soos, G., da Silva, J. A. T., & Menyhart, L. (2021). Water stress modifies canopy light environment and qualitative and quantitative yield components in two soybean varieties. *Irrigation Science*, *39*, 549–566.

Anwar, F., Kamal, G. M., Nadeem, F., & Shabir, G. (2016). Variations of quality characteristics among oils of different soybean varieties. *Journal of King Saud University*, *28*, 332–338.

Assefa, Y., et al. (2019). Assessing variation in US soybean seed composition (protein and oil). *Frontiers in Plant Science*, *10*, 298.

Bandillo, N., Jarquin, D., Song, Q., Nelson, R., Cregan, P., Specht, J., et al. (2015). A population structure and genome-wide association analysis on the USDA soybean germplasm collection. *Plant Genome*, *8*.

Bates, P. D., Stymne, S., & Ohlrogge, J. (2013). Biochemical pathways in seed oil synthesis. *Current Opinion in Plant Biology*, *16*, 358–364.

Bellaloui, N., Gillen, A. M., Mengistu, A., Kebede, H., Fisher, D. K., Smith, J. R., et al. (2013). Responses of nitrogen metabolism and seed nutrition to drought stress in soybean genotypes differing in slow-wilting phenotype. *Frontiers in Plant Science*, *4*, 498.

Bellaloui, N., Hu, Y., Mengistu, A., Kassem, M. A., & Abel, C. A. (2013). Effects of foliar boron application on seed composition, cell wall boron, and seed δ15N and δ13C isotopes in water-stressed soybean plants. *Frontiers in Plant Science*, *4*, 270.

Bellaloui, N., Reddy, K. N., & Mengistu, A. (2015). Drought and heat stress effects on soybean fatty acid composition and oil stability. In *Processing and impact on active components in food* (pp. 377–384). Elsevier.

Bilyeu, K., Palavalli, L., Sleper, D., & Beuselinck, P. (2005). Mutations in soybean microsomal omega-3 fatty acid desaturase genes reduce linolenic acid concentration in soybean seeds. *Crop Science*, *45*, 1830–1836.

Bilyeu, K., Ren, C., Nguyen, H. T., Herman, E., & Sleper, D. A. (2009). Association of a four-Basepair insertion in the P34 gene with the low-allergen trait in soybean. *Plant Genome*, *2*, 141–148.

Bilyeu, K., Skrabisova, M., Allen, D., Rajcan, I., Palmquist, D. E., Gillen, A., et al. (2018). The interaction of the soybean seed high oleic acid oil trait with other fatty acid modifications. *Journal of the American Oil Chemists' Society*, *95*, 39–49.

Blanusa, T., Stikic, R., Vucelic-Radovic, B., Barac, M., & Velickovic, D. (2000). Dynamics of seed protein biosynthesis in two soybean genotypes differing in drought susceptibility. *Biologia Plantarum*, *43*, 55–59.

Bolon, Y. T., et al. (2010). Complementary genetic and genomic approaches help characterize the linkage group I seed protein QTL in soybean. *BMC Plant Biology*, *10*.

Brzostowski, L., & Diers, B. (2017). Agronomic evaluation of a high protein allele from PI407788A on chromosome 15 across two soybean backgrounds. *Crop Science*, *57*(6), 2972–2978. https://doi.org/10.2135/cropsci2017.02.0083.

Cao, Y. C., Li, S. G., Wang, Z. L., Chang, F. G., Kong, J. J., Gai, J. Y., et al. (2017). Identification of major quantitative trait loci for seed oil content in soybeans by combining linkage and genome-wide association mapping. *Frontiers in Plant Science, 8*.

Cardinal, A. J., et al. (2014). Mapping the low palmitate fap1 mutation and validation of its effects in soybean oil and agronomic traits in three soybean populations. *Theoretical and Applied Genetics, 127*, 97–111.

Carrero-Colon, M., Abshire, N., Sweeney, D., Gaskin, E., & Hudson, K. (2014). Mutations in SACPD-C result in a range of elevated stearic acid concentration in soybean seed. *PLoS One, 9*.

Chaudhary, J., Patil, G. B., Sonah, H., Deshmukh, R. K., Vuong, T. D., Valliyodan, B., et al. (2015). Expanding omics resources for improvement of soybean seed composition traits. *Frontiers in Plant Science, 6*, 1–16.

Chebrolu, K. K., Fritschi, F. B., Ye, S. Q., Krishnan, H. B., Smith, J. R., & Gillman, J. D. (2016). Impact of heat stress during seed development on soybean seed metabolome. *Metabolomics, 12*.

Chen, P., Ishibashi, T., Dombek, D. G., & Rupe, J. C. (2011). Registration of R05-1415 and R05-1772 high-protein soybean germplasm lines. *Journal of Plant Registrations, 5*, 410–413.

Chen, P., Sneller, C. H., Ishibashi, T., & Cornelious, B. (2008). Registration of high-protein soybean germplasm line R95-1705. *Journal of Plant Registrations, 2*, 58–59.

Chen, L., Zheng, Y. H., Dong, Z. M., Meng, F. F., Sun, X. M., Fan, X. H., et al. (2018). Soybean (*Glycine max*) WRINKLED1 transcription factor, GmWRI1a, positively regulates seed oil accumulation. *Molecular Genetics and Genomics, 293*, 401–415.

Chung, J., Babka, H. L., Graef, G. L., Staswick, P. E., Lee, D. J., Cregan, P. B., et al. (2003). The seed protein, oil, and yield QTL on soybean linkage group I. *Cell Biology and Molecular Genetics, 43*, 1053–1067.

Ciabotti, S., Juhász, A. C. P., Mandarino, J. M. G., Costa, L. L., Corrêa, A. D., Simão, A. A., et al. (2019). Chemical composition and lipoxygenase activity of soybean (*Glycine max* L. Merrill.) genotypes, specific for human consumption, with different tegument colours. *Brazilian Journal of Food Technology, 22*.

Clemente, T. E., & Cahoon, E. B. (2009). Soybean oil: Genetic approaches for modification of functionality and total content. *Plant Physiology, 151*, 1030–1040.

Cober, E. R., Voldeng, D., & H. (2000). Developing high-protein, high-yield soybean populations and lines. *Crop Science, 40*, 39–42.

de Mejia, E. G., Bradford, T., & Hasler, C. (2003). The anticarcinogenic potential of soybean lectin and lunasin. *Nutrition Reviews, 61*, 239–246.

Dehesh, K., Tai, H., Edwards, P., Byrne, J., & Jaworski, J. G. (2001). Overexpression of 3-ketoacyl-acyl-carrier protein synthase IIIs in plants reduces the rate of lipid synthesis. *Plant Physiology, 125*, 1103–1114.

Demorest, Z. L., et al. (2016). Direct stacking of sequence-specific nuclease-induced mutations to produce high oleic and low linolenic soybean oil. *BMC Plant Biology, 16*.

Di, Q., Piersanti, A., Zhang, Q., Miceli, C., Li, H., & Liu, X. Y. (2022). Genome-wide association study identifies candidate genes related to the linoleic acid content in soybean seeds. *International Journal of Molecular Sciences, 23*.

Dickinson, C. D., Hussein, E. H., & Nielsen, N. C. (1989). Role of posttranslational cleavage in glycinin assembly. *Plant Cell, 1*, 459–469.

Diers, B. W., Keim, P., Fehr, W. R., & Shoemaker, R. C. (1992). RFLP analysis of soybean seed protein and oil content. *Theoretical and Applied Genetics, 83*, 608–612.

Djuricic, I., & Calder, P. C. (2021). Beneficial outcomes of omega-6 and omega-3 polyunsaturated fatty acids on human health: An update for 2021. *Nutrients, 13*, 2421.

Do Wee, C., Hashiguchi, M., Anai, T., Suzuki, A., & Akashi, R. (2017). Fatty acid composition and distribution in wild soybean (*Glycine soja*) seeds collected in Japan. *Asian Journal of Plant Sciences, 16*, 52–64.

Dornbos, D. L., & Mullen, R. E. (1992). Soybean seed protein and oil contents and fatty-acid composition adjustments by drought and temperature. *Journal of the American Oil Chemists' Society*, *69*, 228–231.

Eskandari, M., Cober, E. R., & Rajcan, I. (2013). Genetic control of soybean seed oil: I. QTL and genes associated with seed oil concentration in RIL populations derived from crossing moderately high-oil parents. *Theoretical and Applied Genetics*, *126*, 483–495.

Falco, S. C., Guida, T., Locke, M., Mauvais, J., Sanders, C., Ward, R. T., et al. (1995). Transgenic canola and soybean seeds with increased lysine. *Biotechnology (N. Y.)*, *13*, 577–582.

FAOSTAT. (2020). *FAOSTAT*.

Farhangi-Abriz, S., & Ghassemi-Golezani, K. (2016). Improving amino acid composition of soybean under salt stress by salicylic acid and jasmonic acid. *Journal of Applied Botany and Food Quality*, *89*, 243–248.

Fischer, R. L., & Goldberg, R. B. (1982). Structure and flanking regions of soybean seed protein genes. *Cell*, *29*, 651–660.

FoodData (USDA). (2022). *FoodData*.

Friedman, M., & Brandon, D. L. (2001). Nutritional and health benefits of soy proteins. *Journal of Agricultural and Food Chemistry*, *49*, 1069–1086.

George, M. A., Bhide, S. V., Thengane, R. J., Hosseini, G. H., & Manjaya, J. G. (2008). Identification of low lectin mutants in soybean. *Plant Breeding*, *127*, 150–153.

Ghassemi-Golezani, K., & Farhangi-Abriz, S. (2018). Changes in oil accumulation and fatty acid composition of soybean seeds under salt stress in response to salicylic acid and jasmonic acid. *Russian Journal of Plant Physiology*, *65*, 229–236.

Gillman, J. D., Kim, W. S., & Krishnan, H. B. (2015). Identification of a new soybean Kunitz trypsin inhibitor mutation and its effect on Bowman-Birk protease inhibitor content in soybean seed. *Journal of Agricultural and Food Chemistry*, *63*, 1352–1359.

Grant, D., Nelson, R. T., Cannon, S. B., & Shoemaker, R. C. (2009). SoyBase, the USDA-ARS soybean genetics and genomics database. *Nucleic Acids Research*, *38*, 843–846.

Greenberg, P., & Hartung, H. N. (1998). *The whole soy cookbook: 175 Delicious, nutritious, easy-to-prepare recipes featuring tofu, tempeh, and various forms of nature's healthiest bean*. CA: Three Rivers Press.

Guo, C., Liu, X., Chen, L., Cai, Y., Yao, W., Yuan, S., et al. (2020). Elevated methionine content in soybean seed by overexpressing maize β-zein protein. *Oil Crop Science*, *5*, 11–16.

Harada, J. J., Barker, S. J., & Goldberg, R. B. (1989). Soybean beta-conglycinin genes are clustered in several DNA regions and are regulated by transcriptional and posttranscriptional processes. *Plant Cell*, *1*, 415–425.

Haun, W., et al. (2014). Improved soybean oil quality by targeted mutagenesis of the fatty acid desaturase 2 gene family. *Plant Biotechnology Journal*, *12*, 934–940.

Herman, E. M. (2014). Soybean seed proteome rebalancing. *Frontiers in Plant Science*, *5*.

Huang, J. H., Ma, Q. B., Cai, Z. D., Xia, Q. J., Li, S. X., Jia, J., et al. (2020). Identification and mapping of stable QTLs for seed oil and protein content in soybean [*Glycine max* (L.) Merr.]. *Journal of Agricultural and Food Chemistry*, *68*, 6448–6460.

Huang, S., et al. (2019). Identification of soybean genes related to soybean seed protein content based on quantitative trait loci collinearity analysis. *Journal of Agricultural and Food Chemistry*, *67*, 258–274.

Hwang, E. Y., Song, Q., Jia, G., Specht, J. E., Hyten, D. L., Costa, J., et al. (2014). A genome-wide association study of seed protein and oil content in soybean. *BMC Genomics*, *15*, 1–12.

Ishimoto, M., et al. (2010). Evaluation of amino acid content and nutritional quality of transgenic soybean seeds with high-level tryptophan accumulation. *Molecular Breeding*, *25*, 313–326.

Jokić, S., SudaR, R., Svilović, S., Vidović, S., Bilić, M., Velić, D., et al. (2013). Fatty acid composition of oil obtained from soybeans by extraction with supercritical carbon dioxide. *Czech Journal of Food Sciences, 31*, 116–125.

Joseph, L. M., Hymowitz, T., Schmidt, M. A., & Herman, E. M. (2006). Evaluation of glycine germplasm for nulls of the immunodominant allergen P34/Gly m Bd 30k. *Crop Science, 46*, 1755–1763.

Kalinski, A., Melroy, D. L., Dwivedi, R. S., & Herman, E. M. (1992). A soybean vacuolar protein (P34) related to thiol proteases is synthesized as a glycoprotein precursor during seed maturation. *The Journal of Biological Chemistry, 267*, 12068–12076.

Kanai, M., Yamada, T., Hayashi, M., Mano, S., & Nishimura, M. (2019). Soybean (*Glycine max* L.) triacylglycerol lipase GmSDP1 regulates the quality and quantity of seed oil. *Scientific Reports, 9*.

Karikari, B., Li, S. G., Bhat, J. A., Cao, Y. C., Kong, J. J., Yang, J. Y., et al. (2019). Genome-wide detection of major and epistatic effect QTLs for seed protein and oil content in soybean under multiple environments using high-density bin map. *International Journal of Molecular Sciences, 20*.

Kennedy, A. R. (1998). The Bowman-Birk inhibitor from soybeans as an anticarcinogenic agent. *The American Journal of Clinical Nutrition, 68*, 1406s–1412s.

Khan, A. Z., Shah, P., Khan, H., Nigar, S., Perveen, S., Shah, M. K., et al. (2011). Seed quality and vigor of soybean cultivars as influenced by canopy temperature. *Pakistan Journal of Botany, 43*, 642–648.

Kik, M. J. L., Rojer, J. M., Mouwen, J. M. V. M., Koninkx, J. F. J. G., Vandijk, J. E., & Vanderhage, M. H. (1989). The interaction between plant-lectins and the small intestinal epithelium—A primary cause of intestinal disturbance. *The Veterinary Quarterly, 11*, 108–115.

Kinney, A. J., Jung, R., & Herman, E. M. (2001). Cosuppression of the α subunits of β-conglycinin in transgenic soybean seeds induces the formation of endoplasmic reticulum-derived protein bodies. *Plant Cell, 13*, 1165–1178.

Kita, Y., Nakamoto, Y., Takahashi, M., Kitamura, K., Wakasa, K., & Ishimoto, M. (2010). Manipulation of amino acid composition in soybean seeds by the combination of deregulated tryptophan biosynthesis and storage protein deficiency. *Plant Cell Reports, 29*, 87–95.

Kobayashi, H. (2013). Prevention of cancer and inflammation by soybean protease inhibitors. *Frontiers in Bioscience (Elite Edition), 5*, 966–973.

Koo, S. C., Bae, D. W., Seo, J. S., Park, K. M., Choi, M. S., Kim, S. H., et al. (2011). Proteomic analysis of seed storage proteins in low allergenic soybean accession. *Journal of Applied Biological Chemistry, 54*, 332–339.

Koshiyama, I. (1968). Chemical and physical properties of a 7S protein in soybean globulins. *Cereal Chemistry, 45*, 394–404.

Krishnan, H. B., Kim, W. S., Oehrle, N. W., Smith, J. R., & Gillman, J. D. (2020). Effect of heat stress on seed protein composition and ultrastructure of protein storage vacuoles in the cotyledonary parenchyma cells of soybean genotypes that are either tolerant or sensitive to elevated temperatures. *International Journal of Molecular Sciences, 21*.

Krishnan, H. B., Natarajan, S. S., Mahmoud, A. A., & Nelson, R. L. (2007). Identification of glycinin and beta-conglycinin subunits that contribute to the increased protein content of high-protein soybean lines. *Journal of Agricultural and Food Chemistry, 55*, 1839–1845.

Kwanyuen, P., Pantalone, V. R., Burton, J. W., & Wilson, R. F. (1997). A new approach to genetic alteration of soybean protein composition and quality. *JAOCS, Journal of the American Oil Chemists' Society, 74*, 983–987.

La, T., Large, E., Taliercio, E., Song, Q., Gillman, J. D., Xu, D., et al. (2019). Characterization of select wild soybean accessions in the USDA germplasm collection for seed composition and agronomic traits. *Crop Science, 59*, 233–251.

Lardizabal, K., Effertz, R., Levering, C., Mai, J., Pedroso, M. C., Jury, T., et al. (2008). Expression of Umbelopsis ramanniana DGAT2A in seed increases oil in soybean. *Plant Physiology, 148*, 89–96.

Lau, O. S., & Sun, S. S. M. (2009). Plant seeds as bioreactors for recombinant protein production. *Biotechnology Advances, 27*, 1015–1022.

Lee, J.-D., Bilyeu, K. D., & Shannon, J. G. (2007). Genetics and breeding for modified fatty acid profile in soybean seed oil. *Journal of Crop Science and Biotechnology, 10*, 201–210.

Lee, J. D., Oliva, M. L., Sleper, D. A., & Shannon, J. G. (2008). Irrigation has little effect on unsaturated fatty acid content in soya bean seed oil within genotypes differing in fatty acid profile. *Journal of Agronomy and Crop Science, 194*, 320–324.

Lestari, P., Van, K., Lee, J., Kang, Y. J., & Lee, S. H. (2013). Gene divergence of homeologous regions associated with a major seed protein content QTL in soybean. *Frontiers in Plant Science, 4*, 1–8.

Li, L., & Wurtele, E. S. (2015). The QQS orphan gene of Arabidopsis modulates carbon and nitrogen allocation in soybean. *Plant Biotechnology Journal, 13*, 177–187.

Li, S., Xu, H., Yang, J., & Zhao, T. (2019). Dissecting the genetic architecture of seed protein and oil content in soybean from the Yangtze and Huaihe river valleys using multi-locus genome- wide association studies. *International Journal of Molecular Sciences, 20*.

Li, L., et al. (2015). QQS orphan gene regulates carbon and nitrogen partitioning across species via NF-YC interactions. *Proceedings of the National Academy of Sciences of the United States of America, 112*, 14734–14739.

Li, Q. T., et al. (2017). Selection for a zinc-finger protein contributes to seed oil increase during soybean domestication. *Plant Physiology, 173*, 2208–2224.

Liu, X. Y., Qin, D., Piersanti, A., Zhang, Q., Miceli, C., & Wang, P. W. (2020). Genome-wide association study identifies candidate genes related to oleic acid content in soybean seeds. *BMC Plant Biology, 20*.

Liu, C., Wang, H., Cui, Z., He, X., Wang, X., Zeng, X., et al. (2007). Optimization of extraction and isolation for 11S and 7S globulins of soybean seed storage protein. *Food Chemistry, 102*, 1310–1316.

Liu, A. L., Xiao, Z. X., Li, M. W., Wong, F. L., Yung, W. S., Ku, Y. S., et al. (2019). Transcriptomic reprogramming in soybean seedlings under salt stress. *Plant, Cell & Environment, 42*, 98–114.

Liu, Y. F., et al. (2014). Soybean GmMYB73 promotes lipid accumulation in transgenic plants. *BMC Plant Biology, 14*.

Lu, X., Li, Q. T., Xiong, Q., Li, W., Bi, Y. D., Lai, Y. C., et al. (2016). The transcriptomic signature of developing soybean seeds reveals the genetic basis of seed trait adaptation during domestication. *The Plant Journal, 86*, 530–544.

Ma, J., Sun, S., Whelan, J., & Shou, H. X. (2021). CRISPR/Cas9-mediated knockout of GmFATB1 significantly reduced the amount of saturated fatty acids in soybean seeds. *International Journal of Molecular Sciences, 22*.

Mahmoud, A. A., Natarajan, S. S., Bennett, J. O., Mawhinney, T. P., Wiebold, W. J., & Krishnan, H. B. (2006). Effect of six decades of selective breeding on soybean protein composition and quality: A biochemical and molecular analysis. *Journal of Agricultural and Food Chemistry, 54*, 3916–3922.

Maleki, A., Naderi, A., Naseri, R., Fathi, A., Bahamin, S., & Maleki, R. (2013). Physiological performance of soybean cultivars under drought stress. *Bulletin of Environment, Pharmacology and Life Sciences, 2*, 38–44.

Manan, S., Ahmad, M. Z., Zhang, G., Chen, B., Haq, B. U., Yang, J., et al. (2017). Soybean LEC2 regulates subsets of genes involved in controlling the biosynthesis and catabolism of seed storage substances and seed development. *Frontiers in Plant Science, 8*, 1–16.

McDonald, G. K., & Paulsen, G. M. (1997). High temperature effects on photosynthesis and water relations of grain legumes. *Plant and Soil, 196*, 47–58.

Medic, J., Atkinson, C., & Hurburgh, C. R. (2014). Current knowledge in soybean composition. *Journal of the American Oil Chemists' Society, 91*, 363–384.

Miao, L., Yang, S. N., Zhang, K., He, J. B., Wu, C. H., Ren, Y. H., et al. (2020). Natural variation and selection in GmSWEET39 affect soybean seed oil content. *The New Phytologist, 225*, 1651–1666.

Miransari, M. (2016a). *Environmental stresses in soybean production: Soybean production. Vol. 2.* Academic Press.

Miransari, M. (2016b). Soybean, protein, and oil production under stress. In *Environmental stresses in soybean production* (pp. 157–176). Elsevier.

Nakagawa, A. C. S., Ario, N., Tomita, Y., Tanaka, S., Murayama, N., Mizuta, C., et al. (2020). High temperature during soybean seed development differentially alters lipid and protein metabolism. *Plant Production Science, 23*, 504–512.

Nakagawa, A. C. S., Itoyama, H., Ariyoshi, Y., Ario, N., Tomita, Y., Kondo, Y., et al. (2018). Drought stress during soybean seed filling affects storage compounds through regulation of lipid and protein metabolism. *Acta Physiologiae Plantarum, 40*.

Natarajan, S., Xu, C., Bae, H., Caperna, T. J., & Garrett, W. M. (2006). Proteomic analysis of allergen and antinutritional proteins in wild and cultivated soybean seeds. *Journal of Plant Biochemistry and Biotechnology, 15*, 103–108.

Nielsen, N. C., Dickinson, C. D., Cho, T. J., Thanh, V. H., Scallon, B. J., Fischer, R. L., et al. (1989). Characterization of the glycinin gene family in soybean. *Plant Cell, 1*, 313–328.

Nikiforidis, C. V. (2019). Structure and functions of oleosomes (oil bodies). *Advances in Colloid and Interface Science, 274*.

O'conner, S., Zheng, W., Qi, M., Kandel, Y., Fuller, R., Whitham, S. A., et al. (2021). Gmnf-yc4-2 increases protein, exhibits broad disease resistance and expedites maturity in soybean. *International Journal of Molecular Sciences, 22*.

Oakes, J. L., Bost, K. L., & Piller, K. J. (2009). Stability of a soybean seed-derived vaccine antigen following long-term storage, processing and transport in the absence of a cold chain. *Journal of the Science of Food and Agriculture, 89*, 2191–2199.

Ogawa, T., Tayama, E., Kitamura, K., & Kaizuma, N. (1989). Genetic improvement of seed storage proteins using three variant alleles of 7S globulin subunits in soybean (*Glycine max* L.). *Japanese Journal of Breeding, 39*, 137–147.

Oliva, M. L., Shannon, J. G., Sleper, D. A., Ellersieck, M. R., Cardinal, A. J., Paris, R. L., et al. (2006). Stability of fatty acid profile in soybean genotypes with modified seed oil composition. *Crop Science, 46*, 2069–2075.

Ortiz, A. C., De Smet, I., Sozzani, R., & Locke, A. M. (2022). Field-grown soybean shows genotypic variation in physiological and seed composition responses to heat stress during seed development. *Environmental and Experimental Botany, 195*, 104768.

Paek, N. C., Imsande, J., Shoemaker, R. C., & Shibles, R. (1997). Nutritional control of soybean seed storage protein. *Crop Science, 37*, 498–503.

Pandurangan, S., Pajak, A., Molnar, S. J., Cober, E. R., Dhaubhadel, S., Hernández-Sebasti, C., et al. (2012). Relationship between asparagine metabolism and protein concentration in soybean seed. *Journal of Experimental Botany, 63*, 3173–3184.

Pantalone, V. R., Wilson, R. F., Novitzky, W. P., & Burton, J. W. (2002). Genetic regulation of elevated stearic acid concentration in soybean oil. *Journal of the American Oil Chemists' Society, 79*, 549–553.

Panthee, D. R., Kwanyuen, P., Sams, C. E., West, D. R., Saxton, A. M., & Pantalone, V. R. (2004). Quantitative trait loci for beta-conglycinin (7S) and glycinin (11S) fractions of soybean storage protein. *Journal of the American Oil Chemists' Society, 81*, 1005–1012.

Park, H., Graef, G., Xu, Y. X., Tenopir, P., & Clemente, T. E. (2014). Stacking of a stearoyl-ACP thioesterase with a dual-silenced palmitoyl-ACP thioesterase and Delta 12 fatty acid desaturase in transgenic soybean. *Plant Biotechnology Journal, 12*, 1035–1043.

Patil, G., Mian, R., Vuong, T., Pantalone, V., Song, Q., Chen, P., et al. (2017). Molecular mapping and genomics of soybean seed protein: A review and perspective for the future. *Theoretical and Applied Genetics*, *130*, 1975–1991.

Pham, A. T., Shannon, J. G., & Bilyeu, K. D. (2012). Combinations of mutant FAD2 and FAD3 genes to produce high oleic acid and low linolenic acid soybean oil. *Theoretical and Applied Genetics*, *125*, 503–515.

Phang, T. H., Shao, G. H., & Lam, H. M. (2008). Salt tolerance in soybean. *Journal of Integrative Plant Biology*, *50*, 1196–1212.

Philip, R., Darnowski, D. W., Maughan, P. J., & Vodkin, L. O. (2001). Processing and localization of bovine beta-casein expressed in transgenic soybean seeds under control of a soybean lectin expression cassette. *Plant Science*, *161*, 323–335.

Piper, E. L., & Boote, K. J. (1999). Temperature and cultivar effects on soybean seed oil and protein concentrations. *Journal of the American Oil Chemists' Society*, *76*, 1233–1241.

Priolli, R. H. G., Campos, J. B., Stabellini, N. S., Pinheiro, J. B., & Vello, N. A. (2015). Association mapping of oil content and fatty acid components in soybean. *Euphytica*, *203*, 83–96.

Qin, P., Song, W., Yang, X., Sun, S., Zhou, X., Yang, R., et al. (2014). Regional distribution of protein and oil compositions of soybean cultivars in China. *Crop Science*, *54*, 1139–1146.

Real-Guerra, R., Staniscuaski, F., & Carlini, C. R. (2013). Soybean urease: Over a hundred years of knowledge. In *A comprehensive survey of international soybean research—Genetics, physiology, agronomy and nitrogen relationships* (pp. 317–339).

Rincker, K., Nelson, R., Specht, J., Sleper, D., Cary, T., Cianzio, S. R., et al. (2014). Genetic improvement of US soybean in maturity groups II, III, and IV. *Crop Science*, *54*, 1419–1432.

Rodrigues, F. A., Marcolino-Gomes, J., de Carvalho, J. F. C., do Nascimento, L. C., Neumaier, N., Farias, J. R. B., et al. (2012). Subtractive libraries for prospecting differentially expressed genes in the soybean under water deficit. *Genetics and Molecular Biology*, *35*, 304–314.

Roesler, K., Shen, B., Bermudez, E., Li, C. J., Hunt, J., Damude, H. G., et al. (2016). An improved variant of soybean type 1 diacylglycerol acyltransferase increases the oil content and decreases the soluble carbohydrate content of soybeans. *Plant Physiology*, *171*, 878–893.

Rotundo, J. L., Miller-Garvin, J. E., & Naeve, S. L. (2016). Regional and temporal variation in soybean seed protein and oil across the United States. *Crop Science*, *56*, 797–808.

Rotundo, J. L., & Westgate, M. E. (2009). Meta-analysis of environmental effects on soybean seed composition. *Field Crops Research*, *110*, 147–156.

Russell, D. A., Spatola, L. A., Dian, T., Paradkar, V. M., Dufield, D. R., Carroll, J. A., et al. (2005). Host limits to accurate human growth hormone production in multiple plant systems. *Biotechnology and Bioengineering*, *89*, 775–782.

Sadak, M. S., El-Hameid, A., Asmaa, R., Zaki, F. S. A., Dawood, M. G., & El-Awadi, M. E. (2020). Physiological and biochemical responses of soybean (*Glycine max* L.) to cysteine application under sea salt stress. *Bulletin of the National Research Centre*, *44*, 1–10.

Saio, K., & Watanabe, T. (1978). Differences in functional properties of 7S and 11S soybean proteins. *Journal of Texture Studies*, *9*, 135–157.

Samarah, N. H., Mullen, R. E., Cianzio, S. R., & Scott, P. (2006). Dehydrin-like proteins in soybean seeds in response to drought stress during seed filling. *Crop Science*, *46*, 2141–2150.

Schmidt, M. A., Barbazuk, W. B., Sandford, M., May, G., Song, Z., Zhou, W., et al. (2011). Silencing of soybean seed storage proteins results in a rebalanced protein composition preserving seed protein content without major collateral changes in the metabolome and transcriptome. *Plant Physiology*, *156*, 330–345.

Schmutz, J., et al. (2010). Genome sequence of the palaeopolyploid soybean. *Nature, 463*, 178–183.
Sebolt, A. M., Shoemaker, R. C., & Diers, B. W. (2000). Analysis of a quantitative trait locus allele from wild soybean that increases seed protein concentration in soybean. *Crop Science, 40*, 1438–1444.
Settlage, S. B., Kwanyuen, P., & Wilson, R. F. (1998). Relation between diacylglycerol acyltransferase activity and oil concentration in soybean. *Journal of the American Oil Chemists' Society, 75*, 775–781.
Sharon, N., & Lis, H. (2002). How proteins bind carbohydrates: Lessons from legume lectins. *Journal of Agricultural and Food Chemistry, 50*, 6586–6591.
Shimomura, M., Kanamori, H., Komatsu, S., Namiki, N., Mukai, Y., Kurita, K., et al. (2015). The Glycine max cv. Enrei genome for improvement of Japanese soybean cultivars. *International Journal of Genomics, 2015*. https://doi.org/10.1155/2015/358127.
Silva, E., Perez Da Graça, J., Porto, C., Martin Do Prado, R., Nunes, E., Corrêa Marcelino-Guimarães, F., et al. (2021). Untargeted metabolomics analysis by UHPLC-MS/MS of soybean plant in a compatible response to Phakopsora pachyrhizi infection. *Metabolites, 11*.
Singh, S. K., Barnaby, J. Y., Reddy, V. R., & Sicher, R. C. (2016). Varying response of the concentration and yield of soybean seed mineral elements, carbohydrates, organic acids, amino acids, protein, and oil to phosphorus starvation and CO2 enrichment. *Frontiers in Plant Science, 7*.
Song, S. K., Hou, W. S., Godo, I., Wu, C. X., Yu, Y., Matityahu, I., et al. (2013). Soybean seeds expressing feedback-insensitive cystathionine -synthase exhibit a higher content of methionine. *Journal of Experimental Botany, 64*, 1917–1926.
Song, Q. X., Li, Q. T., Liu, Y. F., Zhang, F. X., Ma, B., Zhang, W. K., et al. (2013). Soybean GmbZIP123 gene enhances lipid content in the seeds of transgenic Arabidopsis plants. *Journal of Experimental Botany, 64*, 4329–4341.
Song, W., Yang, R., Wu, T., Wu, C., Sun, S., Zhang, S., et al. (2016). Analyzing the effects of climate factors on soybean protein, oil contents, and composition by extensive and high-density sampling in China. *Journal of Agricultural and Food Chemistry, 64*, 4121–4130.
SoyStats. (2022). *SoyStats* (p. 2021). St. Louis, Missouri: American Soybean Association (ASA).
Staniak, M., Stepien-Warda, A., Czopek, K., Kocira, A., & Baca, E. (2021). Seeds quality and quantity of soybean [*Glycine max* (L.) Merr.] cultivars in response to cold stress. *Agronomy-Basel, 11*.
Sun, Q., Xue, J., Lin, L., Liu, D., Wu, J., Jiang, J., et al. (2018). Overexpression of soybean transcription factors GmDof4 and GmDof11 significantly increase the oleic acid content in seed of *Brassica napus* L. *Agronomy, 8*, 222.
Tacarindua, C. R. P., Shiraiwa, T., Homma, K., Kumagai, E., & Sameshima, R. (2012). The response of soybean seed growth characteristics to increased temperature under near-field conditions in a temperature gradient chamber. *Field Crops Research, 131*, 26–31.
Turner, N., Davies, S. L., Plummer, J., & Siddique, K. (2005). Seed filling in grain legumes under water deficits, with emphasis on chickpeas. *Advances in Agronomy, 87*, 211–250.
USDA [N.A.S.S.]. (2022). *USDA, National Agricultural Statistics Service*.
Utsumi, S., Damodaran, S., & Kinsella, J. E. (1984). Heat-induced interactions between soybean proteins: Preferential association of 11S basic subunits and β subunits of 7S. *Journal of Agricultural and Food Chemistry, 32*, 1406–1412.
Van, K., & McHale, L. K. (2017). Meta-analyses of QTLs associated with protein and oil contents and compositions in soybean [*Glycine max* (L.) Merr.] seed. *International Journal of Molecular Sciences, 18*, 1180.

VanToai, T. T., Lee, J. D., Goulart, P. F. P., Shannon, J. G., Alves, J. D., Nguyen, H. T., et al. (2012). Soybean (*Glycine max* L. Merr.) seed composition response to soil flooding stress. *Journal of Food, Agriculture and Environment, 10*, 795–801.

Vaughn, J. N., Nelson, R. L., Song, Q., Cregan, P. B., & Li, Z. (2014). The genetic architecture of seed composition in soybean is refined by genome-wide association scans across multiple populations. *G3: Genes, Genomes, Genetics, 4*, 2283–2294.

Wan, C., Shao, G., Chen, Y., & Yan, S. (2002). Relationship between salt tolerance and chemical quality of soybean under salt stress. *Chinese Journal of Oil Crop Sciences, 24*, 67–72.

Wan, T. F., Shao, G. H., Shan, X. C., Zeng, N. Y., & Lam, H. M. (2006). Correlation between AS1 gene expression and seed protein contents in different soybean (*Glycine max* [L.] Merr.) cultivars. *Plant Biology, 8*, 271–276.

Wang, J., Mao, L., Zeng, Z., Yu, X., Lian, J., Feng, J., et al. (2021). Genetic mapping high protein content QTL from soybean 'Nanxiadou 25' and candidate gene analysis. *BMC Plant Biology, 21*, 1–13.

Wang, H. W., Zhang, B., Hao, Y. J., Huang, J., Tian, A. G., Liao, Y., et al. (2007). The soybean Dof-type transcription factor genes, GmDof4 and GmDof11, enhance lipid content in the seeds of transgenic Arabidopsis plants. *The Plant Journal, 52*, 716–729.

Wang, S. D., et al. (2020). Simultaneous changes in seed size, oil content and protein content driven by selection of SWEET homologues during soybean domestication. *National Science Review, 7*, 1776–1786.

Wijewardana, C., Reddy, K. R., & Bellaloui, N. (2019). Soybean seed physiology, quality, and chemical composition under soil moisture stress. *Food Chemistry, 278*, 92–100.

Wilcox, J. R. (1998). Increasing seed protein in soybean with eight cycles of recurrent selection. *Crop Science, 38*, 1536–1540.

Wilson, R. F. (2004). Seed composition. In *Vol. 16. Soybeans Improv. Prod. uses* (pp. 621–677).

Wu, N., Lu, Q., Wang, P. W., Zhang, Q., Zhang, J., Qu, J., et al. (2020). Construction and analysis of GmFAD2-1A and GmFAD2-2A soybean fatty acid desaturase mutants based on CRISPR/Cas9 technology. *International Journal of Molecular Sciences, 21*.

Xu, Y., Yan, F., Liu, Y. J., Wang, Y., Gao, H., Zhao, S. H., et al. (2021). Quantitative proteomic and lipidomics analyses of high oil content GmDGAT1-2 transgenic soybean illustrate the regulatory mechanism of lipoxygenase and oleosin. *Plant Cell Reports, 40*, 2303–2323.

Yang, J., Xing, G. J., Niu, L., He, H. L., Guo, D. Q., Du, Q., et al. (2018). Improved oil quality in transgenic soybean seeds by RNAi-mediated knockdown of GmFAD2-1B. *Transgenic Research, 27*, 155–166.

Yang, A., Yu, X., Zheng, A., & James, A. T. (2016). Rebalance between 7S and 11S globulins in soybean seeds of differing protein content and 11SA4. *Food Chemistry, 210*, 148–155.

Yao, Y., You, Q., Duan, G., Ren, J., Chu, S., Zhao, J., et al. (2020). Quantitative trait loci analysis of seed oil content and composition of wild and cultivated soybean. *BMC Plant Biology, 20*, 1–13.

Yu, Y., Sheng Hou, W., Hacham, Y., Sun, S., Xiang Wu, C., Matityahu, I., et al. (2018). Constitutive expression of feedback-insensitive cystathionine \$γ\$-synthase increases methionine levels in soybean leaves and seeds. *Journal of Integrative Agriculture, 17*, 54–62.

Yung, W. S., Wang, Q. W., Huang, M. K., Wong, F. L., Liu, A. L., Ng, M. S., et al. (2022). Priming-induced alterations in histone modifications modulate transcriptional responses in soybean under salt stress. *The Plant Journal*.

Zhai, Z., Liu, H., Xu, C., & Shanklin, J. (2017). Sugar potentiation of fatty acid and triacylglycerol accumulation. *Plant Physiology, 175*, 696–707.

Zhang, H., Goettel, W., Song, Q., Jiang, H., Hu, Z., Wang, M. L., et al. (2020). Dual use and selection of GmSWEET39 for oil and protein improvement in soybean. *PLoS Genetics*, *16*.

Zhang, Y. Q., Lu, X., Zhao, F. Y., Li, Q. T., Niu, S. L., Wei, W., et al. (2016). Soybean GmDREBL increases lipid content in seeds of transgenic Arabidopsis. *Scientific Reports*, *6*.

Zhang, J., et al. (2014). Overexpression of the soybean transcription factor GmDof4 significantly enhances the lipid content of *Chlorella ellipsoidea*. *Biotechnology for Biofuels*, 7, 128.

Zhang, D. J., et al. (2018). Elevation of soybean seed oil content through selection for seed coat shininess. *Nature Plants*, *4*, 30–35.

Zhang, D., et al. (2019). Artificial selection on GmOLEO1 contributes to the increase in seed oil during soybean domestication. *PLoS Genetics*, *15*.

Zhao, J. Z., Bi, R. R., Li, S. X., Zhou, D., Bai, Y., Jing, G. Q., et al. (2019). Genome-wide analysis and functional characterization of Acyl-CoA:diacylglycerol acyltransferase from soybean identify GmDGAT1A and 1B roles in oil synthesis in Arabidopsis seeds. *Journal of Plant Physiology*, *242*.

Zhao, X., Chang, H., Feng, L., Jing, Y., Teng, W. L., Qiu, L. J., et al. (2019). Genome-wide association mapping and candidate gene analysis for saturated fatty acid content in soybean seed. *Plant Breeding*, *138*, 588–598.

Zhao, Y., et al. (2021). Enhanced production of seed oil with improved fatty acid composition by overexpressing NAD(+)-dependent glycerol-3-phosphate dehydrogenase in soybean. *Journal of Integrative Plant Biology*, *63*, 1036–1053.

Zhou, Z., Lakhssassi, N., Knizia, D., Cullen, M. A., El Baz, A., Embaby, M. G., et al. (2021). Genome-wide identification and analysis of soybean acyl-ACP thioesterase gene family reveals the role of GmFAT to improve fatty acid composition in soybean seed. *Theoretical and Applied Genetics*, *134*, 3611–3623.

Zhou, Z. K., et al. (2015). Resequencing 302 wild and cultivated accessions identifies genes related to domestication and improvement in soybean. *Nature Biotechnology*, *33*, 408–U125.

CHAPTER TEN

Soybean secondary metabolites and flavors: The art of compromise among climate, natural enemies, and human culture

Yee-Shan Ku[a,*], Ming-Sin Ng[a], Sau-Shan Cheng[a], Ching-Yee Luk[a], Ndiko Ludidi[b], Gyuhwa Chung[c], Shwu-Pyng T. Chen[a], and Hon-Ming Lam[a,*]

[a]Centre for Soybean Research of the State Key Laboratory of Agrobiotechnology and School of Life Sciences, The Chinese University of Hong Kong, Shatin, Hong Kong Special Administrative Region, China
[b]Department of Biotechnology and the DST/NRF Centre of Excellence in Food Security, University of the Western Cape, Bellville, South Africa
[c]Department of Biotechnology, Chonnam National University, Yeosu, Korea
*Corresponding authors: e-mail address: ysamyku@cuhk.edu.hk; honming@cuhk.edu.hk

Contents

Advances in Botanical Research, Volume 102
ISSN 0065-2296
https://doi.org/10.1016/bs.abr.2022.03.001

Abstract

Soybean is a food crop in high demand in Northeast Asia. Besides protein and oil, soybean is also a rich source of health-beneficial secondary metabolites such as flavonoids, terpenes, and alkaloids. The long history of soybean domestication resulted in a rich collection of soybean germplasms, which could be generally categorized as wild, landrace, and cultivated soybeans. Previous research has shown that soybean seeds from diverse genetic backgrounds exhibited different metabolite profiles. Germplasms originating from different geographical regions, i.e., at different latitudes and longitudes, probably experienced different selective pressures and evolved different secondary metabolite profiles. Domestication has generally led to a reduction in secondary metabolite contents in seeds since many of these compounds are related to the bitter taste or other agronomic traits that may hinder the ease of farming and harvest. These selection forces have possibly rendered the different flavors of soybean germplasms. Due to the popularity of soy food products, the post-domestication selection of soybean based on flavor is a common phenomenon. In Northeast Asian countries, soy foods such as soy milk, tofu, and fermented soy products are popular. Based on the consumer preference for the flavors of these products, soybean germplasms with different metabolite profiles are selected for different commercial uses. However, the breeding of soybeans for maximizing health benefits and for the preferred flavors of food products may create contradictions. Industrial methods to remove undesirable flavors and molecular breeding to produce cultivars with desired metabolite profiles may be the solution.

1. Introduction

Soybean (*Glycine max* [L.] Merr.) is an important legume crop, the seeds of which serve as human food, animal feed and a source of edible oil (Hartman, West, & Herman, 2011). Soybean is also important in the fossil fuel-free oil production for the biofuel industry (Koc, Abdullah, & Fereidouni, 2011), the production of industrial biomaterials and for the pharmaceutical industry (Chung & Singh, 2008). It is essential in sustainable agriculture because of its ability to enrich soil with nitrogen through biological nitrogen fixation, thus reducing the reliance on chemical fertilizers and their impact on soil quality and greenhouse gas emissions from their production and application (Stagnari, Maggio, Galieni, & Pisante, 2017).

Demands for soybeans are high in Northeast Asian countries including China, Japan, the Republic of Korea, and Russia, where in 2019, the quantity of soybean imports ranked in the top five among major crops (Fig. 1) (FAO, 2021). Besides proteins and oils, soybean is a reservoir of a rich variety of secondary metabolites, some of which are of high nutraceutical/pharmaceutical value to humans. Examples include isoflavones that act as phytoestrogen and are abundantly found in soybean (Messina, 1999; Messina, Persky, Setchell, & Barnes, 1994). Unlike primary metabolites, secondary metabolites are not critical for essential life processes like growth and development. However, these metabolites offer plants adaptive traits and enhance their ecological fitness (Ku, Contador, et al., 2020). The molecular structures and cellular profiles of secondary metabolites in plants are often specific to taxonomically related groups, except for some key enzymes that may have a wider distribution in the plant kingdom through horizontal gene transfer mediated by endophytic bacteria or fungi (Wink, Botschen, Waterman, Schäfer, & Waterman, 2010). By acting both as defence chemicals against natural enemies and signaling molecules mediating communications between plants and symbiotic microorganisms, secondary metabolites promote the physiological fitness of plants within their local communities (Wink et al., 2010). To combat against natural enemies such as herbivores, many of the secondary metabolites have unpleasant flavors.

Being a popular part of the human diet, soybean has undergone domestication with the conscious selections for desirable traits. For legumes in general, since many of the secondary metabolites are related to the bitter taste of seeds or other agronomic traits that may hinder the ease of farming and harvest, domestication has generally led to a reduction in secondary metabolites

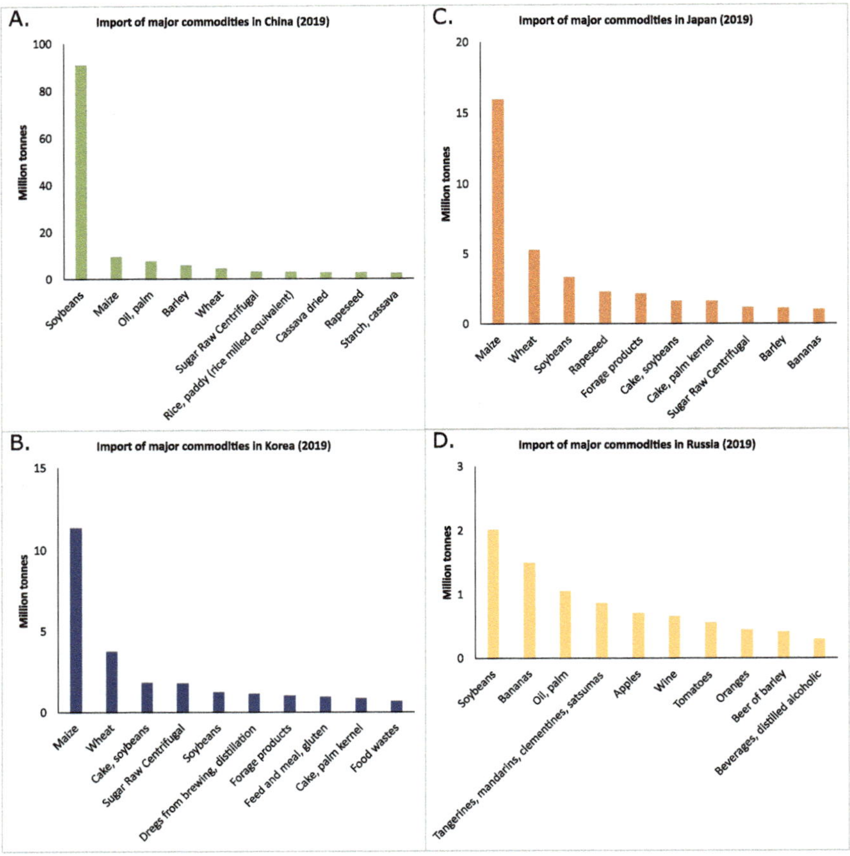

Fig. 1 Import of major commodities in (A) China, (B) Korea, (C) Japan, and (D) Russia in 2019 (FAO, 2021).

in cultivated germplasms and has created a bottleneck in genetic diversity when compared to their wild progenitors (Ku, Contador, et al., 2020). This may have caused the loss of potentially beneficial secondary metabolites among the cultivar populations. On the other hand, the elimination of secondary metabolites that give undesirable tastes from soybean cultivars would be preferable for increasing the acceptance of soy products in the consumer market. Understanding the shifts in secondary metabolite compositions in soybean as influenced by human selection and environmental adaptations, and their underlying genetic basis, might help to build (or re-build) favorable soybean secondary metabolite profiles that fit the needs of environmental adaptations, nutritional value, and flavor. This could be accomplished by exploiting the genetic richness of the wild populations.

2. The diversity of soybean germplasms

Metabolite profiling and genetic analyses of soybean have shown the genetic diversity among germplasms collected from different parts of the world, including China, South Korea, North Korea, Japan, United States, Brazil, Pakistan, Russia, Ghana, South Africa, Zambia, Ethiopia, Cameroon and Nigeria (Chander et al., 2021; Karikari, Bhat, Denwar, & Zhao, 2020; Malik et al., 2017; Valliyodan et al., 2021). The diverse genetic backgrounds of soybean germplasms imply correspondingly diverse metabolite profiles. This speculation is supported by previous findings showing that soybean germplasms with a long history of breeding selection for specific traits have diverse metabolite profiles (Kusano et al., 2015). In another study, it was also demonstrated that the metabolite profiles of soybean seeds from diverse genetic backgrounds showed accession-dependent genetic diversity (Lin et al., 2014). For example, a major difference in seed metabolite profiles was observed between black and yellow soybean seeds (Jo et al., 2021). This seed color-associated difference in metabolite profiles is thought to be a consequence of the differential accumulation of flavonoids and anthocyanins in the epidermal layer of the seed coat. Interestingly, soybean seeds with black seed coats and green cotyledons tend to have higher anthocyanin and phenolics with health benefits than the more commonly cultivated soybean with yellow seed coats and yellow cotyledons (Jo et al., 2021). This creates a niche for black soybean in pharmaceutical applications. The diverse secondary metabolite compositions in cultivated and wild soybean germplasms have been previously reviewed (Ku, Contador, et al., 2020).

3. Major biochemical pathways for the synthesis of secondary metabolites in soybean, and their molecular structures and nutraceutical/pharmaceutical values

Enzymes involved in the biosynthetic pathways of secondary metabolites are possible targets for genetic engineering. The biosynthetic pathways also hint at the nutraceutical values of the secondary metabolites, which are usually related to the molecular structure of these metabolites.

3.1 Flavonoids

Flavonoids are a group of naturally occurring polyphenols that are widely distributed in the plant kingdom. Based on the variations in their molecular structure, flavonoids are classified into several sub-groups including flavones, flavonols, flavanones, flavanonols, anthocyanidins, flavanols, and isoflavones (Ku, Ng, et al., 2020). Different modifications within each sub-group further increase the functional diversity of flavonoids. Among the various sub-groups, isoflavones are most abundant in soybean and render it the main food source of this widely adopted health supplement (Křížová, Dadáková, Kašparovská, & Kašparovský, 2019). There are three major isoflavones in soybean, namely genistein, daidzein and glycitein. The aglycone form, which is also known as the free form without any conjugate, is biologically active while isoflavone glycosides and their glycoconjugates confer stability and solubility for storage in seeds (Ku, Ng, et al., 2020). The amino acids, phenylalanine and tyrosine, are the precursors of secondary metabolite biosynthesis in the phenylpropanoid pathway (Dong & Lin, 2020). Their conversion into various secondary metabolites in this pathway is illustrated in Fig. 2. Phenylalanine is first converted into S-(cinnamoyl)-CoA by phenylalanine ammonia lyase (PAL) together with 4-coumarate-CoA ligase. Naringenin chalcone and isoliquiritigenin are generated by chalcone synthase (CHS) or chalcone reductase (CHR). Chalcone isomerase (CHI) then converts them into naringenin and liquiritigenin, which are then converted into the isoflavonoids, genistein, daidzein and glycitein, which are particularly enriched in soybean seeds (Fig. 2).

Due to the structural similarity between flavonoids (particularly genistein and daidzein) and mammalian estradiol, they are widely recognized as phytoestrogens that display therapeutic and chemopreventive effects on hormone-related cancers, such as ovarian and breast cancers, with epidemiological evidence (Lee, Kim, & Song, 2012; Nagata et al., 2014). Genistein, among the soy isoflavones, has been demonstrated to be the most significant contributor to the anticancer property of soybean (Ahmad et al., 2014). It is found to exert pleiotropic effects in modulating genes in multiple signaling pathways that control cell cycle and apoptosis (Banerjee, Li, Wang, & Sarkar, 2008). Apart from the anti-carcinogenic effects, flavonoids are reported to be antioxidants with free radical-scavenging abilities, exhibiting protective effects against oxidative stress caused by the accumulation of reactive oxygen species (ROS) (Mira et al., 2002). Other beneficial physiological effects of soy isoflavones include anti-inflammatory properties, and preventive effects on many degenerative and chronic diseases such as cardiovascular diseases, obesity, diabetes, and osteoporosis (Křížová et al., 2019). Owing to the high nutraceutical values brought forth by flavonoids, soybean has been regarded as a functional food that can promote human health through dietary consumption.

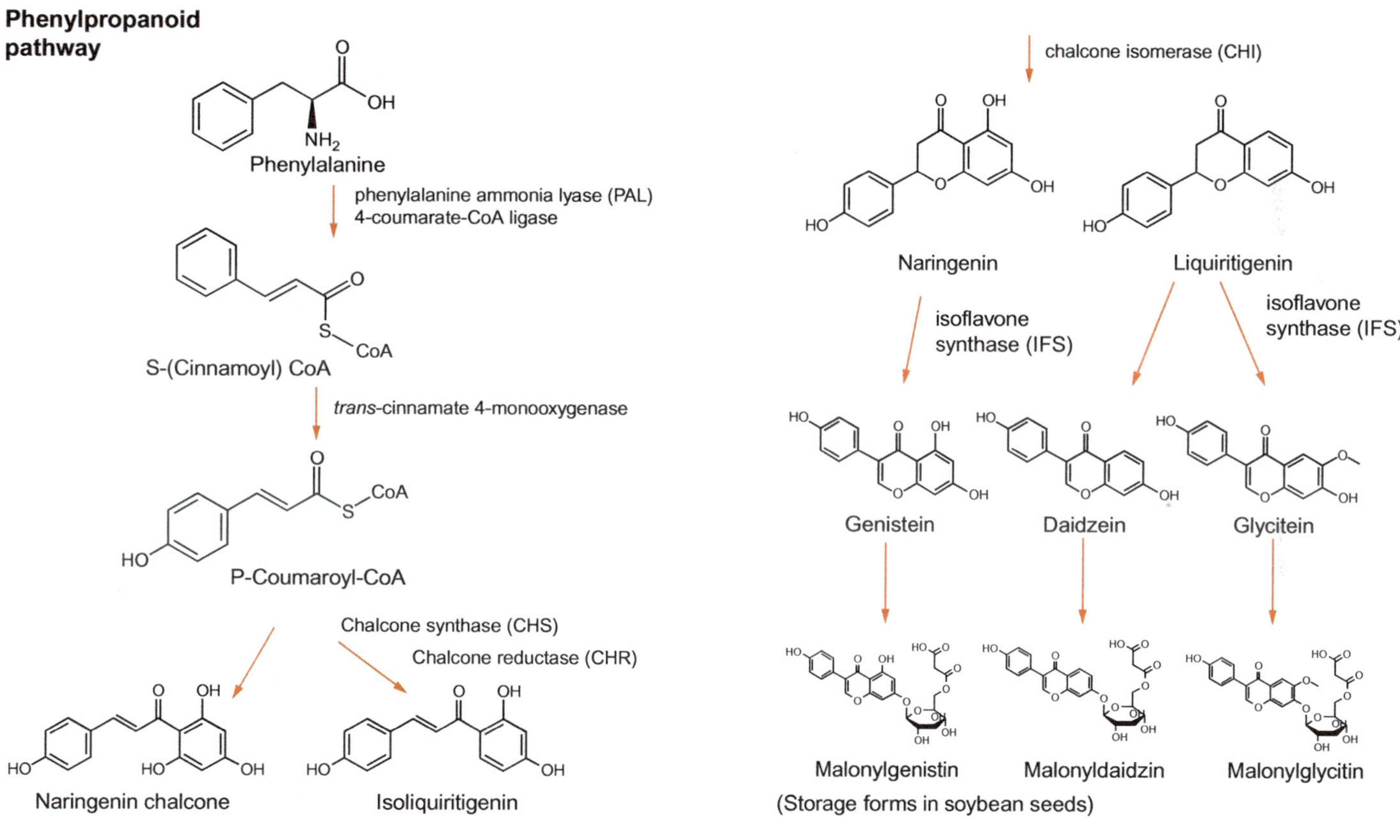

Fig. 2 The phenylpropanoid pathway for flavonoid biosynthesis.

In the plant, flavonoids help mediate interactions with its abiotic and biotic environments. For symbiotic interactions, soy isoflavones act as signaling molecules to initiate legume-rhizobium communications for the formation of nitrogen-fixing nodules (Kosslak, Bookland, Barkei, Paarent, & Appelbaum, 1987). Microbial infections or pathogen attacks also stimulate the biosynthesis and accumulation of soybean isoflavones such as glyceollin, an isoflavonoid phytoalexin, to develop resistance against biotic threats (Araújo et al., 2015; Sepiol, Yu, & Dhaubhadel, 2017). In the face of abiotic stresses, flavonoids protect plants from UV damage by absorbing UV radiation and quenching the resulting reactive oxygen species (ROS) (Kumar & Pandey, 2013). Their antioxidant properties also allow soybean to combat oxidative stresses caused by high salinity (Wu et al., 2008).

In tracking the effect of domestication on flavonoid contents in soybean, the average total seed isoflavone level in wild soybean (*Glycine soja*) was found to be higher than that in the cultivated accessions (Wang, Zhao, & Gai, 2010). In general, wild soybeans possess higher antioxidant content than cultivated ones, and such differences highly correlate to the seed total flavonoid levels (Li et al., 2016). Interestingly, a negative correlation was found between the total isoflavone levels in cultivated soybean and the longitudes and latitudes across different geographic regions of China in which these cultivars were grown, while no such correlation was found in wild accessions, indicating that climate, and possibly human preference, could have contributed to the selection pressure on soybean during the domestication process (Wang et al., 2010).

3.2 Terpenes

3.2.1 Saponins

Saponins are triterpenes or sterol glycosides present in a number of food plants such as legumes, ginseng, garlic, and tea (Oakenfull, 1981), and soybean is one of their major dietary source for both human food and animal feed. Five bioactive saponins have been isolated from soybean, complete with elucidated structures. Soyasaponin I, II and III are group B saponins while soyasaponin A_1 and A_2 are group A saponins (Kitagawa, Saito, Taniyama, & Yoshikawa, 1985; Kitagawa, Yoshikawa, & Yosioka, 1976). Group B saponins are the primary saponins in intact soybeans and generally exist in the 2,3-dihydro-2,5-dihydroxy-6-methyl-4H-pyran-4-one (DDMP)-conjugated form (Kudou et al., 1993). Saponins are derived from the terpenoid (isoprenoid) pathway. Two pathways, the mevalonate pathway and the non-mevalonate pathway (Fig. 4), are involved in isoprenoid biosynthesis.

In the mevalonate pathway, acetyl-CoA is converted into isopentenyl pyrophosphate (IPP), which could be converted to dimethylallyl pyrophosphate (DMAPP) through a reversible reaction. DMAPP is then converted to isoprene (Fig. 3). The non-mevalonate pathway is an alternative IPP and DMAPP biosynthetic pathway (Fig. 4). The mevalonate and non-mevalonate pathways both produce the same products but through different enzymatic reactions. While the former pathway takes place in the cytoplasm of plant cells, the latter pathway occurs in bacteria, eukaryote parasites, and the plastids of plant cells (Banerjee & Sharkey, 2014).

The amphiphilic nature of saponin molecules, which is contributed by the hydrophobic triterpene and the hydrophilic sugar, allows them to act as a natural surfactant to enhance the water and nutrient uptake by plants (Fournier, Erdman, & Gordon, 1998). Moreover, in response to abiotic stresses, the accumulation of saponin would be induced to protect plants from oxidative stress (Wu et al., 2008). For example, the group B saponins were found to be one of the key secondary metabolites that conferred salt tolerance in soybean, making saponins important secondary metabolites in soybean (Wu et al., 2008). The anti-herbivory properties of saponins make them natural insect repellents that protect the plant from biotic threats (De Geyter, Lambert, Geelen, & Smagghe, 2007).

When comparing between wild and cultivated soybean germplasms, clear variety-specific differences could be observed in their saponin profiles. In a study using 800 *Glycine max* and 329 *Glycine soja* germplasms, saponin type AaBc was found predominantly in wild soybeans but it only accounted for 0.3% of total saponin in the cultivated accessions, in which saponin type Aa was predominant (Tsukamoto, Kikuchi, Harada, Kitamura, & Okubo, 1993). In another study on 3025 wild soybean accessions (*Glycine soja*), a large variation in saponin contents was observed (Krishnamurthy, Chung, & Singh, 2013). However, it appeared that group B saponins, especially saponin Bc and Bg, were lost during soybean domestication (Tsukamoto et al., 1993). Therefore, the diverse genetic backgrounds governing saponin compositions in the wild soybean germplasms (*Glycine soja*) could be employed in engineering the cultivated accessions to produce the beneficial group B saponins.

In terms of nutraceutical/therapeutic value, saponins are highly beneficial to humans. Like flavonoids, saponins (particularly those in the DDMP-conjugated form) can act as antioxidants through their high ability to scavenge free radicals (Yoshiki, Kahara, Okubo, Sakabe, & Yamasaki, 2001). Other benefits including anti-inflammatory, anticarcinogenic, hypocholesterolemic,

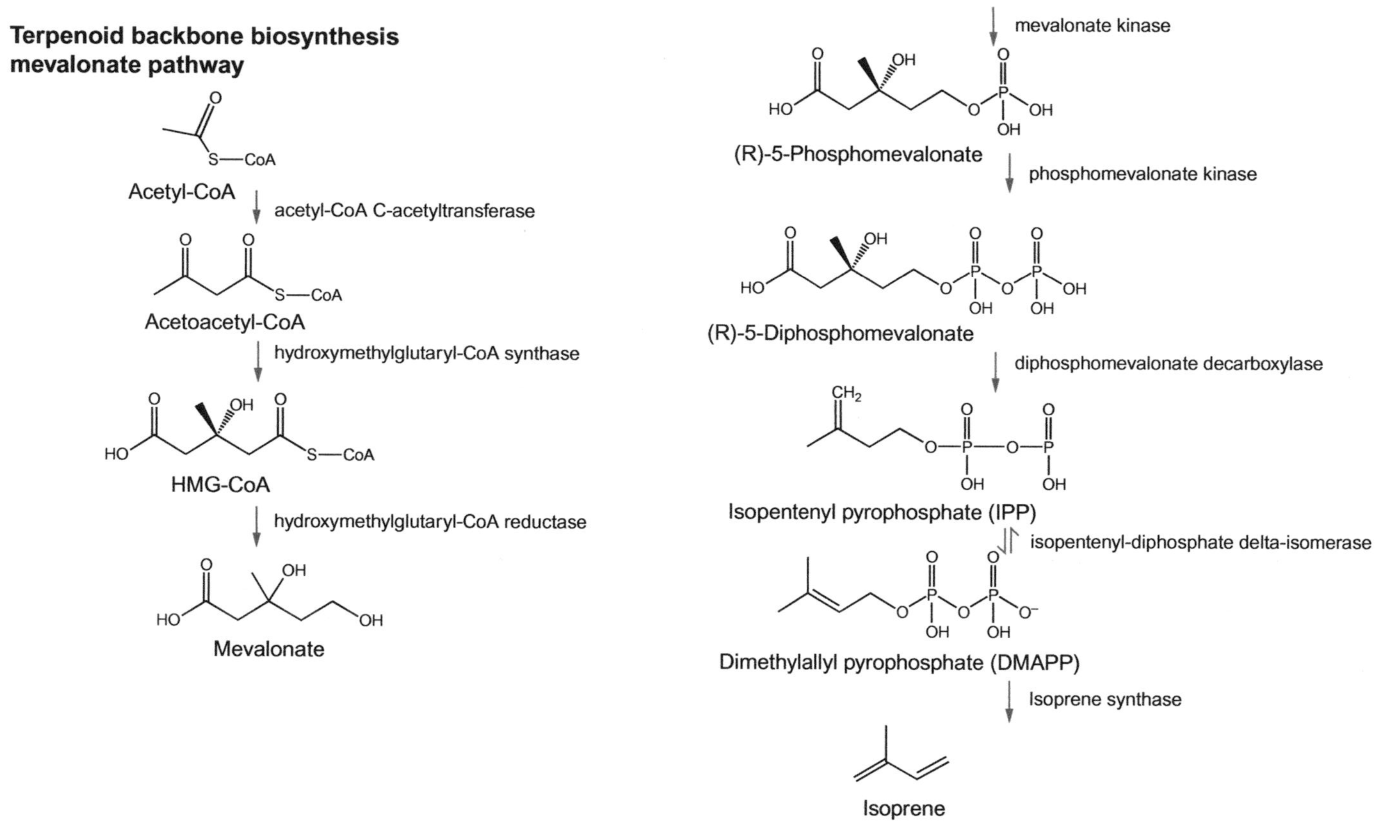

Fig. 3 The mevalonate pathway for isoprene biosynthesis.

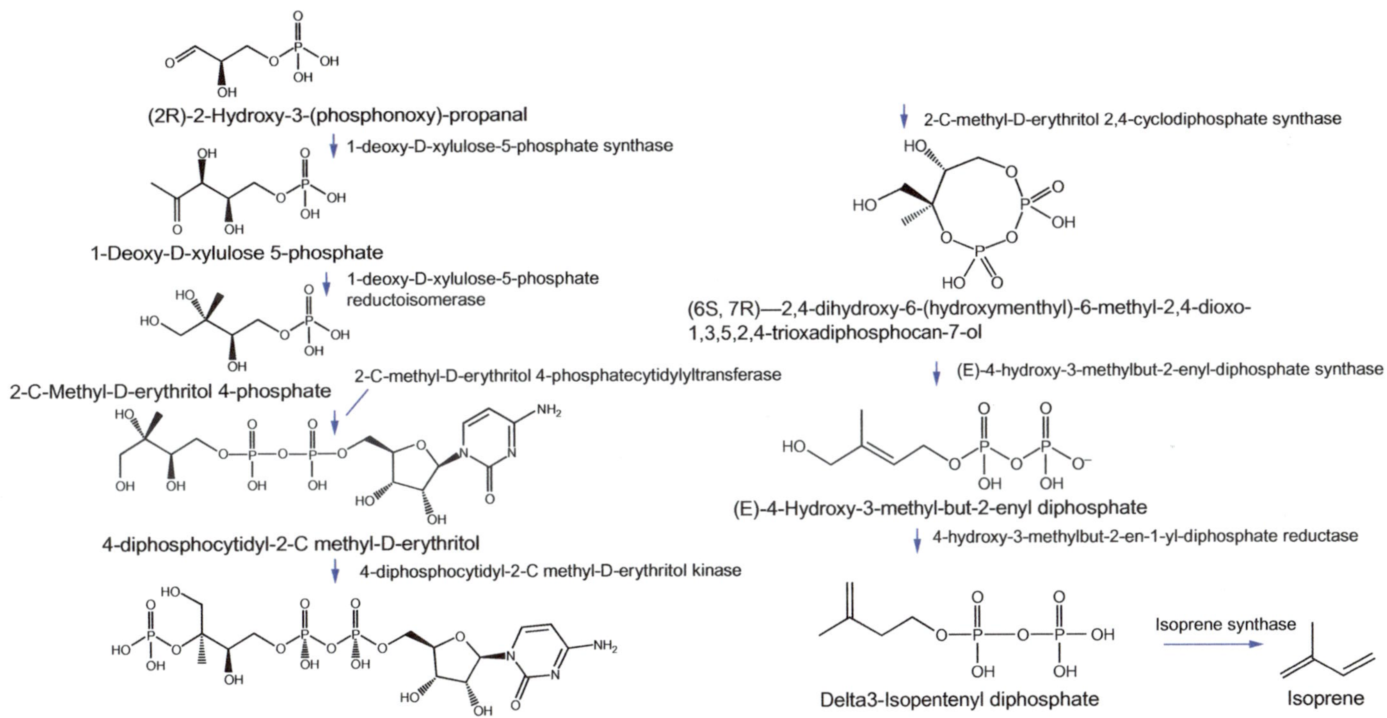

Fig. 4 The non-mevalonate pathway for isoprene biosynthesis.

immunostimulatory, hepatoprotective, and antidiabetic abilities have also been reported with soyasaponins (Fournier et al., 1998; Surh et al., 2001). Although saponins from other plants, such as alfalfa, were found to have strong hemolytic activities and were relatively toxic to some poikilothermic animals such as fish and frog, saponins derived from soybeans were found to be non-toxic and innocuous to mice, chicks, and rats, even when these animals were fed soyasaponins at levels three times higher than the natural level in soybean (Ishaaya, Birk, Bond, & Tencer, 1969). Thus, the dietary intake of soyasaponins should pose no threat to human health (Oakenfull & Sidhu, 1990).

3.2.2 Phytosterols

Phytosterols are a group of steroidal triterpenes naturally occurring in plants (Wang, Yao, & Wei, 2011). The four major types of soybean phytosterols include β-sitosterol, campesterol, stigmasterol, and brassicasterol (Wang et al., 2011). Differences among these phytosterols are mainly due to the presence/absence of a double bond at C22 and the types of substituents at C24 of the molecule (Wang et al., 2011).

Soybean oil deodoriser distillate (SODD) is the main source for commercial phytosterol extraction (Benites, Concha, Reis, & De Oliveira, 2009). The conversion of soybean phytosterols to therapeutic steroidal hormones has long been a practice of the pharmaceutical industry (Wang et al., 2011). The typical A-ring structure with a 5,6-double bond and a 3β-hydroxyl group makes soybean phytosterols an ideal raw material for producing steroidal hormones. Almost all kinds of steroidal hormones, including adrenocortical hormones and sex hormones, could be generated from the intermediates derived from soybean phytosterols (Wang et al., 2011).

Because of the structural similarity between phytosterols and cholesterol, the direct intake of soybean sterols through diet can help reduce the absorption of cholesterol in the intestines, thereby lowering the level of low-density lipoproteins in the serum, and eventually preventing atherosclerosis or other cardiovascular diseases in human (Ostlund, Racette, & Stenson, 2003). Other beneficial properties of phytosterols include anti-inflammatory, antipyretic, antineoplastic, and immunomodulating activities (Shi, Nam, & Ma, 2010).

Like cholesterol in animals, phytosterols are responsible for regulating the fluidity and maintaining the normal functions of the plant cell membrane. This includes cellular proliferation and differentiation (Piironen, Lindsay, Miettinen, Toivo, & Lampi, 2000), which explains the increase in phytosterol content in soybean seeds during germination (Shi et al., 2010).

3.3 Alkaloids

Alkaloids are a group of nitrogen-containing heterocyclic secondary metabolites with great structural diversities. Being in the legume family that possesses nitrogen-fixing ability, soybean can therefore produce more alkaloids than food crops belonging to other families (Wink, 2013).

Alkaloids are mainly derived from lysine, nicotinate, and ornithine, which are all derivatives of L-aspartate. For lysine-derived alkaloids, L-aspartate is first converted to L-lysine, which is then converted to cadaverine and indolizidine by the activities of lysine decarboxylase and lysine 6-dehydrogenase, respectively. Cadaverine is converted to quinolizidine alkaloids or piperideine alkaloids by a series of enzymatic reactions. The quinolizidine ring is then formed from the cyclization of cadaverine (Suzuki, Koike, Murakoshi, & Saito, 1996). Once the quinolizidine skeleton is formed, various alkaloid products can be generated by enzymes such as hydroxylases and dehydrogenases. For nicotinate-derived alkaloids, quinolinate is converted from L-aspartate by quinolinate synthase. Using quinolinate as the precursor, nicotinate is synthesized by nicotinate-nucleotide pyrophosphorylase, and then eventually converted to pyridine alkaloids through a series of enzymatic reactions. In the ornithine pathway, L-glutamate is first converted into L-ornithine, and then through decarboxylation into putrescine. Next, putrescine is converted to pyrrolizidine alkaloids, pyrrolidine alkaloids or tropane alkaloids (Lichman, 2021). The biosynthesis of alkaloids is illustrated in Fig. 5.

Quinolizidine alkaloids (QAs) are mainly found in *Lupin* spp. and have been employed in pharmaceutical uses due to their antibacterial, antiviral, antipyretic, and cytotoxic activities (Saito & Murakoshi, 1995). In bitter *Lupinus angustifolius* accessions, the activity of lysine decarboxylase is high in leaves, stems, pedicels, and pods while the level of QAs is high in the phloem (Otterbach, Yang, Kato, Janfelt, & Geu-Flores, 2019). It was therefore suggested that QAs are synthesized in other tissues and then transported to seeds (Otterbach et al., 2019).

QA biosynthesis in soybean has not been fully characterized. However, both known and novel alkaloids have been isolated from soybean greens (Wang et al., 2016). The known alkaloids include gensenine, (1S,3S)-1-methyl-1,2,3,4-tetrahydro-β-carboline-3-carboxylic acid, (1R,3S)-1-methyl-1,2,3,4-tetrahydro-β-carboline-3-carboxylic acid, and indole-3-carboxylic acid. The newer alkaloids include soyalkaloid A and isoginsenine (Wang et al., 2016). Likewise, lysine decarboxylase has been characterized and isolated from soybean (Kim, Kim, & Cho, 1998). In another study, lysine decarboxylase activities were detected in the extracts of soybean root and hypocotyl (Matsuzaki & Soybean, 2009).

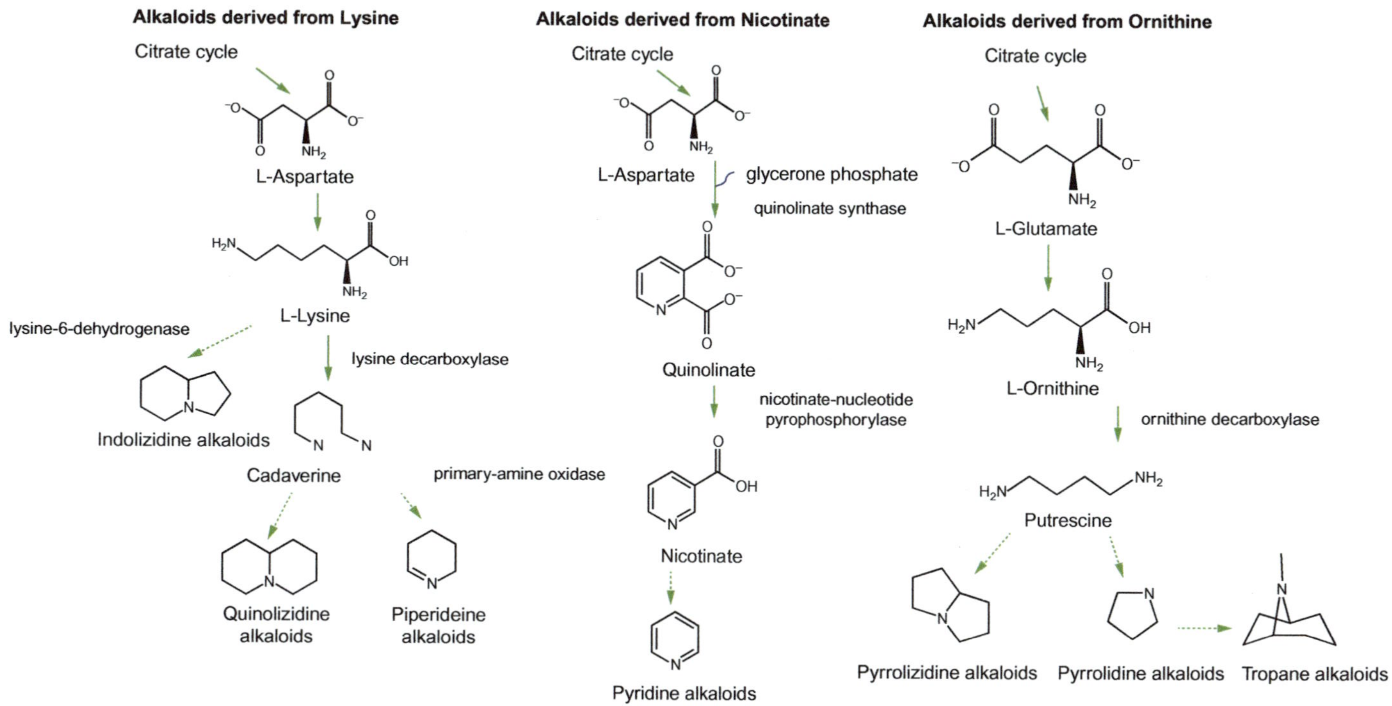

Fig. 5 The pathway for alkaloid biosynthesis.

Many alkaloids are reported to have pronounced pharmaceutical values. They possess anticancer activities (Gupta et al., 2015), anti-malarial activities (Onguéné et al., 2013), neuroprotective potential of preventing stroke by facilitating blood circulation in the brain (Kumar & Khanum, 2012), and potential as therapeutic drugs for targeting neurodegenerative diseases, such as Parkinson's disease and Alzheimer's disease (Amirkia & Heinrich, 2014). In addition, the newly identified soyalkaloid A was found to be able to ameliorate concanavalin A-induced hepatic injury (Wang et al., 2016).

Many alkaloids are neuromodulators or neurotoxins and are thought to have evolved in plants for defence against pathogens and herbivores (Wink, 2013). In one study on the causal constituents of insect resistance in soybean cultivars, quinic acid, which is derived from quinine, is identified to be one of the chemicals with dramatically different levels between the insect-resistant and insect-susceptible lines (Tester, 1977).

4. The influence of climate on soybean secondary metabolite production

4.1 Impacts of latitude on secondary metabolites in soybean: Photoperiod and temperature

Geographical regions at different latitudes have different daylengths and temperatures. Daylength, a measurement of the photoperiod during which plants receive light radiation from the sun, has been known to affect the vegetative period and the time to flowering. Average daily temperature also affects the timing for planting and the rate of growth within a growing season. Evidence showed that planting soybean too early in the season when daylength was short led to premature flowering and reduced yield (Board & Hall, 1984; Islam, Fujita, & Zheng, 2019), while planting too late in the season with shortening daylength was also enough to cause premature flowering and reduced yield (Board & Hall, 1984; Islam et al., 2019). Planting soybean at warmer temperatures was found to reduce the time to first flowering compared to planting at cooler temperatures, and the temperature effects were more prominent under short daylength than under long daylength (Board & Hall, 1984).

Besides influencing the time to flowering and yield, latitude and longitude, which are associated with daylength and temperature, also have a significant effect on secondary metabolite profiles. The diverse seed isoflavone profiles among wild soybean accessions originating in different geographical areas in Korea have been documented (Tsukamoto et al., 2018).

Using geographical data and orthogonal partial least-squares discriminant analyses on soybeans grown at different longitudes and latitudes in China, a study demonstrated that growing soybean at high longitudes yields seeds with higher isoflavone contents. It also showed that soybean grown at higher latitudes had higher accumulation of anthocyanins (Liu et al., 2017). The same study illustrated that enhanced biosynthesis of isoflavones occurred in high-longitude environments whereas a low-longitude environment promoted anthocyanin biosynthesis (Liu et al., 2017). Existing evidence suggests that total phenolic and isoflavone contents in soybean seeds from cooler regions are higher than those from warmer regions (Król-Grzymała & Amarowicz, 2020). However, there is a trade-off between desired metabolite profiles and yield because higher average daytime temperatures within physiological limits are beneficial to soybean yield (Zheng, Chen, & Han, 2009). Higher mean daily maximum temperature during pod filling leads to grain yield increase by a range of 6 − 10% for each 1 °C rise. This can lead to more than 22% of total yield increase per growing season at the optimal growth temperature during the pod filling stage (Zheng et al., 2009).

Taking daylength and temperature into consideration, it is thus important to identify soybean accessions suitable for cultivation in specific geographical regions. Soybean cultivars of different maturity groups have different photoperiod responses for the adaptation to different daylength conditions (Yang et al., 2019). It is possible to breed early-maturing soybean accessions to be grown under short daylength without compromising the yield. Similarly, it will be of interest to exploit the influences of temperature and latitude as a means to tailor soybean breeding to enhance the production of health-beneficial secondary metabolites.

4.2 The role of secondary metabolites in facilitating adaptations to climate change

Climate change is characterized by the increased frequencies and intensities of climate-related disasters including floods, heat waves and drought (Pareek, Dhankher, & Foyer, 2020). These stresses have a detrimental effect on agricultural production and food security, not only because they lead to low crop yields for human food and animal feed, but also because such low yields drive up agricultural commodity prices, which in turn exacerbates the impoverished state of the livelihoods of poor communities. It is thus important to exploit the vast genetic differences between cultivated crops and their wild relatives as a means to re-introduce genes into cultivated crops to enhance their adaptation to the stresses associated with climate change.

Primary and secondary metabolites are differentially accumulated in soybean leaves in response to drought and heat stress (Das, Rushton, & Rohila, 2017). Metabolites associated with carbohydrate, amino acid and peptide metabolism, such as purine and pyrimidine biosynthesis, were impacted by drought and heat stress (Das et al., 2017). Sugar and nitrogen metabolism were also shown to play important roles in regulating the responses to drought and heat stress in soybean (Das et al., 2017).

The consequence of flooding is hypoxia in crops since the water overload limits gas exchange between the atmosphere, the soil and the plant. By contrasting the metabolite profiles of two soybean germplasms with different responses to flooding stress, it was shown that flooding highly affected primary and secondary metabolites in soybean plants. The most affected primary and secondary metabolites were those involved in carbon and nitrogen metabolism; and those involved in the phenylpropanoid pathway, respectively (Coutinho et al., 2018). Integrating the knowledge from metabolomics and proteomics can allow us to accurately pinpoint the key regulators of flooding responses by crop plants, including soybean. Specifically, proteomic analyses have produced evidence strongly suggesting that the most affected proteins due to flooding are related to development, protein synthesis or degradation, and secondary metabolism (Komatsu, Makino, & Yasue, 2013). These proteins were found to be involved in pathways consistent with those uncovered by metabolic profiling of flooding response (Coutinho et al., 2018). Given that genome sequencing of cultivated soybean accessions (*Glycine max*) and their wild relatives (*Glycine soja*) has been proposed as a tool for revealing the mechanisms underlying processes involved in regulating potential pathways for adaptation to climate change (Li et al., 2017), it should be possible to translate this information to comparative proteomic and metabolite profiling to shed light on the genes, proteins and metabolites that are important for the adaptation to climate change. This would require screening contrasting (sensitive vs tolerant) lines and analyzing their genomic variations, as well as the transcriptomic, proteomic and metabolomic responses to drought, heat and flooding stresses.

Hence, such analyses emphasize the relevance of omics approaches to the development of new soybean varieties that can adapt to climate change. Furthermore, it is evident that secondary metabolites in the phenylpropanoid and anthocyanin biosynthesis pathways play a critical role in conferring tolerance to the three abiotic stresses characteristic of climate change, namely drought, heat and flooding.

5. The influence on soybean secondary metabolite production by natural enemies

5.1 Overview of the perception of biotic stress by plants

Plants are sessile and therefore are unable to physically evade microbial infections and herbivore attacks. Instead, plants have evolved a range of defence mechanisms in response to these biotic stimuli, such as the production of antimicrobial or toxic compounds, which are either constitutively produced or synthesized when challenged by biotic stresses. Phytohormones and secondary metabolites such as phenolics and flavonoids play an important role in plant defence responses.

Defence against pathogens starts with the recognition of non-self by plants. Molecules involved in non self-recognition have been extensively reviewed. In summary, fragments of flagellin, DNA, lipoproteins, and lipopolysaccharides from microbes serve as pathogen-associated molecular patterns (PAMPs) or microbe-associated molecular patterns (MAMPs). Recognition of PAMPs or MAMPs by the plant pattern recognition receptors (PRRs) would initiate the downstream defence responses (Ku, Cheng, et al., 2020; Ku, Contador, et al., 2020; Ku, Ng, et al., 2020; Medzhitov & Janeway, 1997). Herbivores produce a suite of herbivore-associated molecular patterns (HAMPs) similar to PAMPs and MAMPs, which induce a set of anti-herbivory defence responses (Felton & Tumlinson, 2008; Mithöfer & Boland, 2008). In plants, the perception of invasions, whether microbial or herbivore-related, rapidly activates ion flow mechanisms, modulates the production of reactive oxygen species (ROS), modifies primary metabolism and the biosynthesis of secondary metabolites, and induces the expressions of defence genes (Romero-puertas, Terrón-camero, Peláez-vico, Molina-moya, & Sandalio, 2021).

5.2 Soybean plants can distinguish between mechanical damage and herbivore attack

Plants have evolved sophisticated signaling networks to perceive and cope with herbivory. The proteomic analyses of the salivary secretion of green stink bug, *Nezara viridula*, identified a significant portion of the active enzymes to be associated with digestion, detoxification, and antioxidation (Giacometti et al., 2020; Lomate & Bonning, 2016). Soybean plants can distinguish between mechanical damage and herbivore attack. When developing soybean seeds are damaged, lignin accumulates in the lacerated area and leads to seed cell wall thickening, but this only occurred in the presence of insect saliva, and not after mechanical damage (Giacometti et al., 2020).

In soybean, it has been shown that the damage by stink bugs is perceived by mitogen-activated protein kinases (MPKs), which triggered jasmonic acid (JA)/ethylene (ET)- and salicylic acid (SA)-regulated defences downstream through the MPK cascades (Giacometti et al., 2020). A similar pathway of defence response activation was also reported in *Nicotiana attenuate* upon herbivory attack (Wu, Hettenhausen, Meldau, & Baldwin, 2007). In soybean, upon stink bug feeding on the pod, the expressions of *MPK3*, *MPK4*, and *MPK6* in the seed were induced (Giacometti et al., 2016). In addition, the MPK3, MPK4, and MPK6 proteins were activated by phosphorylation (Giacometti et al., 2016). On the other hand, mechanical damage alone did not elicit the same effects on the MPK genes and proteins as the damage due to stink bug feeding. Although mechanical damage also induced the expression of *MPK6* and phosphorylation of MPK6, it repressed the expression of *MPK3* and *MPK4* or induce their phosphorylation (Giacometti et al., 2016). In the same study, the genes involved in flavonoid synthesis in developing seeds, *phenylalanine ammonia-lyase 2* (*PAL2*), *chalcone synthase 7* (*CHS7*), and *isoflavone synthase 2* (*IFS2*), had increased transcript levels 72 hours after both stink bug feeding and mechanical damage (Giacometti et al., 2016). At the same time, the transcript levels of *PAL2* and *CHS7* were also inducible by methyl-jasmonate or SA treatment (Giacometti et al., 2016). Such phenomena hinted at the involvement of flavonoids in defence responses. In another study comparing the transcript profiles between aphid infestation-susceptible and aphid infestation-resistant soybean plants after aphid attack, it was found that aphids could suppress phytohormone signaling in the susceptible plants (Studham & MacIntosh, 2013). In the aphid-susceptible soybean plants, a significant induction of abscisic acid (ABA)-related transcripts was observed 7 days after aphid infestation (Studham & MacIntosh, 2013). It was hypothesized that aphids could induce ABA production which suppresses SA- and JA-mediated defence responses (Studham & MacIntosh, 2013).

5.3 Flavonoids play a role in the defence against herbivores

The correlation between flavonoid levels and herbivory deterrence has been investigated (Burden & Norris, 1992; Chiang, Norris, Ciepiela, Shapiro, & Oosterwyk, 1987; Hart, Kogan, & Paxton, 1983; Peruca et al., 2018; Piubelli, Hoffmann-Campo, De Arruda, Franchini, & Lara, 2003;

Zhou et al., 2011). Multiple isoflavones, for example, daidzin, 4′,7-dihyroflavone, daidzein, and formononetin, were induced and accumulated in leaves following the *Spodoptera litura* larvae attack (Murakami et al., 2014; Nakata et al., 2016). In addition, malonyldaidzin, ononin, and malonylononin were accumulated in soybean leaves treated with *S. litura* gut content extracts. By applying ^{13}C-labeled phenylalanine together with gut content extracts to mechanically damaged leaves, ^{13}C was incorporated into the 4′,7-dihydroxyflavone, daidzein, and formononetin found in the leaf extract, supporting the hypothesis that active flavonoid biosynthesis occurs at the site of elicitation (Murakami et al., 2014). The increase in selective *de-novo* biosynthesis of soybean isoflavones upon the gut content elicitation was previously reported (Nakata et al., 2016), suggesting a possible legume-specific defence response against *S. litura*. As the *S. litura* gut content does not have direct β-glucosidase activity, the accumulation of isoflavone aglycones might be a result of the enhanced hydrolysis of isoflavone glycosides. It has been demonstrated that *S. litura* could not metabolize daidzein, which could inhibit the growth of *S. litura* larvae (Zhou et al., 2011). Also, the herbivory-deterrent effect of daidzein was observed in experiments conducted with other Spodoptera members (Peruca et al., 2018). The accumulation of daidzein, without being converted to glyceollin, might serve as an adaptive anti-herbivory defensive strategy of soybean against chewing insects.

The presence of phytoalexins and phenolic compounds in the soybean plant might alter the host selection of Mexican bean beetle, *Epilachna varivestis* Mulsan (Burden & Norris, 1992; Chiang et al., 1987; Hart et al., 1983), and this antiherbivory effect was found to be dose-dependent (Burden & Norris, 1992). Soybean glyceollin reduced the efficiency of food utilization but did not affect the growth, development, and survival of Mexican bean beetle larvae (Hart et al., 1983). The elevated level of total phenolics correlated with the quantifiable and temporal patterns of active PAL and L-tyrosine ammonia-lyase (TAL) enzymes in affected tissues (Chiang et al., 1987). Besides the Mexican bean beetle, the application of glyceollin also showed a dose-dependent antiherbivory effect on the chrysomelids bean leaf beetle, *Cerotoma trifurcata* (Forster), and southern corn rootworm, *Diabrotica undecimpunctata howardi* Barber as well (Fischer, Kogan, & Paxton, 1990). Different species of chewing insects elicit similar defence responses in the soybean plant, where feeding by one species of chewing insect could initiate a broad resistance response by the plant against other chewing insects as well, known as cross-resistance (Srinivas, Danielson, Michael Smith, & Foster, 2001).

Soybean aphid is one of the common herbivores that can colonize soybean for a long term. Strategies must be adopted by the aphid to suppress the soybean defence responses (Kanobe, McCarville, O'Neal, Tylka, & MacIntosh, 2015; Liechti & Farmer, 2002). Metabolomic and transcriptomic analyses on aphid-infested soybean revealed the upregulation of the phenylpropanoid pathway and the accumulation of isoflavonoids, such as daidzein, formononetin, and genistein (Hohenstein et al., 2019). Mass spectrometry imaging showed that isoflavone accumulation due to aphid infestation was localized in mesophyll cells and the epidermis (Hohenstein et al., 2019; Klein et al., 2015). This suggests that the accumulation of isoflavonoids is a part of the non-phloem defence against phloem-sucking aphid colonization. Polyunsaturated fatty acids are the precursors of jasmonic acid (JA) which is involved in plant response to herbivory (Liechti & Farmer, 2002). Herbivore attack might have resulted in the reduced levels of polyunsaturated fatty acids, particularly linoleic and linolenic acids in soybean, with a concomitant increase in palmitic acid in leaves, and stearic and oleic acid in seeds upon aphid colonization (Kanobe et al., 2015). As the remodeling of fatty acid composition was not observed in root colonization by soybean cyst nematode (SCN) and brown stem rot fungus, this observation appeared to be specific to aphid infestation in soybean. Alterations in fatty acid composition might be one of the counter-strategies for aphids to suppress the effectiveness of JA-mediated defence responses by modulating fatty acid desaturase (FAD) 2 and FAD6 activities in soybean (Kanobe et al., 2015). Moreover, SA was also accumulated in the form of pipecolic acid, suggesting the active regulation by aphids of soybean defence via the SA pathway (Klein et al., 2015).

Upon damage to developing seeds due to feeding by the southern green stink bug, elevated levels of daidzein, which is the immediate precursor of glyceollin, and genistein were observed (Piubelli et al., 2003). After the sting bug injury, an increase in the isoflavone concentration was observed. However, glyceollin was not detectable. It was hypothesized that the midgut bacterial community of stink bugs could play a role in deactivating the chemical defences of soybean developing seeds as well. Some of the bacterial community isolates showed *in-vitro* β-glucosidase and raffinase activities, which enable the degradation of galactosyl derivatives and increase the digestibility of soybean plants. Enzymes in midgut isolates might also degrade isoflavonoids and deactivate soybean protease inhibitors, helping aphids tolerate soybean defences and feed on the plant (Medina et al., 2018).

5.4 Phytohormones mediate the accumulation of secondary metabolites during defence responses

The signaling cascade for non-self-recognition can be mediated by various phytohormones, including SA, JA, and ET (Bari & Jones, 2009). These hormones can then trigger biochemical and molecular strategies to deter pathogen colonization and insect herbivory (Bari & Jones, 2009; Ku, Sintaha, Cheung, & Lam, 2018), which often involve the accumulation of secondary metabolites.

The induction of JA upon necrotrophic pathogen infection, insect attack and wounding has been recently reviewed (Ku et al., 2018). In soybean, it was reported that methyl jasmonate treatment on the cotyledon enhanced the light-induced accumulation of malonyl-genistin (Graham & Graham, 1996). When methyl jasmonate was applied together with *Phytophthora sojae* cell-wall glucan elicitor in the light, the accumulation of malonyl-daidzin was increased (Graham & Graham, 1996). However, the induced accumulation of malonyl-genistin and malonyl-daidzin were both inhibited by the application of abscisic acid (ABA) (Graham & Graham, 1996).

Attack by *Aphis glycines* has been shown to increase the accumulation of daidzein and genistein in soybean leaves, and this effect could be further enhanced by the application of beta-aminobutyric acid (BABA), which provided additional upregulation on isoflavone and callose biosynthesis genes, leading to greater accumulation of genistein and callose, and the subsequent increased resistance to *A. glycines* (Yao, Zhong, Wang, Yan, & Wu, 2020).

5.5 The accumulation of secondary metabolites and their role in the defence against pathogenic microbes

In response to the fungus, *Phakopsora pachyrhizi*, that causes Asian soybean rust, soybean produces a vast number of secondary metabolites, such as terpenes and glyceollin. (Parmezan et al., 2020; Silva et al., 2020, 2021; Tremblay, Hosseini, Alkharouf, Li, & Matthews, 2010; van de Mortel et al., 2007). For example, *P. pachyrhizi* inoculation was shown to induce daidzein and genistein accumulation in soybean plants, and the application of quercetin or kaempferol could inhibit *P. pachyrhizi* spore germination (Lygin et al., 2009). It is hypothesized that terpene production might also contribute to the resistance against *P. pachyrhizi* in soybean, and its accumulation is temporally regulated by the *Rpp* genetic background (Parmezan et al., 2020). Temporal variations in the expressions of soybean genes in the flavonoid biosynthetic pathway and of the WRKY family of transcription factors upon *P. pachyrhizi* infection were observed (van de Mortel et al., 2007). In a global gene expression analysis, in the resistant soybean plants, 12h after *P. pachyrhizi* infection when fungal germination and penetration were initiated, a number of leaf genes peaked in expression, then dropped back down to levels similar to those of the mock-inoculated control (van de Mortel et al., 2007). However, in the susceptible plants, the expression levels of this set of genes remained unchanged until 96 hours after *P. pachyrhizi* infection. Some of the genes displaying such differential expression pattern between the resistant plants and the susceptible plants upon *P. pachyrhizi* infection are involved in the flavonoid biosynthetic pathway (van de Mortel et al., 2007).

The necrotrophic plant pathogen, *Sclerotinia sclerotiorum*, induces oxidative stress in the host, resulting in pathogenesis (Chen, Ullah, Reichelt, Gershenzon, & Hammerbacher, 2019). For example, *S. sclerotiorum* was found to secrete quercetin dioxygenase to degrade flavonols in Arabidopsis (Chen et al., 2019), and in soybean, wounding and *S. sclerotiorum* infection in stems induced the accumulation of isoflavones, especially daidzein and genistein, in leaves (Wegulo, Yang, Martinson, & Murphy, 2005). By integrating the results from transcriptomic and metabolomic analyses, it was found that the antifungal activities in soybean upon *S. sclerotiorum* infection were achieved by reprogramming the phenylpropanoid pathway to modulate the redox capacity (Ranjan et al., 2019). In particular, fluctuations in the cellular levels of dehydroascorbic acid (DHA) and trans-4-hydroxy-L-proline suggested the involvement of antioxidants in the resistance response to *S. sclerotiorum* (Ranjan et al., 2019). Moreover, during *S. sclerotiorum* infection, the accumulation of jasmonoyl-L-isoleucine (JA-Ile), which is the bioactive form of jasmonate, was observed earlier in the resistant soybean lines then in the sensitive line. At 6 and 12h after inoculation, the level of JA-Ile was higher in the resistant line than in the sensitive line (Ranjan et al., 2019). However, at 72h after inoculation, the level of JA-Ile was higher in the sensitive line than in the resistant line (Ranjan et al., 2019). Such a phenomenon indicates the importance of JA signaling to the resistance against *S. sclerotiorum* (Ranjan et al., 2019).

5.6 Secondary metabolites play a role in beneficial plant-microbe interactions

Flavonoids and isoflavonoids play a critical role in the interactions between plants and microbes, including both pathogens and beneficial symbionts (Phillips & Kapulnik, 1995; Stafford, 1997). In contrast to the inhibitory effect on the pathogenic fungi *P. pachyrhizi*, flavonoids stimulated the germination of arbuscular mycorrhizal fungus spores and enhanced hyphal growth to facilitate nutrient transports between the rhizosphere and soybean roots (Abdel-Lateif, Bogusz, & Hocher, 2012). Soybean plants use flavonoids and isoflavonoids as chemoattractants and inducers to attract rhizobia toward their roots and to activate the nod genes cassette for nodulation (Spaink, 1995). Plant growth-promoting rhizobacteria (PGPR) could trigger soybean secondary metabolism through microbe-associated molecular patterns (MAMP) systemic induction. PGPR was shown to have successfully primed the plant against *Xanthomonas axonopodis* pv. *glycines*. Although not all PGPR strains could trigger a significant induction of isoflavones, MAMPs from PGPR had shown a cross-resistance defensive metabolism in soybean, suggesting the involvement of other non-isoflavone elements in regulating the defence against this pathogen (Algar et al., 2014). PGPR could also enhance the plant's resistance against soybean cyst nematode (SCN) through reprogramming the biosynthesis of secondary metabolites, such as phenylpropanoids, alkaloids, and fatty acids (Kang, Zhu, Wang, Chen, & Duan, 2018).

5.7 The negative roles of secondary metabolites in plant defence

The susceptibility of soybean plants is associated with the types and amounts of secondary metabolites induced in the plant according to the particular type of stress and to specific herbivores. Compared to the *Euschistus heros*-resistant soybean cultivar (Dowling), the susceptible cultivar (Williams 82) had a higher level of sugars and isoflavones, and was preferred by *Euschistus heros* for oviposition (Silva, O-Panizzi, Blassioli-Moraes, & Panizzi, 2013). In a feeding choice test, the same amount of the phytoalexin, glyceollin, had a greater deterring effect on Mexican bean beetle larvae and adults than on soybean looper (Hart et al., 1983). Transcriptomic and biochemical differences were observed and reported between the resistant and susceptible plants, suggesting the contribution of genetic variations to the resistance against herbivores and pathogens (Chiang et al., 1987; Chiozza, O'Neal, & MacIntosh, 2010; Parmezan et al., 2020; Silva et al., 2013; Studham & MacIntosh, 2013; van de Mortel et al., 2007).

Several genetic variations between the resistant and susceptible soybean plants were comprehensively discussed in insect feeding studies. Aphids feed on the soybean phloem that provides carbohydrates and nitrogenous compounds. A positive correlation between available phloem nitrogen in soybean and aphid abundance was demonstrated (Chiozza et al., 2010; Walter & DiFonzo, 2007). In particular, susceptible soybean plants had a higher level of α-aminobutyric acid, asparagine, glutamine, glutamic acid, histidine, proline, and serine (Chiozza et al., 2010). Such nutritional quality could have influenced the determination of host suitability by the insect (Chiozza et al., 2010). In another study, transcriptome analyses revealed that many defence-related genes were constitutively expressed in the *Rag1*-carrying resistant soybean line, but were activated only during aphid infestation in the susceptible soybean line (Studham & MacIntosh, 2013).

6. Secondary metabolites in soy food products: The balance between nutrients and flavor

6.1 Secondary metabolites as anti-nutrients

Despite being the "super food" that provides valuable nutrients, soybean also contains several anti-nutritional factors that may counteract its nutritional quality.

6.1.1 Trypsin inhibitor

Trypsin inhibitors (TIs) are considered a key anti-nutritional factor in soybean. There are two major protease inhibitors in soybean, the Bowman-Birk trypsin inhibitor (BBI) and the Kunitz trypsin inhibitor (KTI) (Liener, 1995). Consumption of these TIs in relatively high concentrations from raw soy flour was found to cause pancreatic hyperplasia and hypertrophy in rats, weakening their growth (Liener, Nitsan, Srisangnam, Rackis, & Gumbmann, 1985). In rats, prolonged feeding with soy flour containing high TIs eventually led to the development of neoplastic and hyperplastic nodules, including carcinomas, in pancreas (McGuinness et al., 1980).

Fortunately, both TIs could be inactivated by heat or other preprocessing methods such as fermentation and germination (Dixit, Kumar, Rani, Manjaya, & Bhatnagar, 2011). However, valued soy proteins and essential amino acids could become insolubilized by the heat treatment and therefore lost in the final soya products (Dixit et al., 2011). New treatments, for example, gamma irradiation, could be considered for better conservation of nutritional attributes (Dixit et al., 2011). Although it has been recognized as safe to treat food with gamma irradiation (Ravindran & Jaiswal, 2019), such treatment may bring forth other concerns. It has been reported that gamma irradiation leads to the reduction of phenolic compounds and tannin besides the trypsin inhibiting activity in raw and cooked soybean (de Toledo, Canniatti-Brazaca, Arthur, & Piedade, 2007). In general, gamma irradiation is also known to reduce the vitamin contents in food products, such as vitamins A, B_1, C, D, E, and K (Kilcast, 1994; Lima, Vieira, Santos, & Mendes de Souza, 2018; Ravindran & Jaiswal, 2019). Using irradiation rather than heat treatment needs careful consideration as radiation does destroy nutrients like vitamins.

In contrast to KTI, BBI survives in gastric juice (Messina & Barnes, 1991). Despite being an anti-nutritional factor, BBI was found to exhibit anticarcinogenic activities through its ability to inhibit chymotrypsin, in addition to the original substrate trypsin (Fournier et al., 1998). The lowest effective concentration for its chemoprotective activity can be easily achieved through dietary intake of soybean, based on the level used in animal studies (0.1% of total diet) (Messina & Barnes, 1991). Therefore, it is a fine balance between destroying the anti-nutritional properties and retaining the dosage required for beneficial effects when treating soy products to remove TIs.

TIs protect plants against insect feeding and soil pathogens (Marchetti et al., 2000). To understand the effect of domestication on TI contents in soybean, genomic and proteomic profiles of wild, landrace and cultivated soybean accessions were studied (Natarajan, Xu, Bae, & Bailey, 2007). The levels and variations of the protein profiles of KTIs were found to be higher in wild soybean germplasms than cultivated ones (Natarajan et al., 2007). During soybean domestication, the levels and the complexity of KTIs in seeds may have been reduced or lost.

6.1.2 Phytic acid

Phosphorus (P) in soybean seed is mostly stored in the form of phytic acid, an organic phosphate with the structural formula, myo-inositol 1,2,3,4,5,6-hexaphosphate (Raboy, 1997). Due to the strong ability of phytic acid to chelate metal ions, it is primarily deposited as a mixture of phytic salts of various mineral cations, particularly potassium and magnesium (Raboy, 1997). In developing seeds, the concentrations of inorganic P and minerals are important for regulating enzymatic activities involved in critical metabolic processes, such as protein and starch metabolism, for early seed growth and germination. Since the reserve levels of P and metal ions in the seed are often severalfolds greater than the optimal levels, the deposition and remobilization of phytic acid represent important mechanisms for the storage and retrieval of P and mineral ions (Raboy, 1997). It was found that metal cations were deposited to phytic acid at the aleurone layer after they had completed their physiological roles in seed development (Ogawa, Tanaka, & Kasai, 1979).

Phytic acid concentration is high in oilseeds, averaging 1.5% dry weight (Erdman, 1979). Unfortunately, phytic acid P is nutritionally unavailable to nonruminants (Erdman, 1981). It has therefore been regarded as an anti-nutritional factor in the diet, for its excretion as the salts of essential mineral ions might lead to both mineral and P deficiencies in human populations (Erdman, 1981). Conflicting results were often produced regarding the anti-nutritional property of phytic acid. In a study that fed rats with soy-derived meals, a reduction in zinc and iron availability was detected (Liener, 1995), while an experiment with human subjects conducted by USDA did not find such any significant reduction in these minerals (Miles et al., 1987). Despite being an anti-nutrient, the dietary intake of phytic acid was found to suppress iron-catalyzed oxidative damage in the colon, hence there is growing interest in phytic acid being an agent against colon cancer (Graf & Eaton, 1993). Considering the potential benefits and harm that dietary phytic acid might bring to human, its quantitative variations among cultivated soybean lines deserve much attention.

In a study on 38 cultivated soybean germplasms, the proportion of seed total P represented by phytic acid varied from 67% to 77%, with no significant difference among the germplasms (Raboy, Dickinson, & Below, 1984). As for seeds of wild soybean germplasms, the concentrations of phytic acid were generally 30% greater than those in cultivated soybeans (Raboy et al., 1984). Since phytic acid accounted for a relatively constant portion of the total P among all the germplasms, and the total P in seeds was found to be influenced by both heritable multi-genic controls and the environmental availability of P to the developing plants, these factors also accounted for the variations in phytic acid P among the germplasms (Raboy, 1997). Phytic acid accounts for more than 70% of total P in the seed of maize (Raboy et al., 2000). In a study that searched for non-lethal homozygous maize mutants with low phytic acid contents, two mutants named *lpa-1* and *lpa-2* were identified (Raboy et al., 2000). Both of them displayed a wild-type kernel phenotype under a reduction in phytic acid P and an increase in inorganic P (Raboy, Young, & Gerbasi, 1994). This indicates that the sequestration of P as phytic acid is not essential for P homeostasis, and the highly correlated relationship between seed total P and phytic acid P is not an absolute requirement for the normal development and function of the seed (Raboy, 1997). This study might provide insights for engineering low-phytic acid mutants of soybean with wild-type viability, or provide a genetic marker for screening desirable phytic acid P contents in soybean.

6.2 Secondary metabolites and soybean flavor

6.2.1 Lipoxygenases as off-flavor generators

Since lipoxygenases initiate the production of off-flavor compounds, they are regarded as anti-nutritional factors. Lipoxygenases (LOXs) belong to a group of non-heme iron-containing dioxygenases widely distributed in animals, plants, and fungi (Lenis, Gillman, Lee, Shannon, & Bilyeu, 2010). In soybean, up to 2% of seed proteins are represented by lipoxygenases (Lenis et al., 2010). Lox-1, Lox-2, and Lox-3 are the three major isozymes present as storage proteins in the mature soybean seed, and each one is encoded by a unique gene (Wang, Croft, Järlfors, & Hildebrand, 1999). Lox-1 is more active on linoleic acid (LA) and is categorized as Type I LOX with an optimal pH around 9; Lox-2 and Lox-3 are more active on methyl linoleate and are categorized as Type II LOX with an optimal pH between 6.5 and 7 (Dixit et al., 2011). The pH-dependent activities of LOXs give hints on the design of soy food processing methods. Lox-2, although classified in a different category from Type I LOX, can also oxygenate esterified polyunsaturated fatty acid (PUFA) units in biomembranes (Robinson, Wu, Domoney, & Casey, 1995). All soybean lipoxygenase isozymes are monomeric and contain one iron atom per molecule of protein for their enzymatic function (Donnelly & Robinson, 1995).

Lipoxygenases catalyze the oxidation of PUFAs that contain a *cis, cis*-1, 4-pentadiene moiety to produce an initial free radical, which mediates subsequent oxygenation and conjugates to form hydroperoxyl unsaturated acids (Donnelly & Robinson, 1995). These initial products would then be degraded by hydroperoxide lyases and isomerases to generate a variety of aldehydes, ketones and alcohols, also known as the green leaf volatiles (GLVs). Many of them are of low flavor thresholds and give an objectionable rancid flavor and aroma (Robinson et al., 1995).

The physiological role of LOX in soybean seed development remains unclear as no adverse effects were found on the crop performance of the LOX1/2/3 triple null mutant (Wang et al., 1999). The three isoforms disappear during the first day of seed germination, and the absence of substantial oxygenation events on PUFAs suggests that they are not involved in lipid mobilization during germination (Wang et al., 1999). Three new LOX isozymes, LOX 4-6, appear in the cotyledon of germinating soybean (Wang et al., 1999). These LOX isozymes are believed to be involved in plant growth, wounding, and defence against pathogens and herbivores (Wang et al., 1999).

The ability of lipoxygenases in catalyzing free radical formation also makes them an undesirable component in soybean products

(Donnelly & Robinson, 1995). The initial free radical might also trigger chain reactions such as the subsequent oxidation of non-PUFA constituents such as vitamin A and β-carotene, resulting in altered sensory quality (flavor and aroma), loss of essential nutrients and overall reduction in the quality of food (Donnelly & Robinson, 1995). Since most lipoxygenases are able to withstand thermal processing, they remain sufficiently active to affect the quality of soy products after long-term storage (Donnelly & Robinson, 1995). In intact cells, LA and LOX are separated in different subcellular compartments. However, the processing of soy products requires crushing and homogenization of soybean seeds and subsequently results in the mixing of the enzyme with its substrate (Mellor, Bligh, Chandler, & Hodgman, 2010). This allows the endogenous LOX to initiate a series of lipid oxidation reactions and eventually produce a range of volatile aldehydes (hexanal and nonenal) and ketones, the off-flavor compounds that give soybean product an unpleasant grassy-beany flavor and aroma (Robinson et al., 1995).

The presence of these objectionable flavors and aroma greatly reduces product desirability, particularly in beverages made from crushed soybeans (Mellor et al., 2010). N-hexanal, among the off-flavor compounds, is the one of particular interest to food manufacturers in their attempt to improve soy product flavor (Mellor et al., 2010). Some processing methods have been developed to reduce the off-flavor of soy milk, but the trade-off is the introduction of an undesirable chalky texture to soybean products (Iassonova, Johnson, Hammond, & Beattie, 2009). Soybean cultivars with low LOX content are therefore in demand in the food market. However, removing LOX will not totally resolve the off-flavor issue because soybean unsaturated fatty acids have self-oxidative mechanisms that do not require participation of LOX but need oxygen presence (Vercellotti, St. Angelo, & Spanier, 1992). Therefore, removing oxygen in the vessel by deaeration (Chew & Nyam, 2020) during processing will greatly improve the off-flavor issue.

6.2.2 The bitter tastes of flavonoids, saponins, and alkaloids

Despite being beneficial phytonutrients that promote health in human populations, plant-based flavonoids, saponins and alkaloids are mostly bitter. They trigger the bitter rejection response in humans and are routinely removed in debittering processes by the food industry (Drewnowski & Gomez-Carneros, 2000).

In a study on the non-volatile minor constituents responsible for the undesirable bitter taste in soy flour, three isoflavones, glycitein 7-β-O glucoside, daidzein and genistein, were identified to additively contribute to the objectionable flavor of soybean products (Huang, Hsieh, & Chang, 1982). Glycitein 7-β-O glucoside, in particular, possesses herb-like bitterness and astringency (Huang et al., 1982).

In another similar study on ground soybean, both group A and group B saponins, together with isoflavones, contributed significantly to the undesirable bitter and astringent taste (Okubo et al., 1992). Among these molecules, group A soybean saponins had the most intense undesirable bitter taste when conjugated to glycosides (Okubo et al., 1992). While the degradation of soybean saponins to the aglycone form reduced such an undesirable taste, the reverse trend was observed in isoflavone glycosides (Okubo et al., 1992).

Being toxic secondary metabolites that confer herbivory resistance on plants, alkaloids also contribute to the bitter taste of plants (Drewnowski & Gomez-Carneros, 2000). The instinctive rejection by humans due to the bitter taste of alkaloid-containing crops could help to avoid suffering from acute toxicity from consuming an excessive amount of the alkaloids (Wink, 2013).

The physiological roles, nutritional values, and flavors of secondary metabolites in soybean are summarized in Table 1.

7. Natural and human selections of soybean

Considering their physiological roles in soybean plants and nutritional values and effects on flavors in soy foods, secondary metabolites have experienced heavy natural and human selective pressures. While natural selection has worked on improving the adaptations and ecological fitness of the soybean plant to its environment, human selection has focused on enhancing the ease of cultivation, harvest, improving yield, catering to consumer preferences and, ultimately, profitability (Ku, Contador, et al., 2020). Since secondary metabolites are linked to many domestication-related traits, human selection on the crop performance of soybean could indirectly mean the selection of secondary metabolite profiles (Ku, Contador, et al., 2020).

7.1 Natural selection: Adaptations to the environment

In terms of environmental adaptations, both natural and artificial selections favor a more effective response to climate change or abiotic stresses and more favorable interactions with other organisms in the local community, including symbiosis with beneficial rhizobia and fungi, and defence responses against pathogens and herbivores. However, the growing of soybean in fields with favorable cultivation conditions and protection from farmers against crop diseases and herbivory have largely diminished the pressure of natural

Table 1 Summary of the physiological roles, nutritional values, and flavors of secondary metabolites in soybean.

Secondary metabolites	Physiological role in plant	(Anti-)Nutritional value for humans	Flavor	References
Flavonoids	– Initiate symbiotic interactions with rhizobia – Abiotic and biotic stress resistance	– Phytoestrogen with anticarcinogenic effects – Antioxidant – Anti-inflammatory effect – Prevention of degenerative and chronic diseases	Bitter and astringent	Ahmad et al. (2014), Banerjee et al. (2008), Carter et al. (2018), Huang et al. (1982), Kosslak et al. (1987), Křížová et al. (2019), Kumar and Pandey (2013), Lee, Kim, and Song (2012), Mira et al. (2002), Nagata et al. (2014), Sepiol et al. (2017), Wu et al. (2008)
Saponins	– Natural surfactant – Abiotic and biotic stress resistance	– Antioxidant – Anti-inflammatory, anticarcinogenic, hypocholesterolemic, immunostimulatory, hepatoprotective, and antidiabetic effects	Bitter and astringent	Fournier et al. (1998), De Geyter et al. (2007), Okubo et al. (1992), Surh et al. (2001) Wu et al. (2008), Yoshiki et al. (2001)
Alkaloids	– Defence against pathogens and herbivores	– Anticancer activities – Anti-malarial activities – Preventing stroke by facilitating blood circulation in brain – Therapeutic effects in neurodegenerative diseases	Bitter	Gupta et al. (2015), Huang et al. (1982), Kumar and Khanum (2012), Onguéné et al. (2013), Wink (2013)

Continued

Table 1 Summary of the physiological roles, nutritional values, and flavors of secondary metabolites in soybean.—cont'd

Secondary metabolites	Physiological role in plant	(Anti-)Nutritional value for humans	Flavor	References
Phytosterols	– Stabilizing plant cell membrane – Controlling cellular proliferation and differentiation	– Precursors of steroidal hormones – Reducing serum cholesterol levels – Anti-inflammatory, antipyretic, antineoplastic, and immunomodulating activities	Not determined	Ostlund et al. (2003); Piironen et al. (2000), Shi et al. (2010), Wang et al. (2011)
Trypsin inhibitor	– Defence against insects and pathogens	– BBI exhibits anticarcinogenic activities – As anti-nutrients: Prolonged consumption of high concentration of raw soybean may lead to development of neoplastic and hyperplastic nodules, including carcinomas, in pancreas	Not determined	Fournier et al. (1998), Marchetti et al. (2000), McGuinness et al. (1980)
Phytic acid	– Phosphorus storage in developing seed	– Anti-colonic cancer – As anti-nutrients: May lead to mineral and P deficiencies	Not determined	Erdman (1981), Graf and Eaton (1993), Raboy (1997)
Lipoxygenase	– Storage protein	– As anti-nutrients: Leading to loss of essential nutrients	Generating off-flavor compounds such as aldehydes	Donnelly and Robinson (1995), Wang et al. (1999)

selection from abiotic and biotic stresses (Ku, Contador, et al., 2020). Since the secondary metabolites that facilitate ecological adaptions were no longer needed to enhance crop yield or performance in agricultural lands, they were no longer prioritized in the human selection process and might instead be lost due to the undesirable properties they confer on the soybean plant or soy products. For example, the bitter-tasting genistin and group B saponins are found to be key secondary metabolites that were associated with increased salt tolerance and a distinguishing feature between wild and cultivated soybeans (Wu et al., 2008). In cases where it is necessary to cultivate soybean in areas with adverse environmental conditions, such as high latitudes or high salinity, or where the threat from diseases is serious, the wild progenitors can provide the genetic resources for reintroducing the ability to produce these favorable secondary metabolites into cultivars.

7.2 Human selection: Seed traits, nutritional values, and flavors

In some instances, human selection might introduce additional pressure or have a completely opposite direction of selection to that of natural selection during soybean domestication. Domestication-related traits mainly concern the yield and harvestability of crops. Yield-related parameters include leaf length, leaf width, pod length, number of seeds per pod, 100-seed weight, days to flowering, and floral color, and those related to harvestability include pod shattering and germination rate (Lo et al., 2018). Other agronomic traits selected for due to human preferences include seed size, appearance, taste, and digestibility (Ku, Contador, et al., 2020). Many of these agronomic traits are regulated by secondary metabolites. For example, as discussed in Section 6.2, the flavor of soybean seeds is affected by the off-flavor generating lipoxygenases and the bitter-tasting flavonoids, saponins and alkaloids. Other examples of agronomic traits controlled by flavonoids include tissue color and seed coat color, which are related to the ease of sowing and religious concerns (Abbas et al., 2017; Heiser, 1988). Floral pigmentation, a trait related to both insect pollination and human esthetic preferences, is also determined by flavonoids (Tanaka, Sasaki, & Ohmiya, 2008). The digestibility of seeds is influenced by all flavonoids, saponins, and alkaloids (Gupta, 1987).

7.2.1 Simultaneous selection for large seeds with high oil but low protein contents

The focus of crop domestication has been on the improvement of yield and harvestability (Dehaan et al., 2016). Larger seeds are directly related to higher yield and easier harvest. Domestication-related loci associated with seed size were identified in soybean (Zhou, Jiang, et al., 2015; Zhou, Wang, et al., 2015), and It has been suggested that soybean domestication is directly related to the increase in seed size (Wang et al., 2020).

Soybean has been subjected to the simultaneous selection for larger seed size, higher oil content, and lower protein content, and that this was achieved by the selection for the gene, *GmSWEET10a*, which encodes a sugar transporter (Wang et al., 2020). An increased expression of *GmSWEET10a* leads to an enhanced flux of sugars from the maternal tissue to the embryo, thus facilitating cell division and expansion in the embryo and consequently an increased seed size. The increase in sugar supply to the embryo also enhances lipid synthesis. However, the nitrogen supply may become a limiting factor for these processes. The result is a reduction in the protein content (Wang et al., 2020). The selection for *GmSWEET39* for higher oil but lower protein seed contents was suggested to have followed a similar path (Zhang et al., 2020). These findings are consistent with a previous report highlighting the negative correlation between oil and protein contents in soybean seed (Patil et al., 2018).

The selection for desirable levels of oil content in soybean seeds during domestication has been discussed (Zhou, Jiang, et al., 2015; Zhou, Wang, et al., 2015; Zong et al., 2017). The human selection for soybeans with high oil contents was suggested to begin as early as 7500 years B.P. (C-14 ages where the year 1950 is used as "present"), with the high-oil content soybean cultivars established by 4000 years B.P. (Zong et al., 2017). Later, it was found that the increase in soybean seed oil content as a result of domestication had resulted from the artificial selection for the gene *GmOLEO1*, which encodes the oleosin protein for oil body formation (Zhang et al., 2019). It was also found that the quantitative trait loci (QTLs) related to protein, oil, and sucrose contents were found to overlap with those QTLs governing domestication-related traits such as plant maturity and seed color (Patil et al., 2018).

While soybean cultivars have been reported to have a wide range of seed sugar levels (Hou et al., 2009), the process of selecting for seed sugar contents during domestication is less clear than that for seed oil content. In a study attempting to identify the molecular footprints of soybean domestication, a gene related to carbohydrate metabolism was found to be related to domestication (Li et al., 2013). However, genes related to carbohydrate metabolism were found to exhibit significant variations in their DNA methylation levels during soybean domestication (Shen et al., 2018). Although sugar transporters, such as *GmSWEET10a* (Wang et al., 2020) and *GmSWEET39* (Zhang et al., 2020), have been reported to be selected for during domestication, it has been suggested that such selections were indirect results of the selection for lipid synthesis and therefore the seed lipid content (Wang et al., 2020).

7.2.2 Selection for seeds with reduced flavonoids in the seed coat

Seed color, including seed coat color and hilum color, is a domestication related trait (Zhou, Jiang, et al., 2015; Zhou, Wang, et al., 2015). Wild soybean varieties tend to have black seed coats while cultivated varieties tend to have yellow seed coats (Wang et al., 2018). It has been suggested that seeds with light colors were selected during domestication for easier sowing and religious reasons (Heiser, 1988). The loss of seed coat color is a result of the selection for the dominant allele, *I*, which silences *Chalcone Synthase* (*CHS*) alleles and thus inhibits the flavonoid pathway in the seed coat (Paauw, Koes, & Quattrocchio, 2019; Tuteja, Zabala, Varala, Hudson, & Vodkin, 2009). In a study comparing among the metabolic profiles of a cultivated, a landrace and a wild accession, epicatechin was only found in the landrace and the wild accessions but not the cultivated one (Yun, Kang, Kim, Kim, & Kim, 2020). Since epicatechin is related to the hardness of the seed coat, it was suggested that the existence of epicatechin in the landrace and wild accessions renders the hardness and the decreased germination rates of the seeds (Yun et al., 2020).

Besides "domestication," the concept of "post-domestication selection" has also been introduced (Abbo et al., 2014). The appearance and taste of crops could be regarded as being subjected to post-domestication improvement (Abbo et al., 2014).

7.3 Controversy on the selection for off-flavor-producing metabolites in soybean

Glycosides, including those of saponins and isoflavones, have been identified to contribute to the bitterness and astringency of soybean seeds, with group A saponins being the major contributors (Okubo et al., 1992). The *Sg-1* locus was found to regulate the structural diversity of saponins in soybean (Sayama et al., 2012), and is situated in a genomic region under selection during soybean domestication (Zhou, Jiang, et al., 2015; Zhou, Wang, et al., 2015). Nevertheless wild soybeans do not necessarily have higher saponin contents than cultivated ones (Ku, Contador, et al., 2020). During domestication, it is not clear whether soybeans with low levels of saponins were actively selected for. From the perspective of the food industry, it is reasonable to select soybean varieties with low seed saponin levels to rid soybean seeds of bitterness and astringency. However, from the perspective of nutraceutical benefits, there may be reasons to select soybean varieties with high seed saponin levels.

In short, domestication has selected for a more pleasant taste, a better yield, and a more esthetically favorable appearance in the soybean seed. However, such alterations in the characters of the plant often contradict with natural selection for better defence against natural enemies, and higher rates of dispersal and survival of seeds. This genomic evolution due to artificial selection results in a shift in secondary metabolite profiles and generally reduces the genetic diversity and the availability of secondary metabolites in soybean cultivars. However, sometimes due to the close proximity of certain QTLs or genes for secondary metabolite biosynthesis to those responsible for other domestication-related traits, concomitant selections for secondary metabolites may also have happened (Heiser, 1988).

8. Possible post-domestication selection of soybean seeds based on their uses in food products

As discussed above, commercial considerations possibly play a role in human selection on soybeans. Isoflavones in soybeans are related to the bitter taste of soy food products. Specifically, it has been suggested that daidzein and genistein are responsible for the objectionable flavor of soy milk (Matsuura, Obata, & Fukushima, 1989). Genistein was found to stimulate the human bitter receptors hTAS2R14 and hTAS2R39 (Roland et al., 2011). Besides isoflavones, other contents, such as protein, lipid, and sugar, also influence the taste of the soy food products.

Generally, high protein and high sucrose are desirable characteristics for food-grade soybeans (Jegadeesan & Yu, 2020), although most soybeans are used for animal feed (Voora, Larrea, & Bermudez, 2020). Soybean provides the second most important source of vegetable cooking oil (Voora et al., 2020) as well as being consumed in various cuisines. When soybean seeds are made into different food products, there are different considerations for the seed nutrient content. For example, in European countries, the most commonly consumed soybean-based foods are in the form of soy milk and bean sprouts (Keinan-Boker et al., 2002). Soy milk has been used as a major alternative to cow's milk, due to the prevalence of allergies toward cow's milk protein (casein), as well as lactose intolerance (Vanga & Raghavan, 2018). Lactose intolerance has been recognized as a common phenomenon among Asians since 1960s, and therefore there have been concerns about the consumption of cow's milk and other milk products (Davis & Bolin, 1967). Lactose intolerance refers to lactose malabsorption combined with symptoms (Jellema, Schellevis, van der Windt, Kneepkens, & van der Horst, 2010). In a global analysis, the world lactose malabsorption prevalence in children and adults aged 10 years or older was predicted to be 68% (Storhaug, Fosse, & Fadnes, 2017). Besides lactose intolerance, cow's milk allergy (CMA) also raises health concerns. Generally, about 2–3% infants and 0.5–3% adults are allergic to cow's milk, although the percentages vary among different populations (Jooste, Anelich, & Motarjemi, 2014). In the early 1900s, soy milk, which is technically an aqueous extract of soybean, was introduced as an "artificial milk" (Berk, 1992). In the 1930s, Dr. Harry W. Miller, an American physician-missionary, promoted the consumption of soy milk in China in the attempt to improve the nutritional status of the population, especially infants and small children, during food shortages. However, soy milk was not popularized in Hong Kong until after World War II. A beverage company, Hong Kong Soya Bean Products Co. Ltd, promoted soy milk as a soft beverage instead of as an "artificial milk" (Berk, 1992). Despite its popularity in Hong Kong, soy milk was not widely accepted by customers in the West (Berk, 1992). The "beany taste" of soy milk has been considered to be a major hurdle of soy milk promotion (Berk, 1992). Soy protein allergy is another hurdle despite of high quality protein with nine essential amino acids (Cordle, 2004; Singer, Zhang, Mian, & Huang, 2019). Besides the preference of soy milk to cow's milk in Asian countries, soybean is in high demand in Asian countries as a common ingredient in their cuisines. Examples of soybean-based foodstuffs in Asian countries include tofu, soy sauce, *jang*, and *natto*.

In the following sections, common soybean-based food products will be used as examples to illustrate the artificial selection for specific seed content compositions.

8.1 Soy milk

Ketones and aldehydes, which are produced by the oxidation of soybean oil, have been suggested to be the source of the "beany taste" in soy milk (Berk, 1992). Specifically, these compounds include hexanal, hexanol, *trans*-2-nonenal, 1-octen-3-ol, *trans,trans*-2,4-decadienal, benzaldehyde, 2-pentyl furan, 1-octen-3-one, *trans,trans*-2,4-nonadienal, *trans*-2-hexenal, *trans*-2-nonenal, and dimethyl trisulfide (Li et al., 2019; Yu, Liu, Hu, & Xu, 2018; Yuan & Chang, 2007), and their levels in soy milk are influenced by processing methods and soybean varieties.

Different soybean cultivars give different levels of off-flavor compounds in soy milk (Yu et al., 2018). Among 26 soybean cultivars from China processed using the same technology, the total off-flavor compounds in the soy milk, including hexanal, hexanol, benzaldehyde, 1-octen-3-ol, 1-octen-3-one, 2-pentylfuran, *trans*-2-nonenal, *trans,trans*-2-4-nonadienal, and *trans,trans*-2-4-decadienal, were found to range from 8.37 to 39.25 mg/L (Yu et al., 2018). The cultivar having the lowest level of total off-flavor compounds lacks lipoxygenases 1 and 2 (Yu et al., 2018). Lipoxygenase activity, protein content, and linoleic acid are positively correlated with the "beany taste" of soy milk (Yuan & Chang, 2007). It was demonstrated that soy milk produced from soybeans lacking lipoxygenases had less undesirable aroma and bitterness than that made from soybeans having functional lipoxygenases (Yang, Smyth, Chaliha, & James, 2016). A positive correlation between soybean protein content and hexanal in soy milk was shown. This is consistent with a previous study demonstrating that the protein level in soybean was positively correlated with the level of lipids, which was in turn positively correlated with that of volatile compounds formed from the lipids in soy milk (Min, Yu, Yoo, & St. Martin, S., 2005). Another report found that the concentrations of "beany taste"-causing compounds, including hexanal, *trans*-2-hexanal, and 3-methylbutyraldehyde, are significantly correlated with the contents of fatty acids specifically and not with the contents of total protein or total fat (Shi et al., 2015). The rationale behind such correlations remained unclear, but overall, for use in the food industry to produce soy milk with low levels of off-flavor compounds, soybean varieties with low levels of lipoxygenases, proteins and lipids tend to be selected.

8.2 Tofu

Tofu is a popular soy food and is usually consumed as a meat alternative (Sadler, 2004). Similar to soy milk, tofu produced from soybeans lacking lipoxygenases had reduced undesirable aromas and bitterness than that produced from soybeans having lipoxygenases (Yang et al., 2016). In addition to the taste, the firmness of tofu is also an important property (Lim, DeMan, Deman, & Buzzell, 1990). Coagulation is thus a critical procedure for tofu production, and tofu processing leads to 44% reduction of isoflavones although the distribution of different forms of isoflavones was not affected (Wang & Murphy, 1996). Some protein-associated isoflavones may have lost in the whey when Ca^{2+} in the coagulant reacts with the proteins during coagulation (Wang & Murphy, 1996). The lower levels of isoflavones, including free, glycosylated and malonylated isoflavones, in tofu compared to soybean flour was also demonstrated in another report (Liu, Chang, & Wiesenborn, 2005). The lower level of soysaponin I in tofu compared to soybean flour was also reported (Hu, Lee, Hendrich, & Murphy, 2002). Tofu may thus be a desirable soy food product if the flavors of isoflavones and saponins are of concern.

8.3 Fermented soybean products

Fermentation has been used as a means of food preservation (Ross, Morgan, & Hill, 2002), as well as for providing other benefits such as the generation of unique flavor and aroma (Ray & Joshi, 2014) and the production of antibiotics (Ross et al., 2002). Fermented soy products also bring forth health benefits such as alleviating the effects of diabetes mellitus, hypertension, cardiac disorders, and cancer-related issues (Jayachandran & Xu, 2019). Fermented soybean products have been widely adopted in Asian countries. Although these products have been regarded as ethnic foods in Asian countries, they are becoming more widely available in western countries.

8.3.1 Soy sauce

Soy sauce is produced by the fermentation of a mixture of defatted soybean meal and roasted wheat grits, using microorganisms such as *Aspergillus*, *Pediococcus*, *Saccharomyces*, and *Candida* spp. (Sugiyama, 1984; Wilson, 1995). Soy sauce is also called *shoyu*, *jiang-you*, and *kanjang* in Japan, China and Korea, respectively (Wilson, 1995). Soybeans with high protein and unsaturated fatty acid contents are preferred for soy sauce production (Wan, Wu, Wang, Wang, & Hou, 2013). During fermentation, the enzymatic activities on starch and protein result in the production of flavor compounds including alcohols, acids, esters, aldehydes, furans, and pyrazines (Zhao et al., 2018). The degradation of starch and protein in the mixture by microorganisms under high salt concentrations results in the unique aroma of soy sauce (Sugiyama, 1984; Wilson, 1995).

8.3.2 Jang

In Korea, the fermentation of soybean has been recorded since the Three Kingdom period (57 B.C.–A.D. 668) (Cwiertka & Moriya, 2008). A few centuries later, kanjang (soy sauce) and doenjang (fermented soybean paste) were recorded (Cwiertka & Moriya, 2008). After the introduction of red peppers to Korea at the end of the sixteenth century, gochujang (fermented red pepper paste) was recorded (Cwiertka & Moriya, 2008). The aroma of kanjang is mainly contributed by glutamic acid and other free amino acids from the breakdown of soybean proteins by *Bacillus* spp. (Shin & Jeong, 2015). Doenjang is famous for its health benefits such as liver detoxification and tumor control (Shin & Jeong, 2015). These health beneficial effects are due to secondary metabolites including trypsin inhibitors, isoflavones, vitamin E, and linoleic acid (Shin & Jeong, 2015).

The long history of using soybean to produce these food products with very distinctive tastes and aromas may have contributed to the selection for soybean varieties having particular seed content compositions. For the production of *jang*, besides the yield (Ha et al., 2013; Kim, Baek, Han, et al., 2010; Kim, Baek, Ko, et al., 2010), other qualities including the levels of protein, sugar, total nitrogen content, total free amino acid level and glutamic acid content in the seed are also important considerations (Choi et al., 1999).

8.3.3 Natto

The unique tastes and aromas of fermented soybean products are the results of different combinations of metabolites. For natto production, soybean seeds having a pale yellow seed coat, a low ammonia-like flavor, as well as a high carbohydrate content for promoting microbial activities and sweetness are desired (Escamilla, Rosso, Holshouser, Chen, & Zhang, 2019). The awareness of the health benefits due to the isoflavone contents in natto also drives the selection for soybean varieties with high seed isoflavone contents (Ikeda et al., 2006).

9. Industrial methods and the selection of soybean varieties for soy food production

Industrial methods have been developed to minimize the unpleasant flavors of soy food products. Efforts have been paid to minimize the adverse effects from off-flavor components such as flavonoids and lipoxygenase. It was found that soaking, commonly the first step of soy milk production, induced an increase in daidzein and genistein in the soaked soybeans (Matsuura et al., 1989). The level of increase was dependent on the soaking temperature and pH, with 50 °C and pH6 being the most favorable conditions for the daidzein and genistein production (Matsuura et al., 1989).

Such a phenomenon is due to the activity of β-glucosidases which hydrolyze the isoflavones from the glucoside form to the aglycone form (Matsuura et al., 1989). To minimize the production of daidzein and genistein during soybean soaking, glucono-δ-lactone was proposed as an additive in the soaking water to inhibit the activities of β-glucosidases (Matsuura et al., 1989). Similarly, in another study comparing four different processing techniques, namely soaked soybean-boiling before filtration, soaked soybean-boiling after filtration, dry soybean-boiling (no soaking before grinding soybeans with water for boiling) before filtration, dry soybean-boiling (no soaking before grinding soybeans with water for boiling) after filtration, it was found that DS-BBF produced the least beany flavor in soy milk (Yu et al., 2018). Also, the levels of these compounds in soy milk were positively correlated with the soaking temperature (Li et al., 2019). Strategies have been designed to inactivate lipoxygenase during soy food processing. One example is employing thermal inactivation by superheated steam (Chong, Mah, Easa, & Tan, 2019). When applying these strategies to minimize the adverse effects on the flavors of soy products by the off-flavor components, attention must be paid to make sure other nutritional components, such as crude protein, shall not be negatively affected (Chong et al., 2019). With the advancement of soy food processing technologies enhances, there is more flexibility in choosing soybean germplasms for industrial uses. However, the processing technologies have to be tailor-made for each end-product in order to produce a favorable secondary metabolite profile as well as retaining the desirable nutritional components.

10. Conclusion

The composition of secondary metabolites in soybean is a result of both natural and human selections. While natural selection is mostly influenced by environmental factors such as adaptations to local climate and the prevalence of natural enemies, human selection has been focused on the appearance, yield, and harvestability of the seeds. Secondary metabolites, including flavonoids, terpenes, and alkaloids, play important roles in all these aspects. Meanwhile, these metabolites contribute to the flavors of soybean. As soybean is widely employed to produce food products such as soy milk, tofu and fermented soy products, the need for the post-domestication selection of secondary metabolites associated with the flavors of soybean cannot be ignored. Industrial methods have been developed to minimize the effects of off-flavor metabolites on soy food products.

However, the soybean processing conditions have to be tailored to each specific food product. Wild soybean accessions have diverse secondary metabolite compositions in the seed due to the diverse genetic backgrounds. Thus, wild soybean germplasms could serve as a huge gene pool for molecular breeding to develop soybean cultivars with desirable secondary metabolite profiles that fit different uses.

Acknowledgments

J.Y. Chu copy-edited this manuscript. Any opinions, findings, conclusions or recommendations expressed in this publication do not reflect the views of the Government of the Hong Kong Special Administrative Region or the Innovation and Technology Commission.

Funding

This research was funded by the Hong Kong Research Grants Council General Research Fund (14143916) to H.-M.L.

References

Abbas, M., Saeed, F., Anjum, F. M., Afzaal, M., Tufail, T., Bashir, M. S., et al. (2017). Natural polyphenols: An overview. *International Journal of Food Properties*, *20*, 1689–1699.

Abbo, S., Van-Oss, P. R., Gopher, A., Saranga, Y., Ofner, I., & Peleg, Z. (2014). Plant domestication versus crop evolution: A conceptual framework for cereals and grain legumes. *Trends in Plant Science*, *19*, 351–360.

Abdel-Lateif, K., Bogusz, D., & Hocher, V. (2012). The role of flavonoids in the establishment of plant roots endosymbioses with arbuscular mycorrhiza fungi, rhizobia and Frankia bacteria. *Plant Signaling and Behavior*, *7*, 636–641.

Ahmad, A., Hayat, I., Arif, S., Masud, T., Khalid, N., & Ahmed, A. (2014). Mechanisms involved in the therapeutic effects of soybean (*Glycine max*). *International Journal of Food Properties*, *17*, 1332–1354.

Algar, E., Gutierrez-Mañero, F. J., Garcia-Villaraco, A., García-Seco, D., Lucas, J. A., & Ramos-Solano, B. (2014). The role of isoflavone metabolism in plant protection depends on therhizobacterial MAMP that triggers systemic resistance against *Xanthomonas axonopodis* pv. *glycines* in *Glycine max* (L.) Merr. cv. Osumi. *Plant Physiology and Biochemistry*, *82*, 9–16.

Amirkia, V., & Heinrich, M. (2014). Alkaloids as drug leads—A predictive structural and biodiversity-based analysis. *Phytochemistry Letters*, *10*, xlviii–liii.

Araújo, S. S., Beebe, S., Crespi, M., Delbreil, B., González, E. M., Gruber, V., et al. (2015). Abiotic stress responses in legumes: Strategies used to cope with environmental challenges. *Critical Reviews in Plant Sciences*, *34*, 237–280.

Banerjee, S., Li, Y., Wang, Z., & Sarkar, F. H. (2008). Multi-targeted therapy of cancer by genistein. *Cancer Letters*, *269*, 226–242.

Banerjee, A., & Sharkey, T. D. (2014). Methylerythritol 4-phosphate (MEP) pathway metabolic regulation. *Natural Product Reports*, *31*, 1043–1055.

Bari, R., & Jones, J. D. G. (2009). Role of plant hormones in plant defence responses. *Plant Molecular Biology*, *69*, 473–488.

Benites, C. I., Concha, V. O. C., Reis, S. M. P. M., & De Oliveira, A. C. (2009). Physiochemical characterization of soybean oil deodorizer distillate. *Chemical Engineering Transactions*, *17*, 903–908.

Berk, Z. (1992). Soymilk and related products. In *Technology of production of edible flours and protein products from soybeans*. Rome: Food and Agriculture Organization of the United Nations.

Board, J. E., & Hall, W. (1984). Premature flowering in soybean yield reductions at nonoptimal planting dates as influenced by temperature and photoperiod. *Agronomy Journal*, *76*, 700–704.

Burden, B. J., & Norris, D. M. (1992). Role of the isoflavonoid coumestrol in the constitutive antixenosic properties of "Davis" soybeans against an oligophagous insect, the mexican bean beetle. *Journal of Chemical Ecology*, *18*, 1069–1081.

Carter, A., Tenuta, A., Rajcan, I., Welacky, T., Woodrow, L., & Eskandari, M. (2018). Identification of quantitative trait loci for seed isoflavone concentration in soybean (*Glycine max*) against soybean cyst nematode stress. *Plant Breeding*, *137*, 721–729.

Chander, S., Garcia-Oliveira, A. L., Gedil, M., Shah, T., Otusanya, G. O., Asiedu, R., et al. (2021). Genetic diversity and population structure of soybean lines adapted to sub-Saharan Africa using single nucleotide polymorphism (SNP) markers. *Agronomy*, *11*, 604.

Chen, J., Ullah, C., Reichelt, M., Gershenzon, J., & Hammerbacher, A. (2019). Sclerotinia sclerotiorum circumvents flavonoid defenses by catabolizing flavonol glycosides and aglycones. *Plant Physiology*, *180*, 1975–1987.

Chew, S. C., & Nyam, K. L. (2020). Refining of edible oils. In C. M. Galanakis (Ed.), *Lipids and edible oils: properties, processing and applications* (pp. 213–241). Cambridge, MA: Elsevier Inc.

Chiang, H. S., Norris, D. M., Ciepiela, A., Shapiro, P., & Oosterwyk, A. (1987). Inducible versus constitutive PI 227687 soybean resistance to mexican bean beetle, *Epilachna varivestis*. *Journal of Chemical Ecology*, *13*, 741–749.

Chiozza, M. V., O'Neal, M. E., & MacIntosh, G. C. (2010). Constitutive and induced differential accumulation of amino acid in leaves of susceptible and resistant soybean plants in response to the soybean aphid (Hemiptera: Aphididae). *Environmental Entomology*, *39*, 856–864.

Choi, C., Choi, J.-D., Chung, H.-C., Kwon, K.-I., Im, M.-H., Seo, J.-S., et al. (1999). Quality characteristics of kanjang prepared with meju cultivated on different soybean cultivars with Bacillus subtilis var. globigii seed culture. *Journal of Korean Society of Agricultural Chemistry and Biotechnology*, *42*, 283–287.

Chong, W.-K., Mah, S.-Y., Easa, A. M., & Tan, T.-C. (2019). Thermal inactivation of lipoxygenase in soya bean using superheated steam to produce low beany flavour soya milk. *Journal of Food Science and Technology*, *56*, 4371–4379.

Chung, G., & Singh, R. J. (2008). Broadening the genetic base of soybean: A multidisciplinary approach. *Critical Reviews in Plant Sciences*, *27*, 295–341.

Cordle, C. T. (2004). Soy protein allergy: Incidence and relative severity. *The Journal of Nutrition*, *134*, 1213S–1219S.

Coutinho, D. I., Henning, L. M. M., Döpp, S. A., Nepomuceno, A., Moraes, L. A. C., Marcolino-Gomes, J., et al. (2018). Identification of primary and secondary metabolites and transcriptome profile of soybean tissues during different stages of hypoxia. *Data in Brief*, *21*, 1089–1100.

Cwiertka, K. J., & Moriya, A. (2008). Fermented soyfoods in South Korea. In C. M. Du Bois, C.-B. Tan, & S. W. Mintz (Eds.), *The world of soy* (pp. 161–181). Singapore: NUS Press.

Das, A., Rushton, P. J., & Rohila, J. S. (2017). Metabolomic profiling of soybeans (*Glycine max* L.) reveals the importance of sugar and nitrogen. *Plants*, *6*, 21.

Davis, A. E., & Bolin, T. (1967). Lactose intolerance in Asians. *Nature*, *216*, 1244–1245.

De Geyter, E., Lambert, E., Geelen, D., & Smagghe, G. (2007). Novel advances with plant saponins as natural insecticides to control pest insects. *Pest Technology*, *1*, 96–105.

de Toledo, T. C. F., Canniatti-Brazaca, S. G., Arthur, V., & Piedade, S. M. S. (2007). Effects of gamma radiation on total phenolics, trypsin and tannin inhibitors in soybean grains. *Radiation Physics and Chemistry*, *76*, 1653–1656.

Dehaan, L. R., Van Tassel, D. L., Anderson, J. A., Asselin, S. R., Barnes, R., Baute, G. J., et al. (2016). A pipeline strategy for grain crop domestication. *Crop Science*, *56*, 917–930.

Dixit, A. K., Kumar, V., Rani, A., Manjaya, J. G., & Bhatnagar, D. (2011). Effect of gamma irradiation on lipoxygenases, trypsin inhibitor, raffinose family oligosaccharides and nutritional factors of different seed coat colored soybean (*Glycine max* L.). *Radiation Physics and Chemistry*, *80*, 597–603.

Dong, N.-Q., & Lin, H.-X. (2020). Contribution of phenylpropanoid metabolism to plant development and plant -environment interactions. *Journal of Integrative Plant Biology*, *63*, 180–209.

Donnelly, J. K., & Robinson, D. S. (1995). Invited review free radicals in foods. *Free Radical Research*, *22*, 147–176.

Drewnowski, A., & Gomez-Carneros, C. (2000). Bitter taste, phytonutrients, and the consumer: A review. *The American Journal of Clinical Nutrition*, *72*, 1424–1435.

Erdman, J. W. (1979). Oilseed phytates: Nutritional implications. *Journal of the American Oil Chemists' Society*, *56*, 736–741.

Erdman, J. W., Jr. (1981). Bioavailability of trace minerals from cereals and legumes. *Cereal Chemistry*, *58*, 21–26.

Escamilla, D. M., Rosso, M. L., Holshouser, D. L., Chen, P., & Zhang, B. (2019). Improvement of soybean cultivars for natto production through the selection of seed morphological and physiological characteristics and seed compositions: A review. *Plant Breeding*, *138*, 131–139.

FAO. (2021). *FAOSTAT-Food and Agriculture Organization of the United States*. Retrieved from https://www.fao.org/faostat/en/#rankings/commodities_by_country.

Felton, G. W., & Tumlinson, J. H. (2008). Plant-insect dialogs: Complex interactions at the plant-insect interface. *Current Opinion in Plant Biology*, *11*, 457–463.

Fischer, D. C., Kogan, M., & Paxton, J. (1990). Effect of glyceollin, a soybean phytoalexin, on feeding by three phytophagous beetles (Coleoptera: Coccinellidae and Chrysomelidae): Dose versus response. *Environmental Entomology*, *19*, 1278–1282.

Fournier, D. B., Erdman, J. W., Jr., & Gordon, G. B. (1998). Soy, its components, and cancer prevention: A review of the *in vitro*, animal, and human data. *Cancer Epidemiology, Biomarkers & Prevention*, 7, 1055–1065.

Giacometti, R., Barneto, J., Barriga, L. G., Sardoy, P. M., Balestrasse, K., Andrade, A. M., et al. (2016). Early perception of stink bug damage in developing seeds of field-grown soybean induces chemical defences and reduces bug attack. *Pest Management Science*, *72*, 1585–1594.

Giacometti, R., Jacobi, V., Kronberg, F., Panagos, C., Edison, A. S., & Zavala, J. A. (2020). Digestive activity and organic compounds of *Nezara viridula* watery saliva induce defensive soybean seed responses. *Scientific Reports*, *10*, 15468.

Graf, E., & Eaton, J. W. (1993). Suppression of colonic cancer by dietary phytic acid. *Nutrition and Cancer*, *19*, 11–19.

Graham, T. L., & Graham, M. Y. (1996). Signaling in soybean phenylpropanoid responses: Dissection of primary, secondary, and conditioning effects of light, wounding, and elicitor treatments. *Plant Physiology*, *110*, 1123–1133.

Gupta, Y. P. (1987). Anti-nutritional and toxic factors in food legumes: A review. *Plant Foods for Human Nutrition*, *37*, 201–228.

Gupta, A. P., Pandotra, P., Kushwaha, M., Khan, S., Sharma, R., & Gupta, S. (2015). Alkaloids: A source of anticancer agents from nature. In Atta-ur-Rahman (Ed.), *Vol. 46. Studies in natural products chemistry* (pp. 341–445). Amsterdam, Netherlands: Elsevier B. V.

Ha, K.-S., Choi, J.-K., Cho, S.-H., Byun, H.-S., Kim, S.-W., Jang, J.-S., et al. (2013). A new tofu and soy-paste soybean cultivar, 'Cheonga' with disease resistance, lodging tolerance and high yield. *Korean Journal of Breeding Science*, *45*, 452–456.

Hart, S. V., Kogan, M., & Paxton, J. D. (1983). Effect of soybean phytoalexins on the herbivorous insects mexican bean beetle and soybean looper. *Journal of Chemical Ecology*, *9*, 657–672.

Hartman, G. L., West, E. D., & Herman, T. K. (2011). Crops that feed the world 2. Soybean—worldwide production, use, and constraints caused by pathogens and pests. *Food Security*, *3*, 5–17.

Heiser, C. B. (1988). Aspects of unconscious selection and the evolution of domesticated plants. *Euphytica*, *37*, 77–81.

Hohenstein, J. D., Studham, M. E., Klein, A., Kovinich, N., Barry, K., Lee, Y. J., et al. (2019). Transcriptional and chemical changes in soybean leaves in response to long-term aphid colonization. *Frontiers in Plant Science*, *10*, 1–18.

Hou, A., Chen, P., Alloatti, J., Li, D., Mozzoni, L., Zhang, B., et al. (2009). Genetic variability of seed sugar content in worldwide soybean germplasm collections. *Crop Science*, *49*, 903–912.

Hu, J., Lee, S.-O., Hendrich, S., & Murphy, P. A. (2002). Quantification of the group B soyasaponins by high-performance liquid chromatography. *Journal of Agricultural and Food Chemistry*, *50*, 2587–2594.

Huang, A. S., Hsieh, O. A. L., & Chang, S. S. (1982). Characterization of the nonvolatile minor constituents responsible for the objectionable taste of defatted soybean flour. *Journal of Food Science*, *47*, 19–23.

Iassonova, D. R., Johnson, L. A., Hammond, E. G., & Beattie, S. E. (2009). Evidence of an enzymatic source of off flavors in "Lipoxygenase-Null" soybeans. *Journal of the American Oil Chemists' Society*, *86*, 59–64.

Ikeda, Y., Iki, M., Morita, A., Kajita, E., Kagamimori, S., Kagawa, Y., et al. (2006). Intake of fermented soybeans, *Natto*, is associated with reduced bone loss in postmenopausal women: Japanese population-based osteoporosis (JPOS) study. *The Journal of Nutrition*, *136*, 1323–1328.

Ishaaya, B. I., Birk, Y., Bond, A., & Tencer, Y. (1969). SOYABEAN SAPONINS IX. Studies of their effect on birds, mammals and cold-blooded organisms. *Journal of the Science of Food and Agriculture*, *20*, 433–436.

Islam, M. R., Fujita, D., & Zheng, S.-H. (2019). Genotypic variation of sensitivities to photoperiod and temperature in different growth stages in soybean world mini-core collections (GmWMC). *Tropical Agriculture and Development*, *63*, 69–78.

Jayachandran, M., & Xu, B. (2019). An insight into the health benefits of fermented soy products. *Food Chemistry*, *271*, 362–371.

Jegadeesan, S., & Yu, K. (2020). Food grade soybean breeding, current status and future directions. In M. Hasanuzzaman (Ed.), *Legume crops-prospects, production and uses*. London: IntechOpen.

Jellema, P., Schellevis, F. G., van der Windt, D. A. W. M., Kneepkens, C. M. F., & van der Horst, H. E. (2010). Lactose malabsorption and intolerance: A systematic review on the diagnostic value of gastrointestinal symptoms and self-reported milk intolerance. *QMJ*, *103*, 555–572.

Jo, H., Lee, J. Y., Cho, H., Choi, H. J., Son, C. K., Bae, J. S., et al. (2021). Genetic dversity of soybeans (Glycine max (L.) Merr.) with black seed coats and green cotyledons in Korean germplasm. *Agronomy*, *11*, 581.

Jooste, P. J., Anelich, L., & Motarjemi, Y. (2014). Safety of food and beverages: Milk and dairy products. *Encyclopedia of Food Safety*, *3*, 285–296.

Kang, W., Zhu, X., Wang, Y., Chen, L., & Duan, Y. (2018). Transcriptomic and metabolomic analyses reveal that bacteria promote plant defense during infection of soybean cyst nematode in soybean. *BMC Plant Biology*, *18*, 1–14.

Kanobe, C., McCarville, M. T., O'Neal, M. E., Tylka, G. L., & MacIntosh, G. C. (2015). Soybean aphid infestation induces changes in fatty acid metabolism in soybean. *PLoS ONE*, *10*, 1–20.

Karikari, B., Bhat, J. A., Denwar, N. N., & Zhao, T. (2020). Exploring the genetic base of the soybean germplasm from Africa, America and Asia as well as mining of beneficial allele for flowering and seed weight. *3 Biotech*, *10*, 195.

Keinan-Boker, L., Peeters, P. H. M., Mulligan, A. A., Navarro, C., Slimani, N., Mattisson, I., et al. (2002). Soy product consumption in 10 European countries: The European prospective investigation into cancer and nutrition (EPIC) study. *Public Health Nutrition*, *5*, 1217–1226.

Kilcast, D. (1994). Effect of irradiation on vitamins. *Food Chemistry*, *49*, 157–164.

Kim, H.-T., Baek, I.-Y., Han, W.-Y., Ko, J.-M., Park, K.-Y., Oh, K.-W., et al. (2010). A new soy-paste soybean cultivar, "Daeyang" with disease resistance, large seed and high yielding. *Korean Journal of Breeding Science*, *42*, 690–695.

Kim, H., Baek, I., Ko, J., Han, W., Park, K., Oh, K., et al. (2010). A New Soy-paste soybean cultivar, "Nampung" with disease resistance, good combining adaptability and high yielding. *Korean Journal of Breeding Science*, *42*, 721–726.

Kim, H. S., Kim, B. H., & Cho, Y. D. (1998). Purification and characterization of monomeric lysine decarboxylase from soybean (*Glycine max*) axes. *Archives of Biochemistry and Biophysics*, *354*, 40–46.

Kitagawa, I., Saito, M., Taniyama, T., & Yoshikawa, M. (1985). Saponin and Sapogenol. XXVIII. Structure of soyasaponin A2, a bisdesmoside of soyasapogenol A, from soybean, the seeds of Glycine max MERRILL. *Chemical and Pharmaceutical Bulletin*, *33*, 598–608.

Kitagawa, I., Yoshikawa, M., & Yosioka, I. (1976). Saponin and Sapogenol. XIII. Structures of three soybean saponins: Soyasaponin I, soyasaponin II, and soyasaponin III. *Chemical and Pharmaceutical Bulletin*, *24*, 121–129.

Klein, A. T., Yagnik, G. B., Hohenstein, J. D., Ji, Z., Zi, J., Reichert, M. D., et al. (2015). Investigation of the chemical interface in the soybean-aphid and rice-bacteria interactions using maldi-mass spectrometry imaging. *Analytical Chemistry*, *87*, 5294–5301.

Koc, A. B., Abdullah, M., & Fereidouni, M. (2011). Soybeans processing for biodiesel production. In T.-B. Ng (Ed.), *Soybean applications and technoology*. London: IntechOpen.

Komatsu, S., Makino, T., & Yasue, H. (2013). Proteomic and biochemical analyses of the cotyledon and root of flooding-stressed soybean plants. *PLoS ONE*, *8*, e65301.

Kosslak, R. M., Bookland, R., Barkei, J., Paarent, H. E., & Appelbaum, E. R. (1987). Induction of *Bradyrhizobium japonicum* common *nod* genes by isoflavones isolated from *Glycine max*. *Proceedings of the National Academy of Sciences of the United States of America*, *84*, 7428–7432.

Krishnamurthy, P., Chung, G., & Singh, R. J. (2013). Saponin polymorphism in the Korean wild soybean (*Glycine soja* Sieb. and Zucc.). *Plant Breeding*, *132*, 121–126.

Křížová, L., Dadáková, K., Kašparovská, J., & Kašparovský, T. (2019). Isoflavones. *Molecules*, *24*, 1076.

Król-Grzymała, A., & Amarowicz, R. (2020). Phenolic compounds of soybean seeds from two European countries and their antioxidant properties. *Molecules*, *25*, 2075.

Ku, Y. S., Cheng, S. S., Gerhardt, A., Cheung, M. Y., Contador, C. A., Poon, L. Y. W., et al. (2020). Secretory peptides as bullets: Effector peptides from pathogens against antimicrobial peptides from soybean. *International Journal of Molecular Sciences*, *21*, 1–26.

Ku, Y.-S., Contador, C. A., Ming-Sin, N., Yu, J., Chung, G., & Lam, H.-M. (2020). The effects of domestication on secondary metabolite composition in legumes. *Frontiers in Genetics*, *11*, 581357.

Ku, Y.-S., Ng, M.-S., Cheng, S.-S., Lo, A. W.-Y., Xiao, Z., Shin, T.-S., et al. (2020). Understanding the composition, biosynthesis, accumulation and transport of flavonoids in crops for the promotion of crops as healthy sources of flavonoids for human consumption. *Nutrients*, *12*, 1717.

Ku, Y.-S., Sintaha, M., Cheung, M.-Y., & Lam, H.-M. (2018). Plant hormone signaling crosstalks between biotic and abiotic stress responses. *International Journal of Molecular Sciences*, *19*, 3206.

Kudou, S., Tonomura, M., Tsukamoto, C., Uchida, T., Sakabe, T., Tamura, N., et al. (1993). Isolation and structural elucidation of DDMP-conjugated soyasaponins as genuine saponins from soybean seeds. *Boiscience, Biotechnology, and Biochemistry*, *557*, 546–550.

Kumar, G. P., & Khanum, F. (2012). Neuroprotective potential of phytochemicals. *Pharmacognosy Reviews*, *6*, 81–90.

Kumar, S., & Pandey, A. K. (2013). Chemistry and biological activities of flavonoids: An overview. *The Scientific World Journal*, *2013*, 162750.

Kusano, M., Baxter, I., Fukushima, A., Oikawa, A., Okazaki, Y., Nakabayashi, R., et al. (2015). Assessing metabolomic and chemical diversity of a soybean lineage representing 35 years of breeding. *Metabolomics*, *11*, 261–270.

Lee, J., Kim, H. S., & Song, Y. (2012). Genistein as a potential anticancer agent against ovarian cancer. *Journal of Traditional and Complementary Medicine*, *2*, 96–104.

Lenis, J. M., Gillman, J. D., Lee, J. D., Shannon, J. G., & Bilyeu, K. D. (2010). Soybean seed lipoxygenase genes: Molecular characterization and development of molecular marker assays. *Theoretical and Applied Genetics*, *120*, 1139–1149.

Li, X., Liu, X., Hua, Y., Chen, Y., Kong, X., & Zhang, C. (2019). Effects of water absorption of soybean seed on the quality of soymilk and the release of flavor. *RSC Advances*, *9*, 2906–2918.

Li, M. W., Muñoz, N. B., Wong, C. F., Wong, F. L., Wong, K. S., Wong, J. W. H., et al. (2016). QTLs regulating the contents of antioxidants, phenolics, and flavonoids in soybean seeds share a common genomic region. *Frontiers in Plant Science*, *7*, 854.

Li, M., Xin, D., Gao, Y., Li, K., Fan, K., Muñoz, N. B., et al. (2017). Using genomic information to improve soybean adaptability to climate change. *Journal of Experimental Botany*, *68*, 1823–1834.

Li, Y., Zhao, S., Ma, J., Li, D., Yan, L., Li, J., et al. (2013). Molecular footprints of domestication and improvement in soybean revealed by whole genome re-sequencing. *BMC Genomics*, *14*, 579.

Lichman, B. R. (2021). The scaffold-forming steps of plant alkaloid biosynthesis. *Natural Products Reports*, *38*, 103–129.

Liechti, R., & Farmer, E. E. (2002). The Jasmonate pathway. *Science*, *296*, 1649–1650.

Liener, I. E. (1995). Possible adverse effects of soybean anticarcinogens. *The Journal of Nutrition*, *125(Suppl)*, 744S–750S.

Liener, I. E., Nitsan, Z., Srisangnam, C., Rackis, J. J., & Gumbmann, M. R. (1985). The USDA trypsin inhibitor study. II. Timed related biochemical changes in the pancreas of rats. *Plant Foods for Human Nutrition*, *35*, 243–257.

Lim, B. T., DeMan, J. M., Deman, L., & Buzzell, R. I. (1990). Yield and quality of tofu as affected by soybean and soymilk characteristics. Calcium sulfate coagulant. *Journal of Food Science*, *55*, 1088–1107.

Lima, F., Vieira, K., Santos, M., & Mendes de Souza, P. (2018). Effects of radiation technologies on food nutritional quality. In V. Díaz, & R. M. García-Gimeno (Eds.), *Effects of radiation technologies on food nutritional quality*. IntechOpen: Descriptive Food Science.

Lin, H., Rao, J., Shi, J., Hu, C., Cheng, F., Wilson, Z. A., et al. (2014). Seed metabolomic study reveals significant metabolite variations and correlations among different soybean cultivars. *Journal of Integrative Plant Biology*, *56*, 826–836.

Liu, J., Chang, S. K. C., & Wiesenborn, D. (2005). Antioxidant properties of soybean isoflavone extract and tofu *in vitro* and *in vivo*. *Journal of Agricultural and Food Chemistry, 53*, 2333–2340.

Liu, X., Wu, J., Ren, H., Qi, Y., Li, C., Cao, J., et al. (2017). Genetic variation of world soybean maturity date and geographic distribution of maturity groups. *Breeding Science, 67*, 221–232.

Lo, S., Muñoz-Amatriaín, M., Boukar, O., Herniter, I., Cisse, N., Guo, Y.-N., et al. (2018). Identification of QTL controlling domestication-related traits in cowpea (Vigna unguiculata L. Walp). *Scientific Reports, 8*, 6261. https://doi.org/10.1038/s41598-018-24349-4.

Lomate, P. R., & Bonning, B. C. (2016). Distinct properties of proteases and nucleases in the gut, salivary gland and saliva of southern green stink bug, Nezara viridula. *Scientific Reports, 6*, 1–10.

Lygin, A. V., Li, S., Vittal, R., Widholm, J. M., Hartman, G. L., & Lozovaya, V. V. (2009). The importance of phenolic metabolism to limit the growth of Phakopsora pachyrhizi. *Phytopathology, 99*, 1412–1420.

Malik, M. F. A., Tariq, K., Qureshi, A. S., Khan, M. R., Ashraf, M., Naz, G., et al. (2017). Analysis of genetic diversity of soybean germplasm from five different origins using RAPD markers. *Acta Agriculturae Scandinavica, Section B—Soil & Plant Science, 67*, 148–154.

Marchetti, S., Delledonne, M., Fogher, C., Chiaba, C., Chiesa, F., Savazzini, F., et al. (2000). Soybean Kunitz, C-II and PI-IV inhibitor genes confer different levels of insect resistance to tobacco and potato transgenic plants. *Theoretical and Applied Genetics, 101*, 519–526.

Matsuura, M., Obata, A., & Fukushima, D. (1989). Objectionable flavor of soy milk developed during the soaking of soybeans and its control. *Journal of Food Science, 54*, 602–605.

Matsuzaki, S., & Soybean, Á. P.Á. (2009). Putative occurrence of lysine decarboxylase isoforms in soybean (*Glycine max*) seedlings. *Amino Acids, 36*, 65–70.

McGuinness, E. E., Morgan, R. G. H., Levison, D. A., Frape, D. L., Hopwood, D., & Wormsley, K. G. (1980). The effects of long-term feeding of soya flour on the rat pancreas. *Scandinavian Journal of Gastroenterology, 15*, 497–502.

Medina, V., Sardoy, P. M., Soria, M., Vay, C. A., Gutkind, G. O., & Zavala, J. A. (2018). Characterized non-transient microbiota from stinkbug (*Nezara viridula*) midgut deactivates soybean chemical defenses. *PLoS ONE, 13*, 1–23.

Medzhitov, R., & Janeway, C. A. (1997). Innate immunity: The virtues of a nonclonal system of recognition. *Cell, 91*, 295–298.

Mellor, N., Bligh, F., Chandler, I., & Hodgman, C. (2010). Reduction of off-flavor generation in soybean homogenates: A mathematical model. *Journal of Food Science, 75*, R131–R138.

Messina, M. J. (1999). Legumes and soybeans: Overview of their nutritional profiles and health effects. *American Journal of Clinical Nutrition, 70*, 439S–450S.

Messina, M., & Barnes, S. (1991). The role of soy products in reducing risk of cancer. *Journal of the National Cancer Institute, 83*, 541–546.

Messina, M. J., Persky, V., Setchell, K. D. R., & Barnes, S. (1994). Soy intake and cancer risk: A review of the *in vitro* and *in vivo* data. *Nutrition and Cancer, 21*, 113–131.

Miles, C. W., Bodwell, C. E., Morris, E., Ziyad, J. A., Prather, E. S., Mertz, W., et al. (1987). Long-term consumption of beef extended with soy protein by men, women and children. *Plant Foods for Human Nutrition, 37*, 341–376.

Min, S., Yu, Y., Yoo, S., & St. Martin, S. (2005). Effect of soybean varieties and growing locations on the flavor of soymilk. *Journal of Food Science, 70*, C1–C7.

Mira, L., Fernandez, M. T., Santos, M., Rocha, R., Florêncio, M. H., & Jennings, K. R. (2002). Interactions of flavonoids with iron and copper ions: A mechanism for their antioxidant activity. *Free Radical Research, 36*, 1199–1208.

Mithöfer, A., & Boland, W. (2008). Recognition of herbivory-associated molecular patterns. *Plant Physiology, 146*, 825–831.
Murakami, S., Nakata, R., Aboshi, T., Yoshinaga, N., Teraishi, M., Okumoto, Y., et al. (2014). Insect-induced daidzein, formononetin and their conjugates in soybean leaves. *Metabolites, 4*, 532–546.
Nagata, C., Mizoue, T., Tanaka, K., Tsuji, I., Tamakoshi, A., Matsuo, K., et al. (2014). Soy intake and breast cancer risk: An evaluation based on a systematic review of epidemiologic evidence among the Japanese population. *Japanese Journal of Clinical Oncology, 44*, 282–295.
Nakata, R., Kimura, Y., Aoki, K., Yoshinaga, N., Teraishi, M., Okumoto, Y., et al. (2016). Inducible *de novo* biosynthesis of isoflavonoids in soybean leaves by *Spodoptera litura* derived elicitors: Tracer techniques aided by high resolution LCMS. *Journal of Chemical Ecology, 42*, 1226–1236.
Natarajan, S., Xu, C., Bae, H., & Bailey, B. A. (2007). Proteomic and genomic characterization of Kunitz trypsin inhibitors in wild and cultivated soybean genotypes. *Journal of Plant Physiologylant, 164*, 756–763.
Oakenfull, D. (1981). Saponins in food—A review. *Food Chemistry, 7*, 19–40.
Oakenfull, D., & Sidhu, G. S. (1990). Could saponins be a useful treatment for hypercholesterolaemia? *European Journal of Clinical Nutrition, 44*, 79–88.
Ogawa, M., Tanaka, K., & Kasai, Z. (1979). Accumulation of phosphorus, magnesium and potassium in developing rice grains: Followed by electron microprobe X-ray analysis focusing on the aleurone layer. *Plant and Cell Physiology, 20*, 19–27.
Okubo, K., Iijima, M., Kobayashi, Y., Yoshikoshi, M., Uchida, T., & Kudou, S. (1992). Components responsible for the undesirable taste of soybean seeds. *Bioscience, Biotechnology and Biochemistry, 56*, 99–103.
Onguéné, P. A., Ntie-Kang, F., Lifongo, L. L., Ndom, J. C., Sippl, W., & Mbaze, L. M. A. (2013). The potential of anti-malarial compounds derived from African medicinal plants. Part I: A pharmacological evaluation of alkaloids and terpenoids. *Malaria Journal, 12*, 449.
Ostlund, R. E., Jr., Racette, S. B., & Stenson, W. F. (2003). Inhibition of cholesterol absorption by phytosterol-replete wheat germ compared with phytosterol-depleted wheat germ. *The American Journal of Clinical Nutrition, 77*, 1385–1389.
Otterbach, S. L., Yang, T., Kato, L., Janfelt, C., & Geu-Flores, F. (2019). Quinolizidine alkaloids are transported to seeds of bitter narrow-leafed lupin. *Journal of Experimental Botany, 70*, 5799–5808.
Paauw, M., Koes, R., & Quattrocchio, F. M. (2019). Alteration of flavonoid pigmentation patterns during domestication of food crops. *Journal of Experimental Botany, 70*, 3719–3735.
Pareek, A., Dhankher, O. P., & Foyer, C. H. (2020). Mitigating the impact of climate change on plant productivity and ecosystem sustainability. *Journal of Experimental Botany, 71*, 451–456.
Parmezan, T. R., Júnior, S. L. B., De Carvalho, K., De Aquino, M., Birkett, M., Pickett, J., et al. (2020). Transcriptional profile of genes involved in the production of terpenes and glyceollins in response to biotic stresses in soybean. *Genetics and Molecular Biology, 43*, 1–12.
Patil, G., Vuong, T. D., Kale, S., Valliyodan, B., Deshmukh, R., Zhu, C., et al. (2018). Dissecting genomic hotspots underlying seed protein, oil, and sucrose content in an interspecific mapping population of soybean using high-density linkage mapping. *Plant Biotechnology Journal, 16*, 1939–1953.
Peruca, R. D., Coelho, R. G., da Silva, G. G., Pistori, H., Ravaglia, L. M., Roel, A. R., et al. (2018). Impacts of soybean-induced defenses on Spodoptera frugiperda (Lepidoptera: Noctuidae) development. *Arthropod-Plant Interactions, 12*, 257–266.

Phillips, D. A., & Kapulnik, Y. (1995). Plant isoflavonoids, pathogens and symbionts. *Trends in Microbiology*, *3*, 58–64.

Piironen, V., Lindsay, D. G., Miettinen, T. A., Toivo, J., & Lampi, A. M. (2000). Plant sterols: Biosynthesis, biological function and their importance to human nutrition. *Journal of the Science of Food and Agriculture*, *80*, 939–966.

Piubelli, G. C., Hoffmann-Campo, C. B., De Arruda, I. C., Franchini, J. C., & Lara, F. M. (2003). Flavonoid increase in soybean as a response to Nezara viridula injury and its effect on insect-feeding preference. *Journal of Chemical Ecology*, *29*, 1223–1233.

Raboy, V. (1997). Accumulation and storage of phosphate and minerals. In B. A. Larkins, & I. K. Vasil (Eds.), *Cellular and molecular biology of plant seed development* (pp. 441–477). Dordrecht: Springer.

Raboy, V., Dickinson, D. B., & Below, F. E. (1984). Variation in seed total phosphorus, phytic acid, zinc, calcium, magnesium, and protein among lines of *Glycine max* and *G. soja*. *Crop Science*, *24*, 431–434.

Raboy, V., Gerbasi, P. F., Young, K. A., Stoneberg, S. D., Pickett, S. G., Bauman, A. T., et al. (2000). Origin and seed phenotype of maize *low phytic acid 1-1* and *low phytic acid 2-1*. *Plant Physiology*, *124*, 355–368.

Raboy, V., Young, K., 4t. I. C. of P. M. B, & Gerbasi, P. (1994). Maize low phytic acid (Lpa) mutants. In *4th International Congress of Plant Molecular Biology* (pp. 3–18).

Ranjan, A., Westrick, N. M., Jain, S., Piotrowski, J. S., Ranjan, M., Kessens, R., et al. (2019). Resistance against *Sclerotinia sclerotiorum* in soybean involves a reprogramming of the phenylpropanoid pathway and up-regulation of antifungal activity targeting ergosterol biosynthesis. *Plant Biotechnology Journal*, *17*, 1567–1581.

Ravindran, R., & Jaiswal, A. K. (2019). Wholesomeness and safety aspects of irradiated foods. *Food Chemistry*, *285*, 363–368.

Ray, R. C., Joshi, V. K., & Didier, M. (2014). Fermented foods: Past, present and future. In R. C. Ray (Ed.), *Microorganisms and fermentation of traditional foods (1st)* (pp. 1–36). Boca Raton: CRC Press.

Robinson, D. S., Wu, Z., Domoney, C., & Casey, R. (1995). Lipoxygenases and the quality of foods. *Food Chemistry*, *54*, 33–43.

Roland, W. S. U., Vincken, J.-P., Gouka, R. J., van Buren, L., Gruppen, H., & Smit, G. (2011). Soy isoflavones and other isoflavonoids activate the human bitter taste receptors hTAS2R14 and hTAS2R39. *Journal of Agricultural and Food Chemistry*, *59*, 11764–11771.

Romero-puertas, M. C., Terrón-camero, L. C., Peláez-vico, M.Á., Molina-moya, E., & Sandalio, L. M. (2021). An update on redox signals in plant responses to biotic and abiotic stress crosstalk: Insights from cadmium and fungal pathogen interactions. *Journal of Experimental Botany*, *72*, 5857–5875.

Ross, R. P., Morgan, S., & Hill, C. (2002). Preservation and fermentation: Past, present and future. *International Journal of Food Microbiolgy*, *79*, 3–16.

Sadler, M. J. (2004). Meat alternatives- market developments and health benefits. *Trends in Food Science & Technology*, *15*, 250–260.

Saito, K., & Murakoshi, I. (1995). Chemistry, biochemistry and chemotaxonomy of lupine alkaloids in the leguminosae. In Atta-ur-Rahman (Ed.), *Vol. 15. Studies in natural products chemistry* (pp. 519–549). Amsterdam: Elsevier Science B. V.

Sayama, T., Ono, E., Takagi, K., Takada, Y., Horikawa, M., Nakamoto, Y., et al. (2012). The Sg-1 gycosyltransferase locus regulates structural diversity of triterpenoid saponins of soybean. *The Plant Cell*, *24*, 2123–2138.

Sepiol, C. J., Yu, J., & Dhaubhadel, S. (2017). Genome-wide identification of *Chalcone Reductase Gene* family in soybean: Insight into root-specific *GmCHRs* and *Phytophthora sojae* resistance. *Frontiers in Plant Science*, *8*, 2073.

Shen, Y., Zhang, J., Liu, Y., Liu, S., Liu, Z., Duan, Z., et al. (2018). DNA methylation footprints during soybean domestication and improvement. *Genome Biology*, *19*, 128.

Shi, X., Li, J., Wang, S., Zhang, L., Qiu, L., Han, T., et al. (2015). Flavor characteristic analysis of soymilk prepared by different soybean cultivars and establishment of evaluation method of soybean cultivars suitable for soymilk processing. *Food Chemistry, 185*, 422–429.

Shi, H., Nam, P. K., & Ma, Y. (2010). Comprehensive profiling of isoflavones, phytosterols, tocopherols, minerals, crude protein, lipid, and sugar during soybean (*Glycine max*) germination. *Journal of Agricultural and Food Chemistry, 58*, 4970–4976.

Shin, D., & Jeong, D. (2015). Korean traditional fermented soybean products: Jang. *Journal of Ethnic Foods, 2*, 2–7.

Silva, E., da Graça, J. P., Porto, C., Martin do Prado, R., Hoffmann-Campo, C. B., Meyer, M. C., et al. (2020). Unraveling Asian soybean rust metabolomics using mass spectrometry and molecular networking approach. *Scientific Reports, 10*, 1–11.

Silva, F. A. C., O-Panizzi, M. C. C., Blassioli-Moraes, M. C., & Panizzi, A. R. (2013). Influence of volatile and nonvolatile secondary metabolites from soybean pods on feeding and on oviposition behavior of euschistus heros (Hemiptera: Heteroptera: Pentatomidae). *Environmental Entomology, 42*, 1375–1382.

Silva, E., Perez Da Graça, J., Porto, C., Martin Do Prado, R., Nunes, E., Corrêa Marcelino-Guimarães, F., et al. (2021). Untargeted metabolomics analysis by UHPLC-MS/MS of soybean plant in a compatible response to Phakopsora pachyrhizi infection. *Metabolites, 11*. https://doi.org/10.3390/metabo11030179.

Singer, W. M., Zhang, B., Mian, M. A. R., & Huang, H. (2019). Soybean amino acids in health, genetics, and evaluation. In A. Sudarić (Ed.), *Soybean for human consumption and animal feed* (pp. 7–21). London: IntechOpen.

Spaink, H. P. (1995). The molecular basis of infection and nodulation by rhizobia: The ins and outs of sympathogenesis. *Annual Review of Phytopathology, 33*, 345–368.

Srinivas, P., Danielson, S. D., Michael Smith, C., & Foster, J. E. (2001). Cross-resistance and resistance longevity as induced by bean leaf beetle, Cerotoma trifurcata and soybean looper, Pseudoplusia includens herbivory on soybean. *Journal of Insect Science, 1*, 1–5.

Stafford, H. A. (1997). Roles of flavonoids in symbiotic and defense functions in legume roots. *Botanical Review, 63*, 27–39.

Stagnari, F., Maggio, A., Galieni, A., & Pisante, M. (2017). Multiple benefits of legumes for agriculture sustainability: An overview. *Chemical and Biological Technologies in Agriculture, 4*, 2.

Storhaug, C. L., Fosse, S. K., & Fadnes, L. T. (2017). Country, regional, and global estimates for lactose malabsorption in adults: A systematic review and meta-analysis. *Lancet Gastroenterol Hepatol, 2*, 738–746.

Studham, M. E., & MacIntosh, G. C. (2013). Multiple phytohormone signals control the transcriptional response to soybean aphid infestation in susceptible and resistant soybean plants. *Molecular Plant-Microbe Interactions, 26*, 116–129.

Sugiyama, S. (1984). Selection of micro-organisms for use in the fermentation of soy sauce. *Food Microbiology, 1*, 339–347.

Surh, Y., Chun, K., Cha, H., Su, S., Keum, Y., Park, K., et al. (2001). Molecular mechanisms underlying chemopreventive activities of anti-inflammatory phytochemicals: Down-regulation of COX-2 and iNOS through suppression of NF- $κ$B activation. *Mutation Research, 480–481*, 243–268.

Suzuki, H., Koike, Y., Murakoshi, I., & Saito, K. (1996). Subcellular localization of acyltransferases for quinolizidine alkaloid biosynthesis in *Lupinus*. *Phytochemistry, 42*, 1557–1562.

Tanaka, Y., Sasaki, N., & Ohmiya, A. (2008). Biosynthesis of plant pigments: Anthocyanins, betalains and carotenoids. *The Plant Journal, 54*, 733–749.

Tester, C. F. (1977). Constituents of soybean cultivars differing in insect resistance. *Phytochemistry, 16*, 1899–1901.

Tremblay, A., Hosseini, P., Alkharouf, N. W., Li, S., & Matthews, B. F. (2010). Transcriptome analysis of a compatible response by Glycine max to Phakopsora pachyrhizi infection. *Plant Science, 179*, 183–193.

Tsukamoto, C., Kikuchi, A., Harada, K., Kitamura, K., & Okubo, K. (1993). Genetic and chemical polymorphisms of saponins in soybean seed. *Phytochemistry*, *34*, 1351–1356.

Tsukamoto, C., Nawaz, M. A., Kurosaka, A., Le, B., Lee, J. D., Son, E., et al. (2018). Isoflavone profile diversity in Korean wild soybeans (*Glycine soja* Sieb. & Zucc.). *Turkist Journal of Agriculture and Forestry*, *42*, 248–261.

Tuteja, J. H., Zabala, G., Varala, K., Hudson, M., & Vodkin, L. O. (2009). Endogenous, tissue-specific short interfering RNAs silence the chalcone synthase gene family in *Glycine max* seed coats. *The Plant Cell*, *21*, 3063–3077.

Valliyodan, B., Brown, A. V., Wang, J., Patil, G., Liu, Y., Otyama, P. I., et al. (2021). Genetic variation among 481 diverse soybean accessions, inferred from genomic. *Scientific Data*, *8*, 50.

van de Mortel, M., Recknor, J. C., Graham, M. A., Nettleton, D., Dittman, J. D., Nelson, R. T., et al. (2007). Distinct biphasic mRNA changes in response to asian soybean rust infection. *Molecular Plant-Microbe Interactions*, *20*, 887–899.

Vanga, S. K., & Raghavan, V. (2018). How well do plant based alternatives fare nutritionally compared to cow's milk? *Journal of Food Science and Technology*, *55*, 10–20.

Vercellotti, J. R., St. Angelo, A. J., & Spanier, A. M. (1992). Lipid oxidation in foods: An overview. In A. J. St. Angelo (Ed.), *Lipid oxidation in food* (pp. 1–11). Washington: ACS Publications.

Voora, V., Larrea, C., & Bermudez, S. (2020). *Global market report: Soybeans*. International Institute for Sustainable Development (IISD).

Walter, A. J., & DiFonzo, C. D. (2007). Soil potassium deficiency affects soybean phloem nitrogen and soybean aphid populations. *Environmental Entomology*, *36*, 26–33.

Wan, S., Wu, Y., Wang, C., Wang, C., & Hou, L. (2013). The development of soy sauce from organic soy bean. *Agricultural Sciences*, *4*, 116–121.

Wang, C., Croft, K. P., Järlfors, U., & Hildebrand, D. F. (1999). Subcellular localization studies indicate that lipoxygenases 1 to 6 are not involved in lipid mobilization during soybean germination. *Plant Physiology*, *120*, 227–236.

Wang, M., Li, W., Fang, C., Xu, F., Liu, Y., Wang, Z., et al. (2018). Parallel seelction on a dormancy gene during domestication of crops from multiple families. *Nature Genetics*, *50*, 1435–1441.

Wang, S., Liu, S., Wang, J., Yokosho, K., Zhou, B., Yu, Y.-C., et al. (2020). Simutaneous changes in seed size, oil content and protein content driven by selection of *SWEET* homologues during soybean domestication. *National Science Review*, 7, 1776–1786.

Wang, H., & Murphy, P. A. (1996). Mass balance study of isoflavones during soybean processing. *Journal of Agricultural and Food Chemistry*, *44*, 2377–2383.

Wang, F., Yao, K., & Wei, D. (2011). From soybean phytosterols to steroid hormones. In H. El-Shemy (Ed.), *Soybean and health* (pp. 231–252). InTech: Rijeka.

Wang, C., Zhao, T., & Gai, J. (2010). Genetic variability and evolutionary peculiarity of isoflavone content and its components in soybean germplasm from China. *Scientia Agricultura Sinica*, *43*, 3919–3929.

Wang, T., Zhao, J., Li, X., Xu, Q., Liu, Y., Khan, I. A., et al. (2016). New alkaloids from green vegetable soybeans and their inhibitory activities on the proliferation of concanavalin A-activated lymphocytes. *Journal of Agricultural and Food Chemistry*, *64*, 1649–1656.

Wegulo, S. N., Yang, X. B., Martinson, C. A., & Murphy, P. A. (2005). Effects of wounding and inoculation with Sclerotinia sclerotiorum on isoflavone concentrations in soybean. *Canadian Journal of Plant Science*, *85*, 749–760.

Wilson, L. A. (1995). Soy foods. In D. R. Erickson (Ed.), *Practical handbook of soybean processing and utilization* (pp. 428–459). Amsterdam: Elsevier.

Wink, M. (2013). Evolution of secondary metabolites in legumes (Fabaceae). *South African Journal of Botany*, *89*, 164–175.

Wink, M., Botschen, F., Waterman, P. G., Schäfer, H., & Waterman, P. G. (2010). Chemotaxonomy seen from a phylogenetic perspective and evolution of secondary

metabolism. In M. Wink (Ed.), *Annual plant reviews: Vol. 40. Biochemistry of plant secondary metabolism* (2nd ed., pp. 364–433). Oxford: Blackwell Publishing Ltd.
Wu, J., Hettenhausen, C., Meldau, S., & Baldwin, I. T. (2007). Herbivory rapidly activates MAPK signaling in attacked and unattacked leaf regions but not between leaves of Nicotiana attenuata. *Plant Cell, 19*, 1096–1122.
Wu, W., Zhang, Q., Zhu, Y., Lam, H.-M., Cai, Z., & Guo, D. (2008). Comparative metabolic profiling reveals secondary metabolites correlated with soybean salt tolerance. *Journal of Agriculture and Food Chemistry, 56*, 11132–11138.
Yang, A., Smyth, H., Chaliha, M., & James, A. (2016). Sensory quality of soymilk and tofu from soybeans lacking lipoxygenases. *Food Science & Nutrition, 4*, 207–215.
Yang, W., Wu, T., Zhang, X., Song, W., Xu, C., Sun, S., et al. (2019). Critical photoperiod measurement of soybean genotypes in different maturity groups. *Crop Science, 59*, 2055–2061.
Yao, L., Zhong, Y., Wang, B., Yan, J., & Wu, T. (2020). BABA application improves soybean resistance to aphid through activation of phenylpropanoid metabolism and callose deposition. *Pest Management Science, 76*, 384–394.
Yoshiki, Y., Kahara, T., Okubo, K., Sakabe, T., & Yamasaki, T. (2001). Superoxide- and 1,1-diphenyl-2-picrylhydrazyl radical-scavenging activities of soyasaponin $β$ g related to gallic acid. *Bioscience, Biotechnology and Biochemistry, 65*, 2162–2165.
Yu, H., Liu, R., Hu, Y., & Xu, B. (2018). Flavor profiles of soymilk processed with four different processing technologies and 26 soybean cultivars grown in China. *International Journal of Food Properties, 20*, S2887–S2898.
Yuan, S., & Chang, S. K.-C. (2007). Selected odor compounds in soymilk as affected by chemical composition and lipoxygenases in five soybean materials. *Journal of Agricultural and Food Chemistry, 55*, 426–431.
Yun, D.-Y., Kang, Y.-G., Kim, M., Kim, D., & Kim, E.-H. (2020). Metabotyping of different soybean genotypes and distinct metabolism in their seeds and leaves. *Food Chemistry, 330*, 127198.
Zhang, H., Goettel, W., Song, Q., Jiang, H., Hu, Z., Wang, M. L., et al. (2020). Selection of GmSWEET39 for oil and protein improvement in soybean. *PLoS Genetics, 16*, e1009114.
Zhang, D., Zhang, H., Hu, Z., Chu, S., Yu, K., Lv, L., et al. (2019). Artificial selection on *GmOLEO1* contributes to the increase in seed oil during soybean domestication. *PLoS Genetics, 15*, e1008267.
Zhao, G., Ding, L.-L., Yao, Y., Cao, Y., Pan, Z.-H., & Kong, D.-H. (2018). Extracellular proteome analysis and flavor formation during soy sauce fermentation. *Frontiers in Microbiology, 9*, 1872.
Zheng, H., Chen, L., & Han, X. (2009). Response of soybean yield to daytime temperature change during seed filling: A long-term field study in Northeast China. *Plant Production Science, 12*, 526–532.
Zhou, Z., Jiang, Y., Wang, Z., Gou, Z., Lyu, J., Li, W., et al. (2015). Resequencing 302 wild and cultivated accessions identifies genes related to domestication and improvement in soybean. *Nature Biotechnology, 33*, 408–414.
Zhou, H., Lin, J., Johnson, A., Morgan, R. L., Zhong, W., & Ma, W. (2011). *Pseudomonas syringae* type III effector HopZ1 targets a host enzyme to suppress isoflavone biosynthesis and promote infection in soybean. *Cell Host and Microbe, 9*, 177–186.
Zhou, L., Wang, S.-B., Jian, J., Geng, Q.-C., Wen, J., Song, Q., et al. (2015). Identification of domestication-related loci associated with flowering time and seed size in soybean with the RAD-seq genotyping method. *Scientific Reports, 5*, 9350.
Zong, Y., Yao, S., Crawford, G. W., Fang, H., Lang, J., Fan, J., et al. (2017). Selection for oil content during soybean domestication revealed by X-ray tomography of ancient beans. *Scientific Reports, 7*, 43595.

CHAPTER ELEVEN

Seed morphology of soybean

Lianjun Sun* and Zhihui Yuan
College of Agronomy and Biotechnology, China Agricultural University, Beijing, China
*Corresponding author: e-mail address: sunlj@cau.edu.cn

Contents

Abstract

Seed as a major target for breeding is a considerably important source of food. External and internal structure, nutrition value, and chemical composition are normally referred as seed morphology. This chapter will give a description of the structure, physical properties, and composition of soybean seeds and their regulation and domestication process.

Advances in Botanical Research, Volume 102
ISSN 0065-2296
https://doi.org/10.1016/bs.abr.2022.03.004

1. A general description of seed morphology of soybean

The seed of soybean consists of two major components: seed coat and embryo. As dicot seed, soybean seed does not contain endosperm. The embryo consists of three major embryonic organ systems: a plumule with two well-developed primary leaves enclosing one trifoliolate leaf primordium, two large fleshy cotyledons that functions as a storage organ for macromolecular reserves in soybean seeds, and a hypocotyl-radicle axis that rests in a shallow depression formed by the cotyledons. The tip of the radicle, often visible through the seed coat, is located just below the micropyle and surrounded by an envelope of tissue formed by the seed coat.

2. A general description of shape and size of soybean seed

The seed shape and size of different soybean varieties is extremely variable, varying from round to elliptical in shape and 40–550 mg per seed in size by weight. The seeds of most cultivars are oval in outline. These traits are in general controlled by both genetic and environmental factors. The traits related to seed shape and size are of considerable importance because they are directly related to the quality, yield and commercial property of soybean.

Seed weight, measured as individual seed weight or 100-seed weight, is an important yield component of soybean and is positively correlated with the seed yield. Soybean seed is non-endospermic, which means the endosperm is almost degraded in the mature soybean seed. It is because cotyledons absorb the food reserves from the endosperm completely during embryo development, so the cotyledons of soybean account for almost all of the soybean seed mass.

Seed weight largely determines what the seed will be used for. Small-seeded soybeans lighter than 80 mg per seed are often used to produce sprouts and fermented food such as natto. Soybean seeds larger than 250 mg per seed are often used in the production of miso and edamame. In addition, soybean that possess large seed size and high protein content are used for the production of tofu.

Except for the seed weight, researchers also use seed length, seed width and seed thickness as the indicators of seed size, which have much to do with seed germination rate, germination potential and sowing depth

(Schaal, 1980; Weis, 1982). The ratios of length/thickness, length/width and width/thickness are used as indicators of the shape of soybean seeds.

Studies have reported that, under the same sowing depth, seed size is significantly positively correlated with seed germination rate and germination potential, which means large seeds have higher germination rate than small seeds (Simons & Johnston, 2000). In addition, large seeds have a wider range of planting depth than small seeds. As large seeds store more energy and nutrition, there is enough energy for them to form longer stems, which can ensure the emergence of seedling in deeper soil.

Relevant studies have shown that, under stress conditions, large seeds have more advantages than small seeds at the seedling establishment stage because they contain more nutrients to support the seedling development, which allows them to reach the autotrophic state (Moles & Westoby, 2004).

3. A general description of soybean seed coat

The development of highly differentiated seed coat is a general feature of soybean and a characteristic often used as an identification and classification indicator because the seed coat morphology is easy to be evaluated. The seed coat, or testa, of mature soybean seed is derived from the inner integument of the ovule and consists of several cell layers, accounting for 4–8% of the total individual seed weight.

The soybean seed coat has three distinct layers, namely, epidermis, hypodermis and inner parenchyma tissue (Qutob, Ma, Peterson, Bernards, & Gijzen, 2008). The epidermal tissue consists of closely packed, thick-walled palisade cells called macrosclereids. These cells are elongated perpendicular to the surface of the seed and have thickened, pitted walls in the outer part of the cell. A cuticle is present on the outer wall of the macrosclereids. The hypodermis tissue consists of a single layer of sclerified cells variously elongated and separated from each other. The inner parenchyma tissue consists of 6–8 layers of thin-walled, flattened cells that lack contents, and essentially uniform throughout the entire seed coat except the hilum.

At the time of fertilization, the inner integument consists of two or three cell layers. After fertilization, periclinal division occurs, resulting in an increase in thickness of the inner integument to about 10-cell layers. The inner layer of the inner integument will become more densely stained and differentiate into endothelium or integumentary tapetum 10–14 days after flowering, which presumably serves as nutritive function. There is a layer of multicellular tubules between the seed coat and the cotyledons in soybean seeds, which help nutrients transfer from the endothelium to the embryo (Qutob et al., 2008).

The outer integument at fertilization has 2–4 cell layers except in the region of the micropyle and hilum, where it is considerably thicker (Thorne, 1981). The epidermis of the outer integument consists of isodiametric cells. Periclinal division occurs after fertilization and the outer integument becomes approximately 12–15 cell layers thick. During the development of the soybean seeds, these cells elongate radially, especially near the hilum (Thorne, 1981).

In mature soybean seed, the embryo is enclosed solely by the seed coat, which is strictly of maternal origin and forms a barrier between the embryo and its immediate environment, so the seed permeability and its resistance to seed-borne diseases are mainly determined by seed coat and its associated tissues (Aldesuquy, 2012). In addition, the seed composition, texture, and food reserves are also influenced by seed coat. This part gives a description of seed coat color and luster.

3.1 Seed coat and cotyledons color

Color is one of the earliest characteristics being evaluated in genetic research, originating from the research of Mendel. In mature soybean seed, the embryo is enclosed solely by the seed coat, which forms a barrier between the embryo and its external environment. Therefore, as an important morphological marker, the seed coat color is a trait of considerable importance.

Soybean seed coat is rich in color such as yellow, green, brown, and black and it can be of one color, bicolored or variegated. The variety of soybean seed coat color and the color depth are caused by the content of different pigments content. The darker the seed coat color, the higher the content of seed coat pigments. While these pigments usually possess antioxidative activities, dietary supplements based on soybean seed coat pigments can be used as functional food or pharmaceutical raw materials. The black and brown seed-coat varieties may be a better natural source of antioxidants than the common soybean varieties (i.e., yellow and green seed-coat soybeans) (Zhang et al., 2020).

The seed coat pigments are located mainly in the palisade layer such as anthocyanin in the vacuole, chlorophyll in the plastids, proanthocyanidin, flavonoid, carotenoid and various combinations of breakdown products of these pigments (Todd & Vodkin, 2003). Among these, anthocyanin is an important pigment, which provides not only color but also UV protection to soybean seeds, protecting the embryo from radiation damage. Also, proanthocyanidin (also known as condensed tannins) can influence the seed coat color as an unstable intermediate in the synthesis of anthocyanin, although it is soluble and colorless. Besides, it is agriculturally and environmentally important for reducing bloating in ruminants.

The cotyledons of the mature embryo may be green, yellow or chalky yellow. The various combinations of pigments in the seed coat and cotyledons are responsible for the wide range of colors of soybean seeds.

3.2 Seed coat luster

Seed coat is highly variable in their luster, including shiny, intermediate, dull, light bloom, bloom and dense bloom.

3.2.1 Seed coat brightness

Like seed coat color, the seed coat brightness is also closely related to the pigments content (Zhang et al., 2020). No matter what color of seed coat is, the content of different pigments has a significant negative correlation with the seed coat brightness (Zhang et al., 2020).

3.2.2 Seed coat bloom

Seed coat bloom, present in most wild soybean varieties (Fig. 1), is synthesized in the maternal pod endocarp and adhere to the surface of seed coat during late stages of seed development. It can make the seeds less visible and contains hazardous allergens such as hydrophobic protein, offering duel protection against potential predators (Enstone, Peterson, & Gijzen, 2015; Zhang, Sun, et al., 2018; Zhang, Wang, et al., 2018).

Fig. 1 Morphological differences in seed coat bloom of *G. max* and *G. soja* Accessions (Zhang, Sun, et al., 2018; Zhang, Wang, et al., 2018). (A) PI 468916 with seed coat bloom. (B) PI 479252 with seed coat bloom. (C) Williams 82 without seed coat bloom.

As an abundant seed constituent and a potentially hazardous allergen, hydrophobic protein from soybean (HPS), which can cause asthma in human's allergic to soybean dust from the seed coat bloom. HPS is an 8.3 kD alpha-helical bundle with disulfide bridges consisted of eight cysteine residues; it belongs to the prolamin super-family of plant proteins and is most closely related to the lipid transfer proteins (Baud, Pebay-Peyroula, Cohen-Addad, Odani, & Lehmann, 1993; Sodano & Ptak, 1995). Previous work has shown that the amount of HPS present on the seed coat bloom varies among cultivars and is associated with dull seed luster, with greater amounts of HPS in on dull-seeded soybean cultivars (Enstone et al., 2015; Gijzen, Miller, Kuflu, Buzzell, & Miki, 1999). The function of HPS is not clearly known, but there is evidence indicating that it can influence the seed coat luster possibly by mediating the adherence of endocarp secretions to the seed surface (Gijzen et al., 1999, 2003).

Initially, HPS was isolated and studied because of its extraordinary property of spontaneously crystallizing out of solution (Odani, Koide, Ono, Seto, & Tanaka, 1987), which demonstrate its strong self-associative properties (Dhaubhadel, Kuflu, Romero, & Gijzen, 2005; Odani et al., 1987). HPS is deposited along all the pod endocarp from the funiculus to the ventral suture in a heterogeneous secreted matrix. These secretions may play an important role in the seed expansion by allowing the seed surface to slide more easily over the inner surface of the pod endocarp (Enstone et al., 2015).

The installation of air handling and filtration equipment is an effective precaution of asthma in human's allergy to soybean dust from the seed coat bloom. However, airborne soybean dust is still present in soybean-related workplaces, such as some port cities where soybeans are transferred and in regions where soybeans are grown and harvested (Rodrigo et al., 2004). For example, from 1981 to 1987, the community-wide outbreaks of asthma in Barcelona and Cartagena (Spain) were caused by the release of soybean seed coat bloom dust through the unloading of seed from container vessels. Hundreds of individuals were affected and several deaths were reported (Antó, Sunyer, Rodriguez-Roisin, Suarez-Cervera, & Vazquez, 1989). Soybean seed coat bloom is also a probable cause of earlier asthma outbreaks in other cities, including New Orleans (Weill, Ziskind, & Derbes, 1964). Therefore, the shiny seed coat without bloom is necessary for human consumption and health.

3.2.3 Cracks in soybean seed coat

Seed coat integrity is necessary for seed development, but soybean seed coat may crack under different stress conditions and result in the exposure of cotyledons to the external environment, which can easily cause damage to cotyledons and other structures inside the seed coat. In addition, seeds of soybean germinate very poorly and show reduced growth or vigor when hydrated with an injured seed coat (Shelar, Shaikh, & Nikam, 2008). For the same variety, the emergence rate of soybean seeds with seed coat cracks is significantly lower than that of the seeds without seed coat cracks. The number of abnormal seedlings after emergence of seeds with intact seed coat is significantly higher than that of the seeds without seed coat cracks. Therefore, the cracks have negative effects on not only the external appearance but the storage capacity and vigor of soybean seeds, and it must be strictly limited to reduce the deterioration of soybean seed quality.

Two types of cracks have been reported that may occur in soybean seed coat: irregular cracks and net-like cracks (Fig. 2). During the development of seeds, it is uncertain when the seed coat may get broken. The cracking of the seed coat is influenced by many factors such as the soybean variety, external environment and the physiological condition of the plant. Most soybean cultivars occasionally exhibit seed coat cracking to some extent when exposed to adverse environmental conditions such as alternate wetting and drying, or chilling temperatures (about 15°C) at flowering phase (Yamaguchi et al., 2015).

4. A general description and the physiology of soybean seed texture

Soybean seed is non-endospermic that all the seed food storage is contained in the cotyledons such as lipids and proteins to fuel early seedling development. Those food reserves are used for respiration and growth in the period immediately following germination, before the photosynthesis and uptake from the roots.

Soybean is one of the most important vegetable oil and protein feed crops. Dry soybean seed is composed of approximately 40% protein and 20% oil. Thus, the research to improve soybean seed oil and protein content not only contributes to crop quality but also effectively promotes the entire soybean industry all over the world.

Food-grade soybeans are grouped into two major classes on the basis of food manufacture. One is for whole-soy foods including natto, vegetable soybeans and soy nuts. The other is for cracked soybean foods such as tofu, miso (fermented soup-base paste), soymilk and soy sauce. Natto is a typical and popular Japanese fermented soybean food with soft and sticky texture. It has higher isoflavone (genistein and genistin) content than soymilk or tofu.

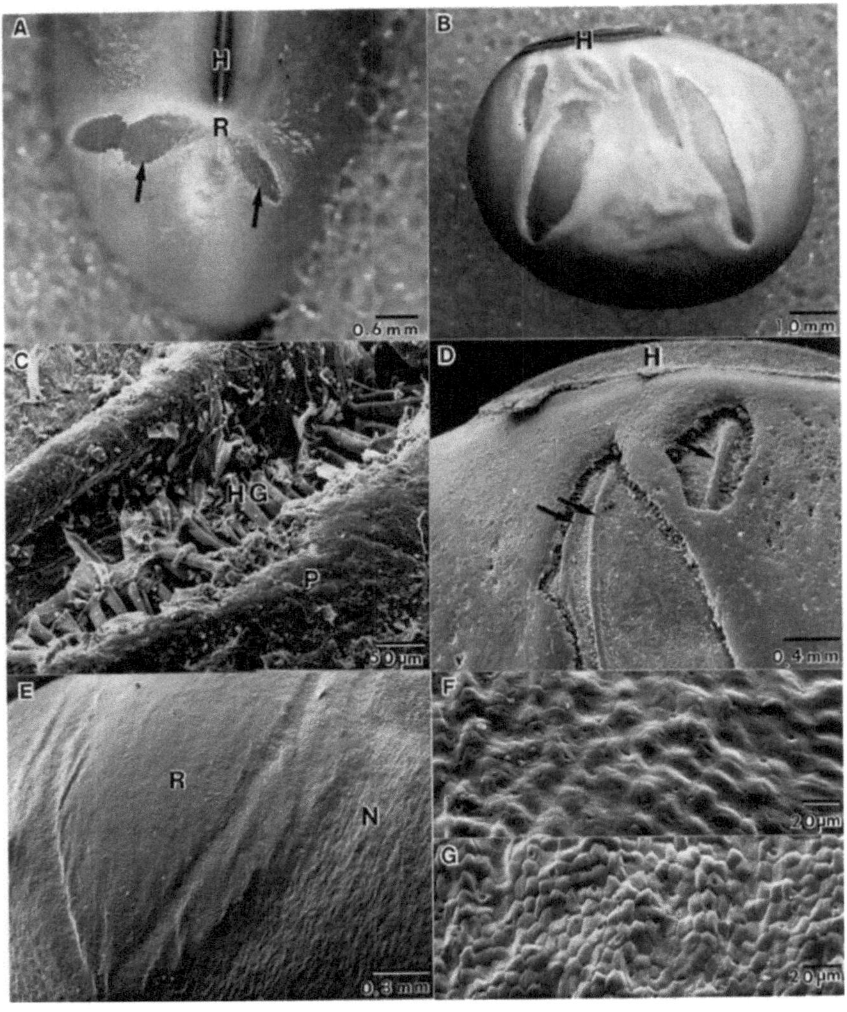

Fig. 2 Light and scanning electron micrographs of soybean seed coat cracking (Yaklich & Barla-Szabo, 1993). (A) Separation of surface cells (arrow) during seed development. H = hilum, R = raphe (×15). (B) Cracked areas forming on seed coat during desiccation of the seed (×10). (C) Palisade (P) and hourglass cells (HG) bordering the cracked area are torn and distorted (×200). (D) Cracked surface of mature seed showing rolling of parenchyma tissue (arrows) (×30). (E) Cotyledon tissue beneath the cracked area. N = normal surface, R = raised surface (×40). (F) Cotyledon surface cells beneath the cracked area (×400). (G) Cotyledon surface cells adjacent to a cracked area (×400).

Vegetable soybean, also called edamame in Japan and Mao Dou in China, is a kind of green and large-seeded soybean with a sweet, nutty flavor, usually harvested at the R6 stage. The important quality properties of vegetable soybean are seed color, texture, and size. Soy nuts are roasted soybean seeds requiring soft texture after roasting and a low percentage of antinutritional factors such as trypsin inhibitors. Apparently, seed texture is an important factor in sensory evaluation of seed characteristics for whole-soy foods consumption. Seed texture is also important for cracked-soy foods because the texture of raw seeds and soybean food products will have an impact on the manufacturing process and consumer acceptance.

4.1 Fats and oils

The cultivated soybean is one of the major oil-seed crops in the world and a predominant plant source of cooking oil. Soybean oil is mainly composed of five fatty acids, including palmitic acid (16:0), stearic acid (18:0), oleic acid (18:1), linoleic acid (18:2) and linolenic acid (18:3), which are present at approximate average concentrations of 10%, 4%, 18%, 55%, and 13%, respectively (Sangwan, Gupta, & Dhindsa, 1986). Among these five fatty acids, palmitic acid is the most common one, but the predominant fatty acids in soybean seeds are unsaturated acids. The quality of soybean oil depends on the composition of fatty acids, which affects the nutritional value, flavor, and its stability.

The fatty acids are stored in soybean seeds as triacylglycerol, which are found within the cells of storage organs and confined to the organelles called lipid bodies. Fatty acids are released from triacylglycerol, and participate in further catabolic pathways to provide energy and carbon skeletons. The mobilization of triacylglycerol involves several consecutive pathways in various cellular compartments. First, triacylglycerols are hydrolyzed into fatty acids by lipases, then the fatty acids pass into another single-membrane-bound organelles called glyoxysomes, where the β-oxidation and glyoxylate cycle occur (Bradford & Hiroyuki, 2007). This process is necessary for soybean storage lipid to breakdown. β-Oxidation converts fatty acids to acetyl-CoA, which is subsequently condensed into four-carbon sugars via the glyoxylate cycle. Then, they are transported to mitochondria. In mitochondria, they can be converted to malate and transported to the cytosol for gluconeogenesis, or used as substrates for respiration.

In addition, soybean seed oils of plant grown in cool climates generally tend to be more unsaturated than those grown in warm climates (Bradford & Hiroyuki, 2007). However, the precise nature of the climatic influence on fatty acid composition remains to be elucidated.

4.2 Protein

Soybean seeds contain higher level (with protein contents of approximately 30–46%) and better-quality plant protein than any other legume varieties.

Because of the rich nutritive components, soybean is originally cultivated for a source of high-quality protein for human and animal consumption. In recent years the consumption of soybean is increasing, providing more than 25% of the world's protein for food and animal feed (Wang et al., 2020).

Soybean seed storage proteins are deposited within special cellular organelles called protein bodies. The protein bodies range in diameter from 0.1 to 25 μm and are surrounded by a single membrane (Kim & Krishnan, 2004). Nutritionally, soybean seed proteins are composed of 18 amino acids, including all the 10 essential amino acids. In addition, soybean seeds are generally deficient in the sulfur amino acids (cysteine and methionine). However, unlike the cereal grains, the lysine content is adequate in soybean seed protein (Zhang, Sun, et al., 2018; Zhang, Wang, et al., 2018). In addition, many of the storage proteins in soybean seed are deposited in insoluble form. To make them fully susceptible to hydrolysis, they must first be modified. In soybean seed cotyledons, there might be modifications brought about through the removal by deaminases of amido groups form storage proteins prior to their hydrolysis. This activity could provide an early source of ammonia to the developing seedling.

Soybean storage protein can be grouped into two types, conglycinin and glycinin, based on the sedimentation coefficients. In addition, the soybean storage protein also contains a 2S fraction, which includes α-conglycinin, several enzymes and trypsin inhibitor (Natarajan, Luthria, Bae, Lakshman, & Mitra, 2013).

Glycinin, a 11S globulin (360 kDa), consists of four acidic and four basic subunits (Natarajan et al., 2013). Glycinin is the major component of the storage protein in soybean cotyledons, accounting for about 70–80% of the total protein in soybean seeds. Glycinin is closely related to the nutritional level of soybean seed and the quality of soybean products. Therefore, glycinin has become an important component of soybean protein products because of its food functional properties.

The 7S storage fraction consists of β-conglycinin, γ-conglycinin and a heterogeneous group of proteins (Fukuda et al., 2005). β-Conglycinin, with a molecular mass of 180 kDa, is the predominant vicilin form consisting of three types of nonidentical but homologous polypeptide subunits, α and α′ of 57 kDa and β of 42 kDa. The α′ subunit consists of extension and core regions, whereas the β subunit only has the core region. All the core regions of three subunits' exhibit high sequence identity with each another. The extension regions of α and α′ subunits have lower identities than the core regions. The subunits of β-conglycinin are encoded by about 20 genes

(Maruyama et al., 2001). These subunits can combine into many different isomeric forms, such as $\alpha'\beta$, $\alpha\beta$ and $\alpha\alpha'\beta$, each varying slightly in amino acid and carbohydrate composition and resulting in subunit variation among different cultivars.

In addition, soybean seed contains high level of trypsin inhibitor proteins, which account for 2–6% of the total protein in soybean seeds. The concentration is the highest found in a range of legumes (Natarajan, Xu, Bae, & Bailey, 2007). These proteins are protective measures to against herbivores and bacterial infections. However, it can reduce the nutritional value of the beans as food and/or animal feed.

4.3 Soybean seed carbohydrates synthesis

The primary constituents of carbohydrates are starch, oligosaccharides (raffinose and stachyose), sucrose and other soluble sugars. Starch is the predominate carbohydrate early in soybean seed development, during which the deposition peaks near mid-seed fill and then declines and is nearly absent in mature soybean seed (Zhang, Sun, et al., 2018; Zhang, Wang, et al., 2018). In conjunction with starch hydrolysis, soluble sugars such as sucrose, fructose and glucose begin to accumulate as a result of the elevated invertase and sucrose synthase activity. Raffinose and stachyose accumulate later in seed development.

During soybean seed development, there is a substantial increase in carbohydrate content in the cotyledons and then followed by a decline, and little carbohydrate remains in the mature soybean seed. This is because the carbohydrate is re-mobilized in the later period of seed development to provide carbon skeletons for oil and protein synthesis. Presumably the products of carbohydrate hydrolysis supplement the restrict flow of assimilates coming into the cotyledons from the mother plant.

4.4 Soybean seed water status during seed development and germination

The seed water status and water relations play regulatory roles in the comprehension of seed biology, particularly in seed development and germination process.

4.4.1 Imbibition of water

Germination begins with the rapid imbibition of water by quiescent dry seeds to activate metabolic processes. Imbibition by soybean seeds is a physical process driven by matric potential difference of seed constituents and can occur in both viable and dead seeds. This imbibition process can be classified into two phases based on the seed imbibition constituents. The first phase is dominated by hydration of seed coat and the second phase is dominated by hydration of cotyledons, which is rate-limited by seed coat permeability. Water enters rapidly through the permeable regions of the seed coat, which is named micropylar region in soybean seeds. Then, water spreads throughout the seed tissues.

The first external sign of the permeable seed's hydration is the wrinkling of seed coat and the wrinkles will be enlarged during imbibition process. Wrinkles formed when water is absorbed by the cells of the seed coat to increase their volumes, and the force of this expansion is sufficient to separate the coat from the cotyledons in wrinkled regions. Therefore, the initial stage of cotyledon hydration in intact soybean seeds could be predicted from external observations of the wrinkles in the seed coat, as the position of the hydrated cotyledon in each seed correlated with the wrinkled area of the seed coat. In addition, water also passes through the hilum to wet the cotyledons, and the hydrated hilum will bulge outward during this process.

However, in most cases, living soybean seeds show a three-phase pattern of water uptake according to the imbibition speed. First is a rapid uptake (phase I), followed by a plateau phase of variable duration (phase II) and ends with the resumption of water uptake associated with embryo growth (phase III) (Meyer, Steudle, & Peterson, 2007). Initial rapid absorption of water during Phase I is primarily a physical process, but physiological activities such as respiration, protein synthesis, and DNA repair are activated well before all tissues are completely hydrated. During Phase II of imbibition, water uptake rate slows and seed water content is either relatively constant or only slowly increasing. Metabolic activities transit from those of a developing seed to those associated with germination. Some stored mRNAs are translated, but most are degraded as new genes are transcribed.

4.4.2 Seed desiccation

During the late stage of soybean seed development, the seeds lose water, desiccate, and mature, leading to a quiescent metabolism status that is placed between development and germination. Desiccation plays an important role in switching seeds from a developmental mode to germinative mode. Seeds that desiccate and become quiescent are known as the orthodox seeds.

In the dry state, most of the metabolic processes, including respiration, are reduced to very low or even undetectable levels, which not only allows seeds to keep longer viability under extreme environmental conditions, but also protect seeds against the attack by bacteria and fungi.

4.5 Soybean seed hardness

In mature soybean seeds, the seed coat protects the seed against detrimental agents from the environment. As a result of this protective function, the seed coat may impose dormancy on the seeds due to its impermeability to water and/or gas, and its mechanical strength which resist the radicle protrusion. This character is called hard-seededness, which exists in most species of the Leguminosae, including soybean, especially some wild varieties. The seeds that have impermeable seed coat are called impermeable or stone seeds. Those seeds with variable seed-coat impermeability or hard-seededness as a mechanism for maintaining seed dormancy to keep viability for long periods until weathering or scarification of the seed coat allow water and/or gas to penetrate. Therefore, impermeability of seed coat is beneficial for germplasm preservation. However, the water can enter the impenetrable seed through the lens, a local area of modified palisade epidermal cells on the opposite side of the hilum from the micropyle.

Although there are several potential advantages, hard-seededness also leads to many problems that impede the utilization of wild germplasms for cultivar enhancement and affects the germination rate, viability, and quality of stored seeds. Therefore, in soybean seeds, the presence of more than 1–2% impermeable seed is undesirable. Soybean seeds with undesirable texture cause processing complications in soybean food production. Stone seeds are not a concern in processing soybeans for vegetable oil. However, seed hardness is an important quality attribute for food grade soybeans because it affects cook-ability, overall texture and results in adverse quality and cost factors. In the manufacture of traditional soy foods, such as soy milk, tofu, miso, natto, or tempeh, the soybean seeds are first washed in water to remove dirt and foreign matter. They are soaked at or below ambient temperature for 12–16 h and then cooked under slight pressure to eliminate endogenous enzyme activity and to soften the seeds. Stone seeds do not absorb water or soften during the cooking phase. If there are significant amounts of stone seeds, they have to be removed, usually after the soaking phase, or else the organoleptic qualities and the texture of the product may be adversely affected. For example, seed hardness was found to be significantly correlated to steamed seed hardness in natto quality-trait study (Mullin & Xu, 2001; Zhang et al., 2008). However, the testing for seed hardness is not only costly and time-consuming but often not practical for the selection at the early stages of soybean breeding program. The suitability of potential cultivars or seed product for the natto market is usually determined by professional testers based on sensory evaluations. At least 0.91 kg of soybean seeds is required for each sensory test in which 10 well-trained researchers taste natto products in a 4-week period with four replications of each product (Mullin & Xu, 2001; Zhang et al., 2008). In addition, there is no standard method in testing soybean seed hardness using scientific instruments. Puncture test is one of the simplest and most commonly used instrumental measurements for soybean texture, but it can give rise to large experimental errors if only a single sample is tested (Bourne, 2002). The shear test also has limited utility without an accepted standard testing procedure. A compression test is rarely used because gases are easily trapped in most products. All these methods require a texture analyzer and are not feasible for breeders to use during selection.

There is a correlation between cell structure of the seed coat and seed hardness. Developing macrosclereids have dense cytoplasm with organelles, which is a typical feature of actively metabolic cells. The macrosclereids of the hard-seeded soybean variety stain intensely with $FeSO_4$, indicating tannins. These phenolic substances are represented by electron-dense deposits when viewed with transmission electron microscopy (Harris, 1987).

The soybean seed coat structure is similar to that of other legumes. Under the cuticle, there is an outer macrosclereid layer underlaid by an osteosclereid layer and the remainder part is a parenchymatous region containing scattered vascular bundles. The impermeability of hard-seeded soybean seeds may involve the lack of pores, the thick phenolic layer and the cutin in the hilum region (Harris, 1987). The major difference between permeable and impermeable seed coat of soybean is the continuity of the outermost cuticle. Small cracks, visible with scanning electron microscopy, are present in cuticles of permeable seeds but absent in impermeable seeds so that water is not admitted to any part of the impermeable seed. Scanning electron microscopy also revealed pores in seed coats of the soft-seeded soybean variety but not in the hard-seeded soybean variety (Harris, 1987).

It has been reported that hard-seededness is associated with calcium content in the seed coat, which can potentially enhance the nutritional value of the soybean seeds. A study of a Chinese variety of food grade soybeans found that the resistance to water and/or gas absorption was consistent with a higher calcium content in the seed coat and with the micropyle being covered with palisade cells (Zhu, Huang, & Wang., 2012). In addition, flavonoids have been reported to reinforce the coat-imposed dormancy by increasing the seed coat thickness and mechanical strength, restricting the permeability of seed coat. The anionic peroxidase may also play a role in the hardening of soybean seed coat and the peroxidase activity is mainly detected in the testa. Also, the phytate content in the seed coat may influence the seed hardness and its cooking quality.

5. Hilum: The gateway of sink-source

The hilum, also known as the seed scar on seed coat, varies in shape from linear to oval. Hilum is the place of connection between seed and pod, which provides a pathway for nutrients and photosynthates to deliver to the developing embryo. In addition, the seed hilum is also the channel for water to influx and efflux during seed development, maturation drying and germination. After the seed maturation, hilum is considered to be a hygroscopically activated valve in the impermeable epidermis of the seed coat, which play an important role in seed dormancy.

Developing seeds are attached to two large vascular bundles in the pod placenta by funiculus. Along the two large bundles, there are numerous smaller bundles depart from each, anastomosing repeatedly within the seed coat. Photosynthate exiting in these smaller bundles must apparently move intercellularly through seed coat to the surface of embryo. Secondary branching from these large bundles supplies photosynthate from pod walls and then to the seeds. However, only a single branch bundle can enter developing seed, which can pass through the funiculus and enter the seed at the posterior end of hilum.

The hilum color of soybean is yellow, brown or black, and its shape varies from oval to triangle. Hilum color derives mainly from the anthocyanin synthesis, which is a slow process near the end of the seed development and maturation phase and have much to do with temperature (Barion, Mosca, & Vamerali, 2016). In addition, during flowering phase, chilling temperatures (about 15 °C) may lead to the development of brown pigmentation of the hilum region (Nakamura, Yang, Kalaiselvi, Uematsu, & Takahashi, 2003). Both the palisade layer and the stellate parenchyma are often pigmented in the hilum, resulting in a more intense coloration to that region.

The hilum region of the seed coat is composed of several distinct tissues, few of which are involved in photosynthate import. At the hilum region, the seed coat becomes considerably thicker and exhibits marked cellular differences from the remainder region. There are two layers of palisade at the hilum of a detached seed, which are composed of the true epidermis of the seed plus a similar layer torn from the funiculus during separation of the seed from the pod placenta. During growth and maturation of the soybean seed, the epidermal cells of the hilum region are elongate radially, resulting in a double layer of elongate thick-walled epidermal cells of the hilum region. A longitudinal cleft in the hilum divides the two layers of palisade and exposes the tracheid bar. This bar is an elongated cluster of nonvascular tracheid-like cells, which function as a water reservoir during seed development and enhance the efficiency of the gas exchange. The tracheid bar may extend along the entire hilum and separate into two groups, one near the micropyle and the other near the chalazal end of the seed. Parenchyma cells of hilum region are considerably smaller than the remainder region of seed coat. There are also two lateral vascular bundles in the funiculus, composed of both phloem and xylem elements. The xylem elements in these bundles are approximately the same number as in the median bundle as it enters the seed, while the amount of phloem is much less. These large lateral bundles are deep within the seed coat, just below a layer of interconnected parenchyma with irregularly shaped. As the seed coat continues to mature, the inner parenchyma cells become crushed and flattened. Eventually, there is little evidence of a functional vascular system (Thorne, 1981). Presumably, the mature soybean seed coat has no functional vascular system. Above this layer, the cells of the hypodermis are hourglass-shaped and elongated approximately 120 μm perpendicular to the seed surface. The unevenly thickened cell walls are thin at the ends of the cell and become thick gradually in the constricted portion of the central region, thus forming a strong supporting layer with considerable inter-cellular space. Finally, the epidermal layer consists of closely packed palisade cells. These spindle-shaped cells are elongated approximately 50 μm perpendicular to the seed surface, pitted walls in the upper part of the cell have thickened and contain numerous starch-containing chloroplasts. During maturation, these cells are fully cutinized, providing a strong, gas-impermeable surface.

At one end of the hilum is the micropyle, a tiny hole formed by the integuments during the seed development, and then covered by cuticle after maturity. At the other end of the hilum is the raphe, a small groove extending to the chalaza, where the integuments are attached to the ovule proper. Near the end of seed filling, the connection between seed and pod become nonfunctional or broken. Usually, the seed separates from the pod cleanly, leaving the hilum with a smooth surface, except in some cases, the funiculus remains attached to the hilum by a core of parenchyma. When the funiculus finally separates, the hilum is rough and has a wide, white central scar formed by the parenchyma tissue. Previous studies have found that hilum morphology and vitality have significant influences to seed weight and quality (Zhao et al., 2021) and the major and minor hilum axes lengths are positively correlated with seed protein content and seed weight. In addition, soybean seeds with an intact hilum have relatively higher seed viability, whereas an incomplete hilum produces poor quality seed that leads to significant yield loss, which is possibly due to bacterial infections or reduced nutrient flows.

In some soybean varieties, hilum can also control the drying process. When the external environmental relative humidity is low, the cells bordering the hilum fissure shrink owing to the loss of water, thus the tissue opening is wider and allows even more drying to occur. When the seeds are under damp conditions, however, the cells bordering the hilum fissure expand to close this fissure. Drying of soybean seeds is thus regulated.

6. Genetic control of seed morphology

Soybean seeds are complex structures that consist of two major components: embryo and seed coat, each with a distinct genotype. The diversity of genotypes suggests that distinct genetic programs underlie the development of each seed component. The embryo is diploid, possessing one paternal and one maternal genome equivalent. The other component of soybean seed is seed coat that is derived from the integuments of the ovule, so the seed coat is of maternal origin.

For decades, genetic diversity has been estimated by phenotypic diversity and this is still an important aspect of germplasm management and evaluation. Phenotypic diversity in soybean is extensive and is under both qualitative and quantitative genetic control.

6.1 Inheritance of soybean seed shape and size

Seed size and shape in soybeans are inherited as quantitative traits. These traits are controlled by different extent of maternal and cytoplasmic effects simultaneously. The inheritance of seed weight and seed length are mainly controlled by cytoplasmic effect (Liang, Dong, Wang, & Fang, 2005), while seed width and seed thickness are mainly controlled by maternal effect (Liang et al., 2005).

In addition, some reports concluded that seed size was primarily controlled by additive gene action (Weber & Moorthy, 1952). The heritability of soybean seed size of soybean was ranged from 27% to 89%, depending on the unit of evaluation and the population studied and marker-assisted selection will be the most effective way for seed size traits (Hoeck et al., 2003).

6.2 Genetic control of seed coat luster

Three complementary loci, B1, B2, and B3 have been reported to control the bloom formation, but only the B1 locus on chromosome 13 was detected by multiple studies (Gijzen et al., 2003; Zhang, Sun, et al., 2018; Zhang, Wang, et al., 2018) and no genetic model account for all of the luster phenotypes was observed. The seed coat bloom is mainly controlled by gene *B1*, which encodes a transmembrane transporter protein for biosynthesis of the seed bloom (Zhang, Sun, et al., 2018; Zhang, Wang, et al., 2018). And HPS genetic locus controls the amount of HPS protein so that it corresponds to the bloom gene that result in dull seeded phenotypes (Gijzen et al., 2003).

Judging from the cracks in different soybean varieties, the seed coat crack is controlled by some genotypes, such as pigment genes, chalcone synthase genes and genes related to maturity and individual seed weight (Nakamura et al., 2003). There are two kinds of seed coat cracking that occur in some cultivars regardless of environmental conditions: irregular cracks and net-like cracks. Irregular crack is controlled by *T* and *I* loci (Ralph & John, 1930). Soybean lines with double recessive genotypes produce severely cracked seeds, while no such symptom was observed with the other allelic combinations. In contrast, genotypes at *T* and *I* loci have no influence on net-like cracks (Liu, 1949).

6.3 Genetic control of seed coat and hilum color

Soybean exhibits natural variation in seed coat and hilum color via the deposition of different anthocyanin pigments. Although these pigments have been well dissected at molecular level, the genes controlling of different soybean seed coat and hilum color traits are not completely understood. Genetic research of soybean seed coat and hilum color date from the beginning of last century. So far, 5 loci (*I*, *T*, *W1*, *R*, *O*) related to the formation of seed coat color have been located, and among these loci, *T* and *I* are responsible for hilum color (Nakamura et al., 2003). The *I* locus and the *R* locus along with the pleiotropic *T* locus have major effects on seed coat color (Nakamura et al., 2003).

I locus is located at a region of chromosome 8 (linkage group A2) and it has four alleles that determine the absence or presence of pigment along with the spatial distribution of the pigments (Todd & Vodkin, 1996). Unlike other species in which the presence of pigment is usually the dominant phenotype, the absence of anthocyanin pigments in soybean seed coats is dominant because of the *I* allele. Within the *I* locus, there is a cluster chalcone synthase (*CHS*) genes with highly conserved sequences, controlling anthocyanin and proanthocyanidin biosynthesis in seed coat (Clough et al., 2004; Todd & Vodkin, 1996; Tuteja, Clough, Chan, & Vodkin, 2004). It was reported that the dominant allele of *I* locus in cultivated soybeans contains a complex structural rearrangement of the *CHS* gene cluster, triggering post-transcriptional gene silencing (PTGS) and inhibits the expression of *CHS* gene family members in the seed coat (Tuteja, Zabala, Varala, Hudson, & Vodkin, 2009; Xie et al., 2019). Research found that the inversion could bring the promoter and the first four exons of a subtilisin gene near to the *CHS* gene cluster. This promoter may drive a chimeric transcript that reads through the subtilisin gene fragment and anti-*CHS1* gene region to cause PTGS of the *CHS* genes (Xie et al., 2019).

Locus *T* and *R* are located on chromosome 6 (linkage group C2) and chromosome 9 (linkage group K), respectively (Owen, 1928; Song et al., 2004). The *R* and *T* loci control specific seed coat colors by controlling the correspond types of the anthocyanin and proanthocyanidin pigments, including black, imperfect black, brown and buff. *R* locus controls the color of the self-colored (*i*) seed coats (Yang et al., 2010). In the presence of the dominant *R* allele, seed coat is black. In the homozygous recessive condition (*r*), seed coat is brown (Song et al., 2004). The pubescence color locus, *T*, exhibits a pleiotropic effect on seed coat color. Self-colored seed and tawny pubescence cultivars have black (*i*, *R*, *T*) or brown (*i*, *r*, *T*) seed. Cultivars with gray pubescence have imperfect-black (*i*, *R*, *t*) or buff (*i*, *r*, *t*) seed-coat color (Owen, 1928; Song et al., 2004).

6.4 Interaction and genetic control of soybean seed protein and oil content

Soybean oil and protein content are very complex quantitative traits, which are controlled by multiple genes. They are genetically negatively correlated and affected by environmental conditions. Such a negative correlation is the major obstacle that hinders the simultaneous improvement of oil and protein content for commercial use.

To date, numerous quantitative trait loci (QTL) that control seed oil and protein content in soybean have been reported, and they were distributed on all the 20 chromosomes in soybean according to the SoyBase (http://www.soybase.org). However, although numerous QTLs have been identified for seed oil and protein content in soybean, only a few stable QTLs have been confirmed and cloned to verify their function.

Glycinin is encoded by five nonallelic genes: *Gy1*, *Gy2*, *Gy3*, *Gy4*, and *Gy5* (Nielsen et al., 1989). These genes code for five precursor protein molecules, G1, G2, G3, G4, and G5, respectively. The five subunits are classified into two distinct major groups, I and II, based on their physical properties, including the identity of their amino acid sequences. Group I subunits consist of G1, G2 and G3 proteins, which contain more methionine residues, and group II contains G4 and G5 subunits.

6.5 Genetic control of hard-seededness

The genetic basis of hard-seededness in soybean remains unclear. Hard-seededness varies widely among seed batches of the same genotype, indicating large environmental influences. This, together with other factors such as polygenic inheritance, makes hard-seededness a typical quantitative trait, which is often described as strong or weak hard-seededness. Because of the integumental origin of the testa, its properties are mainly controlled by the genetics of the maternal plant.

Kilen and Hartwig studied the inheritance mode of seed coat permeability in a cross between a permeable soybean cultivar and an impermeable breeding line. Only 1.4% of F_2 plants had impermeable seed coats and it was estimated that three major genes may be involved in the permeable-impermeable response (Kilen & Hartwig, 1978). Keim used 72 restriction fragment length polymorphism (RFLP) markers to identify five independent genomic regions associated with hard-seededness in a segregating population of a cross between wild and cultivated soybeans. Among them, a genomic region associated with an inhibitor locus of seed coat color accounted for 32% of the observed variation (Keim, Diers, & Shoemaker, 1990). The variation in hard-seededness might be determined by the segregation of alleles at *I* locus, as seed coat color is occasionally correlated with hardness (Thseng, Hou, & Takeda, 1991).

7. Domestication of seed morphology

Domestication is an extremely complex process. It is the earliest form of plant breeding and one of the most important technological innovations in human history. The early stage of domestication was slow, but it has been faster after human exerted strong selection pressures on the varieties. In addition, domestication is a continuous process over time. Domesticated varieties often exhibit convergent phenotypic evolution, referred to as the domestication syndrome.

Soybean was domesticated from its wild relative *Glycine soja* in China 5000 years ago (Shelar et al., 2008), resulting in dramatic morphological modifications, which means wild soybeans exhibit many agronomic traits that are different to cultivated soybean. This is due primarily to the development of soybean landraces in East Asia. Individual farm has grown soybean containing specific traits generation after generation for use as food, feed, medicinal, religious or ceremonial value. Today, these landraces provide the major sources of genetic diversity within soybean germplasm collections. However, these landraces or other unusual strains should not be considered exotic. The term exotic applies only to the wild perennial and annual species.

Wild soybean grows in the form of creepers with many lateral branches, flowers later than cultivated soybean, produces tiny black seeds rather than large yellow seeds, and its pod shatters easily, promoting long-distance seed dispersal. Variations in seed morphology have an impact on biological characteristics and chemical quality. Therefore, the study of seed domestication is helpful to the development of plant identification, classification, and breeding.

7.1 Domestication of seed shape and size

Domestication has enlarged the soybean seeds much beyond the size of their wild-type relatives. The shape of soybean seeds gradually changes from hollow and small to large and plump.

During the process of domestication, seed length, width, thickness and 100-seed weight were all gradually increased, while the ratio of length to width, length to thickness and width to thickness were gradually decreased, with the variation coefficients of all the seed shape and size indicators were gradually decreased, too (Yuan et al., 2012).

At present, the widely accepted classification is wild soybean with 100-seed weight less than 3.0 g, semi-wild soybean with 100-seed weight more than 3.0 g and cultivated soybean seeds with 100-seed weight more than 15.0 g (Yuan et al., 2012).

7.2 Domestication of seed coat

Soybean seed coat color was gradually changed from black to yellow and dark to light during the domestication from wild to cultivated soybean.

The pigment content of different seed coat colors is in significant difference, especially the anthocyanins, proanthocyanidins and total flavonoids content (Todd & Vodkin, 2003). In varieties with dark seed coat color such as black or brown, the pigments concentrated more than the light-colored seed coat varieties, indicating that the contents of the three pigments have been reduced during the domestication of soybean seed coat color.

A shiny seed surface without bloom is desirable for human consumption and health, so the bloom was largely eliminated from the domesticated soybeans (Fig. 1) because it contains hazardous allergens. The domestication from bloom to no-bloom is associated with the artificial selection of a nucleotide mutation that naturally occurred in the coding region of *B1* (Zhang, Sun, et al., 2018; Zhang, Wang, et al., 2018). Studies have shown that the nucleotide mutation occurred in the coding region of B1 not only shined the seed surface, but also elevated seed oil content in domesticated soybeans (Zhang, Sun, et al., 2018; Zhang, Wang, et al., 2018).

7.3 Domestication of seed oil and protein content

One of the most important considerations in connection with the selection of the soybean is that of modifications of seed composition, especially the oil and protein content. Modern cultivars of soybean store significantly higher quantities of food material than either earlier cultivars or their wild progenitors. In general, there was an increase in seed oil content, and a concomitant decrease in protein content to meet the demand of human consumption and health. Studies have shown that there is a negative correlation between lipid and protein content of soybean seeds (Wilcox, 1998).

During domestication, soybean seed oil content was elevated by the nucleotide mutation occurred in the coding region of B1, which also shined the soybean seed coat (Zhang, Sun, et al., 2018; Zhang, Wang, et al., 2018). Furthermore, there was a significant difference in oil content and fatty acid composition between wild and cultivated soybeans. The content of oleic acid in cultivated soybean was significantly higher than that in wild soybean (Sangwan et al., 1986). On the contrary, the content of linolenic acid was lower. In cultivated soybean, the oil content is positively correlated with oleic acid content, and is negatively correlated with other fatty acids (Sangwan et al., 1986). Moreover, the oleic acid content had a negative correlation with all other fatty acids, especially, with linoleic acid and linolenic acid. Therefore, the content of oil and fatty acid compositions are related to the domestication level in soybean, and can be used as a reference index for the evolutionary classification in soybean (Chen et al., 2019).

Unsaturated fatty acids play an important role in human life, including immune system regulation, blood clotting, neurotransmission, cholesterol metabolism, and the structure of membrane phospholipids in the brain and retina. However, unsaturated fatty acids are susceptible to oxidation, resulting in an off-flavor and reduced oil shelf life. Soybeans are thus currently bred to have higher amounts of monounsaturated fatty acids (oleic acid) and lower amounts of polyunsaturated fatty acids (linoleic and linolenic acid), which increases the oxidative stability and is also better for human health (Chen et al., 2019).

The primary goals for enhancing soybean protein quality are to improve essential amino acid balance, increase digestibility of the meal and reduce the environmental impact of animal production. Essential amino acid balance may be augmented via regulation of the expression of genes in particular amino acid pathways, or by increasing the concentration of total crude protein in soybean meal. Digestibility and total metabolizable energy of soybean meal in livestock rations could be improved by reducing the level of raffinose and stachyose in soybean seed. Environmental impact of nutrition waste may be minimized by genetic regulation of phytic acid accumulation during soybean seed development.

7.4 Domestication of hard-seededness

Changes in soybean hard-seededness were selected during the domestication of soybean. Plant breeders have selected against this characteristic and to promote the seed coat permeability, which allows rapid and uniform seed germination. Now, it is widely believed that soybean was domesticated from its wild relative *Glycine soja* in China 5000 years ago, resulting in a diversity of landraces with permeable seed coat (Sakamoto, Abe, Kanazawa, & Shimamoto, 2004). The transition from impermeability to permeability in domesticated soybean was caused by artificial selection of a point mutation. However, many soybean landraces evaded permeability selection because of an alternative selection for seed coat cracking, which also promote the seed imbibition.

8. Conclusion

Seed morphology plays an important role in providing critical information on systematics and can be used to identify the species. The morphology differences among soybean seeds is so complicated so that there are many difficulties in classification. Through the researches of seed morphology, we can have a deeper understanding of seed physiology, seed domestication and the influence of environment.

References

Aldesuquy, H. (2012). *Seed physiology* (1th ed.). Lambert Academic Publishing (Chapter 2).

Antó, J. M., Sunyer, J., Rodriguez-Roisin, R., Suarez-Cervera, M., & Vazquez, L. (1989). Community outbreaks of asthma associated with inhalation of soybean dust. Toxicoepidemiological committee. *The New England Journal of Medicine*, *320*(17), 1097–1102.

Barion, G., Mosca, G., & Vamerali, T. (2016). Estimation of cotyledon isoflavone abundance by a grey luminance-based model in variously hilum-coloured soybean varieties. *Journal of the Science of Food and Agriculture*, *96*(12), 4126–4134.

Baud, F., Pebay-Peyroula, E., Cohen-Addad, C., Odani, S., & Lehmann, M. S. (1993). Crystal structure of hydrophobic protein from soybean; a member of a new cysteine-rich family. *Journal of Molecular Biology*, *231*(3), 877–887.

Bourne, M. C. (2002). *Food texture and viscosity* (2nd ed.). Elsevier Inc.

Bradford, K. J., & Hiroyuki, N. (2007). *Annual plant reviews volume 27: Seed development.* Dormancy and Germination: Blackwell Publishing Ltd (Chapter 11).

Chen, Y., Zhang, S., Wang, L., Wang, L., Li, B., & Sun, J. (2019). Characteristics of oil components and its relationship with domestication of oil components in wild and cultivated soybean accessions. *Acta Agronomica Sinica*, *45*(7), 1038–1049.

Clough, S. J., Tuteja, J. H., Li, M., Marek, L. F., Shoemaker, R. C., & Vodkin, L. O. (2004). Features of a 103-kb gene-rich region in soybean include an inverted perfect repeat cluster of CHS genes comprising the I locus. *Genome*, *47*(5), 819–831.

Dhaubhadel, S., Kuflu, K., Romero, M. C., & Gijzen, M. (2005). A soybean seed protein with carboxylate-binding activity. *Journal of Experimental Botany*, *56*(419), 2335–2344.

Enstone, D. E., Peterson, C. A., & Gijzen, M. (2015). Soybean hydrophobic protein is present in a matrix secreted by the endocarp epidermis during seed development. *Scientific Reports*, *5*(1), 1–12.

Fukuda, T., Maruyama, N., Kanazawa, A., Abe, J., Shimamoto, Y., Hiemori, M., et al. (2005). Molecular analysis and physicochemical properties of electrophoretic variants of wild soybean *Glycine soja* storage proteins. *Journal of Agricultural and Food Chemistry*, *53*(9), 3658–3665.

Gijzen, M., Miller, S. S., Kuflu, K., Buzzell, R. I., & Miki, B. L. (1999). Hydrophobic protein synthesized in the pod endocarp adheres to the seed surface. *Plant Physiology*, *120*(4), 951–959.

Gijzen, M., Weng, C., Kuflu, K., Woodrow, L., Yu, K., & Poysa, V. (2003). Soybean seed lustre phenotype and surface protein cosegregate and map to linkage group E. *Genome*, *46*(4), 659–664.

Harris, W. (1987). Comparative ultrastructure of developing seed coats of hard-seeded and soft-seeded varieties of soybean, *Glycine max* (L.) Merr. *International Journal of Plant Sciences*, *148*(3), 324–331.

Hoeck, J., Fehr, W., Shoemaker, R., Welke, G., Johnson, S., & Cianzio, S. (2003). Molecular marker analysis of seed size in soybean. *Crop Science*, *43*(1), 68–74.

Keim, P., Diers, B. W., & Shoemaker, R. C. (1990). Genetic analysis of soybean hard-seededness with molecular markers. *TAG. Theoretical and applied genetics. Theoretische und angewandte Genetik*, *79*(4), 465–469.

Kilen, T. C., & Hartwig, E. E. (1978). An inheritance study of impermeable seed in soybeans. *Field Crops Research*, *1*(1), 65–70.

Kim, W., & Krishnan, H. B. (2004). Expression of an 11 kDa methionine-rich delta-zein in transgenic soybean results in the formation of two types of novel protein bodies in transitional cells situated between the vascular tissue and storage parenchyma cells. *Plant Biotechnology Journal, 2*(3), 199–210.

Liang, H. Z., Dong, L. W., Wang, H., & Fang, X. J. (2005). Genetic effects on seed traits in soybean. *Journal of Genetics and Genomics, 32*(11), 1199–1204.

Liu, H. (1949). Inheritence of defective seed coat in soybeans. *Journal of Heredity, 40*(12), 317–322.

Maruyama, N., Adachi, M., Takahashi, K., Yagasaki, K., Kohno, M., Takenaka, Y., et al. (2001). Crystal structures of recombinant and native soybean beta-conglycinin beta homotrimers. *European journal of biochemistry / FEBS, 268*(12), 595–3604.

Meyer, C. J., Steudle, E., & Peterson, C. A. (2007). Patterns and kinetics of water uptake by soybean seeds. *Journal of Experimental Botany, 58*(3), 717–732.

Moles, A., & Westoby, M. (2004). Seedling survival and seed size: A synthesis of the literature. *Journal of Ecology, 92*(3), 372–383.

Mullin, W. J., & Xu, W. (2001). Study of soybean seed coat components and their relationship to water absorption. *Journal of Agricultural and Food Chemistry, 49*(11), 5331–5335.

Nakamura, T., Yang, D., Kalaiselvi, S., Uematsu, Y., & Takahashi, R. (2003). Genetic analysis of net-like cracking in soybean seed coats. *Euphytica, 133*(2), 179–184.

Natarajan, S., Luthria, D., Bae, H., Lakshman, D., & Mitra, A. (2013). Transgenic soybeans and soybean protein analysis: An overview. *Journal of Agricultural and Food Chemistry, 61*(48), 11736–11743.

Natarajan, S., Xu, C., Bae, H., & Bailey, B. A. (2007). Proteomic and genomic characterization of Kunitz trypsin inhibitors in wild and cultivated soybean genotypes. *Journal of Plant Physiology, 164*(6), 756–763.

Nielsen, N. C., Dickinson, C. D., Cho, T. J., Thanh, V. H., Scallon, B. J., Fischer, R. L., et al. (1989). Characterization of the glycinin gene family in soybean. *The Plant Cell, 1*(3), 313–328.

Odani, S., Koide, T., Ono, T., Seto, Y., & Tanaka, T. (1987). Soybean hydrophobic protein. Isolation, partial characterization and the complete primary structure. *European Journal of Biochemistry, 162*(3), 485–491.

Owen, F. V. (1928). Inheritance studies in soybeans. III. Seed-coat color and summary of all other mendelian characters thus far reported. *Genetics, 13*(1), 50–79.

Qutob, D., Ma, F., Peterson, C. A., Bernards, M. A., & Gijzen, M. (2008). Structural and permeability properties off the soybean seed coat. *Botany, 86*(3), 219–227.

Ralph, T. S., & John, B. W. (1930). A defective seed-coat character in soybeans 1. *Agronomy Journal, 22*(7), 658–662.

Rodrigo, M., Cruz, M., García, M., Antó, J., Genover, T., & Morell, F. (2004). Epidemic asthma in Barcelona: An evaluation of new strategies for the control of soybean dust emission. *International Archives of Allergy and Immunology, 134*(2), 158–164.

Sakamoto, S., Abe, J., Kanazawa, A., & Shimamoto, Y. (2004). Marker-assisted analysis for soybean hard-seededness with isozyme and simple sequence repeat loci. *Breeding Science, 54*(2), 133–139.

Sangwan, N. K., Gupta, K., & Dhindsa, K. S. (1986). Fatty acid composition of developing soybeans. *Journal of Agricultural and Food Chemistry, 34*(3), 415–417.

Schaal, B. (1980). Reproductive capacity and seed size in Lupinus texensis. *American Journal of Botany, 67*(5), 703–709.

Shelar, V. R., Shaikh, R. S., & Nikam, A. S. (2008). Soybean seed quality during storage: A review. *Agricultural Reviews, 29*(2), 125–131.

Simons, A. M., & Johnston, M. O. (2000). Variation in seed traits of Lobelia inflata (Campanulaceae): Sources and fitness consequences. *American Journal of Botany, 87*(1), 124–132.

Sodano, P., & Ptak, M. (1995). Secondary structure in solution of the hydrophobic protein of soybean (HPS) as revealed by 1H NMR. *Journal of Biomolecular Structure & Dynamics*, *12*(5), 1009–1022.

Song, Q. J., Marek, L. F., Shoemaker, R. C., Lark, K. G., Concibido, V. C., Delannay, X., et al. (2004). A new integrated genetic linkage map of the soybean. *TAG. Theoretical and applied genetics. Theoretische und angewandte Genetik*, *109*(1), 122–128.

Thorne, J. H. (1981). Morphology and ultrastructure of maternal seed tissues of soybean in relation to the import of photosynthate. *Plant Physiology*, *67*(5), 1016–1025.

Thseng, F., Hou, F., & Takeda, K. (1991). Studies on the flooding tolerance of soybean seed: Varietal differences. *Euphytica*, *57*(2), 169–173.

Todd, J. J., & Vodkin, L. O. (1996). Duplications that suppress and deletions that restore expression from a chalcone synthase multigene family. *The Plant Cell*, *8*(4), 687–699.

Todd, J. J., & Vodkin, L. O. (2003). Pigmented soybean (*Glycine max*) seed coats accumulate Proanthocyanidins during development. *Planta*, *102*(2), 663–670.

Tuteja, J. H., Clough, S. J., Chan, W., & Vodkin, L. O. (2004). Tissue-specific gene silencing mediated by a naturally occurring chalcone synthase gene cluster in *Glycine max*. *The Plant Cell*, *16*(4), 819–835.

Tuteja, J. H., Zabala, G., Varala, K., Hudson, M., & Vodkin, L. O. (2009). Endogenous, tissue-specific short interfering RNAs silence the chalcone synthase gene family in *Glycine max* seed coats(W)(OA). *Plant Cell*, *21*(10), 1647–1648.

Wang, S., Liu, S., Wang, J., Yokosho, K., Zhou, B., Yu, Y., et al. (2020). Simultaneous changes in seed size, oil content and protein content driven by selection of SWEET homologues during soybean domestication. *National Science Review*, 7(11), 1776–1786.

Weber, C. R., & Moorthy, B. R. (1952). Heritable and nonheritable relationships and variability of oil content and agronomic characters in the F2 generation of soybean crosses 1. *Agronomy Journal*, *44*(4), 202–209.

Weill, H., Ziskind, M. M., & Derbes, V. J. (1964). Epidemic asthma in New Orleans. *JAMA: The. Journal of the American Medical Association*, *190*(9), 811–814.

Weis, M. (1982). The effects of propagule size on germination and seedling growth in Mirabilis hirsuta. *Canadian Journal of Botany*, *60*(10), 1868–1874.

Wilcox, J. (1998). Increasing seed protein in soybean with eight cycles of recurrent selection. *Crop Science*, *38*(6), 1536–1540.

Xie, M., Chung, C. Y., Li, M., Wong, F., Wang, X., Liu, A., et al. (2019). A reference-grade wild soybean genome. *Nature Communications*, *10*(1), 1–12.

Yaklich, R. W., & Barla-Szabo, G. (1993). Seed coat cracking in soybean. *Crop Science*, *33*(5), 1016–1019. https://doi.org/10.2135/cropsci1993.0011183X003300050028x.

Yamaguchi, N., Taguchi-Shiobara, F., Sayama, T., Miyoshi, T., Kawasaki, M., Ishimoto, M., et al. (2015). Quantitative trait loci associated with tolerance to seed cracking under chilling temperatures in soybean. *Crop Science*, *55*(5), 2100–2107.

Yang, K., Jeong, N., Moon, J., Lee, Y., Lee, S., Kim, H. M., et al. (2010). Genetic analysis of genes controlling natural variation of seed coat and flower colors in soybean. *The Journal of Heredity*, *101*(6), 757–768.

Yuan, N., Yu, X., Guang, J. L., Yun, Q. W., Xiao, F. L., He, N. L., et al. (2012). Domestication of seed size and shape traits in soybean. *Soybean Science*, *31*(04), 522–528.

Zhang, B., Chen, P., Chen, C. Y., Wang, D., Shi, A., Hou, A., et al. (2008). Quantitative trait loci mapping of seed hardness in soybean. *Crop Science*, *48*(4), 1341–1349.

Zhang, D., Sun, L., Li, S., Wang, W., Ding, Y., Swarm, S. A., et al. (2018). Elevation of soybean seed oil content through selection for seed coat shininess. *Nature plants*, *4*(1), 30–35.

Zhang, J., Wang, X., Lu, Y., Bhusal, S. J., Song, Q., Cregan, P. B., et al. (2018). Genome-wide scan for seed composition provides insights into soybean quality improvement and the impacts of domestication and breeding. *Molecular Plant*, *11*(3), 460–472.

Zhang, H., Wu, G., Yang, W., Du, W., Zhao, J., Yue, A., et al. (2020). Analysis of pigment content in different color soybean seed coats. *Journal of Shanxi Agricultural Sciences*, *48*(04), 492–496.

Zhao, Q., Shi, X., Yan, L., Yang, C., Liu, C., Feng, Y., et al. (2021). Characterization of the common genetic basis underlying seed hilum size, yield, and quality traits in soybean. In *Frontiers in plant science*. Retrieved February 25, 2021, from https://doi.org/10.3389/fpls.2021.610214.

Zhu, L., Huang, Y., & Wang, Q. (2012). Nondestructive identification of hard seeds of three legume plants using near infrared spectroscopy. *Editorial Office of Transactions of the Chinese Society of Agricultural Engineering*, *28*(s2), 237–242.

FSC
www.fsc.org
MIX
Papier | Fördert
gute Waldnutzung
FSC® C083411

Kolibri 360 GmbH
Ferdinand-Jühlke-Straße 7
99095 Erfurt, Deutschland
produktsicherheit@kolibri360.de